Applied Mathematical Sciences

Volume 202

More information about this series at http://www.springer.com/series/34

Martin Feinberg

Foundations of Chemical Reaction Network Theory

Springer

Martin Feinberg
Chemical & Biomolecular Engineering
The Ohio State University
Columbus, OH, USA

ISSN 0066-5452 ISSN 2196-968X (electronic)
Applied Mathematical Sciences
ISBN 978-3-030-03857-1 ISBN 978-3-030-03858-8 (eBook)
https://doi.org/10.1007/978-3-030-03858-8

Library of Congress Control Number: 2018961742

Mathematics Subject Classification: 37C10, 37C75, 37N25, 92C42, 80A30, 92B05, 92C37, 92C40, 92C45

This Springer imprint is published by the registered company Springer Nature Switzerland AG.
The registered company address is: Gewerbestrasse 11, 6330 Cham, Switzerland

This book is dedicated to Gail—my friend, teacher, and wife.

Preface

When I am called upon to teach fluid mechanics, I always show students a copy of Newton's *Principia*. I do this for a number of reasons, not least of which is the connection I hope they might feel to antiquity and to genius. But I also want them to know that, although the first volume of *Principia* is nominally about the motion of bodies, the book is without doubt a work of *mathematics*. It is explicitly organized as such, very much like a book of geometry, with definitions, axioms, lemmas, propositions, and theorems. Students should, I think, know that this is how Newton wrote about mechanics. They should also understand that at least some parts of science greatly profit from, and often require, formal mathematical deduction leading to deeper truths about the physical universe.

Although this book is about chemistry, it too is intended as a work of mathematics, very much in the classical tradition. So that others might build upon some of what is now known, I wanted the arguments laid out carefully and in a reasonably self-contained way. Especially when things are complicated, the classical style has served us well over several centuries, and I have not hesitated to invoke it.

However, the book tells a certain story that transcends the mathematics. It is a delicate and beautiful story that, I think, is of special importance to people who need to understand the role that intricate chemical reaction networks exert in biological settings. Nowadays such people are often the most remarkable of scientists, but their stunning achievements are generally experimental and usually unconnected to higher mathematics.

The writing of this book, then, has been something of a challenge. Temptations about just what to do pulled in seemingly contradictory directions. I wanted a fairly complete exposition of the mathematics on record. At the same time, I wanted the central story told in such a way that the narrative—or at least important parts of it—could be understood by people who are interested in reaction network behavior but maybe less so in why certain far-from-obvious assertions are true.

It is fortuitous—and remarkable—that *statements* of some of the deepest theorems about reaction networks can be understood without much mathematical preparation at all or, in some cases, with only a rudimentary knowledge of linear algebra. This is true even though the theorems themselves are about intricate systems of nonlinear equations. Proofs of these theorems are not easy, for they often involve complicated chains of assertions, each building on earlier ones. Even so, the main arguments do not invoke heavy mathematical machinery and are generally accessible to those willing to follow along.

With all of this in mind, I have attempted to write a book that might be of value to two audiences: first, to mathematicians who want to learn about and contribute to chemical reaction network theory and, second, to chemical and biological scientists who want to know some of the deeper things that can be said about reaction networks, especially mass action systems. To read this book, what one needs to know about basic chemistry is not as great as what one needs to know about mathematics. As I have said, though, the mathematical background one needs to understand the theorem *statements* is not great at all. I have tried to place statements of some "big theorems" fairly early in the book, well before the mathematics gets complicated, so that chemical and biological scientists can see what they are. Especially in biology, there is a growing community of people who are at least conversant with modern mathematics, and I hope that this book will be beneficial to them.

The very first chapter contains some thoughts about the strangely distant relationship between the world of mathematics and the world of chemistry—including, of course, biological chemistry. I have traveled in both worlds and have enormous admiration for people I have met in each. In writing a book in which two very different audiences might perhaps find something of value, I hope that I will have helped to soften the boundary between them.

The Book's Scope

This book is not intended to be a survey of all that can be said about the mathematics of chemical reaction networks. The emphasis has been on wide-ranging theorems that connect behavior to reaction network structure, theorems that hold true for broad classes of kinetics without regard to fine parametric details. These, it seems to me, are the kinds of theorems that will be of most use in biology. Moreover, attention is restricted to mixtures at constant temperature, for it is in such systems that reaction network structure exerts itself in the purest way, unobscured by thermal effects.

I have tried to tell a coherent story without going off in directions which, while important, would in my judgment cause an already-challenging narrative path to become highly bifurcated and less straightforward. I regret, therefore, the omission of research that would have afforded a fuller picture. There are indeed pioneers, unrepresented here, whose work readers of this book should pursue.

I want to mention just a few. One is Bruce Clarke, a physical chemist who in the 1970s introduced reaction network ideas admired and used by many. He speaks for himself in a book-length 1980 article published in *Advances in Chemical Physics* [37] and in some shorter survey articles as well [35, 36, 38]. The Russian literature, in particular work by A. I. Vol'pert and A. Ivanova, is not so readily available in English, apart from a chapter in [170]. It is fortunate that their ideas have been used and made more accessible in work by Érdi and Tóth [66], Mincheva and Roussel [128–130], Gnacadja [90–92], and D. Siegel [161]. I will say more about people associated with the early days of reaction network theory in the Acknowledgments. For a sampling of seminal work by my generation of mathematically oriented chemical engineers, I urge readers to see a paper by Fred Krambeck [119] and the monograph by George Gavalas [88].

Although this book is not explicitly biological, it has some intersection with a subject that has come to be called *systems biology*. Readers of this book might find inspiration in any of the rapidly growing collection of systems biology books published recently. An excellent starting point is the very engaging book by Uri Alon [1]. It is simply written but provides much to think about. A much earlier book by Michael Savageau [146] was an important precursor of the systems biology books appearing now.

How This Book Is Organized

This book is divided into three parts, all concerned with connections between reaction network behavior and reaction network structure. Part I, which consists of Chapters 1–5, is largely introductory. To some extent, it is a tutorial for mathematicians. There I indicate how reaction networks give rise to differential equations, and I discuss some elementary geometric connections between solutions of those differential equations and reaction network structure. Part I also contains a small reaction network zoo that serves to exhibit varieties of behavior even simple networks can display. Part II, consisting of Chapters 6–12, is meant to be a tour of some of the main theorems of chemical reaction network theory. In Part III I revisit Part II with an eye toward expanding and deepening the narrative, this time from a perspective that is more overtly mathematical.

Parts I and II form something like a self-contained unit, meant for readers less interested in proofs than in statements of results. Taken together, the two parts are also meant to tell a coherent story with a beginning and an end: Chapter 1, which begins Part I, contains no mathematics at all. Instead it contains a discussion of some conundrums about the qualitative behavior of reaction networks or, more accurately, about the behavior of people for whom reaction networks are an everyday concern. In Chapter 12, which concludes Part II, I return to those conundrums in light of theorems accumulated in the chapters that intervene. There I also examine some mathematical reasons for the richness of behavior found in biological chemistry, a richness that is largely absent from "regular" chemistry.

Although Parts I and II are less mathematical than Part III, they do require of the reader at least some familiarity with modern linear algebra and, to a lesser extent, ways in which linear algebra makes contact with calculus. Nevertheless, I have tried to keep the tone of Parts I and II somewhat informal. They do contain technical material, including some proofs, but those are usually relegated to appendices.

As I indicated earlier, in Part III I revisit several theorems stated in Part II. My aim in Part III is to provide mathematicians not only with some proofs but also with some occasional commentary on the thinking and intuition that gave rise to the proofs. In Part III I also expand on some important ideas, such as detailed balancing and complex balancing, that are mentioned only briefly in Parts I and II. I haven't tried to prove in Part III all theorems stated in Parts I and II. In some cases the journal literature is reasonably accessible and self-contained. My emphasis has been instead on theory for which the existing literature is not so self-contained and could benefit from a friendlier, more consolidated presentation.

Mathematical Prerequisites and Some Suggestions

The mathematical prerequisites for most of this book are not great. Beyond a nodding acquaintance with the language of the elementary qualitative theory of ordinary differential equations (e.g., the meaning of asymptotic stability), a working knowledge of modern linear algebra and, to a lesser extent, advanced calculus should suffice—this despite the fact that the differential equations of chemical reaction networks are highly nonlinear.

Linear algebra is sometimes identified with the mathematics of matrices, row vectors, column vectors, and the like. By linear algebra I mean here something more fundamental: the mathematics of vector spaces and of linear maps from one vector space to another.

There are very few matrices in this book. Important linear transformations that arise for reaction networks have as their domain a linear subspace (the stoichiometric subspace) of an ambient vector space resembling $\mathbb{R}^N$. Because the stoichiometric subspace rarely has a natural basis of its own, there is rarely a canonical representation of such linear transformations in matrix terms. For this reason, it is best to deal with linear transformations directly and naturally, without the imposition of artificial choices having nothing at all to do with the chemistry. Besides, matrix representations as rectangular arrays often require an artificial (and sometimes intrusive) numbering of everything in sight—Which is the ith species or the jth reaction?—again having nothing to do with the chemistry.

Two classical books on linear algebra in the sense I described are the ones by Halmos [102] and by Hoffman and Kunze [106] . However, readers coming from biology, physics, chemistry, or engineering backgrounds will almost certainly find the text by Axler [13] to be more congenial. The book by Greub [98] is much more advanced than any of these, but it is very beautiful in many sections. Determinants are usually presented in an ugly and off-putting way, but Greub's sublime chapter on the subject is enough to make one weep.

I also want to mention a book that starts from scratch and provides most of the mathematics required for a reading of this one: *Advanced Calculus* by Nickerson, Spencer, and Steenrod [133], in particular Chapters 1–8. I studied from it when I began my career as an assistant professor of chemical engineering. The book is actually a reissue of a set of lecture notes provided to advanced calculus honors undergraduates in Princeton University just after the middle of the 20th century. The first five chapters are an introduction to modern linear algebra in the sense I described. The next three contact linear algebra with calculus. For me, those three chapters made clear why, even for the study of nonlinear mathematics, a good knowledge of linear algebra is indispensable. The book is not so friendly, but it is succinct and very powerful in what it teaches. Of all the mathematics texts I've studied, this one had the greatest effect.

A Remark About Computer Implementation

The lifetime of a book—for example, *Anna Karenina*—can be very long. The lifetime of a computer program is short.

I know. Many years ago, to implement some aspects of chemical reaction network theory, I wrote the first version of the Chemical Reaction Network Toolbox. It was written for a personal computer operating system called DOS, which few people remember today. The DOS version can be implemented on current computers only with the support of special DOS emulation software. The *Toolbox* was eventually ported to Windows by Phillipp Ellison. It was also expanded every once in a while (by Phillipp Ellison, Haixia Ji, and Daniel Knight) to provide computer implementation of new results for which such implementation might be desirable.

The current Windows *Toolbox* is freely available for download [62]. It can also operate on contemporary Macintosh computers with the help of a Windows emulation program called *WineBottler*, also freely available. Reference to the *Toolbox* is made occasionally throughout this book.

I would like to think, however, that the usable lifetime of this book will be longer than that of any particular piece of software or operating system. Theorems are permanent. In fact, many of the theorems in this book require no computer assistance at all in deciding, for a particular network, that their hypotheses are satisfied, except perhaps if the network is large. When those hypotheses are satisfied, the consequences are usually powerful and far from obvious.

Nevertheless, there are some theorems in this book for which, in applications, computer assistance will often be needed. As I write today, the current *Toolbox* will, for the specific purposes of this book, usually be the program of choice. But that won't be true years from now when new computer technology leaves the *Toolbox* inoperative. With this in mind, I have tried not to tie this book too tightly to the *Toolbox*. For readers of the future, the value of my references to the *Toolbox* will derive not from its utility but, rather, from knowledge that claims about computer implementations were not far-fetched. At one time, those implementations existed and were used.

Martin Feinberg

Acknowledgments

I regard myself as fortunate to have entered academic life when I did. My education was in chemical engineering, but it came at a singular time. On October 4, 1957, the Soviet Union successfully launched the Sputnik 1 satellite. Although it was only the size of a beach ball, the fact that the satellite successfully orbited Earth caused a profound shock to the psyche of the United States. There were a number of effects of this, superimposed on the earlier effects of World War II. Both the war and Sputnik caused at least some people who might normally devote their lives to mathematics and physics to seek education and careers in engineering.

They brought their mathematical sensibilities with them. In the 1960s and early 1970s, post-Sputnik mathematically inclined chemical engineering students of my age were taught by or collaborated with mathematicians and theoretical physicists such as Neal Amundson, Rutherford Aris, Fritz Horn, and Roy Jackson, people who migrated to chemical engineering in circumstances following World War II. Horn and Jackson play a central role in this book, Aris and Amundson far less so. But that is only because of my choice of subject matter. Indeed, Aris set out quite early [10, 11] to make chemical kinetics a mathematical subject with its own coherence.[1] Even more important than that, he was for so many of us a source of support and inspiration—inspiration not only as a scholar but also as a person. He was a saint.

There were other influences too. It might seem strange that some of the most important early papers cited in this book, including ones by Aris, appeared in the *Archive for Rational Mechanics and Analysis*, a journal whose title does not suggest chemistry. This warrants an explanation and an acknowledgment, both related to the period I just described. During that time the editors of the Archive were Clifford Truesdell and James Serrin, each a separate recipient of the highly prestigious Birkhoff Prize in mathematics. The journal became a meeting ground for, on one

[1] James Wei was another early and very important contributor to efforts in this direction [173–175]. Unlike the others mentioned here, he was educated in chemical engineering and, interestingly, worked on foundational questions while employed in industry, at the Socony Mobil Oil Company.

hand, analysts of the highest caliber and, on the other hand, scientists and engineers interested in examining the foundations of continuum mechanics in the spirit of Euler. Most articles contained proved theorems.

The Archive caught the attention of academic chemical engineering, which of necessity is wide-ranging in the subject matter it embraces.[2] At the same time, Archive contributors such as B. Coleman, W. Noll, M. Gurtin, and R. Bowen had interest in bringing mixtures with reaction and diffusion into the developing mathematical framework of continuum mechanics.[3] As a result, significant cross talk emerged between these Archive authors and chemical engineers: the Archive authors were interested in mixtures, and at least some of the engineers were interested in the extent to which the theorem-proof spirit of the Archive might illuminate chemistry.[4] My contact with Archive editors and contributors helped make this book possible. Although the book is far removed from continuum mechanics, their work and friendship encouraged me to learn mathematics in a serious way and to try to think broadly and deeply about a subject's structure.

My academic career began in 1967 at the University of Rochester, which in the early years was a wonderful place to be, for several reasons. At the time it was, in terms of endowment, the third wealthiest private university in the United States. The endowment supported a climate in which a great faculty could be built and given great freedom. Although my Ph.D. research was in fluid mechanics, I thought (perhaps foolishly in retrospect) that I could, just after my arrival in Rochester, take chances on a research direction—reaction network theory—that was not only new to me but somewhat new to science. The ambient environment was, in many ways, idyllic. Because the university was small and had a wonderful faculty club, I was intellectually enriched by many friendships made there: Richard Lavine and Adrian Nachman in mathematics, Lionel McKenzie and Stanley Engerman in economics, and Bruce Bueno de Mesquita in political science. In the faculty club, Marxists got along well with Chicago School economists. (I am less sure that the Marxists got along well with each other.)

It was in 1970 that Fritz Horn was recruited to Rochester from Rice University, where he had begun work with Roy Jackson on complex balancing [109]—a generalization of detailed balancing—in mass action kinetics. At the time, I was working on questions about the extent to which an underlying detailed balanced mass action system could be inferred uniquely from near-equilibrium experiments [68]. What resulted was an immediate and intense collaboration with Fritz and, indirectly, with Roy. Although this collaboration was one of the greatest joys of my academic life, it was tragically brief: Fritz suffered a major stroke in 1973 and the loss of a son not long after. He retired to Vienna in 1977 and died there in 1978 at age 51.

---

[2] For example, my Ph.D. thesis was in fluid mechanics, extending a paper by Serrin to viscoelastic liquids.

[3] For a sampling of their writings about the thermodynamics of mixtures, not all from the Archive, see [29, 39, 101, 134].

[4] Rutherford Aris was asked to join the editorial board. As a result, the Archive became a venue for early articles on reaction network theory.

In the University of Rochester and at The Ohio State University, to which I moved in 1997, I worked on reaction network theory with superb, highly gifted graduate students. The ones in Rochester were all in chemical engineering; at Ohio State some were in chemical engineering, others in mathematics. Their Ph.D. theses were highly challenging. The technical problems considered were difficult in themselves, and they also resided at the juncture of mathematics, chemistry, and biology.

Although much is said about the importance of interdisciplinary research, little account is taken of the courage it requires. Chemical engineering students such as Paul Schlosser, Phillipp Ellison, and Daniel Knight not only had to learn new mathematics, they pushed the mathematics to a depth that, I think, would give great pride of achievement to formally trained mathematicians—even ones of very high caliber. One of my mathematics students, Gheorghe Craciun, on reading Schlosser's thesis was astonished that Paul was not a mathematician by training. But Gheorghe's own very deep mathematical contributions to reaction network theory required that he, in turn, acquire styles of thinking natural to chemists and biochemists. There is a strong self-selection in students who choose this kind of difficult research at the boundaries. It is not so far-fetched, then, to say that *all* my students, not just the sampling I just mentioned, have been a great gift.

I was not immune from problems associated with trying to work across disciplinary lines. Although the difficulties were certainly not major, I remain grateful for the help I received from others. This came from the chemical engineering community, from mathematics, and from biology. Rutherford Aris and Neal Amundson were always staunch supporters. When, very early in my career, I received venomous proposal reviews from within chemical engineering, Michael Reed—a mathematician at Duke University whom I hardly knew—helped me get a fairer hearing at the US National Science Foundation. He has been a friend ever since. On the mathematics side, I also benefited greatly from very long-term encouragement offered by Chris Jones and Bernold Fiedler. Encouragement from biology came somewhat later, and for that I am grateful to Stan Leibler and, especially, to Uri Alon.

I mentioned the idyllic climate at the University of Rochester during my early years there. I want to say something about Ohio State too. By the time I moved to Ohio State in 1997, mathematical work in chemical engineering, especially in reference to chemical reactors, became largely dormant. This was no longer the post-Sputnik era I described. Large laboratories funded by large grants seemed to become the objective of academic life. Nevertheless, I found Ohio State an extremely congenial place to be. For that I thank L.-S. Fan, the chair of chemical engineering when I arrived. He has the largest of laboratories with grants to match, but his sole requirement for me was that I do good work. Mathematics at Ohio State has also been extremely welcoming, as a department and in terms of close personal friends such as Avner Friedman, Marty Golubitsky, Barbara Keyfitz, and Dave Terman.

Three sabbaticals over many years—one in chemistry, one in mathematics, and one in biology—contributed in different ways to the work appearing in this book. I want to thank my hosts. The first was in 1974 in the Physical Chemistry Laboratory of Oxford University, where my host was Sir John Rowlinson. There I had the time to think more deeply about reaction networks and about the mathematics I was still learning. In 1979 Charles Conley called to ask if I could spend a six months at the Mathematics Research Center of the University of Wisconsin. He said they would pay my salary if their budget could afford it. When I told him my salary, he asked if I could spend a full year. The occasion was an in-gathering of several people having interests in the mathematics of chemistry, with visits of varying duration. It was there, to that wonderful audience, that I gave nine lectures on chemical reaction networks. I wrote up about half of those; the written versions circulated privately and eventually found their way to the Internet [71]. Details of the lectures reverberate in this book, especially in Part III.[5] While I was in Wisconsin, the daily fine-scale interest in the lectures shown by Charles Conley, Chris Jones, and Neil Fenichel was immensely encouraging to me as I continued to find my way into mathematics.[6] My biological sabbatical was in 2007, at Harvard Medical School's Department of Systems Biology. There I was in the presence of both very gifted people and very large laboratory frogs. What can be done now in experimental biology is astonishing. For my wonderful experiences at Harvard, I am especially grateful to Marc Kirschner, Tim Mitchison, Jeremy Gunawardena, and Rebecca Ward.

One of the best things that happened during my visit to Harvard was my first meeting with Guy Shinar. He studied physics in Israel as an undergraduate, began a successful company after the army, and then decided he should study graduate-level biology with Uri Alon at the Weizmann Institute of Science. Just as I was packing for my sabbatical, Uri sent a manuscript of Guy's and asked what I thought. I was busy with the move, so I declined comment. During my stay in Harvard, Guy visited friends there and gave an informal lecture. Some of it, because of my long experience, seemed obvious and easy. But some of it, if true, seemed deep and far from obvious, at least to me. We had a nice talk immediately afterward. I then studied the manuscript that Uri sent. It had rough edges, normal for a graduate student, but the deep things were, I came to believe, correct. With Uri's encouragement, Guy and I began to collaborate long distance, polishing and extending significantly what he had done[154]. At the same time, Guy developed an appetite for learning modern mathematics. I suggested some books, which he devoured with blazing speed. We now had a common language, and our collaboration continued over several years. The result permeates much of the material that appears in Chapters 9–11. That collaboration was, for me, a treasure, not only scientifically but also personally. Guy is brilliant and wise.

I am grateful for financial support from the National Science Foundation, the National Institutes of Health, the Petroleum Research Fund, the Richard Morrow Chair at Ohio State, and The Camille and Henry Dreyfus Foundation in the form of a Teacher-Scholar Grant. Thanks too to Uri Alon for support in connection with visits to the Weizmann Institute of Science.

---

[5] I am grateful to Jennifer Adams for porting the 1979 typewritten lecture text to electronic form. The originals were typed by the magnificent Donna Porcelli.

[6] For an expository summary of some material taken from the full set of nine Wisconsin lectures see [72]. The same volume contains written versions of excellent talks by others at a 1979 Mathematical Research Center symposium on reactive systems.

Great books sometimes result from personal anguish. My wife, Gail, and my daughters, Donna and Sarah, provided no help in this regard. They brought me only great happiness, always.

Martin Feinberg

Contents

Part I

Preliminaries

Part I is intended to be an introduction to the most basic mathematics of reaction networks and their differential equations. For mathematicians, Part I starts at the very beginning and presumes very little in the way of prior knowledge of chemistry. Readers coming from chemical, biological, or engineering backgrounds will, I hope, find in Part I a unifying framework for thinking mathematically—and, in particular, geometrically—about how reactors behave. Part I provides a foundation for Part II, in which we will examine deeper connections between the reaction network structure and reaction network dynamics.

Chapter 1, unlike the other chapters, is not at all mathematical. It contains somewhat surprising observations about attitudes and assumptions in communities of scientists and engineers for whom reaction networks have importance. The behavior of people in these communities provides clues and raises questions about how reaction networks themselves behave. To a great extent, this book is aimed at answering questions raised in Chapter 1. Chapter 2 provides a tutorial about how a reaction network, taken with a kinetics, gives rise to a system of differential equations. There is also a sampling of the kinds of questions these differential equations beget. Chapter 2 concludes with an important section about notation, which readers would do well to study and absorb.

In Chapter 3 we begin to erect a mathematical framework for the study of reaction network behavior, a framework that permeates the entire book. It is here that connections between reaction network structure and dynamics begin to become much more geometrical in character. The material in Chapter 3 is motivated by consideration of a closed reaction system to which no material is added and from which none is removed. In Chapter 4, however, we show how open systems can be incorporated naturally into the mathematical framework constructed in Chapter 3, provided we admit for consideration reaction networks containing pseudoreactions such as $A \to 0$ and $0 \to A$ or, more generally, reaction networks that are "nonconservative." In this way, the geometric insights afforded by Chapter 3 can be brought to the study of systems that are both closed and open.

Chapter 5 amounts to a little zoo of toy networks selected to exhibit a variety of behaviors. The small animals in the zoo not only motivate some questions, they also provide examples (and counterexamples) that are re-examined in light of theorems in Part II.

1

Anticipating the Big Picture: Some Clues

This book is about chemical reaction networks and the differential equations they induce. As such, it is primarily a work of mathematics. This opening chapter, however, contains no equations, nor does it contain any mathematical symbols. Instead, it contains several observations that, when taken together, tell us something about reaction network behavior viewed broadly.

These observations are not about the behavior of chemical systems themselves. Rather, they are about the behavior of *people*—chemists, engineers, biologists, mathematicians, and their communities—who might be called upon to study complex chemical systems. These are highly intelligent people, informed by wide-ranging experiences and a long observational record. Their attitudes and operational habits should, then, reflect informed expectations about how complex chemical systems are likely to act. That is, *their* behavior can carry clues about the behavior—or at least the *anticipated* behavior—of the very systems they study.

Behavioral clues of this kind are, I think, worth noting before we immerse ourselves in the mathematics of reaction networks. Especially with respect to issues concerning stability and instability, these clues will present certain mysteries that a comprehensive theory of reaction networks should help resolve. Even more, they will suggest strongly that there is, in fact, a broad and deep reaction network theory awaiting discovery.

Each of the following sections, except for the last, contains a discussion of one such clue. (They are not completely independent of one another.) Each clue is in the form of a seeming anomaly. Taken together, these anomalies suggest a certain picture of the kinds of behavior we should expect across the broad landscape of all reaction networks. That picture is discussed in the last section of this chapter. Much of this book is aimed at filling in its details.

M. Feinberg, *Foundations of Chemical Reaction Network Theory*,
Applied Mathematical Sciences 202,
https://doi.org/10.1007/978-3-030-03858-8_1

1.1 The Strange Relationship of Mathematics and Chemistry

Hilbert's scientific program at this time, according to Ewald, could be succinctly summarized as follows: "We have reformed mathematics, the next thing to reform is physics, and then we'll go on to chemistry." The chemistry of the day was "somewhat like cooking in a girls' high school." That was the way Hilbert described it.

From Constance Reid's *Hilbert* [140]

1.1.1 The Differential Equations of Chemistry

A reaction network serves to indicate a set of molecular events—the reactions deemed feasible for a particular set of chemical species. More often than not, molecules of a certain species will simultaneously be produced by certain reactions in the network and consumed by still others. Given a reaction network and rules for computing the occurrence rates of the various reactions, the problem then becomes one of understanding how, from a specified initial condition, the number of molecules of each kind evolves in time as these events unfold.

According to a precise formalism, each reaction network gives rise to its own first-order system of differential equations that governs the evolution of the various species concentrations.[1] Systems of differential equations that derive from reaction networks are often large, and they are almost always nonlinear. In fact, when a standard (i.e., mass action) hypothesis for reaction rates is invoked, each reaction network gives rise to its own system of polynomial differential equations in the species concentrations. These systems resemble (but are almost always larger than) the chaotic three-variable Lorenz system [125].

Systems of polynomial differential equations admit a wealth of interesting phenomena. Because polynomials can admit multiple roots, multiple equilibria are common. For systems of polynomial differential equations, there is nothing remarkable about an unstable equilibrium, nor are sustained periodic solutions unexpected. The innocuous Lorenz system of three polynomial differential equations in three variables admits wild, chaotic solutions.

In fact, systems of polynomial differential equations, even small ones, are notoriously difficult to study. In 1900, David Hilbert posed 23 problems he hoped might be solved over the course of the 20th century. One of these, Problem 16, was about the number of distinct limit cycles that might be admitted by polynomial differential equations in just *two* variables. As the 21st century dawned, Hilbert's Problem 16 remained unsolved.

---

[1] We have in mind here a well-stirred mixture, maintained at constant temperature and volume, with instantaneous species concentrations independent of spatial position.

1.1.2 Is There a Mathematical Tradition in Chemistry?

There are thousands of essentially distinct reaction networks that might present themselves for study at one time or another, each with its own system of differential equations. Reaction networks arise naturally in biology, in the manufacture of new materials, in energy production, in environmental dynamics, and in commerce generally. In view of the enormous variety and profound importance of reaction networks in science and industry and in view of the historical interest that mathematicians have had in nonlinear (especially polynomial) systems of differential equations, it is natural to suppose that chemistry, like mechanics, should have a rich and long-standing mathematical tradition.

I want to examine the strength of that tradition a little because the relationship of chemistry and mathematics presents a puzzle that, ultimately, carries important clues about the nature of reaction networks themselves. For the purposes of the discussion, it should be understood that, when I refer to *chemistry*, I do not have in mind that part of chemistry that centers on the study of atomic theory, molecular structure, statistical mechanics or even that part of chemistry that tells us *why* certain reactions occur. Rather, I am thinking about chemistry at an older, more macroscopic level, one that would be familiar to Lavoisier—that is, chemistry as a study of how chemical reactions contribute to the evolution of mixture composition.

For the purpose of determining whether Subject X has a strong mathematical tradition, it is one thing to seek the opinion of practitioners of Subject X and quite another to ask mathematicians. Often these opinions will coincide, but sometimes they won't. I am confident that practitioners of classical (or quantum) mechanics would say that, yes, their subject has a rich mathematical tradition, and I am certain that most mathematicians would agree. There would also be agreement that the tradition has been important for both mechanics and mathematics. The case of economics is a little different, but mostly in degree. Although practicing economists might differ about the importance of a mathematical tradition in their subject, there would, I think, be agreement among economists and mathematicians that such a tradition exists and that it is serious.

For chemistry, the situation is, in my judgment, very different and not just in degree. I am confident that many chemists would argue that there is a strong mathematical tradition in chemistry, but the general opinion of mathematicians is probably not the same. (Remember that here I am not thinking about quantum chemistry or statistical mechanics.)

An indication of the perception held by mathematicians about chemistry can be discerned from the major subject classifications of *Mathematical Reviews*, which abstracts the most recent developments in mathematics. The classification begins with very fundamental mathematics, such as set theory and topology, but then it turns to major categories in applied mathematics. Among those major categorical entries, we find explicit mention of computer science, several branches of mechanics, optics, astronomy, geophysics, economics, and circuits. Although there is an entry labeled "biology and other natural sciences," *there is no major categorical entry that is clearly identifiable with chemistry*. For *Mathematical Reviews*, chemistry—one of the great branches of physical science—is not a mathematical subject, at least not one deserving of a major categorical entry.[2]

1.1.3 A Historical Puzzle

If thousands of distinct reaction networks exert themselves in subjects as crucial and diverse as cell biology, petroleum refining, and atmospheric chemistry, and if each such network gives rise, in a precise way, to its own system of nonlinear (typically polynomial) differential equations, how can it be that the mathematical tradition of chemistry is sufficiently weak as to escape the notice of *Mathematical Reviews* (at the level of the major topical categories)? Shouldn't the great variety of intricate behaviors exhibited by polynomial systems have attracted to chemistry a Newton or a Poincaré?

1.2 Fruit Flies and Platypuses

There is another, related puzzle I want to consider: the existence of what an anonymous journal reviewer called "the BZ Cult." (The reviewer wrote that the BZ Cult might perhaps find certain results of mine interesting.) By the BZ Cult, the reviewer certainly had in mind a group of chemists, chemical engineers, and mathematicians (many of them my friends) who studied, in various contexts and in various ways, a chemical system discovered the middle of the 20th century by Belousov and Zhabotinsky [18, 178]. (See also [64, 65].)

The "BZ system" is a solution of malonic acid, potassium bromate, and cerium sulfate in dilute sulfuric acid. It has the property that, in a test tube, an indicator oscillates in color between blue and red for a considerable period of time. (The color change is triggered by the rise and fall in concentration of a certain reaction product.) In the test tube, these composition oscillations eventually cease, but they can be made to persist, with a definite period, in a reactor to which fresh reagents are fed continuously and from which reacting mixture is removed continuously. When it was first observed, this periodic behavior was of great interest. Later, it became evident that, in the continuously fed reactor, the observed composition oscillations would sometimes be sustained but with no apparent repetitive pattern. In other words, the composition fluctuations were chaotic. Note, however, that the BZ system was of great interest and was subject to much study and modeling even *before* chaotic composition oscillations were observed. *Regular, periodic behavior was already considered remarkable.*

---

[2] I am referring here to the Mathematics Subject Classification 2000. Strangely, there is a *chemistry* subcategory under *biology and other natural sciences*, but it has the same rank there as the subcategory *genetics and population dynamics*. *Mathematical Reviews* does indeed abstract articles about the differential equations of chemistry, but—again strangely—these are often classified under the major applied mathematics category *classical thermodynamics, heat transfer*.

What is interesting to me here is not the BZ system itself but, rather, the great attention it attracted. If there are thousands of distinct reaction networks in nature and if each is governed by its own system of nonlinear equations, why should periodic behavior in the BZ system have been considered so extraordinary? Why is the observational record, centuries old, so lacking in other examples? Why should there be a BZ Cult at all?

There are different reasons for studying a particular example. In biology one studies the fruit fly because it is simple and, in many ways, typical. One studies the platypus because it is not typical and, in fact, bizarre. I can tell you that I had a BZ kit to entertain colleagues and students. This was a platypus.

1.3 Chemical Education: The Tacit Doctrine of Stable Reactor Behavior

I was born thirty years too late to enjoy the era of more horse sense and less mathematics in chemical engineering.

Olaf Hougen,
University of Wisconsin,
ca. 1958

I have thought for some years that chemical engineering is weighted down with more mathematics than it can support.

P. V. Danckwerts,
Cambridge University,
1982

I was a student of chemical engineering, and I now teach the subject as a professor. Chemical reactors lie at the heart of chemical engineering, and real reactors generally involve many reactions occurring simultaneously. Because reactors are governed by systems of nonlinear differential equations, it is peculiar that chemical engineering, even in the universities, did not begin to look like a mathematical discipline until the late 1950s or early 1960s.[3] In fact, much of the change in the discipline was not driven by chemistry at all. Rather, it was driven by the needs of fluid mechanics and heat transfer.

Even now, there is something very odd about the questions chemical engineering students are taught to ask—and *not* to ask—about reactor stability. They fully understand that *nonisothermal* homogeneous open reactors (e.g., adiabatic continuous-flow stirred-tank reactors) can admit multiple steady states, unstable steady states, and sustained composition-temperature oscillations even when the chemistry is

[3] Economics became a serious mathematical subject much earlier. As an example, Paul Samuelson's *Foundations of Economic Analysis* was published in 1948. In its mathematical sophistication, the Samuelson book surpasses much of what is written about chemical reactors even today.

simple. Yet, stability questions for *isothermal* homogeneous reactors seem never to arise, even when the chemistry is highly intricate.

Regardless of complexity in the underlying reaction network, the tacit assumption seems to be that an isothermal homogeneous continuous-flow stirred-tank reactor will behave in the most mundane way: there will be precisely one steady state, and that steady state will eventually be approached from any initial composition within the reactor. In fact, the very language with which such reactors are discussed is prejudicial. Textbook talk is always of *the* steady state, independent of initial conditions, as if there could not be more than one. In the isothermal homogeneous case, questions about stability are rarely asked.[4]

There is something else that is highly puzzling—indeed remarkable—about chemical education, at least as it stood some years ago. Always hungry for excellent graduate students, chemical engineering departments would be very open to the admission of candidates trained not in chemical engineering but, rather, in chemistry. A frequent stumbling block, though, was such a student's mathematics preparation. A question that always needed to be asked was whether the chemistry student had, during his or her undergraduate years, taken *any* course in differential equations—not necessarily an advanced course, merely *any course at all!*

If each new (isothermal) system gives rise to its own peculiar system of differential equations, and if these are generally large polynomial systems, why then do stability questions about them remain unasked? And if differential equations such as these underlie chemical systems generally, how can it be that chemistry students, unlike physics students, might pass through their curriculum unencumbered by any knowledge of differential equations?

1.4 Biochemistry Is Different from "Regular" Chemistry

In at least certain biological circles, the attitudes and expectations regarding complex (isothermal) chemistry seem somewhat different from those one finds, say, in chemical engineering circles. Although in biochemistry there is a residual predisposition to the same doctrine of stable dull behavior that permeates the engineering texts, it is clear that it is losing its grip.

No wonder. The cell is a remarkably dynamic and complex piece of chemical machinery, filled with a variety of switches that respond to a variety of signals. It is true that cellular chemistry is, in many ways, very different from normal engineering chemistry, for the cell is filled with very large molecules crowded into a very small space. Certainly some cellular switching events involve chemical entities (e.g.,

[4] It is interesting, however, to recall a 1955 chemical engineering paper by Bilous and Amundson [22], which was seminal in its consideration of stability questions. It too focused primarily on simple *nonisothermal* reactors and set in motion a burgeoning research effort about them, lasting decades and spanning many universities. Little noticed, though, was a section of the paper in which the authors examined a four-species, three-reaction *isothermal* system. They found it to be stable for all parameter values. Nevertheless, Bilous and Amundson understood that the stability question was one worth asking.

genes) that are too few in number to be governed by differential equations one would normally write for homogenous solutions. But still other switching components are likely to be found entirely in the cytoplasm and at the more traditional level of solution chemistry, involving enzymes, their substrates, and their products.

In fact, there is now a burgeoning literature addressed to the rich dynamics that *biochemical* reaction networks might exhibit, but it will be useful to focus on a much-cited paper by Huang and Ferrell [110] because it is unusually explicit in laying bare the full set of elementary reactions and differential equations thought to be operative in an important signaling network, the mitogen-activated protein kinase cascade. The paper was not overtly concerned with stability issues, but in a subsequent paper by Qiao and Shvartsman et al. [137], the Huang-Ferrell network and the resulting differential equations were shown to be rich in their capacity for multiple steady states and sustained composition oscillations.

There was no hint of "vitalism" in any of this, no invocation of new mechanisms fashioned to explain switch-like or clock-like behavior in living systems. *Instead, the presumed kinetics governing the multienzyme Huang-Ferrell reaction network was of the same (mass action) kind commonly invoked in the chemical engineering texts, and passage to the resulting system of polynomial differential equations proceeded in the same way.*

Nor was there any hint of surprise that exotic dynamics could result from classical modeling precepts *applied to biochemistry*. Indeed, it now seems to be "in the air"—if not yet in the weighty cell biology books—that bistability or oscillations arising naturally in enzyme-driven reaction networks might underlie important aspects of cell function. If these phenomena are not entirely expected in biochemical networks, they are not unexpected either. The Huang-Ferrell network, modeled in the usual way by polynomial differential equations, is considered to be interesting in its behavior, but there are no intimations that it is bizarre. It is not a platypus.

If, however, biochemical networks are governed by differential equations formulated according to the same rules that apply to reaction networks generally, and if there is a wide-ranging theoretical basis for the commonly held (and empirically motivated) doctrine that even complex reaction networks should behave stably, then *what is it about the particular mechanisms of biochemistry that seems to encourage circumvention of that doctrine*?

1.5 The Big Picture, Veiled

What might we infer from all of this? It appears that the vast landscape of all reaction networks, including highly complex ones, contains huge regions in which dull, stable behavior prevails. That behavior is sufficiently prevalent and sufficiently dull that it has rendered chemistry uninteresting to mathematicians and mathematics unimportant to at least some chemists. For the purposes of chemical engineering applications, belief in the prevalence of stable behavior is so compelling that in the case of isothermal reactors, it seems safe to leave stability questions unconsidered, even when the chemistry involves many species and many reactions.

Yet, in the reaction network landscape, there are isolated places where a platypus, such as the Belousov-Zhabotinsky example, will raise its head above ground. And there are broader patches too in which instability is not so alien and where some—perhaps much—of biochemistry resides.

The fact is that, in the roughly isothermal chemical world we inhabit, things *are* very dull. At least on the surface, we are not surrounded by dynamical surprises that are hard to anticipate. Dependable stable behavior is almost certainly important to us: it would be difficult for our lives to play out against a wildly unpredictable chemical background.

At the same time, life—literally speaking—is not dull. The chemical soup that comprises each of us does not behave as the chemical engineer's intuition (or the doctrine of stable reactor behavior) might suggest. Against a largely stable and reliable chemical background, biochemistry seems to provide an exquisite overlay of dynamism, with switches and oscillators playing starring roles. Yet, as in the Huang-Ferrell example, passage from a (mass action) biochemical network to the resulting polynomial differential equations proceeds just as it does for the thousands of networks that chemical engineers might confront at one time or another and for which they would presumably intuit boring dynamics (almost always correctly).

What is it, then, that imparts dull behavior across vast regions of the reaction network landscape but yet permits in that same landscape occasional eruptions of more exotic behavior? And what is it about biochemical networks that causes them to reside in that part of the landscape more prone to exotica? Can we somehow make sense of this landscape by drawing connections between the structure of a reaction network and the kinds of behavior its differential equations might admit?

A primary goal of this book is to do just that. In Part II, meant for mathematicians and non-mathematicians alike, I will present some theorems that take us a long way toward understanding the preponderance of dull behavior across the reaction network landscape. One of these theorems will also indicate how, remarkably, some of the most fundamental mechanisms of enzyme catalysis circumvent the prevalence of stable behavior found more generally.[5] The puzzles and clues of this chapter will be discussed once again in the very last chapter of Part II, at which point we will know much more.

The next chapter is meant to provide some rudimentary preliminaries.

[5] The same is true of catalysis on metal surfaces, but it is in biochemistry that richer, more exotic behavior seems evolutionarily purposeful.

2

Chemical and Notational Preliminaries

In this chapter I will provide some chemical preliminaries—essentially, a description of the very natural way that a chemical reaction network induces a system of differential equations—and I will also provide an indication of the kinds of questions we will ask. Finally, at the end of the chapter, I will discuss the notational scheme—one common in graph theory—that is used throughout the book. Although for some readers the scheme might be a little unfamiliar, it is simple and will serve us well over the long run.

2.1 How the Differential Equations of Chemistry Come About

Here I shall try to explain in an informal way how chemists and chemical engineers arrive at the differential equations they work with and how these differential equations are tied to reaction network structure. Once this is done, we can begin to understand why a reasonably general theory of chemical reaction networks is necessary. Moreover, we can begin to understand why, despite the great complexity of the differential equations involved, such a theory should even be possible in principle.

In order that I might provide some sense of how the equations of chemical dynamics come about, it will be useful if I write down an example of a reaction network and indicate informally how it induces a system of ordinary differential equations. Then I can discuss the kinds of problems we will want to consider.

Suppose that A, B, C, D and E are chemical species, and suppose I believe that the chemical reactions occurring among these species are reasonably well-reflected in the display shown as (2.1). What I have written down is a diagram of a chemical

$$\begin{array}{ccc} A & \rightleftarrows & 2B \\ A+C & \rightleftarrows & D \\ \nwarrow & & \swarrow \\ & B+E & \end{array} \tag{2.1}$$

M. Feinberg, *Foundations of Chemical Reaction Network Theory*,
Applied Mathematical Sciences 202,
https://doi.org/10.1007/978-3-030-03858-8_2

reaction network. It indicates that a molecule of A can decompose into two molecules of B, that two molecules of B can react to form one molecule of A, that a molecule of A can react with a molecule of C to form a molecule of D, and so on.

Now suppose that I throw various amounts of my species into a pot. I am going to presume that the pot is stirred constantly so that its contents remain spatially homogeneous for all time, and I shall also suppose that the contents of the pot are forever maintained at fixed temperature and total volume. This, of course, is not to say that the chemical composition of the mixture within the pot will remain constant in time, for the occurrence of chemical reactions will serve to consume certain species and generate others. In fact, it is the temporal evolution of the composition that we wish to investigate. With this in mind, we denote the values of the molar concentrations of the species at time t by $c_A(t), c_B(t), c_C(t), c_D(t)$, and $c_E(t)$, and we abbreviate this list of numbers by the "composition vector" $c(t)$.[1] We shall be somewhat more precise about what we mean by the "composition vector" in Section 2.2. Thus the picture we are thinking about, at least for the moment, looks something like that shown in Figure 2.1.[2]

Fig. 2.1. A simple well-stirred reactor

[1] A molar concentration, say c_A, specifies the number of A molecules per unit volume of mixture. More precisely, c_A is the number of A molecules per unit volume divided by Avogadro's number, 6.023×10^{23}.

[2] The reactor depicted in Figure 2.1 is closed with respect to the exchange of matter with its environment. Our focus on such reactors is *temporary* and is merely intended to illustrate in a simple context how chemists and engineers formulate differential equations based upon a set of reactions believed to approximate the true chemistry. We will begin to consider "open" reactors in Chapter 4. There we shall indicate how open reactors can be modeled in terms of reaction networks and how the appropriate differential equations, like those for closed reactors, bear a definite relationship to reaction network structure.

2.1.1 General Kinetics

We would like to write down differential equations that describe the evolution of the five molar concentrations. Since chemical reactions are the source of composition changes, the key to understanding how to write down differential equations lies in knowing how rapidly each of the several reactions occurs. What is generally assumed is that the instantaneous occurrence rate of each reaction depends in its own way on the instantaneous mixture composition vector, c. Thus, we presume, for example, the existence of a nonnegative real-valued rate function $\mathscr{K}_{A \to 2B}(\cdot)$ such that $\mathscr{K}_{A \to 2B}(c)$ is the instantaneous occurrence rate of reaction $A \to 2B$ (per unit volume of mixture) when the instantaneous mixture composition is given by the vector c.[3] Similarly, we presume the existence of a rate function $\mathscr{K}_{2B \to A}(\cdot)$ for the reaction $2B \to A$, a rate function $\mathscr{K}_{A+C \to D}(\cdot)$ for the reaction $A + C \to D$, and so on. A *kinetics* for a reaction network is an assignment of a rate function to each reaction in the network.

Once we presume that network (2.1) is endowed with a kinetics, we are in a position to write down the system of differential equations that govern our reactor. Suppose that, at some instant, the reactor is in some composition state c. Let us begin by thinking about the instantaneous rate of change of c_A. Every time the reaction $A \to 2B$ occurs, we *lose* a molecule of A, and that reaction has an occurrence rate $\mathscr{K}_{A \to 2B}(c)$. On the other hand, every time the reaction $2B \to A$ occurs, we *gain* a molecule of A, and that reaction occurs at rate $\mathscr{K}_{2B \to A}(c)$. Similarly, the reactions $B + E \to A + C$ and $D \to A + C$ produce a molecule of A with each occurrence, while each occurrence of the reaction $A + C \to D$ results in the loss of a molecule of A. Thus we write

$$\dot{c}_A = -\mathscr{K}_{A \to 2B}(c) + \mathscr{K}_{2B \to A}(c) - \mathscr{K}_{A+C \to D}(c) + \mathscr{K}_{D \to A+C}(c) + \mathscr{K}_{B+E \to A+C}(c).$$

If we turn our attention to species B, we notice that whenever the reaction $A \to 2B$ occurs, we gain *two* molecules of B, and whenever $2B \to A$ occurs, we lose *two* molecules of B. When $D \to B + E$ occurs, we gain *one* B, and when $B + E \to A + C$ occurs, we lose *one* B. With this in mind, we write

$$\dot{c}_B = 2\mathscr{K}_{A \to 2B}(c) - 2\mathscr{K}_{2B \to A}(c) + \mathscr{K}_{D \to B+E}(c) - \mathscr{K}_{B+E \to A+C}(c).$$

Continuing in this way, we can write down equations for c_C, c_D, and c_E to generate the full system of differential equations that govern our reactor:

$$\begin{aligned}
\dot{c}_A &= \mathscr{K}_{A \to 2B}(c) + \mathscr{K}_{2B \to A}(c) - \mathscr{K}_{A+C \to D}(c) + \mathscr{K}_{D \to A+C}(c) + \mathscr{K}_{B+E \to A+C}(c).\\
\dot{c}_B &= 2\mathscr{K}_{A \to 2B}(c) - 2\mathscr{K}_{2B \to A}(c) + \mathscr{K}_{D \to B+E}(c) - \mathscr{K}_{B+E \to A+C}(c).\\
\dot{c}_C &= -\mathscr{K}_{A+C \to D}(c) + \mathscr{K}_{D \to A+C}(c) + \mathscr{K}_{B+E \to A+C}(c)\\
\dot{c}_D &= \mathscr{K}_{A+C \to D}(c) - \mathscr{K}_{D \to A+C}(c) - \mathscr{K}_{D \to B+E}(c)\\
\dot{c}_E &= \mathscr{K}_{D \to B+E}(c) - \mathscr{K}_{B+E \to A+C}(c)
\end{aligned} \tag{2.2}$$

---

[3] More precisely, $\mathscr{K}_{A \to 2B}(c)$ is the number of occurrences of $A \to 2B$ per unit time per unit volume divided by Avogadro's number.

2.1.2 Mass Action Kinetics

Thus far we haven't said anything about the nature of rate functions, and that is what we shall do now. More often than not, chemists and engineers presume the kinetics to be of *mass action* type [169]. With mass action kinetics, one can merely look at a reaction and write down its rate function up to a multiplicative positive constant.

Here's the way things work: For the reaction $A \to 2B$, we presume that the more A there is in the reactor, the more occurrences of the reaction there will be. In fact, we presume that the instantaneous occurrence rate (per unit mixture volume) of $A \to 2B$ is proportional to the instantaneous value of c_A. Thus, we write

$$\mathscr{K}_{A\to 2B}(c) \equiv \alpha c_A.$$

where α is a positive constant. For the reaction $A + C \to D$, the situation is a little different. An occurrence requires that a molecule of A meet a molecule of C in the reactor, and we take the probability of such an encounter to be proportional to the product $c_A c_C$. Although we do not presume that every such encounter yields a molecule of D, we nevertheless take the occurrence rate of $A + C \to D$ to be given by

$$\mathscr{K}_{A+C\to D}(c) \equiv \gamma c_A c_C,$$

where γ is a positive constant. Similarly, an occurrence of the reaction $2B \to A$ requires that two molecules of B have an encounter, and we take

$$\mathscr{K}_{2B\to A}(c) \equiv \beta (c_B)^2,$$

where β is a positive constant.

Thus, with mass action kinetics, the rate functions for network (2.1) take the form (2.3).

$$\begin{array}{lll} \mathscr{K}_{A\to 2B}(c) \equiv \alpha c_A & \mathscr{K}_{2B\to A}(c) \equiv \beta (c_B)^2 & \mathscr{K}_{A+C\to D}(c) \equiv \gamma c_A c_C \\ \mathscr{K}_{D\to B+E}(c) \equiv \varepsilon c_D & \mathscr{K}_{D\to A+C}(c) \equiv \delta c_D & \mathscr{K}_{B+E\to A+C}(c) \equiv \xi c_B c_E \end{array} \tag{2.3}$$

The positive numbers $\alpha, \beta, \gamma, \delta, \varepsilon$, and ξ—called *rate constants* for the corresponding reactions—are sometimes estimated on the basis of chemical principles or else one makes an attempt to deduce them from experiments. When a reaction network is presumed to be endowed with mass action kinetics, it is the custom to indicate the rate constants (or symbols for them) alongside the corresponding reaction arrows in the network diagram. Thus, for our example, we might have a display such as that shown in (2.4).

$$\begin{array}{ccc} A & \underset{\beta}{\overset{\alpha}{\rightleftarrows}} & 2B \\ & & \\ A+C & \underset{\delta}{\overset{\gamma}{\rightleftarrows}} & D \\ \xi \nwarrow & & \swarrow \varepsilon \\ & B+E & \end{array} \tag{2.4}$$

If we assume mass action kinetics for the network we have been studying, the appropriate differential equations are those shown in (2.5), obtained by inserting (2.3) into (2.2):

$$\begin{aligned}
\dot{c}_A &= -\alpha c_A + \beta(c_B)^2 - \gamma c_A c_C + \delta c_D + \xi c_B c_E \\
\dot{c}_B &= 2\alpha c_A - 2\beta(c_B)^2 + \varepsilon c_D - \xi c_B c_E \\
\dot{c}_C &= -\gamma c_A c_C + \delta c_D + \xi c_B c_E \\
\dot{c}_D &= \gamma c_A c_C - (\delta + \varepsilon) c_D \\
\dot{c}_E &= \varepsilon c_D - \xi c_B c_E
\end{aligned} \tag{2.5}$$

2.1.3 Some Questions

We have arrived at a fairly concrete system of ordinary differential equations, and we can begin to pose questions about them. Here are some of the questions we might like to ask:

(a) Does the system (2.5) admit a positive equilibrium—that is, an equilibrium at which all species concentrations are positive?

(b) Does the system (2.5) admit *multiple* positive equilibria (in a sense[4] made precise in Remark 3.4.10)?

(c) Does the system (2.5) admit an *unstable* positive equilibrium?

(d) Does the system (2.5) admit a nontrivial *periodic* composition trajectory, along which all concentrations are positive?

These are not easy questions, and answers to them—see the next section—might of course depend on the particular values taken by the rate constants α, β, γ, δ, ε, and ξ. Even if we could answer these questions for all positive values of the rate constants, what would we have accomplished? We would have understood *one* model chemical system fairly well, at least with respect to certain qualitative issues. There are, however, *thousands* of distinct reaction networks that might, on one occasion or another, command our attention. Each has its own system of differential equations, perhaps more complicated by far than the system we have been considering.

How, then, are we to proceed? It is clear that we cannot rely forever on purely *ad hoc* studies of whatever systems might present themselves for examination. Even if we cast aside the long-term immensity of such an undertaking, there are still two problems that must be faced in the short run. First, questions of the kind we have posed will, for the most part, be confronted by engineers, chemists, and biologists, not mathematicians. Second, it is by no means clear that mathematicians, say those trained broadly in differential equations, are in a position to provide much help. The fact is that even moderately large systems of nonlinear differential equations—in particular polynomial systems like those displayed in (2.5)—remain poorly understood in general.

[4] We shall want to know whether there can exist multiple positive equilibria within a *stoichiometric compatibility class*. In rough terms, a stoichiometric compatibility class is a certain set of compositions which remains invariant under the flow given by (2.5)

2.1.4 Our Long-Term Objectives

It seems that what's required is a broad-based theory *of those systems of differential equations that derive from reaction networks*, a theory that would in some sense cut across the fine details of individual problems to provide qualitative information about large classes of systems *all at once*. Moreover, *we would like the results of such a theory to be of the kind that engineers, chemists, and biologists can use easily in addressing questions like those we have posed.*

This seems like a lot to ask, and I should try to explain why a theory of the kind I have in mind should even be possible in principle. Although we shall also be interested in the more general situation, let me temporarily restrict my attention to reaction networks endowed with mass action kinetics. Thinking back to the source of the system (2.5), we recall that it derived in an orderly way from the network (2.4). In fact, we knew how to write down the appropriate differential equations (up to values of the rate constants) merely from inspection of the reaction diagram. Had we begun with a different network, we would have arrived at a different system of differential equations, but again the essential shape of those equations (up to values of the rate constants) would have derived from the reaction network in a precise way. Indeed, it is the close connection between reaction network structure and the shape of the induced differential equations that lends the subject of chemical reactor theory its coherency.

This is our source of hope. If reactor behavior is determined by a system of differential equations which, in turn, is determined by the underlying reaction network in a precise way, then perhaps one can prove theorems that tie qualitative aspects of reactor behavior directly to reaction network structure.

Can this in fact be done? I hope that this book will help to demonstrate that one can proceed surprisingly far in this direction. In Part II I will state several theorems, one of which—the Deficiency Zero Theorem—will *immediately* answer all four questions that we posed about the system (2.5). The answers are *yes, no, no,* and *no*; these answers hold for *all* positive values of the rate constants α, β, γ, δ, ε, and ξ. Moreover, *these answers can be obtained merely from inspection of the reaction diagram* (2.4)*; one need not even write out the differential equations.* The fact is that one can delineate a large class of networks—some extremely complicated—for which the corresponding differential equations admit solutions of a very limited variety, regardless of the (positive) values the rate constants take.

Our objectives will be broad, and I should try to make clear what these are. Results of the kind I just described are typical of those we are after. We seek to classify reaction networks according to their capacity to induce differential equations that admit behavior of a specified type. When we restrict our attention to networks endowed with mass action kinetics, we will *not* ask, for example, whether the differential equations for a particular network taken with *specified* rate constants admit periodic orbits.

Rather, we will ask if the network is such that the induced differential equations admit periodic orbits for *at least one set* of rate constants—that is, if the network *has the capacity* to admit periodic orbits. *The network itself will be our object of study, not the network endowed with a particular set of rate constants.*[5]

2.2 About Setting and Notation

Because a reaction network is, at bottom, a graph and because our interest is in the differential equations the network induces, we will be operating in mathematical terrain where dynamical systems theory and graph theory come together. The notational traditions of the two subjects are, however, rather different, and for good reason. The conventional vector space setting, $\mathbb{R}^n$, of most modern books on differential equations can cause difficulties when graphs share the central stage. These difficulties, described next, are best circumvented at the outset.

2.2.1 What's the Problem?

The differential equations induced by a reaction network derive, to a large extent, from the network's graphical structure. I haven't been very explicit about that structure, but we can already see something of it in the example studied earlier, which is displayed again here as (2.6):

$$\begin{array}{ccc} A & \rightleftarrows & 2B \\ A+C & \rightleftarrows & D \\ & \nwarrow \quad \swarrow & \\ & B+E & \end{array} \tag{2.6}$$

We will need a little vocabulary: The objects that sit at the heads and the tails of the reaction arrows—A, $2B$, $A+C$, D, and $B+E$—are called the *complexes* of the reaction network. (This terminology was introduced by Horn and Jackson [109].)

The display shown in (2.6) is an example of a *standard reaction diagram*, in which each complex appears precisely once, with reactions indicated by arrows connecting the various complexes. In this way, a reaction network gives rise to a directed graph, the complexes playing the role of vertices and the reactions playing the role of directed edges. It should be noted, however, that there is more to the diagram (2.6) than just its graphical structure, for the complexes also carry something of an algebraic structure.

---

[5] It should be mentioned here that, in practice, complete sets of rate constants for intricate networks are hardly ever known with great precision. It is often the case that chemists have a very good sense of what reactions are occurring but can estimate or measure rate constants only to within a considerable margin of uncertainty. For a discussion of the relationship between reaction network structure and the extent to which rate constants can be determined uniquely from certain classes of experiments, see [70] and, for more detail, [119] and [68]. See also [51]. In [68] there is also a discussion of how information about the reaction network itself can, in principle, be inferred from near-equilibrium experiments.

A main preoccupation of this book will be with connections between reaction network structure (in both its graphical and algebraic aspects) and properties of the differential equations—especially the mass action differential equations—that the network induces. As I indicated earlier, much that we do will have the spirit of both dynamical systems theory and graph theory.

At least from the dynamical systems side, we will want to work in vector space settings. That is, we will want to have available ways to speak, for example, of a "composition vector" or of a "vector of rate constants." If N is the number of species and p is the number of reactions, it would seem reasonable enough to work—as tradition would have it—in the familiar vector spaces $\mathbb{R}^N$ and $\mathbb{R}^p$. (We will want vector spaces associated not only with the species and the reactions but also with the complexes. Thus, if n is the number of complexes, then $\mathbb{R}^n$ is the candidate that suggests itself first.)

It turns out, however, that $\mathbb{R}^N$, $\mathbb{R}^p$, and $\mathbb{R}^n$ are awkward settings in which to do business: At the very least, these spaces require that we number everything in sight so that we can speak of the "ith species," the "jth reaction," or the "kth complex." Thus, we must impose an artificial ordering on each of the three sets of objects even before we begin to work, and we must carry that order around thereafter, suppressing or rearranging it whenever it becomes intrusive. (Should we have to decide at the outset which complex in network (2.6) should be deemed the first and which the last?)

The same problem arises more generally in the graph-theoretical literature whenever one wants to associate a vector space with either the nodes or the edges of a graph. For example, one might have an assignment of a real number with each edge (perhaps a current), and one might want to represent that assignment vectorially (as a "current vector"). How to proceed?

Here is the sensible graph-theoretical resolution: If $\mathscr{E}$ is the set of edges, then the assignment of numbers to edges is nothing more than a function $f : \mathscr{E} \to \mathbb{R}$, where $\mathbb{R}$ is the set of real numbers. *But f is* already *a vector in the vector space of real-valued functions on $\mathscr{E}$, sometimes denoted $\mathbb{R}^{\mathscr{E}}$*. The vector space $\mathbb{R}^{\mathscr{E}}$, then, becomes the natural medium in which to work, requiring no *a priori* numbering of the edges.

This is the general approach we will take. With just a little experience, the vector space $\mathbb{R}^{\mathscr{E}}$ is no more difficult to understand than is the vector space $\mathbb{R}^p$, where p is the number of edges. As examples in the next section will make clear, the two vector spaces are very similar, but the first is unencumbered by gratuitous ordering.

2.2.2 Core Notation

If $\mathscr{I}$ is a finite set of objects (e.g., species, complexes, or reactions), then $\mathbb{R}^{\mathscr{I}}$ will denote the vector space of real-valued functions on $\mathscr{I}$. That is, a vector $x \in \mathbb{R}^{\mathscr{I}}$ associates a real number, denoted x_i, with each element $i \in \mathscr{I}$. (Addition of functions and multiplication of a function by a real number are defined in the usual way.)

Example 2.2.1. Suppose that we take $\mathscr{I} := \{1,2,\ldots,N\}$, the first N integers. In this case, $\mathbb{R}^{\mathscr{I}}$ is essentially the vector space usually denoted $\mathbb{R}^N$. That is, if x is a member of $\mathbb{R}^{\mathscr{I}}$, then x associates a number x_i with each $i \in \{1,2,\ldots,N\}$. Stated differently, x is the sequence of real numbers $[x_1,x_2,\ldots,x_N]$.

Example 2.2.2. Suppose that $\mathscr{S}$ is the set of species in a reaction network. Suppose also that, in a well-stirred reactor, the molar concentration of each species $s \in \mathscr{S}$ is given by the number c_s. Then that same information is carried by the *composition vector* $c \in \mathbb{R}^{\mathscr{S}}$: It associates with each species $s \in \mathscr{S}$ its corresponding molar concentration, c_s. In this way, the composition state of the mixture can be represented vectorially in $\mathbb{R}^{\mathscr{S}}$ without first numbering the species—a numbering that would be required for a vector representation in $\mathbb{R}^N$.

Remark 2.2.3. Comparison of the two examples should make clear that the traditional use of $\mathbb{R}^N$ derives less from necessity than from habit. In fact, geometric intuition is no more difficult to acquire in $\mathbb{R}^{\mathscr{S}}$ than it is in $\mathbb{R}^N$: In the usual Cartesian vector space depiction, one need only label the coordinate axes with the names of the species rather than with the first N integers.

We denote by $\mathbb{R}_+$ the positive real numbers and by $\overline{\mathbb{R}}_+$ the nonnegative real numbers. By $\mathbb{R}_+^{\mathscr{I}}$ [respectively, $\overline{\mathbb{R}}_+^{\mathscr{I}}$], we mean the set of members of $\mathbb{R}^{\mathscr{I}}$ that take exclusively positive [nonnegative] values.

If x and y are members of $\mathbb{R}^{\mathscr{I}}$, then by xy we mean the member of $\mathbb{R}^{\mathscr{I}}$ defined by

$$(xy)_i = x_i y_i, \quad \forall i \in \mathscr{I}. \tag{2.7}$$

If x is a member of $\mathbb{R}^{\mathscr{I}}$, then by e^x we mean the member of $\mathbb{R}_+^{\mathscr{I}}$ defined by

$$(e^x)_i = e^{x_i}, \quad \forall i \in \mathscr{I}. \tag{2.8}$$

Similarly, for $z \in \mathbb{R}_+^{\mathscr{I}}$ we define $\ln z \in \mathbb{R}^{\mathscr{I}}$ by

$$(\ln z)_i = \ln z_i, \quad \forall i \in \mathscr{I}. \tag{2.9}$$

By the *support* of a vector $x \in \mathbb{R}^{\mathscr{I}}$, denoted supp x, we mean the subset of $\mathscr{I}$ on which x takes nonzero values. That is,

$$\operatorname{supp} x := \{i \in \mathscr{I} : x_i \neq 0\}. \tag{2.10}$$

Thus, if, as in Example 2.2.2, $c \in \mathbb{R}^{\mathscr{S}}$ is vector indicating the composition state of a mixture, then $\operatorname{supp} c$ is the set of species actually present in the mixture.

When i is a member of $\mathscr{I}$, we reserve the symbol ω_i to denote the special vector of $\mathbb{R}^{\mathscr{I}}$ defined by

$$(\omega_i)_j = \begin{cases} 1 & \text{if} \quad j = i \\ 0 & \text{if} \quad j \neq i. \end{cases} \tag{2.11}$$

By the *standard basis for* $\mathbb{R}^{\mathscr{I}}$, we mean the set

$$\{\omega_i\}_{i \in \mathscr{I}}. \tag{2.12}$$

It is evident that this set is indeed linearly independent and that each $x \in \mathbb{R}^{\mathscr{I}}$ has the representation

$$x = \sum_{i \in \mathscr{I}} x_i \omega_i. \tag{2.13}$$

Remark 2.2.4. Note that if, as in Example 2.2.1, $\mathscr{I} = \{1,2,\ldots,N\}$, then the set $\{\omega_i\}_{i \in \mathscr{I}}$ can be identified with the canonical basis for $\mathbb{R}^N$.

If $\mathscr{L}$ is a subset of $\mathscr{I}$, we denote by $\omega_{\mathscr{L}}$ the vector of $\mathbb{R}^{\mathscr{I}}$ that takes the value 1 for every $i \in \mathscr{L}$ and the value 0 for every $i \notin \mathscr{L}$. Stated differently,

$$\omega_{\mathscr{L}} := \sum_{i \in \mathscr{L}} \omega_i. \tag{2.14}$$

In particular,

$$\omega_{\mathscr{I}} := \sum_{i \in \mathscr{I}} \omega_i. \tag{2.15}$$

We define the *standard scalar product* in $\mathbb{R}^{\mathscr{I}}$ as follows: If x and z are vectors in $\mathbb{R}^{\mathscr{I}}$, then

$$x \cdot z := \sum_{i \in \mathscr{I}} x_i z_i. \tag{2.16}$$

With respect to this scalar product, the standard basis for $\mathbb{R}^{\mathscr{I}}$ is orthonormal. Unless stated otherwise we will take $\mathbb{R}^{\mathscr{I}}$ to be endowed with its standard scalar product. Moreover, we will take $\mathbb{R}^{\mathscr{I}}$ to be equipped with the usual finite-dimensional vector space topology given, for example, by the norm derived from the standard scalar product.

2.2.3 A Special Case: The Vector Space Generated by the Species

There is one final matter of notation we need to consider. Suppose that $\mathscr{I}$ is a (finite) set and that $\{\omega_i\}_{i \in \mathscr{I}}$ is the standard basis for $\mathbb{R}^{\mathscr{I}}$. When the set $\mathscr{I}$ carries no algebraic structure, it is sometimes the custom to replace the symbol ω_i by i itself.[6] Thus, the symbol $i + j$ becomes an abbreviation for the vector $\omega_i + \omega_j \in \mathbb{R}^{\mathscr{I}}$. Similarly, $2j$ becomes an abbreviation for the vector $2\omega_j \in \mathbb{R}^{\mathscr{I}}$.

We shall adopt this convention when (and only when) the set in question is the set $\mathscr{S}$ of chemical species. Thus if, as in our example (2.6), $\mathscr{S} = \{A,B,C,D,E\}$, we can regard the symbol $A + C$ as a surrogate for $\omega_A + \omega_C \in \mathbb{R}^{\mathscr{S}}$, the symbol $2B$ as a surrogate for $2\omega_B \in \mathbb{R}^{\mathscr{S}}$, and so on. In this way we can regard what we have called the *complexes* of a network—A, $2B$, $A + C$, D, and $B + E$ in network (2.6)—as vectors in $\mathbb{R}^{\mathscr{S}}$ and, in particular, as vectors in $\overline{\mathbb{R}}_+^{\mathscr{S}}$. This view will permit an easy transition between our formal definition of a reaction network, given at the beginning of the next chapter, and the portrayal of a network in its usual diagrammatic form.

[6] See, for example, pp. 197–199 of [133] or pp. 240–241 of [120].

3

Reaction Networks, Kinetics, and the Induced Differential Equations

In the first three sections of this chapter, we'll make precise some of the ideas that were introduced casually in Chapter 2. Section 3.1 contains our definition of a reaction network along with a small amount of auxiliary terminology. In Section 3.2 we introduce the notion of a kinetics for a network, and we discuss mass action kinetics as the archetypal example. In Section 3.3 we indicate in vectorial terms how a kinetic system—that is, a reaction network endowed with a kinetics—induces a system of differential equations.

In Section 3.4 we begin to examine some elementary geometric connections between solutions of the differential equations for a kinetic system and the structure of the underlying reaction network. These connections are hardly deep, but our awareness of them will help set the stage for the statement of substantive theorems later on. In Section 3.5 we describe a simple structural property a network must have in order that it might give rise to an equilibrium at which all species concentrations are positive or to a cyclic composition trajectory along which all species concentrations are positive.

This chapter closes with an appendix which, while connected intimately to the main material of the chapter, is slightly more technical. This appendix can be skipped for now, but its content will find use later on.

3.1 Reaction Networks

Before placing on record our formal definition of a reaction network, I want to recall and expand upon some ideas discussed at the end of Chapter 2. Recall that the complexes of a network are just the entities that appear at the heads and tails of the reaction arrows—A, 2B, A + C, D, and B + E in network (2.6). We shall regard a network to be specified by its set $\mathscr{S}$ of species, by its set $\mathscr{C}$ of complexes, and by a "reacts to" relation $\mathscr{R}$ that indicates how the complexes are joined by reaction arrows. That is, specification of $\mathscr{R}$ amounts to specification of a subset of $\mathscr{C} \times \mathscr{C}$,

The original version of this chapter was revised: Definition 3.1.1 has been corrected. The correction to this chapter is available at https://doi.org/10.1007/978-3-030-03858-8_20

M. Feinberg, *Foundations of Chemical Reaction Network Theory*,
Applied Mathematical Sciences 202,
https://doi.org/10.1007/978-3-030-03858-8_3

the set of all ordered pairs of complexes. We will require, naturally, that no complex reacts to itself and that no complex is isolated—that is, that each complex lies either at the head or at the tail of some reaction arrow.

Recall also that, in the sense of the discussion in Section 2.2.3, we regard the complexes to be vectors in $\mathbb{R}^{\mathscr{S}}$. (In particular, they lie in $\overline{\mathbb{R}}_+^{\mathscr{S}}$.) As with any vectors of $\mathbb{R}^{\mathscr{S}}$, it makes sense to add two complexes, to subtract one complex from another, to multiply a complex by a number, and to take the scalar product of a complex with any other vector of $\mathbb{R}^{\mathscr{S}}$. We shall always use symbols such as y, y' or y'' to indicate complexes in a reaction network; it should always be kept in mind that these are citizens of $\mathbb{R}^{\mathscr{S}}$.

Definition 3.1.1. *A* **chemical reaction network** *consists of three sets:*

(i) a finite set $\mathscr{S}$, elements of which are the **species** *of the network*
(ii) a finite set $\mathscr{C}$ of vectors in $\overline{\mathbb{R}}_+^{\mathscr{S}}$ called the **complexes** *of the network*
(iii) a set $\mathscr{R} \subset \mathscr{C} \times \mathscr{C}$ such that
  (a) for each $y \in \mathscr{C}$, $(y,y) \notin \mathscr{R}$
  (b) for each $y \in \mathscr{C}$ there is a $y' \in \mathscr{C}$ such that $(y,y') \in \mathscr{R}$ or $(y',y) \in \mathscr{R}$.

Members of $\mathscr{R}$ are the **reactions** *of the network. For each (y,y') in $\mathscr{R}$, we say that complex y* **reacts to** *complex y', and we write the more suggestive $y \to y'$ in place of $(y,y') \in \mathscr{R}$ if and only if y reacts to y'. The vector y is called the* **reactant complex** *of the reaction $y \to y'$, and y' is called its* **product complex**.

Example 3.1.2. For network (2.6)

$$\begin{aligned}
\mathscr{S} &= \{A,B,C,D,E\} \\
\mathscr{C} &= \{A,2B,A+C,D,B+E\} \subset \overline{\mathbb{R}}_+^{\mathscr{S}} \\
\mathscr{R} &= \{A \to 2B, 2B \to A, A+C \to D, D \to A+C, D \to B+E, B+E \to A+C\}
\end{aligned}$$

Remark 3.1.3. We will reserve the symbol n to denote the number of complexes in a given network.

Remark 3.1.4. The component y_s (corresponding to species $s \in \mathscr{S}$) of the complex $y \in \overline{\mathbb{R}}_+^{\mathscr{S}}$ is usually called by chemists the *stoichiometric coefficient* of species s in complex y.[1] For example, in the complex $A+C$ of network (2.6), the stoichiometric coefficient of A is 1, the stoichiometric coefficient of C is 1, and the stoichiometric coefficients of B, D, and E are 0. In the complex $2B$, the stoichiometric coefficient of B is 2, and the stoichiometric coefficients of A, C, D, and E are 0.

[1] Ugly though it is, the word "stoichiometry" has come to occupy an important place in the vocabulary of chemists and chemical engineers. Rutherford Aris, who knew about such things, traced the word back to its Greek roots—in fact, back to Plato's discussion of the material elements. He—Aris, not Plato—asserted, "Stoichiometry literally means the measurement of the elements, but the word is commonly used to refer to all manner of calculations regarding the components of a chemical system ... Stoichiometry is essentially the bookkeeping of the material components of the chemical system" [9]

In preparation for the next section, we note that, because a complex y is a member of $\mathbb{R}^{\mathscr{S}}$, it makes sense to speak about the *support* of y (Section 2.2.2). In fact, supp y is just the set of species that have nonzero stoichiometric coefficients in complex y. Thus, for example, supp $A+B = \{A,B\}$ and supp $2B = \{B\}$. The support of a complex is, in effect, the set of species that "appear in" that complex.

3.2 Kinetics

We introduced the notion of a kinetics for a reaction network in an informal way in Chapter 2. Here we will be a little more precise about what we mean.

3.2.1 General Kinetics Revisited

Definition 3.2.1. *A* **kinetics** *for a reaction network* $\{\mathscr{S},\mathscr{C},\mathscr{R}\}$ *is an assignment to each reaction* $y \to y' \in \mathscr{R}$ *of a continuously differentiable* **rate function** $\mathscr{K}_{y\to y'} : \overline{\mathbb{R}}_+^{\mathscr{S}} \to \overline{\mathbb{R}}_+$ *such that*

$$\mathscr{K}_{y\to y'}(c) > 0 \textit{ if and only if } \operatorname{supp} y \subset \operatorname{supp} c. \tag{3.1}$$

Interpretation: In the context of Chapter 2, $\mathscr{K}_{y\to y'}(c)$ was the occurrence rate of the reaction $y \to y'$ when the mixture under study had composition c. There we required only that the rate functions take nonnegative values. In condition (3.1), we go a little further by delineating those $c \in \overline{\mathbb{R}}_+^{\mathscr{S}}$ for which the function $\mathscr{K}_{y\to y'}(\cdot)$ takes *positive* values.

Note that, for the reaction $y \to y'$, supp y is just the set of species appearing in the reactant complex y. (Thus, for $A+C \to D$, $\operatorname{supp}(A+C) = \{A,C\}$.) Roughly speaking, supp y is the set of "ingredients" required for the occurrence of $y \to y'$. On the other hand, if c is the instantaneous composition state of the mixture, then supp c is the set of species that are actually present—*i.e.*, those species that have nonzero concentrations.

In words, then, condition (3.1) says this: *Reaction* $y \to y'$ *proceeds at nonzero rate, however slowly, if and only if all species appearing in the* reactant *complex* y *are actually present in the mixture.* For example, $A+C \to D$ proceeds at nonzero rate if and only if c_A and c_C are both positive.

Remark 3.2.2. We will say that c is a *positive composition* if it is a member of $\mathbb{R}_+^{\mathscr{S}}$—that is, if all species concentrations are strictly positive. In this case supp $c = \mathscr{S}$, whereupon supp $y \subset$ supp c for all $y \in \mathscr{C}$. Thus, if c is a positive composition, we have

$$\mathscr{K}_{y\to y'}(c) > 0 \quad \text{for all} \quad y \to y' \in \mathscr{R}, \tag{3.2}$$

which is to say that all reactions are "switched on" at composition c.

Remark 3.2.3. We required continuous differentiability of the rate functions only to avoid a more fussy treatment. For much of what we say, continuity will suffice, and for some results we will not even need that. Sometimes when stating one theorem or another, we will stipulate that the rate functions should be continuous or differentiable, even though those attributes are already built into the definition of a kinetics. In such cases, the intent is to indicate, in a particular instance, what specific degree of smoothness is actually required. When the kinetics is mass action, we will have all the smoothness we need, at least when the stoichiometric coefficients are restricted to be nonnegative integers.

Remark 3.2.4. A kinetics for a reaction network $\{\mathscr{S},\mathscr{C},\mathscr{R}\}$ is essentially a function $\mathscr{K}$ that assigns to each reaction $y \to y' \in \mathscr{R}$ the corresponding rate function $\mathscr{K}_{y\to y'}(\cdot)$. Thus, we shall often refer to "the kinetics $\mathscr{K}$."

Definition 3.2.5. *A* **kinetic system** $\{\mathscr{S},\mathscr{C},\mathscr{R},\mathscr{K}\}$ *is a reaction network* $\{\mathscr{S},\mathscr{C},\mathscr{R}\}$ *taken with a kinetics* $\mathscr{K}$.

3.2.2 Mass Action Kinetics Revisited

Next we consider once again the archetypal example, mass action kinetics, this time more formally.

Definition 3.2.6. *A kinetics* $\mathscr{K}$ *for a reaction network* $\{\mathscr{S},\mathscr{C},\mathscr{R}\}$ *is* **mass action** *if, for each* $y \to y' \in \mathscr{R}$, *there is a positive number* $k_{y\to y'}$ *such that*

$$\mathscr{K}_{y\to y'}(c) \equiv k_{y\to y'} \prod_{s\in\mathscr{S}} c_s^{y_s}. \tag{3.3}$$

The number $k_{y\to y'}$ *is the* **rate constant** *for the reaction* $y \to y'$.

Remark 3.2.7. It should be noted that in (3.3), y_s is the stoichiometric coefficient of species s in the *reactant* complex y of the reaction $y \to y'$. Thus, for reaction A + C → D of network (2.6), equation (3.3) becomes

$$\mathscr{K}_{A+C\to D}(c) \equiv k_{A+C\to D}(c_A)^1(c_B)^0(c_C)^1(c_D)^0(c_E)^0 = k_{A+C\to D}c_A c_C.$$

Similarly,

$$\mathscr{K}_{2B\to A}(c) \equiv k_{2B\to A}(c_A)^0(c_B)^2(c_C)^0(c_D)^0(c_E)^0 = k_{2B\to A}(c_B)^2$$

and so on. In the event that a particular molar concentration takes the value 0, we take $0^0 = 1$.

Remark 3.2.8. Because we will be working extensively with mass action kinetics, it will be helpful to introduce the special notation used by Horn and Jackson [109]. For c and y in $\overline{\mathbb{R}}_+^{\mathscr{S}}$, we define c^y as follows:

$$c^y := \prod_{s\in\mathscr{S}} c_s^{y_s}. \tag{3.4}$$

Thus, mass action rate functions take the form

$$\mathscr{K}_{y\to y'}(c) \equiv k_{y\to y'}c^y. \tag{3.5}$$

Remark 3.2.9. A mass action kinetics for a reaction network $\{\mathscr{S},\mathscr{C},\mathscr{R}\}$ is completely specified by an assignment to each reaction $y \to y' \in \mathscr{R}$ of a positive rate constant $k_{y\to y'}$. This is to say that a mass action kinetics is specified by an element $k \in \mathbb{R}_{+}^{\mathscr{R}}$. Thus, notwithstanding a minor abuse of language, we shall speak of a "mass action kinetics k" and of a "mass action system $\{\mathscr{S},\mathscr{C},\mathscr{R},k\}$." In this context it will be understood that k is an element of $\mathbb{R}_{+}^{\mathscr{R}}$.

Definition 3.2.10. *A* **mass action system** $\{\mathscr{S},\mathscr{C},\mathscr{R},k\}$ *is a chemical reaction network* $\{\mathscr{S},\mathscr{C},\mathscr{R}\}$ *taken with a* **rate constant specification** $k \in \mathbb{R}_{+}^{\mathscr{R}}$.

3.3 The Differential Equations for a Kinetic System

In Chapter 2 we saw by means of an example how one writes the differential equations for a kinetic system. Here the differential equations for a general kinetic system will be cast in vectorial terms. We begin with the following definition:

Definition 3.3.1. *Let* $\{\mathscr{S},\mathscr{C},\mathscr{R}\}$ *be a reaction network. The* **reaction vector** *corresponding to reaction* $y \to y' \in \mathscr{R}$ *is the vector* $y' - y \in \mathbb{R}^{\mathscr{S}}$.

Remark 3.3.2. Note that the component[2] of $y' - y$ corresponding to species $s \in \mathscr{S}$ is just $y'_s - y_s$, the difference between the stoichiometric coefficient of s in the product complex y' and its stoichiometric coefficient in the reactant complex y. This difference is the net number of molecules of s produced with each occurrence of the reaction $y \to y'$. Consider, for example, the reaction A → 2B in network (2.6), with corresponding reaction vector $2B - A$. The component corresponding to species A is -1, the component corresponding to B is 2, and the components corresponding to C, D, and E are each zero.

Definition 3.3.3. *For a kinetic system* $\{\mathscr{S},\mathscr{C},\mathscr{R},\mathscr{K}\}$*, the* **species-formation-rate function** $f(\cdot)$ *is defined by the requirement*[3] *that, for all* $c \in \overline{\mathbb{R}}_{+}^{\mathscr{S}}$,

$$f(c) := \sum_{\mathscr{R}} \mathscr{K}_{y\to y'}(c)(y' - y). \tag{3.6}$$

A composition $c^* \in \overline{\mathbb{R}}_{+}^{\mathscr{S}}$ *is an* **equilibrium** *of the kinetic system if* $f(c^*) = 0$*. A* **positive equilibrium** *is an equilibrium in* $\mathbb{R}_{+}^{\mathscr{S}}$.

Interpretation: If, in our homogeneous reactor, the instantaneous composition is c, then, for each $s \in \mathscr{S}$, $f_s(c)$ gives the instantaneous rate of generation (per unit volume of mixture) of moles of species s due to the simultaneous occurrence of all reactions in $\mathscr{R}$. Note that

[2] The components to which we refer are, of course, the components of $y' - y$ relative to the standard basis for $\mathbb{R}^{\mathscr{S}}$.

[3] The symbol $\mathscr{R}$ below a summation sign will always be understood to be an abbreviation for "$y \to y' \in \mathscr{R}$"

$$f_{\mathscr{s}}(c) := \sum_{\mathscr{R}} \mathscr{K}_{y\to y'}(c)(y'_{\mathscr{s}} - y_{\mathscr{s}}) \tag{3.7}$$

so that $f_{\mathscr{s}}(c)$ is obtained by summing all the reaction occurrence rates, each weighted by the net number of molecules of $\mathscr{s}$ produced with each occurrence of the corresponding reaction . This is essentially the idea we used in Chapter 2. At an equilibrium, the generation rate of every species is zero.

Note that the species-formation-rate function for the *mass action* kinetic system $\{\mathscr{S},\mathscr{C},\mathscr{R},k\}$ takes the form

$$f(c) := \sum_{\mathscr{R}} k_{y\to y'} c^y (y' - y). \tag{3.8}$$

By the differential equation for a kinetic system, we mean

$$\dot{c} = f(c), \tag{3.9}$$

where the overdot denotes time differentiation and $f(\cdot)$ is the species-formation-rate function. That is, for a kinetic system $\{\mathscr{S},\mathscr{C},\mathscr{R},\mathscr{K}\}$ the corresponding differential equation is

$$\boxed{\dot{c} = \sum_{\mathscr{R}} \mathscr{K}_{y\to y'}(c)(y' - y).} \tag{3.10}$$

This vector equation of course encodes a system of scalar equations, one for each species:

$$\dot{c}_{\mathscr{s}} = \sum_{\mathscr{R}} \mathscr{K}_{y\to y'}(c)(y'_{\mathscr{s}} - y_{\mathscr{s}}), \quad \forall \mathscr{s} \in \mathscr{S}. \tag{3.11}$$

Written for our sample network (2.1), the system (3.11) reduces to (2.2).

For a mass action system $\{\mathscr{S},\mathscr{C},\mathscr{R},k\}$, the governing vector differential equation becomes

$$\boxed{\dot{c} = \sum_{\mathscr{R}} k_{y\to y'} c^y (y' - y),} \tag{3.12}$$

where, it will be recalled, c^y is given by (3.4). The component form is

$$\dot{c}_{\mathscr{s}} = \sum_{\mathscr{R}} k_{y\to y'} c^y (y'_{\mathscr{s}} - y_{\mathscr{s}}), \quad \forall \mathscr{s} \in \mathscr{S}. \tag{3.13}$$

The reader can confirm that, for our illustrative mass action system (2.4), the equations given by (3.13) reduce to the polynomial system shown in (2.5).

Remark 3.3.4. (When does a polynomial system of differential equations derive from a mass action system?) Because we will be able to say a surprising amount about mass action systems, it is natural to ask when those results might be brought to bear in the study of polynomial systems of differential equations generally. As a first step in this direction, it is important to know which polynomial systems can, in fact, be viewed as having derived from a mass action system in the sense of (3.12).

This question was studied by Hárs and Tóth [103]. Consider the system

$$\dot{x}_i = f_i(x_1, x_2, \ldots, x_N), \quad i = 1, 2, \ldots, N, \tag{3.14}$$

where each $f_i(\cdot,\cdot,\ldots,\cdot)$ is a polynomial in its variables. Then there is mass action system that gives rise to the differential equations (3.14) if and only if, for each $i = 1, 2, \ldots, N$, every negative monomial of $f_i(\cdot,\cdot,\ldots,\cdot)$ contains the variable x_i raised to a nonzero power. See also [66].

3.4 Stoichiometric Compatibility

In this section we will explore an elementary but extremely important property of the differential equations induced by a kinetic system. The essential idea is that, *regardless of the kinetics*, reaction network structure *alone* imposes restrictions on the way that composition trajectories can look. In particular, a trajectory that passes through composition c can eventually reach composition c' only if the pair (c', c) is compatible with certain "stoichiometric" conditions the network itself imposes. In very rough terms, composition trajectories are not generally free to wander in an arbitrary fashion through $\overline{\mathbb{R}}_+^{\mathscr{S}}$ because there are only certain directions in which the species-formation-rate vector can point.

To see this we consider a reaction system $\{\mathscr{S}, \mathscr{C}, \mathscr{R}, \mathscr{K}\}$. Again, the species-formation-rate function is given by

$$f(c) := \sum_{\mathscr{R}} \mathscr{K}_{y \to y'}(c)(y' - y). \tag{3.15}$$

Thus, for each $c \in \overline{\mathbb{R}}_+^{\mathscr{S}}$, $f(c)$ is a nonnegative linear combination of the reaction vectors for the network $\{\mathscr{S}, \mathscr{C}, \mathscr{R}\}$. In particular, for $c \in \mathbb{R}_+^{\mathscr{S}}$, $f(c)$ is a positive linear combination of the reaction vectors. In any case $f(c)$ must point along the cone generated by the reaction vectors and must certainly lie in the linear subspace of $\mathbb{R}^{\mathscr{S}}$ spanned by them. This last idea serves as motivation for our next definition.

Definition 3.4.1. *The* **stoichiometric subspace** *for a reaction network* $\{\mathscr{S}, \mathscr{C}, \mathscr{R}\}$ *is the linear subspace* $S \subset \mathbb{R}^{\mathscr{S}}$ *defined by*

$$S := \operatorname{span}\{y' - y \in \mathbb{R}^{\mathscr{S}} : y \to y' \in \mathscr{R}\}. \tag{3.16}$$

By the span of the set of reaction vectors, we mean of course the linear subspace of $\mathbb{R}^{\mathscr{S}}$ consisting of all real linear combinations of them.

Since the species-formation-rate function for a kinetic system must take values in the stoichiometric subspace for the underlying reaction network, we shall want to know something about the dimension of that subspace. With this in mind, we record the following definition:

Definition 3.4.2. *The* **rank** *of a reaction network is the rank of its set of reaction vectors. That is, the network* $\{\mathscr{S},\mathscr{C},\mathscr{R}\}$ *has rank s if there exists in the set*

$$\{y'-y \in \mathbb{R}^{\mathscr{S}} : y \to y' \in \mathscr{R}\}$$

a linearly independent subset containing s vectors but no linear independent subset containing $s+1$ *vectors. We reserve the symbol s to denote the rank of a network.*

Remark 3.4.3. From elementary considerations in linear algebra, it follows that the dimension of the stoichiometric subspace for a reaction network is just the rank of the network:

$$s = \dim S. \tag{3.17}$$

Example 3.4.4. For network (2.6) the reaction vectors are

$$\{2B-A,\ A-2B,\ D-A-C,\ A+C-D,\ B+E-D,\ A+C-B-E\} \tag{3.18}$$

The three-element subset

$$\{2B-A,\ A+C-D,\ B+E-D\} \tag{3.19}$$

is linearly independent, and any vector in (3.18) can be written as a linear combination of (3.19). Thus, $s = 3$ so that the stoichiometric subspace for network (2.6) is a three-dimensional linear subspace of the five-dimensional ambient vector space $\mathbb{R}^{\mathscr{S}}$.

We have already noted that limitations on the directions in which species-formation-rate vectors can point place restrictions on the way composition trajectories can evolve. The following lemma describes such a restriction.

Lemma 3.4.5. *Let* $\{\mathscr{S},\mathscr{C},\mathscr{R},\mathscr{K}\}$ *be a kinetic system, and let* $c(\cdot): I \to \overline{\mathbb{R}}_+^{\mathscr{S}}$ *be a solution of* (3.10)*, where* $I \subset \mathbb{R}$ *is an interval. Then, for each* $t_1, t_2 \in I$ *with* $t_2 > t_1$*, there exists* $\alpha \in \overline{\mathbb{R}}_+^{\mathscr{R}}$ *such that*

$$c(t_2) - c(t_1) = \sum_{y \to y' \in \mathscr{R}} \alpha_{y \to y'}(y'-y). \tag{3.20}$$

If, for some t^* with $t_1 < t^* < t_2$, $c(t^*)$ *lies in* $\mathbb{R}_+^{\mathscr{S}}$*, then there is an* $\alpha \in \mathbb{R}_+^{\mathscr{R}}$ *that is a solution of* (3.20).

Proof. Integrating (3.10) between t_1 and t_2 along the solution $c(\cdot)$, we obtain

$$c(t_2) - c(t_1) = \sum_{y \to y' \in \mathscr{R}} \left(\int_{t_1}^{t_2} \mathscr{K}_{y \to y'}(c(t))\, dt\right)(y'-y). \tag{3.21}$$

Noting that each integrand in (3.21) is nonnegative, we set

$$\alpha_{y \to y'} := \int_{t_1}^{t_2} \mathscr{K}_{y \to y'}(c(t))\, dt, \quad \forall y \to y' \in \mathscr{R} \tag{3.22}$$

to get $\alpha \in \overline{\mathbb{R}}_+^{\mathscr{R}}$ satisfying (3.20). Finally we note that if for some t^* with $t_1 < t^* < t_2$, $c(t^*)$ lies in $\mathbb{R}_+^{\mathscr{S}}$, then $\mathscr{K}_{y \to y'}(c(t^*))$ is positive for all $y \to y' \in \mathscr{R}$ by virtue of (3.1) and the fact that $\operatorname{supp} c(t^*) = \mathscr{S}$. In this case, α resides in $\mathbb{R}_+^{\mathscr{R}}$. □

Lemma 3.4.5 tells us that a composition c' can follow a composition c along a solution of (3.10) only if $c' - c$ lies in the stoichiometric subspace (and, in particular, the stoichiometric cone[4]) for the network $\{\mathscr{S},\mathscr{C},\mathscr{R}\}$. Thus, if $c(\cdot) : I \to \overline{\mathbb{R}}_+^{\mathscr{S}}$ is a solution of (3.10) that passes through a composition c^0, then, for all $t \in I$, we must have

$$c(t) \in (c^0 + S) \cap \overline{\mathbb{R}}_+^{\mathscr{S}}. \tag{3.23}$$

where S is the stoichiometric subspace and the set

$$c^0 + S := \{c^0 + \sigma \in \mathbb{R}^{\mathscr{S}} : \sigma \in S\} \tag{3.24}$$

is S shifted by addition of the vector c^0. Thus, a composition c can lie on a trajectory passing through c^0 only if c and c^0 are "stoichiometrically compatible." This provides motivation for the following definition, after which we will discuss geometric implications.

Definition 3.4.6. *Let $\{\mathscr{S},\mathscr{C},\mathscr{R}\}$ be a reaction network, and let $S \subset \mathbb{R}^{\mathscr{S}}$ be its stoichiometric subspace. Two vectors c and $c' \in \overline{\mathbb{R}}_+^{\mathscr{S}}$ are* **stoichiometrically compatible** *if $c' - c$ lies in S. Stoichiometric compatibility is an equivalence relation that induces a partition of $\overline{\mathbb{R}}_+^{\mathscr{S}}$ [resp., $\mathbb{R}_+^{\mathscr{S}}$] into equivalence classes called the* **stoichiometric compatibility classes** *[resp.,* **positive stoichiometric compatibility classes***] for the network. In particular, the stoichiometric compatibility class containing $c \in \overline{\mathbb{R}}_+^{\mathscr{S}}$ is the set $(c+S) \cap \overline{\mathbb{R}}_+^{\mathscr{S}}$, and the positive stoichiometric compatibility class containing $c \in \mathbb{R}_+^{\mathscr{S}}$ is the set $(c+S) \cap \mathbb{R}_+^{\mathscr{S}}$.*

Our considerations thus far give some geometric insight into the way phase portraits are structured. Regardless of the kinetics, a composition trajectory passing through composition c^0 must lie entirely within the stoichiometric compatibility class containing c^0. In rough terms, this stoichiometric compatibility class is obtained by shifting S up to c^0 (by parallel translation) and intersecting the resulting parallel of S with $\overline{\mathbb{R}}_+^{\mathscr{S}}$. Some examples might be helpful.

Example 3.4.7. Consider the simple network

$$A \rightleftarrows 2B. \tag{3.25}$$

Here $\mathscr{S} = \{A,B\}$, and the reaction vectors are $\{A - 2B,\ 2B - A\} \subset \mathbb{R}^{\mathscr{S}}$. The rank of the network is clearly one, and the stoichiometric subspace S is one-dimensional: it is the line in $\mathbb{R}^{\mathscr{S}}$ containing the vector $A - 2B$. The stoichiometric compatibility classes are those parts of lines parallel to S that lie in $\overline{\mathbb{R}}_+^{\mathscr{S}}$. Regardless of the kinetics, each composition trajectory compatible with the induced differential equations lies entirely within a stoichiometric compatibility class.

---

[4] The *stoichiometric cone* is the set of all nonnegative linear combinations of the reaction vectors. See Remark 3.4.11.

For illustrative purposes we shall suppose that network (3.25) is endowed with mass action kinetics. In this case the appropriate differential equations are

$$\begin{aligned} \dot{c}_A &= k_{2B \to A}(c_B)^2 - k_{A \to 2B}\, c_A \\ \dot{c}_B &= 2k_{A \to 2B}\, c_A - 2k_{2B \to A}(c_B)^2. \end{aligned} \tag{3.26}$$

The set of equilibrium points for (3.26) is given by those $c \in \overline{\mathbb{R}}_+^{\mathscr{S}}$ that satisfy

$$c_A = \frac{k_{2B \to A}}{k_{A \to 2B}}(c_B)^2. \tag{3.27}$$

The phase portrait for (3.26) is sketched in Figure 3.1.

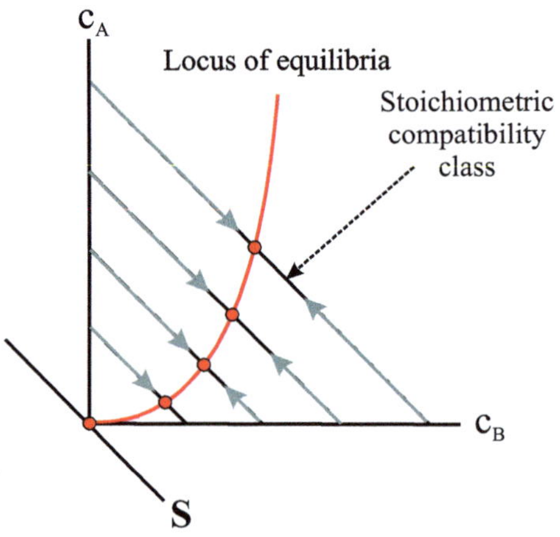

Fig. 3.1. Phase portrait for Example 3.4.7

Example 3.4.8. Consider the network shown in (3.28).

$$\begin{array}{ccc} 2A & \rightleftarrows & B \\ \nwarrow & & \swarrow \\ & C & \end{array} \tag{3.28}$$

Here $\mathscr{S} = \{A, B, C\}$, and the four reaction vectors are

$$\{B - 2A,\ 2A - B,\ C - B,\ 2A - C\} \subset \mathbb{R}^{\mathscr{S}}. \tag{3.29}$$

The rank of the network is readily confirmed to be two so that the stoichiometric subspace S is two-dimensional. The stoichiometric compatibility classes, as indicated in Figure 3.2, are those triangles that are the intersection of parallels of S with $\overline{\mathbb{R}}_+^{\mathscr{S}}$. As the figure is intended to suggest, a composition trajectory must lie entirely within a stoichiometric compatibility class.

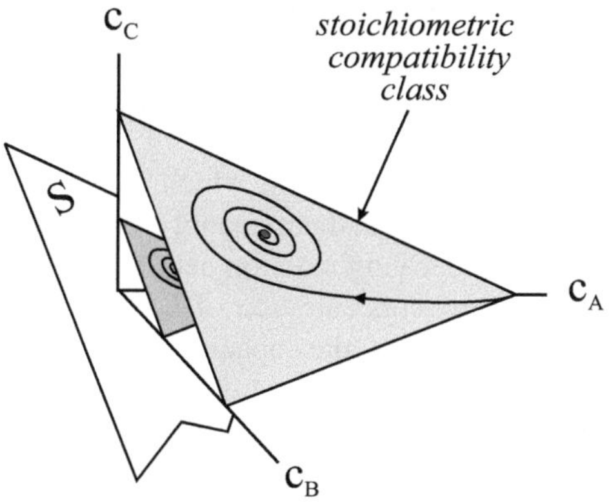

Fig. 3.2. Phase portrait for Example 3.4.8

Remark 3.4.9. In each of these examples, the stoichiometric subspace has dimension one less than the number of species. Of course, this will not always be the case. For example, the simple three-species network $A + B \rightleftarrows C$ has a one-dimensional stoichiometric subspace, and the stoichiometric compatibility classes are line segments.

Remark 3.4.10. Note that in Figure 3.1 there is precisely one equilibrium in each stoichiometric compatibility class (in fact, in each positive stoichiometric compatibility class), and the situation, as depicted, is the same in Figure 3.2. When, for a kinetic system, we ask about the existence of multiple equilibria, we shall always have in mind the existence of multiple equilibria *within the same stoichiometric compatibility class*. Similarly, when we ask about the stability of an equilibrium, we shall always have in mind stability against composition perturbations *within the same stoichiometric compatibility class*.

Remark 3.4.11. The discussion just prior to Definition 3.4.1 suggests that we should also have interest in the *stoichiometric cone* for a network, defined to be the set of all nonnegative linear combinations of its reaction vectors. It is easy to see, however, that the stoichiometric cone and stoichiometric subspace for a network coincide when the zero vector of $\mathbb{R}^{\mathscr{S}}$ is representable as a positive linear combination of the reaction vectors. In the next section, we shall see that this last condition must be satisfied if the network is consistent with the existence of a positive equilibrium. Since we shall focus heavily on situations in which there does exist a positive equilibrium, there will be no distinction between the stoichiometric cone and the stoichiometric subspace in much that we do.

3.5 An Elementary Necessary Condition for the Existence of a Positive Equilibrium or a Cyclic Composition Trajectory Passing Through a Positive Composition

Recall that a positive equilibrium is one in which all species concentrations are strictly positive. It is easy to see that there cannot be a positive equilibrium for a kinetic system in which the underlying reaction network is $A + B \to C \to 2A$: as long as both A and B are present, the concentration of B will decrease. This example is a simple one. In more complex cases, the capacity of a reaction network to admit a positive equilibrium is sometimes less transparent.

Suppose that a kinetic system $\{\mathscr{S},\mathscr{C},\mathscr{R},\mathscr{K}\}$ does admit a positive equilibrium c^*, which is to say that

$$\sum_{\mathscr{R}} \mathscr{K}_{y\to y'}(c^*)(y'-y) = 0. \tag{3.30}$$

Because c^* is a positive equilibrium, we have $\operatorname{supp} c^* = \mathscr{S}$ so that $\operatorname{supp} y \subset \operatorname{supp} c^*$ for all $y \in \mathscr{C}$. From the properties of a kinetics, it follows that $\mathscr{K}_{y\to y'}(c^*)$ is positive for each $y \to y' \in \mathscr{R}$. In this case the reaction vectors are *positively dependent*:

Definition 3.5.1. *A reaction network* $\{\mathscr{S},\mathscr{C},\mathscr{R}\}$ *has* **positively dependent** *reaction vectors if there is a set of positive numbers* $\{\alpha_{y\to y'}\}_{y\to y' \in \mathscr{R}}$ *such that*

$$\sum_{\mathscr{R}} \alpha_{y\to y'}(y'-y) = 0. \tag{3.31}$$

Now suppose that the differential equation for a kinetic system $\{\mathscr{S},\mathscr{C},\mathscr{R},\mathscr{K}\}$ gives rise to a cyclic composition trajectory that passes through a positive composition. That is, suppose that the differential equation admits solution $c(\cdot): I \to \overline{\mathbb{R}}_+^{\mathscr{S}}$, where $I \subset \mathbb{R}$ is an interval, such that for distinct t_1 and t_2 in I we have $c(t_1) = c(t_2) \in \overline{\mathbb{R}}_+^{\mathscr{S}}$ and also $c(t^*) \in \mathbb{R}_+^{\mathscr{S}}$ for some t^* between t_1 and t_2. From Lemma 3.4.5 it follows that the reaction vectors for $\{\mathscr{S},\mathscr{C},\mathscr{R}\}$ must be positively dependent.

These simple considerations are summarized in the following lemma, which will find use later on.

Lemma 3.5.2. *For the differential equation of a kinetic system to admit either a positive equilibrium or a cyclic composition trajectory that passes through a positive composition, it is necessary that the reaction vectors for the underlying reaction network be positively dependent.*

The following lemma tells us that if a reaction network has positively dependent reaction vectors, then there is always an assignment of rate constants such that the resulting mass action system admits a positive equilibrium. In fact, the lemma tells us considerably more.

Lemma 3.5.3. *Suppose that* $\{\mathscr{S},\mathscr{C},\mathscr{R}\}$ *is a reaction network with positively dependent reaction vectors. Then for any choice of positive composition* $c^* \in \mathbb{R}_+^{\mathscr{S}}$, *there is an assignment of rate constants* $k \in \mathbb{R}_+^{\mathscr{R}}$ *such that the mass action system* $\{\mathscr{S},\mathscr{C},\mathscr{R},k\}$ *admits* c^* *as an equilibrium.*

Proof. Let $\{\alpha_{y \to y'}\}_{y \to y' \in \mathscr{R}}$ be as in Definition 3.5.1. For each $y \to y' \in \mathscr{R}$ choose $k_{y \to y'}$ to satisfy $\alpha_{y \to y'} = k_{y \to y'}(c^*)^y$. □

Remark 3.5.4. Note that, when the reaction vectors are positively dependent, the lemma asserts the existence of very many assignments of rate constants such that the resulting mass action systems admit positive equilibria. However, the lemma does *not* assert that for *every* assignment of rate constants there will be a positive equilibrium, nor does the lemma assert that there is always an assignment of rate constants such that there is at least one equilibrium in every positive stoichiometric compatibility class. As we shall see in counterexamples, neither of these last assertions would be true. (See, in particular, Section 5.1.3 and Remark 8.7.1.)

3.6 About the Derivative of the Species-Formation-Rate Function

We shall often be interested in changes in the value of the species-formation-rate function that correspond to small departures from a positive composition c, but only departures toward nearby compositions that are stoichiometrically compatible with c. In this case, for a kinetic system $\{\mathscr{S}, \mathscr{C}, \mathscr{R}, \mathscr{K}\}$ with stoichiometric subspace S, with smooth reaction-rate functions, and with species-formation-rate function[5] $f : \overline{\mathbb{R}}^{\mathscr{S}}_{+} \to S$, we will want to work with the derivative $df(c) : S \to S$, given by

$$df(c)\sigma = \left. \frac{df(c + \theta\sigma)}{d\theta} \right|_{\theta=0}, \qquad \forall \sigma \in S. \tag{3.32}$$

We say that $c^* \in \mathbb{R}^{\mathscr{S}}_{+}$ is a *degenerate equilibrium* if c^* is an equilibrium and if, moreover, $df(c^*)$ is singular; otherwise c^* is a *nondegenerate equilibrium.*

Remark 3.6.1. When we refer to eigenvalues of $df(c) : S \to S$, we will mean, of course, eigenvalues associated with eigenvectors in the stoichiometric subspace.

Remark 3.6.2. For a mass action system $\{\mathscr{S}, \mathscr{C}, \mathscr{R}, k\}$ the derivative takes an interesting and useful form: Let c be a positive composition. Then, for each $\sigma \in S$,

$$df(c)\sigma = \sum_{y \to y' \in \mathscr{R}} \eta_{y \to y'}(y *_c \sigma)(y' - y), \tag{3.33}$$

where, for each $y \to y' \in \mathscr{R}$,

$$\eta_{y \to y'} := k_{y \to y'}(c)^y \tag{3.34}$$

and "$*_c$" indicates the scalar product in $\mathbb{R}^{\mathscr{S}}$ defined by

$$v *_c w := \sum_{s \in \mathscr{S}} \frac{v_s w_s}{c_s}, \qquad \forall v, w \in \mathbb{R}^{\mathscr{S}}. \tag{3.35}$$

---

[5] When the species-formation-rate function was first introduced, we did not have the idea of the stoichiometric subspace at our disposal. Now that we do, we will understand the codomain of the species-formation-rate function to be the stoichiometric subspace. That is, we shall regard species-formation-rate function to be a map $f : \overline{\mathbb{R}}^{\mathscr{S}}_{+} \to S$.

3.7 Elementary Boundary Behavior of the Differential Equations: Where Can Equilibria and Cyclic Composition Trajectories Reside?

Here we consider, for a kinetic system $\{\mathscr{S},\mathscr{C},\mathscr{R},\mathscr{K}\}$, elementary behavior of the differential equations on the boundary of $\overline{\mathbb{R}}_+^{\mathscr{S}}$. This section can be passed over without substantial loss of continuity.

At a composition on the boundary of $\overline{\mathbb{R}}_+^{\mathscr{S}}$, certain species are present, while still others are absent. At such a composition, the absence of one or another species might mean that certain reactions do not proceed at all: they are "switched off" simply because their reactant species are not available. In this section we consider where on the boundary of $\overline{\mathbb{R}}_+^{\mathscr{S}}$ equilibria or cyclic composition trajectories might reside. We shall see that, in either case, only *certain* combinations of species can be present, and only *certain* combinations of reactions can be "switched on."

We begin by showing that, for any kinetic system, the species-formation-rate function inherently possesses a natural property: If, for species s, composition c is such that $c_s = 0$, then, at composition c, the rate of production of s cannot be negative. In geometric terms, this means that on the boundary of the set of all possible compositions—that is, on the boundary of $\overline{\mathbb{R}}_+^{\mathscr{S}}$—$\dot{c}$ does not point outward.

In fact, we can describe regions on the boundary in which the generation rate of at least one absent species must actually be positive. Clearly, no equilibria can reside in such a region. Moreover, no point in such a region could be approached in the limit with increasing time by a solution of (3.10) originating from the interior of $\overline{\mathbb{R}}_+^{\mathscr{S}}$.

Lemma 3.7.1. *Let* $\{\mathscr{S},\mathscr{C},\mathscr{R},\mathscr{K}\}$ *be a kinetic system with species formation rate function* $f(\cdot)$. *Then, for every* $s \in \mathscr{S}$ *and every* $c \in \overline{\mathbb{R}}_+^{\mathscr{S}}$, $c_s = 0$ *implies that* $f_s(c) \geq 0$.

Proof. Let $\tilde{c} \in \overline{\mathbb{R}}_+^{\mathscr{S}}$ be such that for some $s* \in \mathscr{S}$, $\tilde{c}_{s*} = 0$. We want to show that $f_{s*}(\tilde{c}) \geq 0$.

Recall from Definition 3.2.1 that $\mathscr{K}_{y\to y'}(c) = 0$ when $\operatorname{supp} y \not\subset \operatorname{supp} c$. Thus $\mathscr{K}_{y\to y'}(\tilde{c}) = 0$ for all $y \to y' \in \mathscr{R}$ such that $y_{s*} \neq 0$. From this it follows that at composition $\tilde{c}$ the only reactions that might proceed at nonzero rates are those contained in the set

$$\mathscr{R}* := \{y \to y' \in \mathscr{R} : y_{s*} = 0\}. \tag{3.36}$$

Writing (3.7) for species $s*$, we obtain, for the composition $\tilde{c}$,

$$\begin{aligned} f_{s*}(\tilde{c}) &= \sum_{\mathscr{R}} \mathscr{K}_{y\to y'}(\tilde{c})(y'_{s*} - y_{s*}) \\ &= \sum_{\mathscr{R}*} \mathscr{K}_{y\to y'}(\tilde{c}) y'_{s*}. \end{aligned} \tag{3.37}$$

Since $y'_{s*} \geq 0$ for all $y' \in \mathscr{C}$ and since the kinetic rate functions take nonnegative values, we have $f_{s*}(\tilde{c}) \geq 0$. □

Remark 3.7.2. From (3.37) it follows that $f_{s*}(\tilde{c})$ will be *positive* if and only if there exists $y \to y' \in \mathscr{R}$ such that $y'_{s*} > 0$ and $\operatorname{supp} y \subset \operatorname{supp} \tilde{c}$.

Remark 3.7.2 provides information about where on the boundary of $\overline{\mathbb{R}}_+^{\mathscr{S}}$ the generation rate of an absent species must actually be positive and, therefore, information about where on the boundary equilibria might or cannot reside. The following definition derives from [73].[6]

Definition 3.7.3. *For a reaction network* $\{\mathscr{S},\mathscr{C},\mathscr{R}\}$*, a species set* $\mathscr{S}^* \subset \mathscr{S}$ *is* **reaction-transitive** *if, for all* $y \to y' \in \mathscr{R}$,

$$\operatorname{supp} y \subset \mathscr{S}^* \ \Rightarrow\ \operatorname{supp} y' \subset \mathscr{S}^*. \tag{3.38}$$

The idea behind Definition 3.7.3 is simple: Species set $\mathscr{S}^*$ is reaction-transitive if, whenever all the *reactant* species of a reaction are present in $\mathscr{S}^*$, then so too are that reaction's *product* species.

Lemma 3.7.4. *Consider a kinetic system* $\{\mathscr{S},\mathscr{C},\mathscr{R},\mathscr{K}\}$ *with species-formation-rate function* $f(\cdot)$*. If, for composition* $\tilde{c} \in \overline{\mathbb{R}}_+^{\mathscr{S}}$, $\operatorname{supp} \tilde{c}$ *is not reaction-transitive, then there is a species* $s* \in \mathscr{S}$ *such that* $\tilde{c}_{s*} = 0$ *and* $f_{s*}(\tilde{c}) > 0$.

Proof. If $\operatorname{supp} \tilde{c}$ is not reaction-transitive, then there is a reaction $y \to y'$ such that $\operatorname{supp} y \subset \operatorname{supp} \tilde{c}$ while $\operatorname{supp} y' \not\subset \operatorname{supp} \tilde{c}$. In particular, there is a species $s*$ such that $\tilde{c}_{s*} = 0$ and $y'_{s*} > 0$. From Remark 3.7.2 it follows that $f_{s*}(\tilde{c}) > 0$. □

The preceding lemma has an immediate consequence:

Proposition 3.7.5 ([73]). *Suppose that* c^* *is an equilibrium for a kinetic system. Then* $\operatorname{supp} c^*$ *is reaction-transitive.*

Hereafter when we say that "composition c is reaction-transitive," we shall mean that $\operatorname{supp} c$ *is reaction-transitive.* Thus, we can now say, for example, that all equilibria for a kinetic system (not necessarily the positive ones) are reaction-transitive.

Remark 3.7.6 (Reactions that are switched on or switched off at a reaction-transitive composition). Recall that for a kinetic system $\{\mathscr{S},\mathscr{C},\mathscr{R},\mathscr{K}\}$ we have, for reaction $y \to y'$, $\mathscr{K}_{y\to y'}(c) > 0$ if and only if $\operatorname{supp} y$ is contained in $\operatorname{supp} c$. This is to say that reaction $y \to y'$ is "switched on" at composition c precisely when $\operatorname{supp} y$ is contained in $\operatorname{supp} c$.

Now suppose that the underlying network $\{\mathscr{S},\mathscr{C},\mathscr{R}\}$ contains a directed sequence of reactions

$$y_1 \to y_2 \to \cdots \to y_n. \tag{3.39}$$

[6] See in particular Propositions 5.3.1–5.3.2 and Appendices I and II in [73]. The idea behind Definition 3.7.3 also appears in more recent papers, but in different guises. A species set is called *semi-locking* in [2] or a *siphon* in [7] if its complement in the full set of species is reaction-transitive.

If composition c^* is reaction-transitive—in particular if c^* is an equilibrium—and if $\operatorname{supp} y_1$ is contained in $\operatorname{supp} c$, then it must also be the case that $\operatorname{supp} y_\theta$ is contained in $\operatorname{supp} c$ for $\theta = 2,\ldots,n$. Stated differently, *if the reaction $y_1 \to y_2$ is switched on at composition c^*, then all the remaining reactions in the sequence* (3.39) *must also be switched on.*

This has some interesting consequences, among which is the following: Suppose that the underlying network contains a directed reaction cycle

$$y_1 \to y_2 \to \ldots y_n \to y_1. \tag{3.40}$$

In this case, if composition c^* is reaction-transitive—in particular if c^* is an equilibrium—then either all reactions in the cycle are switched on at c^* or none are.

Remark 3.7.7 (Where on the boundary of $\overline{\mathbb{R}}_+^{\mathscr{S}}$ can cyclic composition trajectories reside?). We now know that an equilibrium of a kinetic system $\{\mathscr{S},\mathscr{C},\mathscr{R},\mathscr{K}\}$ can have its support only on a reaction-transitive set of species. This restricts very sharply locations on the boundary of $\overline{\mathbb{R}}_+^{\mathscr{S}}$ where equilibria can reside. In fact, one can say something very similar about cyclic composition trajectories.

Suppose that $c : [t_1,t_2] \to \overline{\mathbb{R}}_+^{\mathscr{S}}$ is a solution of the differential equation for the kinetic system with $c(t_1) = c(t_2)$. We say that *the corresponding cyclic composition trajectory has support $\mathscr{S}^*$* if, for all $t \in [t_1,t_2]$, $\operatorname{sup}, c(t) = \mathscr{S}^*$. In [73] it is argued that *if a cyclic composition trajectory has support $\mathscr{S}^*$, then $\mathscr{S}^*$ must be reaction-transitive.*[7] In this case, the location of a cyclic composition trajectory on the boundary of $\overline{\mathbb{R}}_+^{\mathscr{S}}$ is severely constrained, for *every composition along the trajectory must be reaction-transitive.*

We will be able to make much sharper statements about boundary locations of equilibria and cyclic composition trajectories—and also about combinations of reactions that are switched on and off—when we discuss the special class of *deficiency zero* reaction networks.

Appendix 3.A The Kinetic Subspace

This appendix amounts to a technical digression. Although it can be skipped for now, its content sheds light on issues that will arise later on. The placement of the appendix here results from its intimate connection to ideas in Section 3.4.

For almost all that we will do, our discussion of the stoichiometric subspace, stoichiometric compatibility, and stoichiometric compatibility classes will be more than adequate. There are, however, certain instances in which we will benefit from a sharpening of observations we've already made. We'll begin to lay the groundwork for that sharpening here.

[7] In [73] see Remark 5.3.F, Proposition 5.3.2, and its proof in Appendix II.

For a kinetic system $\{\mathscr{S},\mathscr{C},\mathscr{R},\mathscr{K}\}$ our introduction of the stoichiometric subspace derived from the fact that at every composition $c \in \overline{\mathbb{R}}_+^{\mathscr{S}}$, the corresponding value of the species-formation-rate function, $f(c)$, takes values in the span of the reaction vectors for the underlying reaction network $\{\mathscr{S},\mathscr{C},\mathscr{R}\}$. Thus, with $\operatorname{im} f$ denoting the image of $f(\cdot)$ and S denoting the network's stoichiometric subspace, we invariably have the relation

$$\operatorname{im} f \subset S. \tag{3.A.1}$$

This is to say that values of $f(\cdot)$ can't point just anywhere. They must point along the stoichiometric subspace, and so too must values of the "velocity vector" $\dot{c}$. *This is true regardless of the nature of the kinetics* $\mathscr{K}$.

Clearly, though, we also have the relation

$$\operatorname{im} f \subset \operatorname{span}(\operatorname{im} f), \tag{3.A.2}$$

where $\operatorname{span}(\operatorname{im} f)$ is the smallest linear subspace of $\mathbb{R}^{\mathscr{S}}$ containing $\operatorname{im} f$. Now it might be the case that $\operatorname{span}(\operatorname{im} f)$ is a smaller linear subspace than S, in which case (3.A.2) gives a sharpening of (3.A.1). This possibility serves as motivation for the following definition:

Definition 3.A.1. *The* **kinetic subspace** *K for a kinetic system $\{\mathscr{S},\mathscr{C},\mathscr{R},\mathscr{K}\}$ with species formation function $f(\cdot)$ is the linear subspace of $\mathbb{R}^{\mathscr{S}}$ defined by*

$$K := \operatorname{span}(\operatorname{im} f). \tag{3.A.3}$$

3.A.1 When the Kinetic Subspace Is Smaller than the Stoichiometric Subspace

To see that the kinetic subspace can indeed be smaller than the stoichiometric subspace, we begin with a concrete example that will also play a role later in the book.

Example 3.A.2. Consider the mass action system (3.A.4).

$$\begin{gathered} B \overset{\alpha}{\leftarrow} A \overset{1}{\rightarrow} C \\ B + C \overset{1}{\rightarrow} 2A \end{gathered} \tag{3.A.4}$$

Note that the stoichiometric subspace for the underlying network is given by

$$S = \operatorname{span}\{B - A, C - A, 2A - B - C\}, \tag{3.A.5}$$

the dimension of which is easily seen to be two. The species-formation-rate function $f : \overline{\mathbb{R}}_+^{\mathscr{S}} \to S$ is given by

$$f(c) = c_A(\alpha(B - A) + (C - A)) + c_C c_B(2A - B - C). \tag{3.A.6}$$

It is not difficult to see that the kinetic subspace $K \subset S$ is given by

$$K = \operatorname{span}\{\alpha(B - A) + (C - A),\ 2A - B - C\}. \tag{3.A.7}$$

Note that, for $\alpha \neq 1$, the vectors $\alpha(B-A)+(C-A)$ and $2A-B-C$ are not colinear. In this case the dimension of K is two, whereupon K coincides with S. On the other hand, for the very special value $\alpha = 1$, the vectors $\alpha(B-A)+(C-A)$ and $2A-B-C$ are colinear. In this case the dimension of K is one, and K is actually smaller than S.

Remark 3.A.3. Example 3.A.2 teaches an important lesson: At least for certain mass action systems, the nature and, in particular, the dimension of the kinetic subspace can vary in a jarring way as values of the rate constants change. For the mass action system (3.A.4), even the smallest perturbation away from the rate constant value $\alpha = 1$ causes a sudden explosion in the dimension of the kinetic subspace. The *stoichiometric* subspace, on the other hand, is unaffected by *any* kinetic parameter values, for it depends solely on the set of reactions and not at all on the rate functions associated with the reactions.

Remark 3.A.4. When the kinetic subspace is smaller than the stoichiometric subspace (as in the preceding example for the special case $\alpha = 1$), there can be behavior that is troubling (or at least inconvenient). Consider, for example, a kinetic system $\{\mathscr{S},\mathscr{C},\mathscr{R},\mathscr{K}\}$ with species-formation-rate function $f(\cdot)$ and kinetic subspace K. Moreover, suppose that S is the stoichiometric subspace for the underlying network $\{\mathscr{S},\mathscr{C},\mathscr{R}\}$.

If K is smaller than S, then, at every *composition* $c^* \in \mathbb{R}_+^{\mathscr{S}}$, *the derivative* $df(c^*): S \to S$ *is singular.* In fact, from a standard theorem of linear algebra relating the dimensions of the domain, kernel, and image of a linear transformation, we have

$$\dim\, S = \dim\, \ker\, df(c^*) + \dim\, \mathrm{im}\, df(c^*). \tag{3.A.8}$$

Because $f(\cdot)$ takes values in K, it follows from (3.32) that the linear map $df(c^*)$ takes values in K for each $c^* \in \mathbb{R}_+^{\mathscr{S}}$. Thus, for each $c^* \in \mathbb{R}_+^{\mathscr{S}}$ we have $\dim\, \mathrm{im}\, df(c^*) \leq \dim K < \dim S$, whereupon $\dim\, \ker df(c^*) > 0$ in which case $df(c^*)$ is singular. In particular, in the sense of Section 3.6, *every positive equilibrium is degenerate.*

3.A.2 Should We Focus on the *Kinetic* Rather than the *Stoichiometric* Subspace?

Our discussion in Section 3.4 gave considerable geometric insight into the nature of trajectories for the differential equation $\dot{c} = f(c)$, where $f(\cdot)$ is the species-formation-rate function for the kinetic system under study. The entire discussion was predicated on the fact that $f(\cdot)$ takes values in the stoichiometric subspace for the underlying reaction network, in which case the "velocity vector" $\dot{c}$ must invariably point along the stoichiometric subspace.

As we've noted, however, values of $f(\cdot)$ and, consequently, of $\dot{c}$ must also point along the (perhaps smaller) kinetic subspace for the kinetic system at hand. Because $\dot{c}$ is constrained to point along the kinetic subspace K, the reasoning used in Section 3.4 can be repeated, in almost everything we said there, to replace the stoichiometric subspace S with the kinetic subspace K (and to replace stoichiometric compatibility with kinetic compatibility, defined in the obvious way). Indeed, the dilemma indicated in Remark 3.A.4 could have been mitigated if, for a particular kinetic system under study, we had regarded the derivative $df(c^*)$ not as a map from S to S but instead as a map from K to K.

In view of the fact that K *might* be smaller than S, should we, then, change our focus from the stoichiometric subspace to the kinetic subspace? For us, the answer will be *no*. There are two important, intimately related, reasons for resisting the temptation:

3.A.2.1 Mass Action Systems for Which $K \neq S$ Never *Have That Property Robustly*

Because of their fundamental importance in the mathematical description of reacting mixtures, mass action systems play a central role in this book. Even kinetic systems that are not of mass action type often come about as approximations to more refined mass action models. In the following sense, mass action systems for which $K \neq S$ *never* have that property robustly:

If a mass action system (with specified rate constant values) has a kinetic subspace that is smaller than the stoichiometric subspace, then it is always possible to make certain reactions reversible, *with arbitrarily small rate constants for the reactions added*, such that the new mass action system has the same stoichiometric subspace as the old one *and for which the new kinetic subspace now coincides with the stoichiometric subspace.*[8] This is discussed more fully in an appendix to Chapter 8; see also [73, 80].

Even when the reaction network itself remains unperturbed, tiny perturbations of those rate constants for which $K \neq S$ can make for great changes in qualitative behavior. For example, the mass action system (3.A.4) admits an infinite number of positive equilibria in each positive stoichiometric compatibility class when $\alpha = 1$ (in which case the kinetic subspace is smaller than the stoichiometric subspace). On the other hand, there are no positive equilibria at all when $\alpha \neq 1$ (in which case the kinetic and stoichiometric subspaces coincide). Striking mathematical phenomena that disappear completely with the smallest perturbations of the model are of dubious relevance to the study of real chemistry.

3.A.2.2 The Kinetic Subspace Is Not an Attribute of a Reaction Network; It Is an Attribute of a Particular *Kinetic System*

It should be kept in mind that the main goal of chemical reaction network theory is to make bold statements about the behavior of *reaction networks*, taken perhaps with kinetics circumscribed to lie within a very broad class. This aligns well with what is typically known in chemistry: In all but the simplest networks, knowledge of kinetic parameter values (e.g., rate constants) is almost always poor. Note that the stoichiometric subspace depends only on the reaction network and not at all on kinetics.

[8] Were all reactions in the original system reversible, then K would have already coincided with S [73, 80].

Thus, to know the stoichiometric subspace, one need only know the network, not the fine details of kinetic parameter values.[9] By way of contrast, kinetic parameter values influence the kinetic subspace directly. Indeed, as Example 3.A.2 demonstrates, even the *dimension* of the kinetic subspace might, and sometimes does, hinge on certain exceptional kinetic parameter values. Focus on the kinetic subspace would simply be inappropriate to the larger goals of chemical reaction network theory. Again, the stoichiometric subspace is an attribute of a reaction network, while the kinetic subspace is an attribute of a reaction network taken with a *particular* kinetics.

3.A.3 Thinking (and Not Thinking) About the Kinetic Subspace

In summary, it is important to be aware that, in certain (typically non-robust) cases, the kinetic subspace might be smaller than the stoichiometric subspace and that, when this happens, certain odd behavior might appear. But we shall not let that small tail wag a very large dog. Our focus will remain on the stoichiometric subspace, which derives from the reaction network alone and not at all from peculiarities of one kinetics or another.

One of the fortuitous blessings of chemical reaction network theory is that, at least for mass action systems, there is a very large, highly robust, and easily described class of reaction networks for which the stoichiometric and kinetic subspaces coincide, no matter what values the rate constants take [73, 80]. We will discuss this class in an appendix to Chapter 8, at which point we will have more vocabulary at our disposal.[10] For now it suffices to say that *every network for which each reaction is reversible is a member of this class, but the class is far larger.*[11] For networks in this class, knowledge of the network *alone* suffices for knowledge of both K and S.

[9] In fact, as we shall see in Remark 6.2.2, to know S one need only know certain coarse features of the network.

[10] For a still deeper discussion, see Appendix 16.B in Chapter 16.

[11] In terminology of Chapter 6, the class includes not only all *weakly* reversible networks but also all $t = \ell$ networks—that is, all networks for which the number of linkage classes is identical to the number of terminal strong-linkage classes.

4

Open Systems: Why Study Nonconservative and Otherwise Peculiar Reaction Networks?

Anyone who passed through high school chemistry will remember that, in addition to more urgent adolescent preoccupations, there was the problem of "balancing chemical equations." Not only was the total mass on the two sides of a chemical reaction supposed to be identical, so too were the total number of atoms of each kind and the total charge. In this chapter we will indicate not only why it makes sense to consider nonconservative networks but also why, if reaction network theory is to be wide-ranging in its utility, it is essential that we do so.

4.1 Conservative Networks

When a reaction occurs, a certain collection of molecules is converted into a different collection of molecules. It is generally supposed that the total mass of the first collection is the same as the total mass of the second collection. That is, it is generally supposed that chemical reactions conserve mass. Thus, if $A + B \to C$ is a reaction, and if $M_A, M_B,$ and M_C are the molecular weights of $A, B,$ and C, mass conservation would be reflected in the equation $M_A + M_B = M_C$. Similarly, if $A \to 2B$ is yet another reaction, then mass conservation would be reflected in the equation $M_A = 2M_B$.

More generally, let $\{\mathscr{S}, \mathscr{C}, \mathscr{R}\}$ be a reaction network, and let $M \in \mathbb{R}^{\mathscr{S}}_{+}$ be the vector of molecular weights of the various species. That is, M_s is the molecular weight of species $s \in \mathscr{S}$. If each reaction respects mass conservation, then we would have

$$\sum_{s \in \mathscr{S}} y_s M_s = \sum_{s \in \mathscr{S}} y'_s M_s, \quad \forall\, y \to y' \in \mathscr{R}, \tag{4.1}$$

or, equivalently,

$$(y' - y) \cdot M = 0, \quad \forall\, y \to y' \in \mathscr{R}. \tag{4.2}$$

(Recall that the standard scalar product in $\mathbb{R}^{\mathscr{S}}$ is given by (2.16).) The conservation equation (4.2) is an assertion that M is orthogonal to all of the reaction vectors for the network and, therefore, to their span. Thus, if S is the stoichiometric subspace for the network, then (4.2) is equivalent to the inclusion

M. Feinberg, *Foundations of Chemical Reaction Network Theory*,
Applied Mathematical Sciences 202,
https://doi.org/10.1007/978-3-030-03858-8_4

$$M \in S^{\perp}, \tag{4.3}$$

where $S^{\perp}$ is the orthogonal complement of S relative to the standard scalar product in $\mathbb{R}^{\mathscr{S}}$.

If each of the individual reactions conserves mass, then we should expect the total mixture mass to be conserved in a closed reactor of the kind depicted and discussed in our original motivating Section 2.1. In fact, suppose that the chemistry in such a reactor is described by a kinetic system $\{\mathscr{S},\mathscr{C},\mathscr{R},\mathscr{K}\}$ and that, as before, $M \in \mathbb{R}_+^{\mathscr{S}}$ is the vector of molecular weights. If $c \in \overline{\mathbb{R}}_+^{\mathscr{S}}$ is an instantaneous mixture composition, with c_s the molar concentration of species s, then the instantaneous mass per unit volume in the mixture is given by

$$M \cdot c = \sum_{s \in \mathscr{S}} M_s c_s. \tag{4.4}$$

Recall that evolution of mixture composition is given by the differential equation

$$\dot{c} = \sum_{\mathscr{R}} \mathscr{K}_{y \to y'}(c)(y' - y), \tag{4.5}$$

from which it follows that

$$M \cdot \dot{c} = \sum_{\mathscr{R}} \mathscr{K}_{y \to y'}(c)(y' - y) \cdot M. \tag{4.6}$$

Thus, if the reactions respect mass conservation—that is, if (4.2) obtains—then along composition trajectories we have $M \cdot \dot{c} = 0$, which is to say that mixture mass is conserved.

Motivated by these considerations, we follow Horn and Jackson [109] in saying that a reaction network is *conservative* if it respects a conservation condition of the kind indicated.

Definition 4.1.1. *A reaction network $\{\mathscr{S},\mathscr{C},\mathscr{R}\}$ with stoichiometric subspace S is* **conservative** *if $S^{\perp} \subset \mathbb{R}^{\mathscr{S}}$ contains at least one vector of $\mathbb{R}_+^{\mathscr{S}}$.*

As shown by Horn and Jackson—see Appendix 1 in [109]—conservative networks have particularly pleasant properties, in particular the boundedness of the stoichiometric compatibility classes:

Proposition 4.1.2. *All of the stoichiometric compatibility classes of a conservative reaction network are closed and bounded. A reaction network admits a closed and bounded stoichiometric compatibility class only if it is conservative.*

Example 4.1.3. The very simple network $A \rightleftarrows 2B$ discussed in Example 3.4.7 is conservative: if we choose $M_A = 2$ and $M_B = 1$, then the vector $M \in \mathbb{R}_+^{\mathscr{S}}$ is orthogonal to each of the reaction vectors $2B - A$ and $A - 2B$. Note that the stoichiometric compatibility classes of Figure 3.1 are closed and bounded line segments.

Example 4.1.4. Network (3.28) is conservative. If we choose $M_A = 1$, $M_B = 2$, and $M_C = 2$, then the vector $M \in \mathbb{R}^{\mathscr{S}}_{+}$ is orthogonal to each of the reaction vectors shown in (3.29). The stoichiometric compatibility classes of Figure 3.2 are closed and bounded triangles.

Example 4.1.5. Our very first network example (2.1) is also conservative: Choose $M_A = 2, M_B = 1, M_C = 1, M_D = 3, M_E = 2$. Here the stoichiometric compatibility classes are again closed and bounded, but they are more difficult to visualize.

In the high school problem of "balancing chemical equations," not only was the total mass on the two sides of a chemical equation supposed to be identical, so too were the total number of atoms of each kind and the total charge. For example, if for each species $s \in \mathscr{S}$, the number of phosphorus atoms in a molecule of s was P_s, then the reactions in a network under consideration were required to respect the phosphorus balance,

$$\sum_{s \in \mathscr{S}} y_s P_s = \sum_{s \in \mathscr{S}} y'_s P_s, \quad \forall\, y \to y' \in \mathscr{R}. \tag{4.7}$$

With $P \in \overline{\mathbb{R}}^{\mathscr{S}}_{+}$ denoting the vector of atomic phosphorus contents of the various species, (4.7) is equivalent to

$$(y' - y) \cdot P = 0, \quad \forall\, y \to y' \in \mathscr{R}, \tag{4.8}$$

which in turn is equivalent to the inclusion

$$P \in S^{\perp}. \tag{4.9}$$

From our discussion of mass conservation, it is not difficult to see how a phosphorus balance for each of the individual reactions would be reflected in an overall conservation of phosphorus atoms for a closed reactor of the kind discussed in Section 2.1. In particular, we would expect that $P \cdot \dot{c} = 0$, which is to say that the total number of phosphorus atoms is conserved along every composition trajectory.

4.2 Reaction Network Descriptions of Open Systems

It is the purpose of this section to explain why, in the remainder of the book, we shall admit for consideration networks containing peculiar reactions such as $A \to 2A$ or $0 \to A$ (zero reacts to A) that, at first glance, appear to be incompatible with the conservation of matter. In fact, we shall want to consider nonconservative networks such as

$$\begin{aligned} A &\rightleftarrows 2B \\ A + C &\rightleftarrows B \end{aligned}$$

that offend our sensibilities in a more subtle way: If a molecule of A can decompose into two molecules of B, then the molecular weight of a molecule of B would appear to be half that of a molecule of A. How then can a molecule of B result from a chemical combination of the heavier molecule of A with a molecule of C?

In order to understand why it makes sense to consider such peculiar networks, we must move beyond the simple picture painted at the beginning of Section 2.1, a picture that has motivated all our considerations thus far. There we studied a homogeneous reactor (portrayed in Figure 2.1) that was *closed* with respect to the exchange of matter with the external world. Composition changes resulted *solely* from the occurrence of chemical reactions, and our differential equations were formulated accordingly.

We would also like to study homogeneous (well-stirred) reactors that are open to the influx or efflux of at least certain species. In this case we would expect composition changes to result not only from the occurrence of chemical reactions but also from the transport of various species into and out of the reactor. Consequently, these effects should also manifest themselves in the differential equations that govern the reactor's behavior. For this reason the differential equations (3.10) and (3.12) we have begun to examine might *seem* inappropriate to the study of open reactors, for those equations were developed in consideration of reactors in which composition changes come about by virtue of chemical reactions alone.

The fact is, however, that equations (3.10) and (3.12) are indeed suited to the study of important categories of open reactors, *provided that our conception of a reaction network is suitably broadened to incorporate certain "pseudoreactions" tailored to encode the infusion or effusion of those species supplied to or removed from a particular open reactor under study*. That is, there are varieties of open reactors for which the appropriate differential equations can be viewed as deriving from a reaction network obtained by modifying or augmenting the true chemical network in such a way as to model, by means of "pseudoreactions," various nonchemical effects.

We shall see how this works in a few examples, but before turning to them, it will be useful to make explicit the importance of our ability to subsume open reactors within the framework we have erected so far. Suppose that the possibilities suggested in Chapter 2 could be realized, that there could be developed a theory of the general mass action equation (3.12) which would, for example, indicate that the differential equations for networks of a certain large class cannot admit periodic orbits for any set of rate constants. Suppose further that a particular open system could be *modeled* in terms of a reaction network (composed in part of "pseudoreactions") taken with mass action kinetics.

That is, suppose (3.12), written for the *model* network and its kinetics, would yield precisely those differential equations one would write for the open system from first principles. Then the aforementioned theory of equation (3.12) would connect qualitative properties of these equations with the structure of the *model* network, and the possibility of periodic orbits for the open system under study might, for example, be decided solely on the basis of *model* network structure. In this way the theory would provide information not only about closed reactors but also about open reactors.

We turn now to examples in which we indicate how, by means of appropriately constructed model networks, certain categories of reactors are in fact describable within the framework we have already constructed. So that the basic ideas should not be obscured, I have, in each case, taken the "true" chemistry to be very simple. The examples are easily generalized to situations in which the chemistry is far more complex. Moreover, in each example I have taken the kinetics of the true reactions to be mass action, and I have indicated how each reactor is describable in the context of the mass action equation (3.12). When the kinetics is not mass action, similar considerations enable one to describe each reactor in the context of the more general equation (3.10).

4.2.1 The Continuous-Flow Stirred-Tank Reactor (CFSTR)

Consider the reactor shown in Figure 4.1. The reactor contents, a liquid mixture of species *A* and *B*, are maintained homogeneous, isothermal, and of fixed volume, V cm^3. Feed of fixed composition is continuously supplied to the reactor at a constant volumetric flow rate g cm^3/sec with molar concentrations of A and B in the feed equal to c_A^F and c_B^F moles/cm^3, respectively. The contents of the reactor are continuously removed at volumetric flow rate g cm^3/sec. The concentrations of *A* and *B* in the effluent stream are precisely those in the reactor's interior. In the reactor the only chemical reactions that occur are

$$2A \underset{k'}{\overset{k}{\rightleftarrows}} B. \tag{4.10}$$

The kinetics is mass action with rate constants as indicated in (4.10).

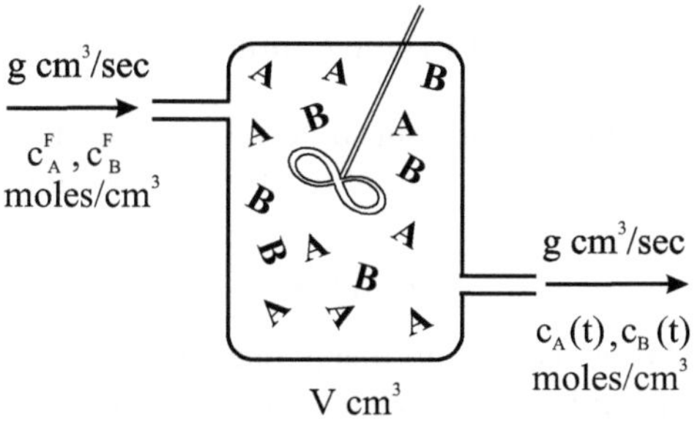

Fig. 4.1. A simple continuous-flow stirred-tank reactor (CFSTR)

Simple mole balances give the following pair of differential equations governing the molar concentrations of *A* and *B*:

$$\begin{aligned} \dot{c}_A &= \alpha - \xi c_A - 2kc_A^2 + 2k'c_B \\ \dot{c}_B &= \beta - \xi c_B + kc_A^2 - k'c_B, \end{aligned} \tag{4.11}$$

where $\alpha = gc_A^F/V$, $\beta = gc_B^F/V$, and $\xi = g/V$. In each of these equations, the first term derives from the addition of *A* or *B* in the feed stream, while the second term derives from removal in the effluent stream. Note that (4.11) is *not* the system of differential equations that correspond, in the sense of (3.12), to the mass action system shown in (4.10): For that simple system, terms corresponding to contributions from the feed and effluent streams would be absent.

Consider, however, the model network

$$2A \underset{k'}{\overset{k}{\rightleftarrows}} B \underset{\beta}{\overset{\xi}{\rightleftarrows}} 0 \underset{\xi}{\overset{\alpha}{\rightleftarrows}} A. \tag{4.12}$$

The entity "0" in (4.12) is the *zero complex*, interpreted as the zero vector of $\mathbb{R}^{\mathscr{S}}$, where, for our example, $\mathscr{S} = \{A, B\}$. The reactions $0 \to A$ and $0 \to B$ are adjoined to the reactions $2A \rightleftarrows B$ to reflect the infusion of A and B in the feed stream, and the reactions $A \to 0$ and $B \to 0$ are similarly adjoined to reflect the effusion of A and B in the exit stream. Now suppose that this model network is assigned mass action kinetics with rate constants as indicated alongside the reaction arrows in (4.12). Then (3.12), written for the mass action system depicted in (4.12), yields a vector differential equation whose component equations are precisely those displayed in (4.11).[1]

Thus, any theory obtained for the general equation (3.12) that draws connections between dynamics and reaction network structure becomes applicable to the flow reactor we have been considering, *provided it is understood that the network of interest is* (4.12) *rather than the simpler* (4.10).

It is not difficult to see how wide varieties of continuous-flow stirred-tank reactors can be codified in reaction network terms by the means described here. Our next example will be one such variation.

4.2.2 Semi-Open Reactors: An Enzyme Example

Here we consider the joining of two substrates, $S1$ and $S2$, to form a product, P, in the presence of an enzyme, E. The overall reaction is sometimes indicated schematically as

$$S1 + S2 \overset{E}{\to} P. \tag{4.13}$$

A display of this kind is actually a shorthand, one that masks a considerable amount of unspecified detail about how the enzyme works its magic. A catalytic *mechanism* serves to fill in some of that detail by postulating, at the level of more elementary reactions, the various steps by which conversion of substrates to products comes about. These elementary reactions are almost always taken to be governed by mass action kinetics. An exhaustive account of the variety of mechanistic possibilities is given in [150].

A classical mechanism for (4.13) is shown in (4.14):

[1] Note that, according to the mass action prescription given by Definition 3.2.6, the reaction $0 \to A$ proceeds at rate per unit volume $\alpha(c_A)^0(c_B)^0 = \alpha = gc_A^F/V$.

$$
\begin{array}{ccc}
S1 + E \underset{k_2}{\overset{k_1}{\rightleftarrows}} ES1 & & S2 + E \underset{k_4}{\overset{k_3}{\rightleftarrows}} ES2 \\
\\
S2 + ES1 \underset{k_6}{\overset{k_5}{\rightleftarrows}} & ES1S2 & \underset{k_8}{\overset{k_7}{\leftrightarrows}} S1 + ES2 \\
& k_9 \downarrow & \\
& P + E &
\end{array}
\tag{4.14}
$$

In the first two steps, $S1$ and $S2$ bind to (and unbind from) the enzyme E, each presumably at its own binding site. Then $S2$ binds reversibly to $ES1$, and, similarly, $S1$ binds reversibly to $ES2$. Once both $S1$ and $S2$ are bound to the enzyme, the product P is formed, and, once formed, it is immediately released from the enzyme. This scheme is sometimes called a *random* or *unordered* mechanism because the substrates are not compelled to bind to the enzyme in a prescribed order.

Now consider a homogeneous (well-stirred) reactor in which the enzyme is operative. The picture will be similar to that shown in Figure 4.1. This time, however, we will suppose that the large-molecule enzyme is entrapped within the reactor chamber by membranes at the feed and effluent ports that deny passage of the enzyme but which, to different extents, permit passage of unbound small-molecule product and substrates through the exit port.

We presume that the substrates $S1$ and $S2$ are added at constant rates (per unit mixture volume) F_{S1} and F_{S2} and that substrates and product are removed at rates (per unit mixture volume) proportional to their respective concentrations in the chamber. If ξ_{S1}, ξ_{S2}, and ξ_P are the proportionality constants and if the operative catalytic mechanism is that depicted in (4.14) with mass action rate constants as indicated, then the species concentrations are governed by the differential equations shown in (4.15).

$$
\begin{aligned}
\dot{c}_{S1} &= -k_1 c_E c_{S1} + k_2 c_{ES1} - k_7 c_{S1} c_{ES2} + k_8 c_{ES1S2} + F_{S1} - \xi_{S1} c_{S1} \\
\dot{c}_{S2} &= -k_3 c_E c_{S2} + k_4 c_{ES2} - k_5 c_{S2} c_{ES1} + k_6 c_{ES1S2} + F_{S2} - \xi_{S2} c_{S2} \\
\dot{c}_P &= k_9 c_{ES1S2} - \xi_P c_P \\
\dot{c}_E &= -k_1 c_E c_{S1} + k_2 c_{ES1} - k_3 c_E c_{S2} + k_4 c_{ES2} + k_9 c_{ES1S2} \\
\dot{c}_{ES1} &= k_1 c_E c_{S1} - k_2 c_{ES1} - k_5 c_{ES1} c_{S2} + k_6 c_{ES1S2} \\
\dot{c}_{ES2} &= k_3 c_E c_{S2} - k_4 c_{ES2} - k_7 c_{ES2} c_{S1} + k_8 c_{ES1S2} \\
\dot{c}_{ES1S2} &= k_5 c_{S2} c_{ES1} + k_7 c_{S1} c_{ES2} - (k_6 + k_8 + k_9) c_{ES1S2}
\end{aligned}
\tag{4.15}
$$

Note that these are *not* the equations that correspond to the mass action system (4.14)—for which terms such as $F_{S1} - \xi_{S1} c_{S1}$ would not appear—but these *are* the differential equations for the mass action system (4.16).

$$
\begin{array}{c}
S1 + E \underset{k_2}{\overset{k_1}{\rightleftarrows}} ES1 \qquad\qquad S2 + E \underset{k_4}{\overset{k_3}{\rightleftarrows}} ES2 \\[1em]
S2 + ES1 \underset{k_6}{\overset{k_5}{\rightleftarrows}} ES1S2 \underset{k_8}{\overset{k_7}{\leftrightarrows}} S1 + ES2 \\
k_9 \downarrow \\
P + E \\[1em]
S1 \underset{F_{S1}}{\overset{\xi_{S1}}{\rightleftarrows}} \; 0 \; \underset{F_{S2}}{\overset{\xi_{S2}}{\leftrightarrows}} S2 \\
\xi_P \uparrow \\
P
\end{array}
\tag{4.16}
$$

Thus, theory that connects reaction network structure to properties of the induced differential equations might be invoked to give information about properties of the system (4.15), provided it is understood that the network of interest is (4.16), not (4.14). Here again, "pseudoreactions" such as $0 \to S1$ or $S1 \to 0$ are invoked to model the supply of $S1$ at fixed rate or its disappearance at a rate proportional to its current concentration.

Remark 4.2.1 (Species synthesis and degradation). We motivated the differential equations (4.15) with a picture of the kind displayed in Figure 4.1, but similar equations emerge in models intended to describe chemical processes in cell biology. There, constant terms such as F_{S1} or linear terms such as $\xi_{S1} c_{S1}$ are often invoked to capture constant-rate synthesis of a particular species or its disappearance via a first-order reaction. Good examples of this can be found in recent models of the *Wnt* pathway [93].

4.2.3 Reactors with Certain Species Concentrations Regarded Constant

Next, we shall consider a reactor studied by Edelstein [58]. The reactor contents are well stirred and are maintained at constant temperature and volume. The reactions occurring are those depicted in network (4.17).

$$
\begin{array}{c}
A + L \underset{\beta}{\overset{\alpha}{\rightleftarrows}} 2A \\[1em]
A + B \underset{\eta}{\overset{\gamma}{\rightleftarrows}} C \underset{\xi}{\overset{\varepsilon}{\rightleftarrows}} M + B
\end{array}
\tag{4.17}
$$

The kinetics is mass action with rate constants denoted by Greek letters alongside the corresponding reaction arrows. We shall suppose, hypothetically, that species L and M are added to or removed from the reactor in such a manner as to keep the molar concentrations of L and M within the reactor fixed at values c_L^* and c_M^*, respectively. (The situation described is far-fetched, but it serves as a metaphor for a practical approximation that is both common and important; see Remark 4.2.2 below.) The differential equations for the five species concentrations within the reactor are then those shown in (4.18).

$$
\begin{aligned}
\dot{c}_A &= \alpha c_L^* c_A - \beta c_A^2 - \gamma c_A c_B + \eta c_C \\
\dot{c}_B &= -\gamma c_A c_B + \eta c_C + \varepsilon c_C - \xi c_M^* c_B \\
\dot{c}_C &= \gamma c_A c_B - \eta c_C - \varepsilon c_C + \xi c_M^* c_B \\
\dot{c}_L &= 0 \\
\dot{c}_M &= 0
\end{aligned}
\tag{4.18}
$$

These are *not* the differential equations one would obtain by writing (3.12) for the mass action system depicted in (4.17). However, the first three equations of (4.17)—the only ones of real interest—*are* induced by the mass action system (4.19).

$$
\begin{gathered}
A \underset{\beta}{\overset{\alpha c_L^*}{\rightleftarrows}} 2A \\
A + B \underset{\eta}{\overset{\gamma}{\rightleftarrows}} C \underset{\xi c_M^*}{\overset{\varepsilon}{\rightleftarrows}} B
\end{gathered}
\tag{4.19}
$$

That is, the evolution of the molar concentrations of species A, B, and C are described by the first three equations of (4.17), and these can be viewed as deriving from the model network (4.19) by means of the usual mass action formalism. Thus, any theory of equation (3.12) that draws connections between dynamics and network structure becomes applicable to the Edelstein system, so long as it is understood that the network of interest is that shown in (4.19) rather than that shown in (4.17).

This example is easily generalized to other reactors of similar type. To obtain the appropriate model network, one merely "strips away" species with time-invariant concentration from the network of true chemical reactions, and one modifies certain rate constants in a manner suggested by the example.[2]

Remark 4.2.2 (Species available in very large supply). The reactor just considered is open: species L and M are added to and removed from the reactor in such a way as to maintain the concentrations of L and M constant in time. In fact, the example—at least in the way I described it—is far-fetched, for it is difficult to see how, in practical terms, control of the concentrations of L and M could be easily managed.

Nevertheless, the example amounts to a formalization of a common situation that chemists really do think about. Imagine that Edelstein's reactions (4.17) are taking place in a closed reactor in which species L and M are present in amounts much larger than those of A, B and C. In such a situation, it is reasonable to suppose that, for a very long time, the concentrations of L and M can be deemed constant in the differential equations for A, B and C; there simply isn't enough A, B or C available to affect the amounts of L and M appreciably, at least not in the short run.

In the context of this supposition and for a reasonably chosen time interval, the closed Edelstein reactor is, insofar as the dynamics of A, B and C are concerned, more or less equivalent to the physical picture painted above, and the differential

[2] If all species in a particular complex have time-invariant concentrations, then the "stripping away" procedure will result in the zero complex.

equations for A, B and C can be viewed as having been induced by the model mass action system depicted in (4.19).

The supposition of constant concentrations for large-supply species is common. It is almost always invoked implicitly for the hydrogen ion, H^+, even when H^+ is a participant in one or more reactions within a network. (In such cases, a reaction rate might regarded to be "pH dependent," for the concentration of hydrogen ion has been folded into the rate constant.) Similarly, in biochemical contexts, the concentration of highly abundant ATP (adenosine triphosphate) is often presumed constant in enzyme-driven phosphorylation reactions, even though in those reactions ATP is converted to ADP (adenosine diphosphate).

Remark 4.2.3 (The Brusselator). Because it will play multiple roles in the toy examples of Chapter 5, we mention here a simple network that derives from one in which certain species concentrations are deemed constant. The toy chemistry begins with the reactions displayed in (4.20).

$$\begin{aligned} L &\to A \\ M + A &\to N + B \\ 2A + B &\to 3A \\ A &\to P \end{aligned} \tag{4.20}$$

The concentrations of L, M, N and P are regarded constant, perhaps because they are available in very large supply. In this case, these species are stripped from (4.20) to produce the network of real interest, (4.21). This two-species network, which in various contexts exhibits surprisingly rich behavior, is called the Brusselator because it was studied originally by a group in Brussels. See, for example,[136].

$$\begin{aligned} 0 \rightleftarrows A &\to B \\ 2A + B &\to 3A \end{aligned} \tag{4.21}$$

4.2.4 Interconnected Cells

As shown in Figure 4.2, we consider a simple system consisting of two "cells," each containing a homogeneous mixture of A and B. The cells are presumed to have the same volume and are maintained at the same constant temperature. The intracellular chemistry is described by the mass action system

$$A \underset{k'}{\overset{k}{\rightleftarrows}} 2B, \tag{4.22}$$

with rate constants as indicated.

Species A and B diffuse from cell to cell at rates proportional to the difference in species concentration between the cells. That is, the net rate, per unit cell volume, of receipt of moles of A in Cell 1 from Cell 2 is $D_A(c_{A2} - c_{A1})$, where c_{A1} and c_{A2} are the molar concentrations of A in Cell 1 and Cell 2, respectively, and D_A is a "diffusion

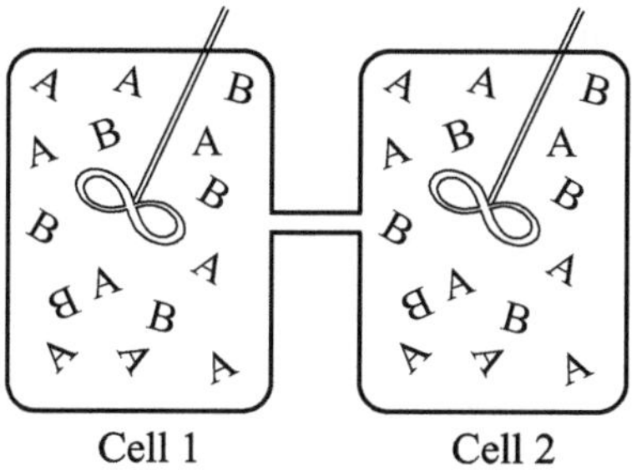

Fig. 4.2. A simple two-cell reactor

coefficient" for species A. Similarly, the net rate, per unit cell volume, of receipt of moles of B in Cell 1 is $D_B(c_{B2} - c_{B1})$. The situation for Cell 2 is similar.

The differential equations governing the concentrations of A and B in the two vessels are those shown in (4.23).

$$\begin{aligned}
\dot{c}_{A1} &= -kc_{A1} + k'c_{B1}^2 + D_A(c_{A2} - c_{A1}) \\
\dot{c}_{B1} &= 2kc_{A1} - 2k'c_{B1}^2 + D_B(c_{B2} - c_{B1}) \\
\dot{c}_{A2} &= -kc_{A2} + k'c_{B2}^2 - D_A(c_{A2} - c_{A1}) \\
\dot{c}_{B2} &= 2kc_{A2} - 2k'c_{B2}^2 - D_B(c_{B2} - c_{B1})
\end{aligned} \tag{4.23}$$

These equations can be viewed as having derived from the mass action system shown in (4.24).

$$\begin{gathered}
2B_1 \underset{k'}{\overset{k}{\leftrightarrows}} A_1 \underset{D_A}{\overset{D_A}{\rightleftarrows}} A_2 \underset{k'}{\overset{k}{\rightleftarrows}} 2B_2 \\
B_1 \underset{D_B}{\overset{D_B}{\rightleftarrows}} B_2
\end{gathered} \tag{4.24}$$

Clearly, larger more complex multicell systems can be described in reaction network terms similarly. In fact, it is an easy extension to include arrangements of several communicating "cells" of the kind considered in Sections 4.2.1–4.2.3. In this way, an incisive reaction network theory could be brought to bear in consideration of intricate arrangements of cells, each housing its own perhaps intricate chemistry. For example, we might want to know what properties the underlying intracellular chemistry (e.g., (4.22)) must have in order that the multicell system might admit a steady state for which compositions within the various cells are *not* the same.

4.3 Going Forward

The examples in Section 4.2 provide only a partial picture of the way in which the framework we have constructed is sufficiently broad as to accommodate a large variety of reactors by means of suitably constructed model networks. These few examples should nevertheless make clear why we would like whatever theory we generate to embrace networks which contain "funny" reactions such as $A \rightarrow 2A$ or $0 \rightarrow A$ and which apparently violate conservation of matter. Indeed, the seemingly strange non-conservative network

$$\begin{aligned} A &\rightleftarrows 2B \\ A + C &\rightleftarrows B \end{aligned} \tag{4.25}$$

with which we began our discussion in Section 4.2 might have resulted from the "stripping away" of species D and E (in the sense Section 4.2.3) from the network

$$\begin{aligned} A + D &\rightleftarrows 2B \\ A + C &\rightleftarrows B + E. \end{aligned} \tag{4.26}$$

Thus, network (4.25) becomes a legitimate object of study in the context of a reactor in which the concentrations of D and E are regarded fixed.

Motivated by these considerations, we shall take the view that *all reaction networks are legitimate objects of study*. In this way we can aspire to a theory that embraces all networks we are likely to encounter, be they "true" chemical networks or model networks that describe reactors like those considered in our examples.

Remark 4.3.1 (Study of mass action systems that do not comply with detailed balance). Chemists often assert that rate constants in a mass action system are not always independent of each other. In particular, they believe that rate constants associated with true chemical reactions should be orchestrated to satisfy a condition that we will study later on, the condition of *detailed balance* discussed in Part III. Note, however, that in some of the examples in Section 4.2, the mass action systems that became objects of study involved rate constants that were not at all "true" rate constants. Instead, they were constructs in mass action system *models*—typically, models of reactors that are open to the transport of material to and from the external world.

For this reason, *we shall also regard all mass action systems as legitimate objects of study, whether or not they comply with detailed balance*. In this way, we can aspire to a theory that has a far broader range of applications. Indeed, in chemical reaction network theory, some of the most important results for mass action systems are theorems that read schematically as follows: *Consider a reaction network that has structural Property X. Then, for* any *choice of (positive) rate constants, the dynamics of the resulting mass action system has Property Y.*

Theorems of this kind are very broad in scope: They tell us that, *by virtue of reaction network structure alone*, we should expect only a certain kind of dynamical behavior, so long as the kinetics is mass action, *no matter what values the rate constants take*. There is no reliance on rate-constant-value constraints that detailed balance might impose.[3] The theorems then become applicable to mass action models broadly, such as those that arise in open-reactor contexts discussed in Section 4.2.

[3] Sometimes, however, there are stronger things that can be said when the supposition of detailed balance is in force. We will discuss these later on, when we also take up the idea of *complex balancing* introduced by Horn and Jackson [109].

Remark 4.3.2 (A vocabulary issue: equilibria vs. steady states). Chemists, especially those who have a strong interest in thermodynamics, like to reserve the word *equilibrium* only for consideration of closed reactors, in particular those constrained to operate at fixed temperature and fixed volume or pressure. They prefer to use the term *steady state* for rest points of open reactors, such as those open reactors discussed earlier in this chapter. Because we are undertaking the study of *all* reaction networks within a common framework, whether they derive from consideration of open or closed reactors, and because this book is heavily a work of mathematics, we will not hesitate to use the word *equilibrium* in its almost universal *mathematical* sense, the one invoked for a kinetic system in Definition 3.3.3.

5

A Toy-Reaction-Network Zoo: Varieties of Behavior and Some Questions

In my local zoo, the most common animals are squirrels (not counting humans, birds, and insects). There is no exhibit called The Squirrel House. People aren't interested in seeing more squirrels, which already roam freely along the walkways. The zoo is meant to exhibit the uncommon.

This chapter is something like a zoo, with exhibits indicating a variety of ways in which reaction networks might behave. Although the behaviors we shall see (e.g., multiple equilibria, bistability, sustained oscillations, symmetry breaking) are not at all exotic from a dynamical systems perspective, they are not commonly observed in real (isothermal) chemical systems, even very complex ones. It is in this sense that these examples constitute a zoo, a display of the uncommon. In fact, they constitute a toy zoo, for the networks considered are small, miniature ones designed to exhibit phenomena that some larger, real networks admit.

Although they are toys, the examples serve a very serious purpose. In the spirit of the introductory chapter, we shall want to understand why vast parts of the reaction network landscape are prone to stable, dull behavior. At the very least, the toy examples tell us that there is nothing *intrinsic* in the differential equations of reaction networks that enforces *universally* mundane dynamics. Rather, the examples tell us that dull behavior must derive from specific structural features that are shared by many, perhaps most, but not all reaction networks. In our search for theorems that tie stable behavior to aspects of reaction network structure, the networks in this chapter will provide small, revealing counterexamples. They set boundaries on how far such stability-enforcing theorems can actually go.

Our concern is with the relationship between the structure of a reaction network and properties of the system of differential equations it induces. Recall that one such elementary relationship was already established in Section 3.4: For a kinetic system $\{\mathscr{S},\mathscr{C},\mathscr{R},\mathscr{K}\}$, a composition trajectory containing a point in a particular stoichiometry compatibility class for the network $\{\mathscr{S},\mathscr{C},\mathscr{R}\}$ lies entirely within that stoichiometric compatibility class. We would like to know more. We would like to know in qualitative terms what happens within the stoichiometric compatibility classes, and we would like to tie that qualitative behavior to reaction network structure.

M. Feinberg, *Foundations of Chemical Reaction Network Theory*, Applied Mathematical Sciences 202,
https://doi.org/10.1007/978-3-030-03858-8_5

The toys will help to motivate some specific questions. Although we shall also be interested in the more general situation, the examples in this chapter will be confined to networks endowed with mass action kinetics. Because we will eventually seek to classify networks according to the kind of behavior the induced differential equations might admit, it should be kept in mind that *the network itself is our object of study*, not the network taken with a *particular* set of rate constants.

We will not, for example, ask whether the differential equations for a *specified* mass action system $\{\mathscr{S},\mathscr{C},\mathscr{R},k^*\}$ admit a periodic solution. We will, however, ask if, for the network $\{\mathscr{S},\mathscr{C},\mathscr{R}\}$, there can exist *some* $k^* \in \mathbb{R}_+^{\mathscr{R}}$ such that the differential equations induced by the mass action system $\{\mathscr{S},\mathscr{C},\mathscr{R},k^*\}$ admit a periodic solution. That is, we will ask whether the network $\{\mathscr{S},\mathscr{C},\mathscr{R}\}$ does or does not *have the capacity* to generate periodic composition oscillations.

With this in mind, we will pose many of the problems considered later in the book in the following way: *Describe the class of networks which, when taken with mass action kinetics, are such that the induced differential equations have Property X regardless of values of the rate constants.* Clearly, a solution to this problem provides a solution to the complementary problem: Describe the class of networks which, when taken with mass action kinetics, are such that the induced differential equations *fail* to have Property X for some values of the rate constants. Thus if we could delineate the full class of networks that generate no periodic composition trajectories for any set of rate constants, we would then know the complementary class of networks which have the capacity to generate periodic composition trajectories for at least one assignment of rate constants.

Before we begin, it will be worth recalling the very simple mass action system considered in Example 3.4.7, the one that derived from the tiny network $A \rightleftarrows 2B$. The phase portrait drawn there, *which is qualitatively the same for all rate constant values*, is repeated below in Figure 5.1.

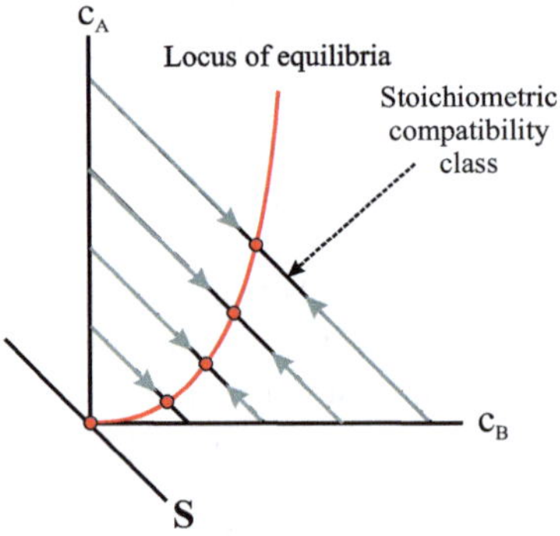

Fig. 5.1. Phase portrait for the mass action system $A \underset{k'}{\overset{k}{\rightleftarrows}} 2B$

The phase portrait is typical of the dull behavior chemists expect, even for much larger and more intricate mass action systems: *Regardless of rate constant values*, in each positive stoichiometric compatibility class there is *precisely one* equilibrium; that equilibrium is *asymptotically stable* against composition perturbations in the same stoichiometric compatibility class; and there are *no cyclic composition trajectories. Again, this is the behavior chemists expect, regardless of rate constant values.*

The examples in our zoo will behave very differently. We will not write out the governing differential equations, but in each case they derive from (3.12) implemented for the particular mass action system at hand.

5.1 Multiple Stoichiometrically Compatible Equilibria and Unstable Equilibria

5.1.1 The Horn and Jackson Example

We begin with a simple two-species mass action system (5.1) studied by Horn and Jackson [109]. Note that both vertical rate constants are taken to be 1, while the horizontal rate constants are taken to be ε. In Figure 5.2 we show phase portraits for two different ε ranges.

$$\begin{array}{ccc} 3A & \overset{\varepsilon}{\rightarrow} & A+2B \\ 1\uparrow & & \downarrow 1 \\ 2A+B & \underset{\varepsilon}{\leftarrow} & 3B \end{array} \tag{5.1}$$

For $\varepsilon \geq \frac{1}{6}$ the phase portrait is very much like that in Figure 5.1; in each positive stoichiometric compatibility class there is precisely one equilibrium, and it is asymptotically stable relative to perturbations within the same stoichiometric compatibility class. For $\varepsilon < \frac{1}{6}$, however, there are three equilibria in each stoichiometric compatibility class; the central one is unstable.

The example is, of course, a toy, and in this case one that bears almost no relation to true chemistry. Nevertheless, Horn and Jackson studied it for very compelling reasons. As we shall see, the example will have much to teach us.

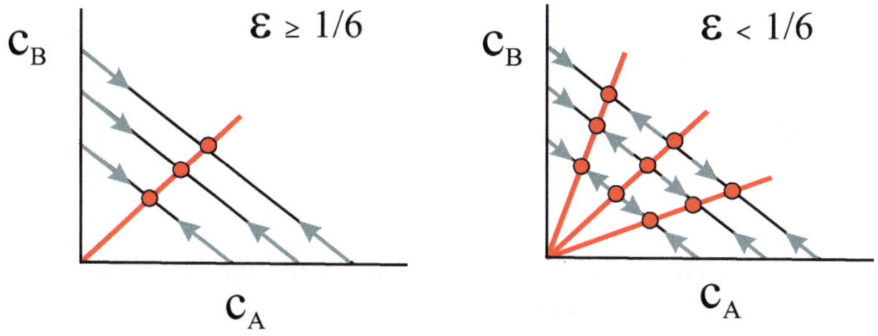

Fig. 5.2. Phase portraits for the Horn and Jackson example

5.1.2 The Edelstein Example

In the Horn-Jackson example, there is a uniformity of behavior from one stoichiometric compatibility class to another. The same is not true of the Edelstein [58] mass action system discussed earlier in Section 4.2.3. Here we consider the Edelstein network taken with fixed values of the rate constants, as indicated in (5.2).

$$
\begin{gathered}
A \underset{1}{\overset{8.5}{\rightleftarrows}} 2A \\
A+B \underset{1}{\overset{1}{\rightleftarrows}} C \underset{0.2}{\overset{1}{\rightleftarrows}} B
\end{gathered}
\tag{5.2}
$$

We show a (not-to-scale) sketch of the equilibrium set in Figure 5.3. In this case, the stoichiometric compatibility classes are unbounded rectangles, parallel to the A-axis, that go off to infinity in the A-direction. As the locus of equilibria snakes through the positive orthant, it intersects the various stoichiometric compatibility classes in one, two, or three equilibria. In those stoichiometric compatibility classes containing three equilibria, one is unstable. For still other sets of rate constants, different from those shown in (5.2), every positive stoichiometric compatibility class contains precisely one equilibrium, and it is stable relative to nearby initial conditions in the same stoichiometric compatibility class.

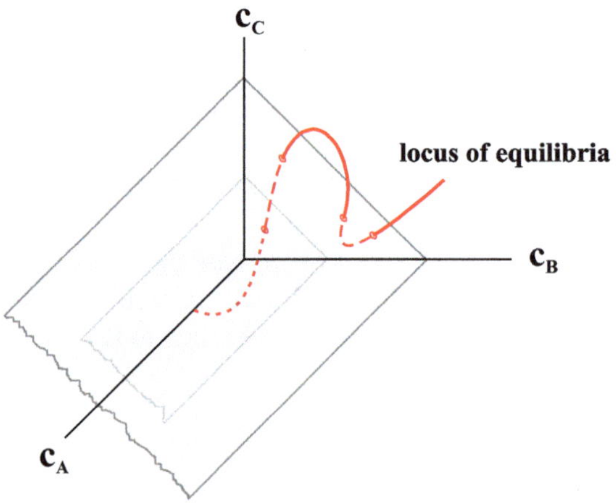

Fig. 5.3. A sketch of the locus of equilibria for the mass action system (5.2)

5.1.3 An Example Having Stoichiometric Compatibility Classes with Zero, One, and Two Positive Equilibria

For the Horn-Jackson and Edelstein networks, there are values of the rate constants for which one gets only dull behavior; in particular, in each positive stoichiometric compatibility class, there is precisely one equilibrium, and it is asymptotically stable. This is not the case, however, for network (5.3),

$$\begin{aligned} B &\to A \\ A + 2B &\to 3B, \end{aligned} \tag{5.3}$$

a network studied in [73] and for which peculiar behavior is present no matter what positive values the rate constants take.

A phase portrait, which is qualitatively the same for all rate constant values, is shown in Figure 5.4. Note that there are some positive stoichiometric compatibility classes that have no positive equilibria at all. There is one stoichiometric compatibility class that has one positive equilibrium. Others have two positive equilibria, with one of these unstable. Note too that a small perturbation away from a positive equilibrium can result in the extinction of one of the species, in this case B.

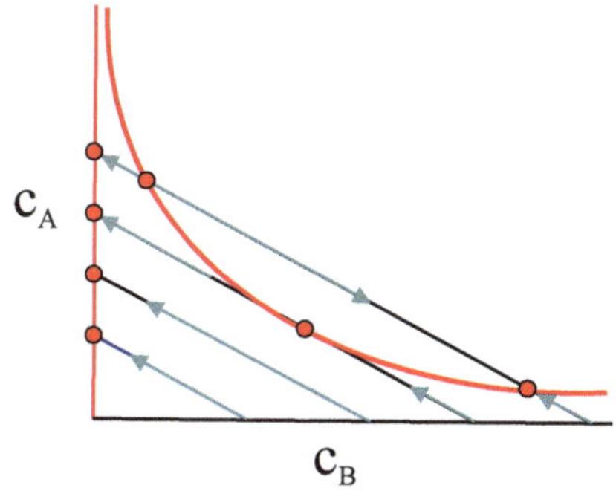

Fig. 5.4. A sketch of the locus of equilibria for the mass action system (5.3)

Again, the qualitative behavior shown in Figure 5.4 persists no matter how the rate constants are assigned to network (5.3). This stands in contrast not only to the situation for the Horn-Jackson and Edelstein networks but also to a remarkable fact that we shall prove later: *For any reversible network—and, indeed, for any weakly reversible network (Definition 6.5.3)—there invariably exists a rate constant assignment such that there is precisely one equilibrium in each positive stoichiometric compatibility class and it is asymptotically stable.*

The Edelstein network is reversible, and the Horn-Jackson network is weakly reversible. Network (5.3) is neither.

5.2 Symmetry Breaking

This section is, in effect, a continuation of the preceding one, for again we shall be concerned with unstable equilibria and the possibility of multiple stoichiometrically compatible equilibria. Here, however, the examples revolve around a common theme: symmetry breaking.

5.2.1 Left-Handed and Right-Handed Molecules: Origins of Enantiomeric Excess

Some sugars and amino acids have "left-handed" and "right-handed" versions, identical in all respects except that, in terms of molecular structure, one is the mirror image of the other. Such pairs are called *enantiomers*. In evolved organisms it sometimes happens that one form is present almost to the exclusion of the other.

The evolutionary origins of strong "enantiomeric excess" are a mystery. It is certainly plausible that life began with a rare event that produced one of the two forms, which then propagated. But there are other conjectures, some relying instead on a tiny perturbation of an initially *racemic* (i.e., 50–50) mixture that results in a runaway asymmetric synthesis in one of two possible directions [23, 86, 142].

In such a case, one can imagine that there is an *unstable* racemic steady state and two mirror-image stable non-racemic steady states. A tiny asymmetric perturbation of the unstable racemic steady state might then result in a composition trajectory that terminates at a stable steady state, one rich in either the right- or left-handed enantiomer. The Horn-Jackson mass action system (5.1), which is symmetric against the interchange of species A and B, behaves in precisely this way for $\varepsilon < \frac{1}{6}$. The two stable mirror-image steady states become increasingly pure in A or B as ε approaches zero.

Here we consider a very small toy model for enantiometric enhancement that does something far more exotic: There are again two mirror-image stable steady states, each highly pure in one of the enantiomers. This time, however, the sole racemic steady state is also *stable*. An initial condition that is just slightly left-right asymmetric can nevertheless give rise to a composition trajectory that evolves not toward the stable racemic steady state but, rather, toward one of the stable enantiomer-pure steady states. The model is thoroughly unrealistic in terms of the plausibility of its chemistry. It is meant only to demonstrate how even very small mass action systems can give rise to behaviors that are surprisingly rich.

We consider the toy chemistry shown as (5.4), in which L and R are left- and right-handed enantiomers. The substance P promotes uniformity. It binds triplets of enantiomers in which one is different from the others and then restructures the odd enantiomer to resemble the majority. In the process, P gets converted to Q. On the other hand, Q promotes diversity. It binds three identical enantiomers and converts one of them to its opposite. Along the way, Q gets converted to P. We presume the kinetics to be mass action, with rate constants as indicated. Note that (5.4) is symmetric against permutation of the enantiomers.

$$L + 2R + P \underset{2.56}{\overset{277\frac{1}{3}}{\rightleftarrows}} 3R + Q \qquad R + 2L + P \underset{2.56}{\overset{277\frac{1}{3}}{\rightleftarrows}} 3L + Q \tag{5.4}$$

That's the toy chemistry. Now, for the toy context, imagine a well-stirred semi-open reactor depicted schematically in Figure 5.5. There are membranes at the inlet and outlet ports that permit free passage of P and Q but which deny passage of enantiomers R and L. Thus, enantiomers within the reactor chamber will never leave it through the ports. We will suppose that the volume of the reactor is 1, that the volumetric flow rates of the liquid feed and effluent streams are 1, and that the concentrations of P and Q in the feed stream are, respectively, 2.5 and 0.1 moles per volume.[1] Moreover, we suppose that the concentrations of P and Q in the exit stream are identical to those in the chamber.

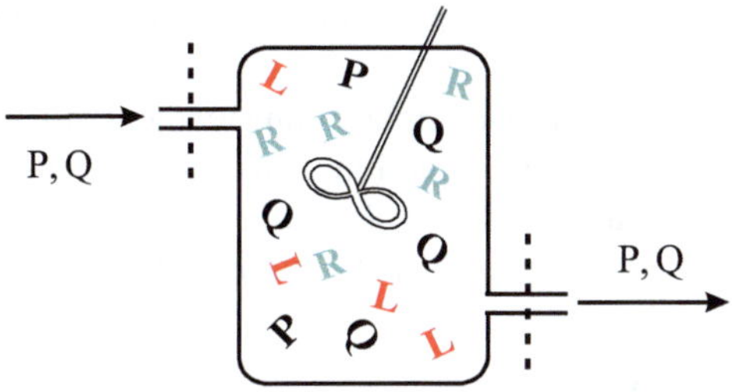

Fig. 5.5. A toy reactor that gives rise to sharp enantiomer enrichment

In the spirit of Sections 4.2.1 and 4.2.2, the differential equations governing the reactor are those that derive from the mass action system (5.5):

$$L + 2R + P \underset{2.56}{\overset{277\frac{1}{3}}{\rightleftarrows}} 3R + Q \qquad R + 2L + P \underset{2.56}{\overset{277\frac{1}{3}}{\rightleftarrows}} 3L + Q \tag{5.5}$$

$$P \underset{2.5}{\overset{1}{\rightleftarrows}} 0 \underset{0.1}{\overset{1}{\leftrightarrows}} Q$$

When the total concentration of enantiomers in the chamber is 1, those equations give rise to *five* positive steady states, displayed in Table 5.1. The racemic steady state is *stable*. Moreover, there are two *stable* mirror-image steady states that are virtually pure in one enantiomer or the other. There are also two mirror-image *unstable* steady states, one R-rich and one L-rich, but those are not nearly as pure.

Table 5.1. Five steady states for the mass action system (5.5)

	Stable R-Rich	Unstable R-Rich	Stable Racemic	Unstable L-Rich	Stable L-Rich
L	0.00036995	0.25	0.5	0.75	0.99963005
R	0.99963005	0.75	0.5	0.25	0.00036995
P	2.49981030	0.1	0.05867	0.1	2.49981030
Q	0.10018970	2.5	2.54133	2.5	0.10018970

Now consider the following thought experiment: By way of experimental preparation, imagine that, while the chamber contains no enantiomer at all, the reactor initially experiences the indicated flow-through of P and Q. In this case, the respective concentrations of P and Q in the feed stream, in the reactor chamber, and in the

[1] Units for volume, concentrations, rate constants, and flow rates are presumed to be self-consistent.

exit stream are 2.5 and 0.1 moles per volume. Now suppose that the reactor chamber is opened momentarily at the top and that *slightly* unequal amounts of L and R crystals are immediately dissolved in the liquid mixture within the chamber. In our thought experiment, the concentrations of R and L are instantaneously brought to 0.51 and 0.49 moles per volume, respectively, while the concentrations of P and Q remain at 2.5 and 0.1.

The result is a striking, almost-complete enhancement of the initially slight $R-L$ asymmetry: Simulations of the mass action differential equations for (5.5) indicate that the near-racemic initial mixture evolves not toward the racemic stable steady state but, rather, toward a stable steady state that is virtually pure in R.

5.2.2 Pattern Formation

Here we consider two interconnected cells such as those discussed in Section 4.2.4. In each cell the network of chemical reactions is that shown in (5.6); the kinetics is mass action with rate constants as indicated. The intercellular diffusion constants, D_A and D_B, described in Section 4.2.4, are both taken to be 1. (Recall from Remark 4.2.3 that the network is the Brusselator, which in turn derives from a conservative network in which certain species concentrations are deemed constant. In fact, the example is taken from [136].)

$$\begin{gathered} 0 \underset{1}{\overset{2}{\rightleftarrows}} A \overset{10}{\rightarrow} B \\ 2A + B \overset{1}{\rightarrow} 3A \end{gathered} \tag{5.6}$$

In the sense of Section 4.2.4, the system of four differential equations governing the concentrations of A and B in the two cells is the one that derives from the mass action system depicted in (5.7).

$$\begin{array}{ccccccc} 2A_1 + B_1 \overset{1}{\rightarrow} 3A_1 & & B_1 & \underset{1}{\overset{1}{\rightleftarrows}} & B_2 & & 2A_2 + B_2 \overset{1}{\rightarrow} 3A_2 \\ & & 10\uparrow & & \uparrow 10 & & \\ & & A_1 & \underset{1}{\overset{1}{\rightleftarrows}} & A_2 & & \\ & & 1 \nwarrow\!\!\searrow 2 & & 2 \nearrow\!\!\swarrow 1 & & \\ & & & 0 & & & \end{array} \tag{5.7}$$

Those differential equations admit an *unstable* uniform steady state shown in Table 5.2, with the concentrations of A and B the same in both cells. However, there is also a *stable asymmetric* steady state with concentrations shown in Table 5.3 and depicted schematically in Figure 5.6.

Table 5.2. An unstable uniform steady state

Cell 1	Species	Cell 2
2	A	2
5	B	5

Table 5.3. A stable asymmetric steady state for an otherwise symmetric two-cell system

Cell 1	Species	Cell 2
3.1547	A	0.8453
3.5179	B	6.9820

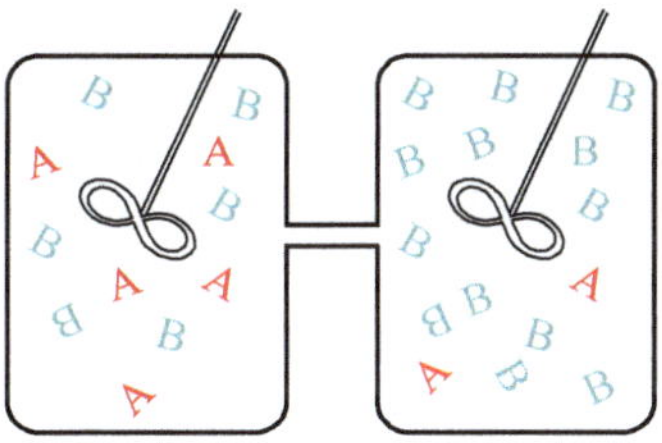

Fig. 5.6. An asymmetric steady state

5.3 Cyclic Composition Trajectories

Here we consider two simple examples that give rise to cyclic composition trajectories.

5.3.1 The Lotka Example: Rabbits and Wolves

We begin with a mass action system first studied in 1910 by Alfred Lotka [126] and displayed here as (5.8). Although Lotka clearly had chemistry in mind, the resulting mass action differential equations are often interpreted in ecological terms:

Imagine an island on which there are only rabbits (A), wolves (B), and a large, constantly replenishing supply of carrots. The rabbits happily eat the carrots and procreate, the procreation rate being proportional to the current concentration of rabbits on the island; this is reflected in the first reaction of (5.8). The wolves are not vegetarians. The wolves eat rabbits, which they need to procreate. However, rabbits, unlike carrots, might be in momentarily short supply. The wolf procreation rate is deemed proportional to the product of the island's rabbit and wolf concentrations; this is reflected in the second reaction. The third reaction reflects the death of wolves at a rate proportional to the current wolf concentration on the island. (Rabbits play no role in these deaths, but the rabbits are not saddened by them.)

$$
\begin{aligned}
A &\xrightarrow{\alpha} 2A \\
A + B &\xrightarrow{\beta} 2B \\
B &\xrightarrow{\gamma} 0
\end{aligned}
\tag{5.8}
$$

The corresponding mass action differential equations are given by (5.9).

$$
\begin{aligned}
\dot{c}_A &= \alpha c_A - \beta c_A c_B \\
\dot{c}_B &= \beta c_A c_B - \gamma c_B .
\end{aligned}
\tag{5.9}
$$

There is a positive equilibrium, $c_A = \frac{\gamma}{\beta}$, $c_B = \frac{\alpha}{\beta}$, at which the rabbit and wolf concentrations remain constant. Suppose, however, that at a certain moment, the rabbit population on the island is moderate and the wolf population is very low. In this case, the rabbits happily procreate, and their concentration on the island increases. The few wolves can then gorge themselves on rabbits, and their population increases. The situation for the rabbits becomes increasingly unpleasant, and their population eventually decreases. But then the wolves experience a rabbit famine, and their population also decreases. When sufficiently many wolves die off, the island becomes a happier place for the rabbits, notwithstanding the pileup of rotting wolf carcasses. In these good times, the rabbit population increases, and the cycle begins anew.

The presence of a cycle is literal: the phase portrait for (5.8) is the one schematically depicted in Figure 5.7. Every positive composition, apart from the positive equilibrium, resides on a nontrivial cyclic composition trajectory. This is the case regardless of the (positive) values the rate constants take.

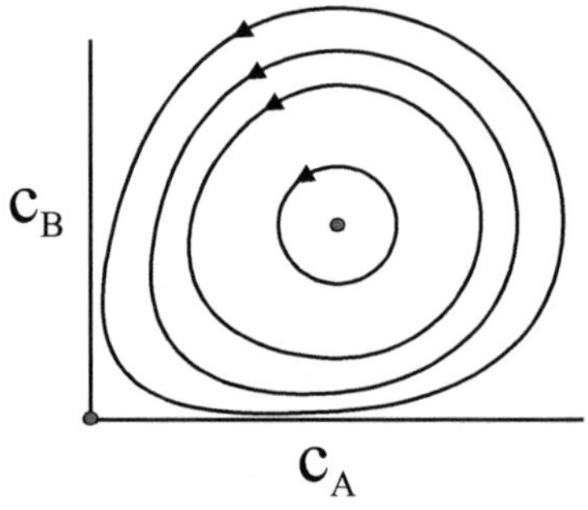

Fig. 5.7. Phase portrait for the Lotka mass action system (5.8)

5.3.2 The Brusselator Example

Although the mass action system (5.10), derived from the Brusselator network (4.21), also gives rise to a cyclic composition trajectory, the dynamics are qualitatively different from those of the Lotka system in three respects.

$$0 \underset{1}{\overset{1}{\rightleftarrows}} A \xrightarrow{3} B \tag{5.10}$$

$$2A + B \xrightarrow{1} 3A$$

First, the phase portrait for (5.10), shown in Figure 5.8, indicates that there is a single (stable) *limit cycle* that is approached by all trajectories originating near it. Second, there is an unstable positive equilibrium. (The positive equilibrium of the Lotka mass action system is stable, but not asymptotically stable.) Finally, unlike the Lotka system, it is not true that, for all choices of rate constants, there is a nontrivial cyclic composition trajectory: there is none when the rate constant for $A \to B$ is changed from 3 to 1. In this case, there is again a single positive equilibrium, and it is asymptotically stable. (As the rate constant for $A \to B$ is increased gradually from 1 to 3, the differential equations give rise to a Hopf bifurcation.)

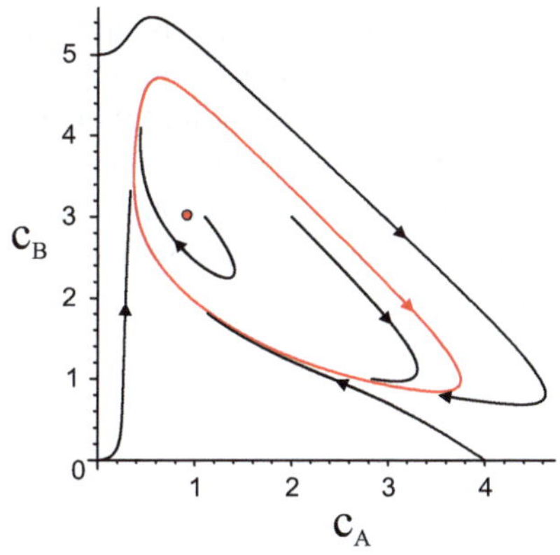

Fig. 5.8. Phase portrait for (5.10)

5.4 Questions: Fruit Flies and Platypuses Again

Here we return to what was said at this chapter's beginning: The networks studied in Sections 5.1–5.3 are meant to comprise a toy zoo, a collection of small examples that exhibit the chemically *uncommon*. Our zoo "animals" are less like the fruit flies of Chapter 1 than they are like the platypuses: Fruit flies hold interest because they are unexceptional; study of them tells us much about what many organisms have in common. Platypuses hold interest because they are odd.

In contrast to the behavior exhibited in the zoo examples, the dynamics shown in Figure 5.1 for the fruit fly mass action system

$$A \underset{k'}{\overset{k}{\rightleftarrows}} 2B \tag{5.11}$$

is far more typical of the behavior that chemists expect: *In each positive stoichiometric compatibility class, there is precisely one equilibrium; it is asymptotically stable; and there are no cyclic composition trajectories. Moreover, when* (5.11) *is the operative chemistry, there is no possibility of an asymmetric two-cell steady state of the kind discussed in Section 5.2.2. These assertions do not depend on the positive values that the rate constants take.*

Yes, (5.11) is a tiny mass action system, but it is not merely the small size of the network that precludes more exotic behavior. The same dullness of dynamics, *regardless of rate constant values*, is extremely common in mass action systems that derive from networks containing hundreds of species and hundreds of reactions. And, again, *it is this parameter-indifferent dull behavior that chemists expect broadly, notwithstanding what might be great complexity in the governing differential equations.*

But why? What is it in the mathematics of chemical reaction networks, and of mass action systems in particular, that causes chemists to harbor such expectations across broad expanses of the reaction network landscape, without regard to parameter values? And what is it in the zoo examples that makes them exceptional, that enables them to exhibit behavior that chemists don't ordinarily expect?

We have clearly returned to mysteries of the kind raised in Chapter 1, but now in the context of a more fully articulated mathematical framework. In Part II of this book, we shall see how some of the principal theorems of chemical reaction network theory help resolve those mysteries.

Part II

Some Principal Theorems: A First Look

Part II is meant to give a somewhat informal sampling of the kind of theorems that chemical reaction network theory provides. At the beginning we will invoke mass action kinetics, but then we will consider kinetic categories that are far more general.

All of the theorems in Part II address the relationship between reaction network behavior and reaction network structure. Part II begins with Chapter 6, devoted to the development of a vocabulary with which aspects of reaction network structure might be discussed. As we shall see, an important structural attribute of a reaction network is its *deficiency*, an easily calculated nonnegative integer index that provides one means of network classification. Thus, there are networks of deficiency zero, of deficiency one, and so on. The deficiency is not a measure of the network's size. Networks of deficiency zero, for example, can contain hundreds of species and hundreds of reactions.

In Chapters 7–9, the deficiency of a network is its central structural feature. In particular, Chapters 7 and 8 contain statements of the Deficiency Zero and Deficiency One Theorems, which give a surprising amount of dynamical information about networks that might be enormously complex, so long as results of very simple deficiency calculations conform to certain requirements. These theorems go a long way toward explaining why vast parts of the reaction network landscape are associated with dull behavior, at least when the kinetics is mass action. Chapter 9 deals with different questions, related to *concentration robustness*—the remarkable ability of certain biological networks to maintain the concentration of a critical network species within very narrow bounds, despite what might be very large externally imposed fluctuations in concentrations of the remaining species. Although questions in Chapter 9 are unlike those in the preceding two chapters, the deficiency remains pivotal.

Chapters 10 and 11, on the other hand, are not at all deficiency-oriented. Instead, the crucial network-structural attribute is *concordance*. A reaction network is either concordant or else it is *discordant*. To know that a network is concordant is to know that, when it is endowed with any kinetics satisfying very weak and natural conditions, behavior of the resulting kinetic system is severely constrained. Whether a network is concordant or discordant can be decided by freely available computer tools [62].

Chapter 10 gives the basic facts about concordance, discordance, and their dynamical consequences. Chapter 11 begins by associating with every reaction network a *Species-Reaction Graph*. The central theorem in Chapter 11 asserts that, when a network's Species-Reaction Graph satisfies very weak conditions, the network itself is concordant (and thereby inherits all the dynamical consequences that concordance confers). As a result, there is a companion theorem—actually a corollary of the first theorem—in which concordance is not mentioned at all. It asserts that, when a network's Species-Reaction Graph satisfies very mild constraints, the variety of dynamics the network can engender is severely limited.

Chapter 12 concludes Part II. It revisits questions posed about reaction networks at the very beginning of the book, in Chapter 1, this time in light of material accumulated in the intervening chapters. In particular, there is an extended discussion of mathematical reasons for the richness of behavior found in biological chemistry, a richness that is largely absent in "regular" chemistry.

Remark. Chapter 10, about reaction network concordance, is the most technical—but perhaps the most powerful—of all the chapters in Part II. It warrants some discussion. Unfortunately, the very definition of concordance is somewhat technical, which to a great extent accounts for the more mathematical tone of Chapter 10. By a simple expedient, I could have omitted Chapter 10 entirely or, indeed, any mention of concordance at all: I could have stated only theorems about dynamical properties of networks having "nice" Species-Reaction Graphs—that is, theorems about those networks having Species-Reaction Graphs for which concordance is ensured.

But then so much of the additional power and nuance contained in Chapter 10 would have been lost. As a compromise, I have placed some advice at the beginning of Chapter 10, in effect a dispensation whereby a reader who wishes to do so can pass over much (or all) of that chapter without serious loss of continuity. Chapter 11, about the Species-Reaction Graph, begins with some further advice and reassurance for readers who might have read Chapter 10 very lightly or skipped it entirely.

6

Aspects of Reaction Network Structure

Our aim in this chapter is to provide a vocabulary with which reaction network structure might be discussed. Although some of the terms will find use well before others, they are all intimately related. For this reason we will find it advantageous to locate their definitions in one place. At the end of the chapter, there is an appendix, in which we discuss a sense in which a reaction network can sometimes be decomposed into *independent subnetworks*. This is an important aspect of reaction network structure, one that belongs naturally in this chapter. Nevertheless, the discussion of the decomposition has been placed in an appendix because it is slightly more technical than the rest and is not essential to an understanding of the chapters immediately following.

In representing reaction networks diagrammatically, we have, without saying so explicitly, always followed a certain procedure. We displayed each complex precisely once, and then we joined the various complexes with the appropriate reaction arrows. By the *standard reaction diagram* for a network, we will mean a display of this kind. The standard reaction diagram has the virtue of making apparent crucial aspects of a reaction network's structure in the form of a directed graph, with complexes playing the role of vertices and reaction arrows playing the role of directed edges.

At least for now, there are only a few ideas we need: what we mean by the *linkage classes* and *strong-linkage classes* of a network, what it means for a network to be *weakly reversible*, and what we mean by the *deficiency* of a network. All but the last of these depend solely on a network's character as a graph, the precise nature of the complexes playing no essential role. The algebraic nature of the complexes will, however, influence a network's deficiency. These basic ideas, especially those associated with the graphical structure of a network, are easy to understand in an intuitive way, and each will be introduced informally before a formal definition is given.

It will be understood throughout this chapter that all definitions refer to structural aspects of a reaction network $\{\mathscr{S},\mathscr{C},\mathscr{R}\}$.

M. Feinberg, *Foundations of Chemical Reaction Network Theory*,
Applied Mathematical Sciences 202,
https://doi.org/10.1007/978-3-030-03858-8_6

6.1 The Linkage Classes of a Reaction Network

One way in which we can begin discussing the structure of a reaction network is by saying some things about its standard reaction diagram's gross features. Consider the reaction diagram (6.1). What strikes one immediately about the diagram is that it is composed of two distinct "pieces," one containing the complexes $A+B$, C, $D+E$, and F, the other containing the complexes $2A$, $B+G$, and H. Disregarding the reaction arrow directions, we can see that complexes in the set $\{A+B, C, D+E, F\}$ are linked to each other, perhaps indirectly, but not to any other complex in the network. The same is true of complexes in the set $\{2A, B+G, H\}$. The sets $\{A+B, C, D+E, F\}$ and $\{2A, B+G, H\}$ are the *linkage classes* of network (6.1). In rough terms, the linkage classes of a network are the sets of complexes in the various "pieces" of which the network is composed.

$$
\begin{array}{ccccccc}
A+B & \rightleftarrows & C & \rightarrow & D+E & \rightleftarrows & F \\
 & & 2A & \rightleftarrows & B+G & & \\
 & & \nwarrow & & \swarrow & & \\
 & & & H & & &
\end{array} \tag{6.1}
$$

We make this precise in the following definition:

Definition 6.1.1. *Two complexes $y \in \mathscr{C}$ and $y' \in \mathscr{C}$ are* **directly linked** *if $y \to y'$ or $y' \to y$. If y and y' are directly linked, we write $y \leftrightarrow y'$. Two complexes $y \in \mathscr{C}$ and $y' \in \mathscr{C}$ are* **linked** *if any of the following conditions is satisfied:*

(i) $y = y'$;
(ii) $y \leftrightarrow y'$;
(iii) There is a sequence of complexes $y(1), y(2), \ldots, y(k)$ such that

$$y \leftrightarrow y(1) \leftrightarrow y(2) \leftrightarrow \ldots \leftrightarrow y(k) \leftrightarrow y'.$$

If y and y' are linked, we write $y \sim y'$. The equivalence relation $\sim$ induces a partition of $\mathscr{C}$ into equivalence classes called the **linkage classes** *of the network. We reserve the symbol ℓ to denote the number of linkage classes in the network.*

Remark 6.1.2. When two complexes, y and y', are directly linked, we shall sometimes use the phrase *the direct link $y \leftrightarrow y'$* to mean the set of reactions—either one or two—that directly connect complexes y and y'. In (6.1), for example, the direct link $B+G \leftrightarrow H$ is synonymous with the single reaction $B+G \to H$; on the other hand, the direct link $2A \leftrightarrow B+G$ is the reaction pair $2A \rightleftarrows B+G$.

Remark 6.1.3. It should be remembered that, for us, linkage classes are *sets of complexes* in a network, not subnetworks of a parent network. Thus, the networks (6.1) and (6.2) have identical linkage classes: $\{A+B, C, D+E, F\}$ and $\{2A, B+G, H\}$.

$$
\begin{array}{ccccccc}
F & \rightarrow & A+B & \rightleftarrows & C & \rightleftarrows & D+E \\
 & & H & \rightarrow & 2A & \rightleftarrows & B+G
\end{array} \tag{6.2}
$$

6.2 The Rank of a Reaction Network Revisited

We met the idea of *rank* of a reaction network before, in Chapter 3, where the rank was identified with the dimension of the stoichiometric subspace. The definition of rank is repeated below, not only because it will play a role in the next section but also because we are now in a position to discuss the idea just a little more deeply.

Definition 6.2.1. *The* **rank** *of a reaction network* $\{\mathscr{S},\mathscr{C},\mathscr{R}\}$ *is the rank of its set of reaction vectors. That is, the network has rank s if there exists in the set*

$$\{y' - y \in \mathbb{R}^{\mathscr{S}} : y \to y' \in \mathscr{R}\}$$

a linearly independent subset containing s vectors but no linearly independent subset containing $s+l$ *vectors. We reserve the symbol s to denote the rank of a network.*

Remark 6.2.2. Because the rank of reaction network is the number of elements in the largest linearly independent set of reaction vectors for the network, it is easy to see that the rank of network (6.1) is the same as the rank of the network (6.3). A little less obvious is that the rank of network (6.3)—and, therefore, of network (6.1)—is the same as the rank of network (6.4).

$$\begin{array}{ccccccc} A+B & \to & C & \to & D+E & \to & F \\ & & 2A & \to & B+G & & \\ & & & \swarrow & & & \\ & & & H & & & \end{array} \qquad (6.3)$$

$$\begin{array}{ccccc} A+B & \leftarrow & C & \to & D+E \\ & & \downarrow & & \\ & & F & & \\ & & & & \\ 2A & \leftarrow & H & \to & B+G \end{array} \qquad (6.4)$$

It is not difficult to show that *two reaction networks having the same complexes and the same linkage classes have the same rank. In fact, both networks have the same stoichiometric subspace.* (Remember that a linkage class is a set of complexes.) The rank of a network is completely insensitive to the precise way in which its reaction arrows join the complexes *within* the various linkage classes. With respect to a network's rank, the reaction arrows exert their influence only to the extent that they serve to partition the complexes into linkage classes.

Thus, to calculate the rank of a network such as (6.1), it suffices to calculate the rank of a far simpler network such as (6.4), one having just enough reactions to give rise to the same linkage classes as those of the original network. Indeed, for this purpose one can, as in (6.4), form a new network in which the complexes in each linkage class of the original network are arranged in a star, with reactions radiating out from one arbitrarily chosen central complex of the linkage class to the remaining ones. (Replacement of the star with any tree-like arrangement will serve the same purpose.) The five reaction vectors for network (6.4) are easily seen to be linearly independent, so the ranks of networks (6.1)–(6.4) are all five.

Remark 6.2.3 (A bound on the rank). These considerations tell us that there is a bound to how large the rank can be for a reaction network having n complexes and ℓ linkage classes: Since the given network has the same rank as one with the star-like arrangement of complexes within linkage classes, and since the starry "galactic network" has just $n-\ell$ reactions (and, therefore, just $n-\ell$ reaction vectors), its rank cannot exceed $n-\ell$.

We have, then, a relation that obtains for every reaction network: If n is the number of complexes, if ℓ is the number of linkage classes, and if s is the rank, we invariably have

$$n-\ell \geq s. \tag{6.5}$$

Equality will obtain in (6.5) precisely when the reaction vectors for the $n-\ell$ reactions in a "galactic" network, constructed as indicated above, are linearly independent. That equality in (6.5) need not obtain is demonstrated by the network shown below as (6.6), for which $n=5$, $\ell=2$, and $s=2$.

$$\begin{gathered} A \rightleftarrows C \rightleftarrows B \\ A+B \rightleftarrows 2C \end{gathered} \tag{6.6}$$

Remark 6.2.4 (The rank of a reaction network: computational considerations). The readily available computational tools of linear algebra offer a variety of ways to calculate the rank of a matrix. These can be exploited to calculate the rank of reaction network in the following way: For a given network with n complexes and ℓ linkage classes, form a galactic network of the kind shown in (6.4), one that retains the linkage classes of the original network. If the species are numbered $1\ldots N$, each of the $n-\ell$ reaction vectors for the galactic network can be identified with a vector of $\mathbb{R}^N$ in the obvious way. These can then be stacked to form a matrix having $n-\ell$ rows and N columns. The rank of the resulting matrix is identical to the rank of the original network.

6.3 The Deficiency of a Reaction Network

For a reaction network with n complexes, ℓ linkage classes, and rank s, it follows from Remark 6.2.3 that $n-\ell-s \geq 0$, with equality holding precisely when the reaction vectors for the network are as linearly independent as the partition of complexes into linkage classes will allow. In this sense, the nonnegative integer $n-\ell-s$ is a measure of how independent the reaction vectors are, given the network's linkage class structure. In the following definition, we give that integer a name.

Definition 6.3.1. *The* **deficiency**, δ, *of a reaction network with n complexes, ℓ linkage classes, and rank s is defined by*

$$\delta := n - \ell - s. \tag{6.7}$$

Thus, the deficiency of the network with which we began Chapter 2, repeated below as (6.8), has a deficiency of zero ($n = 5, \ell = 2, s = 3$). Networks (6.1)–(6.4) all have deficiencies of zero. (For each, $n = 7, \ell = 2, s = 5$.) Network (6.6) has a deficiency of one ($n = 5, \ell = 2, s = 2$).

$$\begin{array}{ccc} A & \rightleftarrows & 2B \\ A+C & \rightleftarrows & D \\ & \nwarrow \quad \swarrow & \\ & B+E & \end{array} \tag{6.8}$$

Remark 6.3.2. The deficiency of a network is not a measure of its size. Networks of deficiency zero can have hundreds of species and hundreds of reactions.

Remark 6.3.3 (Insensitivity of the deficiency to the fine details of reaction arrow structure). Recall that any two reaction networks having the same complexes and the same linkage classes have the same rank. From this it follows that *any two reaction networks having the same complexes and the same linkage classes have the same deficiency. Thus, the deficiency of a network, like the rank, is completely insensitive to the precise way in which its reaction arrows join the complexes within the various linkage classes.* With respect to the deficiency, the reaction arrows exert their influence only to the extent that they serve to partition the complexes into linkage classes.

Remark 6.3.4 (Deficiency zero networks). From the discussion at the beginning of this section, it follows that *deficiency zero networks are ones for which the reaction vectors are as independent as the partition of complexes into linkage classes will allow*. Any subnetwork of a deficiency zero network is also a deficiency zero network.

6.4 The Rank and Deficiency of a Linkage Class

While motivating the idea of linkage classes, we noted that a standard reaction diagram could be regarded to be composed of separate "pieces." Within each piece the various complexes are linked to each other but to no complex not in that piece.[1] Thus, for example, the network (6.8) can be regarded to be composed of two separate

[1] The term "piece" in this informal discussion is meant to connote a subnetwork of the parent network. I have resisted using the graph-theoretical term "connected component" because there is a lack of uniformity in how that term is used. For some authors it refers to a subgraph, and for others it refers to the set of vertices, roughly analogous to what we have called a linkage class—i.e., a set of complexes.

pieces, one associated with the linkage class $\{A, 2B\}$ and the other associated with the linkage class $\{A+C, D, B+E\}$.

Each piece can be viewed as a reaction network unto itself, a reaction network having just one linkage class. With each such single-linkage-class network, we can associate a rank and a deficiency. For reasons given before, that rank and deficiency are determined solely by the complexes within the piece, not by the precise way in which the reaction arrows join the complexes. It makes sense, then, to refer more fundamentally to *the rank and deficiency of a given linkage class* within the parent network.

Remark 6.4.1. Recall that s generally denotes rank of a network as whole. When the linkage classes of a network are denoted $\mathscr{L}_\theta$, $\theta = 1, 2, \ldots, \ell$, we shall denote by s_θ the rank of linkage class $\mathscr{L}_\theta$. Network (6.9) serves to indicate that the sum of the ranks of a network's linkage classes need not be the same as the network's rank: The network has a rank of two. The upper linkage class has a rank of two, and the lower linkage class has a rank of one.

$$\begin{gathered} A \rightleftarrows C \rightleftarrows B \\ A+B \rightleftarrows 2C \end{gathered} \tag{6.9}$$

It follows from simple considerations in linear algebra that we always have the relation

$$\sum_{\theta=1}^{\ell} s_\theta \geq s, \tag{6.10}$$

with equality holding precisely when the stoichiometric subspace for the network as a whole is the direct sum [98] of the stoichiometric subspaces for its ℓ pieces. The implications of equality in (6.10) are discussed in the appendix at the end of this chapter.

As before we denote by n the number of complexes in the network as a whole, and now we denote by n_θ the number of complexes in linkage class $\mathscr{L}_\theta$, $\theta = 1, 2, \ldots, \ell$. Then the deficiency of linkage class $\mathscr{L}_\theta$ is given by

$$\delta_\theta = n_\theta - 1 - s_\theta, \quad \theta = 1, 2, \ldots, \ell. \tag{6.11}$$

Note that

$$\sum_{\theta=1}^{\ell} \delta_\theta = \sum_{\theta=1}^{\ell} (n_\theta - 1 - s_\theta) = n - \ell - \sum_{\theta=1}^{\ell} s_\theta. \tag{6.12}$$

Recall that the deficiency of the network as a whole is given by

$$\delta := n - \ell - s. \tag{6.13}$$

Thus, from (6.10) we always have

$$\sum_{\theta=1}^{\ell} \delta_\theta \leq \delta, \tag{6.14}$$

with equality holding in (6.14) precisely when equality holds in (6.10).

Network (6.9) has a deficiency of one, while each of its linkage classes has a deficiency of zero. On the other hand, equality holds for network (6.8): the deficiency of the network is zero, and each of its linkage classes has a deficiency of zero.

6.5 Reversibility and Weak Reversibility

So far, the directions of reaction arrows have played no role. The arrow directions will, however, play a role in our next idea, weak reversibility.

In standard language of chemistry, a reaction network such as (6.15) is *reversible*: the "reacts to" relation is symmetric in the sense that every reaction is accompanied by its reverse. For the record, we record this in a definition.

$$\begin{array}{ccc} A & \rightleftarrows & 2B \\ A+C & \rightleftarrows & D \\ \nwarrow\!\!\searrow & & \swarrow\!\!\nearrow \\ & B+E & \end{array} \qquad (6.15)$$

Definition 6.5.1. *A reaction network* $\{\mathscr{S},\mathscr{C},\mathscr{R}\}$ *is* **reversible** *if the reacts to relation* $\mathscr{R}$ *is symmetric—that is, if* $y' \to y$ *is a member of* $\mathscr{R}$ *whenever* $y \to y'$ *is a member of* $\mathscr{R}$.

Reaction network (6.16) is not reversible, but it nevertheless has a kind of reversibility: whenever there is a directed arrow path from one complex to another, there is also a directed arrow path from the second complex back to the first. For example, there is a directed arrow path from D to $B+E$ (consisting only of the reaction $D \to B+E$), and there is also a directed arrow path from $B+E$ back to D, via $A+C$. In this case, we say the (6.16) is *weakly reversible*.

$$\begin{array}{ccc} A & \rightleftarrows & 2B \\ A+C & \rightleftarrows & D \\ \nwarrow & & \swarrow \\ & B+E & \end{array} \qquad (6.16)$$

To make this precise, we first define the *ultimately reacts to* relation in the complex set. In effect, a complex ultimately reacts to a different complex if there is a directed arrow pathway connecting the first complex to the second. For technical reasons, we adopt the convention that every complex ultimately reacts to itself.

Definition 6.5.2. *A complex* $y \in \mathscr{C}$ **ultimately reacts to** *complex* $y' \in \mathscr{C}$ *if any of the following conditions is satisfied:*

(i) $y = y'$;
(ii) $y \to y'$;
(iii) *There is a sequence of complexes* $y(1), y(2), \ldots, y(k)$ *such that*

$$y \to y(1) \to y(2) \to \ldots \to y(k) \to y'.$$

If y *ultimately reacts to* y', *we write* $y \Rightarrow y'$.

Definition 6.5.3. *A reaction network* $\{\mathscr{S}, \mathscr{C}, \mathscr{R}\}$ *is* **weakly reversible** *if its ultimately reacts to relation* $\Rightarrow$ *is symmetric—that is, if* $y' \Rightarrow y$ *whenever* $y \Rightarrow y'$.

Remark 6.5.4. It is not difficult to see that a network is weakly reversible precisely when, in the standard reaction diagram, every reaction is contained in a directed cycle; that is, whenever we have $y \to y'$, we also have $y' \Rightarrow y$.

Neither of the networks (6.17) nor (6.18) is weakly reversible. In (6.17) there is a directed arrow path from A to $2B$ but none from $2B$ to A. In (6.18) there is a directed arrow path from $B+E$ to $A+C$ via D, but none from $A+C$ to $B+E$.

$$\begin{array}{ccc} A & \to & 2B \\ A+C & \rightleftarrows & D \\ & \nwarrow \quad \swarrow & \\ & B+E & \end{array} \tag{6.17}$$

$$\begin{array}{ccc} A & \rightleftarrows & 2B \\ A+C & \rightleftarrows & D \\ & \nwarrow \quad \nearrow & \\ & B+E & \end{array} \tag{6.18}$$

Remark 6.5.5. Chemists often insist that every reaction network should be deemed reversible.[2] However, for many—but not all—of the main theorems of chemical reaction network theory, the less stringent presumption of weak reversibility suffices. Clearly, every reversible network is also weakly reversible, so theorems about weakly reversible networks embrace reversible networks as special cases. In fact, there are still milder graphical conditions that can often be invoked in the place of weak reversibility. This will be discussed in Remark 6.6.5.

[2] In this, they usually have in mind "true" chemical reaction networks, not reaction networks, such as those considered in Section 4.2, constructed to model open systems.

6.6 The Strong-Linkage Classes of a Reaction Network and Terminal Strong-Linkage Classes

Recall from Section 6.1 that the "reacts to" relation in a network induces a partition of the network's complexes into subsets called the network's *linkage classes*. Note that the partition into linkage classes is completely insensitive to the precise directions of the reaction arrows; reversal of any or all reaction arrows will leave the linkage classes unchanged.

We shall have interest in a finer partition of the complexes, also related to how the complexes are joined by reaction arrows, but this time the directions of the reaction arrows will matter heavily. The resulting subsets of complexes are the *strong-linkage classes* of the network:

Definition 6.6.1. *Two complexes* $y \in \mathscr{C}$ *and* $y' \in \mathscr{C}$ *are* **strongly linked** *if both* $y \Rightarrow y'$ *and* $y' \Rightarrow y$*—that is, if each ultimately reacts to the other. If* y *and* y' *are strongly linked, we write* $y \approx y'$*. The equivalence relation* $\approx$ *induces a partition of* $\mathscr{C}$ *into equivalence classes called the* **strong-linkage classes** *of the network.*

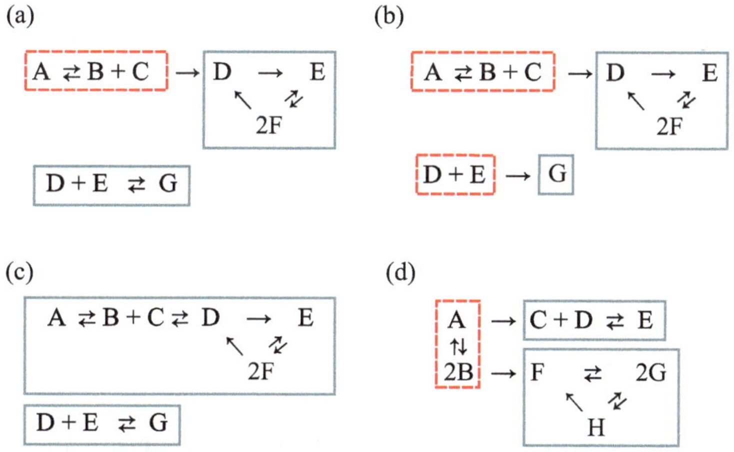

Fig. 6.1. Strong-linkage classes and terminal strong-linkage classes

The ideas in Definition 6.6.1 are illustrated in Figure 6.1, where, in the four networks displayed, the various strong-linkage classes are contained within rectangles. (The distinction between the solid and dashed rectangles will be discussed soon.) In a sense suggested by the figure, the strong-linkage classes can be associated with a network's "weakly reversible parts."

Remark 6.6.2. Clearly $y \approx y'$ implies that $y \sim y'$. Thus, every strong-linkage class lies within a linkage class. In fact, every linkage class is the disjoint union of strong-linkage classes. The linkage classes of a reaction network correspond to what, in graph-theoretical terminology, are sometimes called the *weak components* of a directed graph, and the strong-linkage classes correspond to what are sometimes called the *strong components*. In chemical parlance, however, the word "component" is sometimes used as a synonym for "species."

Figure 6.1 suggests that certain strong-linkage classes—those enclosed in solid rectangles—are terminal, in the sense that no reaction arrow leaves them. The terminal strong-linkage classes will often have a distinguished role to play.

Definition 6.6.3. *A* **terminal strong-linkage class** *is a strong-linkage class Λ with the property that no complex in Λ reacts to a complex not in Λ; that is, $y \in \Lambda$ and $y \to y'$ imply that y' is in Λ. The symbol $\mathfrak{t}$ will be used to indicate the number of terminal strong-linkage classes in a network.*

Remark 6.6.4. By a *terminal complex*, we mean one that resides in a terminal strong-linkage class; otherwise the complex is *nonterminal*.

Remark 6.6.5 (The $\mathfrak{t} = \ell$ condition). Weakly reversible networks are characterized by the fact that, for them, the linkage classes, strong-linkage classes, and terminal strong-linkage classes coincide. As we shall see, the coincidence of the linkage classes and terminal strong-linkage classes makes for a certain pleasantness in dealing with weakly reversible networks. However, some—but not all—of the nice features a weakly reversible network enjoys derive simply from the fact that the *number* of its linkage classes is identical to the *number* of its terminal strong-linkage classes (i.e., $\mathfrak{t} = \ell$) . A network need not be weakly reversible for it to have this property—see Figure 6.1(a),(b)—and it often happens that certain assertions that hold true for weakly reversible networks carry over (perhaps with minor modification) to networks for which $\mathfrak{t} = \ell$.

Here is an example: *For a mass action system, regardless of rate constant values, the kinetic subspace (discussed in Appendix 3.A) coincides with the stoichiometric subspace if, in the underlying reaction network, every linkage class contains just one terminal strong-linkage class.* This is proved in Appendix 16.B.

6.7 Cuts, Trees, and Forests in the Standard Reaction Diagram

We shall sometimes be concerned with a reaction or a pair of reversible reactions whose removal results in a "disconnection" of the standard reaction diagram. Network (6.19) will provide some examples.

$$\begin{array}{ccccccccc} A+B & \rightleftarrows & C & \to & D+E & \rightleftarrows & F \\ & I & \to & 2A & \rightleftarrows & B+G & \\ & & & \nwarrow & & \swarrow & \\ & & & & H & & \end{array} \tag{6.19}$$

Recall from Remark 6.1.2 that when in a reaction network the complexes y and y' are directly linked, we regard *the direct link* $y \leftrightarrow y'$ to be a synonym for the set of reactions, either one or two, that directly connect y and y'. Thus, in network (6.19) the direct link $A+B \leftrightarrow C$ consists of the reaction pair $A+B \rightleftarrows C$; the direct link $B+G \leftrightarrow H$ consists of the single reaction $B+G \to H$.

Definition 6.7.1. *A* **cut-link** *in a reaction network is a direct link* $y \leftrightarrow y'$ *such that removal of* $y \leftrightarrow y'$ *leaves* y *and* y' *unlinked.*

When we say "leaves y and y' unlinked," we mean that the linkage class containing y and y' can be partitioned into two sets of complexes, $\mathscr{W}$ and $\mathscr{W}'$, one containing y and the other containing y', such that there is no direct link connecting a complex of $\mathscr{W}$ with a complex of $\mathscr{W}'$ apart from $y \leftrightarrow y'$.

Example 6.7.2. In network (6.19), the cut-links are $A+B \leftrightarrow C$, $C \leftrightarrow D+E$, $D+E \leftrightarrow F$, and $I \leftrightarrow 2A$. The following direct links are not cut-links: $2A \leftrightarrow B+G$, $B+G \leftrightarrow H$, and $H \leftrightarrow 2A$.

Remark 6.7.3. If $y \leftrightarrow y'$ is a cut-link and y and y' lie in the same strong-linkage class, then the cut-link consists of the *reversible* reaction pair $y \rightleftarrows y'$: Suppose that $y \leftrightarrow y'$ consists of the single reaction $y \to y'$. Then removal of that reaction would leave a directed reaction path connecting y' back to y.

We shall see that linkage classes (and strong-linkage classes) whose direct links are all cut-links often have especially nice properties. This serves as motivation for the following definition.

Definition 6.7.4. *A linkage class (strong-linkage class) in a reaction network is* **tree-like** *if every direct link connecting two complexes in that linkage class (strong-linkage class) is a cut-link. The reaction network is* **forest-like** *if each of its linkage classes is tree-like.*

Thus, the topmost linkage class in network (6.19) is tree-like, as is the terminal strong-linkage class containing the complexes $D+E$ and F. Networks (6.20) and (6.21) are both forest-like.

$$\begin{array}{ccccc} A+B & \to & C & \rightleftarrows & D+E \\ & & \upharpoonleft\downharpoonright & & \\ & & F & & \\ & & & & \\ 2A & \leftarrow & H & \rightleftarrows & B+G \end{array} \tag{6.20}$$

$$\begin{array}{c} A \rightleftarrows C \rightleftarrows B \\ A+B \rightleftarrows 2C \end{array} \tag{6.21}$$

Remark 6.7.5. It is not difficult to see that *every forest-like weakly reversible network is, in fact, reversible*. Moreover, a reversible forest-like network is one for which, in the standard reaction diagram, there are no directed cycles involving more than two complexes.

Appendix 6.A Independent Subnetworks

In Section 6.4 we saw that it is not necessarily true that the rank of a reaction network is the same as the sum of the ranks of its linkage classes, nor is it invariably true that the deficiency of a network is the same as the sum of the deficiencies of its linkage classes. As we shall see, networks for which equality *does* hold often have nice properties, for then certain aspects of network behavior derive from the behavior of subnetworks that, in a sense, act independently of one another. Because this idea will arise later on more than once, it is worthwhile to elaborate on the decomposition of a reaction network into "independent subnetworks" [73] early and in a single place. That is the purpose of this appendix.

We begin with some elementary considerations in linear algebra [98]. Let V be a vector space and let U be a linear subspace of V. When $U_1, U_2, \ldots, U_p$ are linear subspaces of U, we say that U is the *sum* of $U_1, U_2, \ldots, U_p$ if each vector $x \in U$ has at least one representation of the form

$$x = x_1 + x_2 + \cdots + x_p, \quad x_\theta \in U_\theta,\ \theta = 1,2,\ldots,p.$$

In this case we write

$$U = U_1 + U_2 + \cdots + U_p.$$

We say that U is the *direct sum* of $U_1, U_2, \ldots, U_p$ if, in addition, the equation

$$0 = x_1 + x_2 + \cdots + x_p, \quad x_\theta \in U_\theta,\ \theta = 1,2,\ldots,p$$

is satisfied only if $x_\theta = 0,\ \theta = 1,2,\ldots,p$. In this case we write

$$U = U_1 \oplus U_2 \oplus \cdots \oplus U_p.$$

Here is a necessary and sufficient condition for direct summing: *When U is the sum of $U_1, U_2, \ldots, U_p$, the sum is direct if and only if the dimensions of these various subspaces satisfy the equation*

$$\dim U = \sum_{\theta=1}^{p} \dim U_\theta.$$

Now let $\{\mathscr{S},\mathscr{C},\mathscr{R}\}$ be a reaction network with stoichiometric subspace S and rank s (= $\dim S$). Suppose that, for one reason or another, we choose to partition the reaction set $\mathscr{R}$ into subsets $\mathscr{R}_1,\mathscr{R}_2,\ldots\mathscr{R}_p$. In this case, we can think of the partition as giving rise to a partition of the parent reaction network into subnetworks[3] $\{\mathscr{S},\mathscr{C}_\theta,\mathscr{R}_\theta\}, \theta = 1,2,\ldots,p$, where

$$\mathscr{C}_\theta := \{y \in \mathscr{C} : \text{there exists } y \to y' \in \mathscr{R}_\theta \text{ or } y' \to y \in \mathscr{R}_\theta\}. \tag{6.A.1}$$

[3] We nevertheless regard each subnetwork to have the same species set as the parent network even when, in a particular subnetwork, certain species play no role in the subnetwork's reactions.

Each subnetwork $\{\mathscr{S},\mathscr{C}_\theta,\mathscr{R}_\theta\}$ will have its own stoichiometric subspace $S_\theta \subset S$ and its own rank s_θ (= $\dim S_\theta$). Because

$$S = \operatorname{span}\{y' - y \in \mathbb{R}^{\mathscr{S}} : y \to y' \in \mathscr{R}\},$$

and

$$S_\theta = \operatorname{span}\{y' - y \in \mathbb{R}^{\mathscr{S}} : y \to y' \in \mathscr{R}_\theta\}, \quad \theta = 1,2,\ldots,p$$

it is easy to see that

$$S = S_1 + S_2 + \cdots + S_p.$$

In fact, we have direct summing,

$$S = S_1 \oplus S_2 \oplus \cdots \oplus S_p, \tag{6.A.2}$$

precisely when the rank of the network and the ranks of the subnetworks satisfy the equation

$$s = s_1 + s_2 + \cdots + s_p. \tag{6.A.3}$$

In this case we say that the parent network $\{\mathscr{S},\mathscr{C},\mathscr{R}\}$ has been partitioned into *independent subnetworks* $\{\mathscr{S},\mathscr{C}_\theta,\mathscr{R}_\theta\}, \theta = 1,2,\ldots,p$.

Remark 6.A.1. Even when there is no *a priori* specification of a reaction set partition, we shall sometimes find it convenient to say that a subnetwork $\{\mathscr{S},\mathscr{C}_*,\mathscr{R}_*\}$ of a parent network $\{\mathscr{S},\mathscr{C},\mathscr{R}\}$ is an *independent subnetwork of the parent network.* In this case, we mean that the subnetwork and its complement, derived from the reaction subset $\mathscr{R} \setminus \mathscr{R}_*$, constitute a partition of the parent network into two independent subnetworks. This is equivalent to saying that the ranks of the subnetwork and of its complement sum to the rank of the parent network.

Partition of a reaction network into independent subnetworks can sometimes be of considerable help in the study of a kinetic system derived from the network. In fact, suppose that $\{\mathscr{S},\mathscr{C},\mathscr{R},\mathscr{K}\}$ is a kinetic system and that a partition of the reaction set $\mathscr{R}$ induces a partition of the network $\{\mathscr{S},\mathscr{C},\mathscr{R}\}$ into independent subnetworks $\{\mathscr{S},\mathscr{C}_\theta,\mathscr{R}_\theta\}, \theta = 1,2,\ldots,p$. In this case we can think of the kinetic system $\{\mathscr{S},\mathscr{C},\mathscr{R},\mathscr{K}\}$ as having been partitioned into kinetic subsystems $\{\mathscr{S},\mathscr{C}_\theta,\mathscr{R}_\theta,\mathscr{K}_\theta\}, \theta = 1,2,\ldots,p$, where $\mathscr{K}_\theta$ is just the restriction of $\mathscr{K}$ to reactions in $\mathscr{R}_\theta$.

Recall that the species-formation-rate function $f : \overline{\mathbb{R}}_+^{\mathscr{S}} \to S$ for the kinetic system $\{\mathscr{S},\mathscr{C},\mathscr{R},\mathscr{K}\}$ is given by

$$f(c) := \sum_{\mathscr{R}} \mathscr{K}_{y \to y'}(c)(y' - y) = \sum_{\theta=1}^{p} \sum_{\mathscr{R}_\theta} \mathscr{K}_{y \to y'}(c)(y' - y). \tag{6.A.4}$$

We can associate with each kinetic subsystem $\{\mathscr{S},\mathscr{C}_\theta,\mathscr{R}_\theta,\mathscr{K}_\theta\}$ its own species-formation-rate function $f_\theta : \overline{\mathbb{R}}_+^{\mathscr{S}} \to S_\theta$ in a natural way:

$$f_\theta(c) := \sum_{\mathscr{R}_\theta} \mathscr{K}_{y\to y'}(c)(y'-y). \tag{6.A.5}$$

Then we clearly have, for each $c \in \overline{\mathbb{R}}_+^{\mathscr{S}}$,

$$f(c) = f_1(c) + f_2(c) + \cdots + f_p(c) \tag{6.A.6}$$

with $f_\theta(c) \in S_\theta,\ \theta = 1,2,\ldots p$.

Now suppose that $c^* \in \overline{\mathbb{R}}_+^{\mathscr{S}}$ is an equilibrium of the kinetic system under discussion, in which case $f(c^*) = 0$ and, therefore,

$$0 = f_1(c^*) + f_2(c^*) + \cdots + f_p(c^*). \tag{6.A.7}$$

By supposition the subnetworks $\{\mathscr{S},\mathscr{C}_\theta,\mathscr{R}_\theta\}, \theta = 1,2,\ldots,p$ are independent, so we have the direct sum (6.A.2), whereupon we must have

$$f_\theta(c^*) = 0, \quad \theta = 1,2,\ldots,p. \tag{6.A.8}$$

This is to say that *if* c^* *is an equilibrium of the parent kinetic system* $\{\mathscr{S},\mathscr{C},\mathscr{R},\mathscr{K}\}$ *then it must be an equilibrium of each of the kinetic subsystems* $\{\mathscr{S},\mathscr{C}_\theta,\mathscr{R}_\theta,\mathscr{K}_\theta\}$, $\theta = 1,2,\ldots,p$ *separately*. This simple idea will usually find use in the following way: Suppose that $E \subset \overline{\mathbb{R}}_+^{\mathscr{S}}$ is the set of equilibria for the kinetic system $\{\mathscr{S},\mathscr{C},\mathscr{R},\mathscr{K}\}$ and that $E_\theta \subset \overline{\mathbb{R}}_+^{\mathscr{S}}$, $\theta = 1,2,\ldots,p$ are the equilibrium sets of the kinetic subsystems $\{\mathscr{S},\mathscr{C}_\theta,\mathscr{R}_\theta,\mathscr{K}_\theta\}$, $\theta = 1,2,\ldots,p$. Then

$$E = \bigcap_{\theta=1}^{p} E_\theta. \tag{6.A.9}$$

Example 6.A.2 (A variant of the Edelstein network). We show in (6.A.10) an augmented version of the Edelstein mass action network displayed earlier as (5.2) and as (6.A.11) below. The difference is in the addition of the reactions $A \rightleftarrows C$, taken with rate constants k and k' of unspecified value. (The example, as structured, will prove useful—and, in fact, crucial—when we discuss the Deficiency One Theorem.)

$$\begin{array}{ccccc} & & A & \underset{1}{\overset{8.5}{\rightleftarrows}} & 2A \\ & & k \downarrow\uparrow k' & & \\ A+B & \underset{1}{\overset{1}{\rightleftarrows}} & C & \underset{0.2}{\overset{1}{\rightleftarrows}} & B \end{array} \tag{6.A.10}$$

It is easily determined that the rank of the augmented Edelstein network shown in (6.A.10) is three. Note that the network underlying the original Edelstein mass action system, shown below as (6.A.11), has a rank of two, while network underlying the small added mass action system (6.A.12) has a rank of one.

$$A \underset{1}{\overset{8.5}{\rightleftarrows}} 2A \tag{6.A.11}$$

$$A+B \underset{1}{\overset{1}{\rightleftarrows}} C \underset{0.2}{\overset{1}{\rightleftarrows}} B$$

$$A \underset{k'}{\overset{k}{\rightleftarrows}} C \tag{6.A.12}$$

Because the two ranks sum to the rank of the parent network shown in (6.A.10), the two smaller networks constitute a partition of the parent network into *independent* subnetworks. Thus any equilibrium of the parent mass action system (6.A.10) must, in fact, be an equilibrium of *each* of the smaller mass action systems (6.A.11) and (6.A.12) separately.

Recall that the locus of equilibria for the mass action system (6.A.11) was drawn schematically in Figure 5.3. On the other hand, the equilibria for the small mass action system (6.A.12) are those compositions that satisfy the equation

$$c_C = \frac{k}{k'} c_A. \tag{6.A.13}$$

Represented in the context of Figure 5.3, such compositions would constitute that part of the positive orthant consisting of a plane containing the B-axis and passing through the line in the C-A plane given by (6.A.13). Such a plane intersects the curve of equilibria for (6.A.11) in precisely one point. That point is the sole equilibrium of the mass action system (6.A.10).

As we shall see, the fact that there is precisely one positive equilibrium is an immediate consequence of the Deficiency One Theorem, discussed in Chapter 8.

Remark 6.A.3 (When the deficiencies of the linkage classes sum to the deficiency of the network). We shall see that an important condition in the hypothesis of the Deficiency One Theorem is that, for the network under consideration, the deficiencies of its linkage classes sum to the deficiency of the network as a whole. As we pointed out in Section 6.4, this is equivalent to the requirement that the ranks of the individual linkage classes sum to the rank of the network as a whole. If we associate a subnetwork with each of the ℓ linkage classes in the obvious way, then the ranks of the subnetworks and the rank of the entire network will satisfy

$$s = s_1 + s_2 + \cdots + s_\ell,$$

and the stoichiometric subspace of the entire network will be the direct sum of the stoichiometric subspaces associated with the individual linkage classes:

$$S = S_1 \oplus S_2 \oplus \cdots \oplus S_\ell. \tag{6.A.14}$$

In this case, the collection of subnetworks associated with the ℓ linkage classes constitute a partition of the parent network into *independent* subnetworks.

If the network is given a kinetics and c^* is an equilibrium, then we must have

$$f_\theta(c^*) = 0, \quad \theta = 1,2,\ldots,\ell, \tag{6.A.15}$$

where $f_\theta(\cdot) : \overline{\mathbb{R}}_+^{\mathscr{S}} \to S_\theta$ is the species-formation-rate function associated with linkage class $\mathscr{L}^\theta$. That is, equilibration must take place separately in the kinetic system associated with each linkage class.

7

The Deficiency Zero Theorem

Recall that the reaction networks of deficiency zero can contain hundreds of species and hundreds of reactions. In such cases, the corresponding differential equations will be extraordinarily complex. When the kinetics is mass action, these will amount to a large and intricate system of coupled polynomial equations in perhaps 100 species concentrations and in which a large number of perhaps unknown parameters (rate constants) appear. Recall too that not much is known *in general* about systems of polynomial equations, even fairly small ones. Nevertheless, when these derive from deficiency zero networks, quite a lot can be said.

7.1 A Statement of the Deficiency Zero Theorem

What we present below as the Deficiency Zero Theorem had its origins in three papers by F. J. M. Horn, R. Jackson , and me, published in 1972 [67, 107, 109]. As stated, the theorem amounts to a headline, not the full story, but it captures a great deal about the remarkable way in which the behavior of deficiency zero networks is circumscribed. We shall have more to say about that behavior later in this chapter and still more in Part III of this book.

Theorem 7.1.1 (The Deficiency Zero Theorem). *Let $\{\mathscr{S},\mathscr{C},\mathscr{R}\}$ be a reaction network of deficiency zero.*

(i) If the network is not weakly reversible, then, for an arbitrary kinetics $\mathscr{K}$, the differential equations for the kinetic system $\{\mathscr{S},\mathscr{C},\mathscr{R},\mathscr{K}\}$ cannot admit a positive equilibrium, nor can they admit a cyclic composition trajectory containing a composition at which all species concentrations are positive.

(ii) If the network is weakly reversible (in particular, if it is reversible) and the kinetics is mass action, then, regardless of rate constant values, the resulting differential equations have the following properties: There exists within each positive stoichiometric compatibility class precisely one equilibrium; that equilibrium is asymptotically stable; and there cannot exist a nontrivial cyclic composition trajectory along which all species concentrations are positive.

M. Feinberg, *Foundations of Chemical Reaction Network Theory*,
Applied Mathematical Sciences 202,
https://doi.org/10.1007/978-3-030-03858-8_7

Part III of this book contains a proof of Theorem 7.1.1, along the lines of the one in [71]. A preview of the proof is given toward the end this chapter, where we also provide a preliminary discussion of *complex balancing*, an important idea of Horn and Jackson [109] that lies at the heart of the proof.

7.2 Examples

Here we will consider how the Deficiency Zero Theorem works in several examples and, in particular, how it fits the facts in the case of some networks we have already considered.

Example 7.2.1 (Our very first network, revisited). Recall the very first network we discussed, back in Chapter 2. It is repeated yet again as (7.1).

$$\begin{array}{ccc} A & \rightleftarrows & 2B \\ A+C & \rightleftarrows & D \\ \nwarrow & & \swarrow \\ & B+E & \end{array} \tag{7.1}$$

Remember that we gave the network mass action kinetics (with rate constants indicated by Greek letters in (2.4)), and we wrote out the corresponding differential equations (2.5)—*an intricate system of five coupled polynomial equations in five variables (the species concentrations) and in which six parameters (rate constants) appear.* Then, in Section 2.1.3, we asked some difficult questions about those equations.

Despite the complexity of the system (2.5), we can now answer all of them. Network (7.1) is weakly reversible, and we have already calculated its deficiency to be zero. Thus, part (ii) of the Deficiency Zero Theorem gives the following information: *Regardless of values of the rate constants*, there is precisely one equilibrium in each positive stoichiometric compatibility class; each of these is asymptotically stable; and there are no periodic (positive) composition trajectories. Rate constants are rarely known with precision—if they are known at all—*but we do not need to know them to make incisive, nontrivial statements about qualitative behavior.*

Were all reactions in (7.1) made reversible, the resulting differential equations would be still more complicated, and there would be two more parameters (rate constants). But the deficiency would remain zero, and answers to the questions we posed would remain the same.

Example 7.2.2 (Two variants of our very first network). Next we will consider network (7.2) which is very similar to (7.1). Here again the deficiency is zero, but (7.2) is not weakly reversible. Thus, the Deficiency Zero Theorem tells us that, regardless of the kinetics, the induced differential equations can admit neither a positive equilibrium nor a cyclic composition trajectory containing a positive composition. This is more or less obvious: Species E is not produced by any reaction, so as long as species B is present, the supply of E will dwindle.

$$\begin{array}{ccc} A & \rightleftarrows & 2B \\ A+C & \rightleftarrows & D \\ & \nwarrow \quad \nearrow & \\ & B+E & \end{array} \tag{7.2}$$

The situation for network (7.3) is far less transparent. Here again the deficiency is zero, and (7.3) is not weakly reversible. But now *every* species is both produced and consumed by chemical reactions. Nevertheless, part (i) of the Deficiency Zero Theorem tells us that, for any kinetics (mass action or otherwise), the induced differential equations admit no positive equilibrium nor a periodic composition trajectory containing a positive composition.

$$\begin{array}{ccc} A & \rightarrow & 2B \\ A+C & \rightleftarrows & D \\ & \nwarrow \quad \swarrow & \\ & B+E & \end{array} \tag{7.3}$$

Example 7.2.3 (The Edelstein network and a variant). Next we will recall the mass action system studied in effect by Edelstein and discussed earlier in Section 4.2.3:

$$\begin{gathered} A \underset{\beta}{\overset{\theta}{\rightleftarrows}} 2A \\ A+B \underset{\eta}{\overset{\gamma}{\rightleftarrows}} C \underset{\rho}{\overset{\varepsilon}{\rightleftarrows}} B. \end{gathered} \tag{7.4}$$

The induced differential equations are

$$\begin{aligned} \dot{c}_A &= \theta c_A - \beta c_A^2 - \gamma c_A c_B + \eta c_C \\ \dot{c}_B &= -\gamma c_A c_B + (\eta + \varepsilon) c_C - \rho c_B \\ \dot{c}_C &= \gamma c_A c_B - (\eta + \varepsilon) c_C + \rho c_B. \end{aligned} \tag{7.5}$$

Remember that for some values of the rate constants, in particular for those shown in (5.2), the system (7.5) admits three equilibria within a positive stoichiometric compatibility class. (One is unstable.) Network (7.4) is weakly reversible (in fact, reversible), but its deficiency is *one* (n = 5, ℓ = 2, s =2) . Thus, there is no contradiction of the Deficiency Zero Theorem.

Now consider the slightly different mass action system

$$\begin{gathered} A \underset{\beta}{\overset{\theta}{\rightleftarrows}} 2A \\ A+B \underset{\eta}{\overset{\gamma}{\rightleftarrows}} C \underset{\rho}{\overset{\varepsilon}{\rightleftarrows}} 2B, \end{gathered} \tag{7.6}$$

for which the induced differential equations are

$$\begin{aligned}\dot{c}_A &= \theta c_A - \beta c_A^2 - \gamma c_A c_B + \eta c_C \\ \dot{c}_B &= -\gamma c_A c_B + (\eta + 2\varepsilon) c_C - 2\rho c_B^2 \\ \dot{c}_C &= \gamma c_A c_B - (\eta + \varepsilon) c_C + \rho c_B^2.\end{aligned} \tag{7.7}$$

In very rough terms, the system (7.7) is "more nonlinear" than the system (7.5): While the rate of reaction $B \to C$ in network (7.4) is proportional to c_B, the rate of reaction $2B \to C$ in network (7.6) is proportional to c_B^2. At least to the extent that they can be compared, one might expect that phase portraits for the system (7.7) should be at least as interesting as those for (7.5).

But the deficiency of the reversible network (7.6) is readily calculated to be *zero* (n = 5, ℓ = 2, s = 3). Thus, for any choice of positive numbers for the Greek letters in (7.7)), the system admits precisely one positive equilibrium; it is asymptotically stable; and there are no (positive) periodic solutions.

Example 7.2.4 (Rabbits and wolves again). Here we revisit the Lotka mass action system discussed in Section 5.3.1 and repeated below as (7.8).

$$\begin{aligned}A &\xrightarrow{\alpha} 2A \\ A + B &\xrightarrow{\beta} 2B \\ B &\xrightarrow{\gamma} 0.\end{aligned} \tag{7.8}$$

The differential equations were given by (7.9).

$$\begin{aligned}\dot{c}_A &= \alpha c_A - \beta c_A c_B \\ \dot{c}_B &= \beta c_A c_B - \gamma c_B.\end{aligned} \tag{7.9}$$

In Section 5.3.1 we interpreted the Lotka system in ecological terms, with rabbits indicated by A and rabbit-consuming wolves indicated by B. Recall that the reaction $B \to 0$ was taken to indicate the demise and disappearance of wolves due to natural causes, including rabbit famine.

Recall also that, regardless of the positive values that the rate constants take, the phase portrait for the system (7.9) has the qualitative shape shown in Figure 7.1. Note that there is a positive equilibrium, and there are also positive cyclic composition trajectories. Despite the lack of weak reversibility, there is no contradiction to the Deficiency Zero Theorem, for in this case the deficiency of network (7.8) is *one* ($n = 6$, $\ell = 3$, $s = 2$).

Suppose now that we chose to explicitly take full account of wolf carcasses. In this case, we might replace the reaction $B \to 0$ with $B \to C$, with C indicating a dead wolf. This, of course, has no effect on the dynamics of c_A and c_B, the differential equations for which remain precisely those shown in (7.9), and Figure 7.1 remains a faithful depiction of the relative dynamics of c_A and c_B.

However, our mass action system becomes

$$\begin{aligned}A &\xrightarrow{\alpha} 2A \\ A + B &\xrightarrow{\beta} 2B \\ B &\xrightarrow{\gamma} C.\end{aligned} \tag{7.10}$$

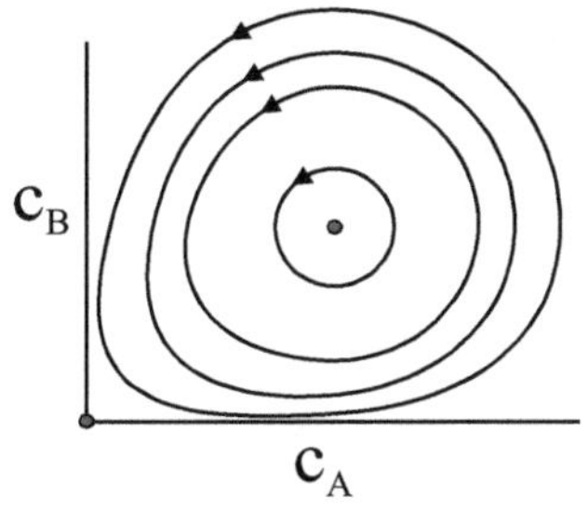

Fig. 7.1. Phase portrait for the Lotka mass action system (7.8)

The deficiency of the network in (7.10) is now *zero* ($n = 6$, $\ell = 3$, $s = 3$). The network is not weakly reversible, *so the Deficiency Zero Theorem precludes both positive equilibria and positive cyclic composition trajectories.*

Does Figure 7.1 contradict the Deficiency Zero Theorem? It doesn't: For the three-species mass action system (7.10), Figure 7.1 is merely a projection of a three-dimensional phase portrait onto the $c_A - c_B$ plane. Keep in mind that, for the mass action system (7.10), the differential equations in (7.9) are supplemented by a third differential equation,

$$\dot{c}_C = \gamma c_B. \tag{7.11}$$

What appears to be a positive equilibrium in Figure 7.1 is, in three dimensions, a trajectory moving along a line perpendicular to the page in the positive C direction. What appear to be cycles are actually helices, also moving outward from the page.

Remark 7.2.5. The Edelstein and Lotka examples were taken from Chapter 5's toy zoo, in which the "animals" on exhibit were exotic in one respect or another. There were multiple positive equilibria, unstable positive equilibria, or periodic composition trajectories. It is easy to confirm that in all such cases, the deficiency of the underlying reaction network exceeds zero. A visit to the zoo, then, might suggest that networks of deficiency zero are uncommon. It should be remembered, though, that the sample there was highly biased: the zoo animals were chosen *because of* their capacity to exhibit behavior that the Deficiency Zero Theorem *precludes*.

7.3 A Derivative Version of the Deficiency Zero Theorem

There is much that can be said about deficiency zero networks that is not already said in Theorem 7.1.1. Especially because it is related to what we shall later call *nondegenerate networks*, we state below a "local" Deficiency Zero Theorem that gives considerable information about the derivative of the species-formation-rate function at a positive equilibrium. (It will help to recall that *forest-like* reaction networks were defined in Section 6.7. In essence, a forest-like network is one in which removal of reaction(s) directly joining a pair of complexes results in a "disconnection" of the network.)

Theorem 7.3.1 (Derivative Version of the Deficiency Zero Theorem). *Let $\{\mathscr{S},\mathscr{C},\mathscr{R}\}$ be a weakly reversible deficiency zero reaction network with stoichiometric subspace S. For any assignment of rate constants $k \in \mathbb{R}_+^{\mathscr{R}}$, the resulting mass action system $\{\mathscr{S},\mathscr{C},\mathscr{R},k\}$ has the following properties: If c^* is a positive equilibrium, then the derivative $df(c^*): S \to S$ of the species-formation-rate function at c^* is nonsingular. In fact, relative to the scalar product "$*$" in $\mathbb{R}^{\mathscr{S}}$ given by*

$$u * w := \sum_{s \in \mathscr{S}} \frac{u_s w_s}{c_s^*}, \tag{7.12}$$

$df(c^*)$ is negative-definite. If the network is not only weakly reversible but also forest-like (and therefore reversible), then $df(c^*)$ is symmetric relative to the "$*$" scalar product. In this case, all roots of the characteristic polynomial of $df(c^*)$ are real and negative; moreover, there exists in S a basis consisting of eigenvectors of $df(c^*)$.

When we say that $df(c^*)$ is negative-definite relative to "$*$," we mean that

$$\sigma * df(c^*)\sigma < 0$$

for all nonzero $\sigma \in S$. When we say that $df(c^*)$ is symmetric relative to "$*$," we mean that

$$\sigma' * df(c^*)\sigma = \sigma * df(c^*)\sigma'$$

for all σ and σ' in S.

Example 7.3.2 (An Edelstein network variant, again). Recall that network (7.6) has a deficiency of zero. It is reversible and therefore weakly reversible. The rank of the network is three, as is the dimension of the stoichiometric subspace. Thus the stoichiometric subspace for this three-species network is all of $\mathbb{R}^{\mathscr{S}}$, and there is just one positive stoichiometric compatibility class, all of $\mathbb{R}_+^{\mathscr{S}}$. For any assignment of rate constants, the Deficiency Zero Theorem tells us that there is precisely one positive equilibrium, that this equilibrium is asymptotically stable, and that there are no cyclic composition trajectories in $\mathbb{R}_+^{\mathscr{S}}$.

Theorem 7.3.1 tells us more: At the unique positive equilibrium c^*, the derivative $df(c^*): S \to S$ is not only nonsingular but also negative-definite relative to the "$*$" scalar product. Moreover, because the network in (7.6) is forest-like, all roots of the characteristic polynomial of $df(c^*)$ are real and negative. The final approach of a composition trajectory to c^* cannot be oscillatory; instead, in some neighborhood of c^*, the trajectory will appear to ride along one of the eigenvectors of $df(c^*)$.

7.4 Boundary Equilibria for Deficiency Zero Networks: Only Certain Combinations of Species Can Coexist

The Deficiency Zero Theorem tells us that an equilibrium for a kinetic system derived from a not-weakly-reversible deficiency zero network cannot be positive. This is to say that, at such an equilibrium, one or more species must be absent. In Section 3.7 we argued that for a general kinetic system—not necessarily one derived from a deficiency zero network—there are only certain combinations of species that can be present or absent at an equilibrium. In this section we will state a theorem that is far more specific when the underlying network does indeed have a deficiency of zero.

Let us review what we know in the general case. If c^* is an equilibrium of a kinetic system $\{\mathscr{S},\mathscr{C},\mathscr{R},\mathscr{K}\}$ then the set of species actually present at that equilibrium—i.e., supp c^*—must be *reaction-transitive*: If there is a directed sequence of reactions

$$y_1 \to y_2 \to \cdots \to y_n \tag{7.13}$$

and if $\text{supp}\, y_1$ is contained in supp c^*, then $\text{supp}\, y_\theta$ is contained in supp c^* for all $\theta = 2,\dots,n$. Recall the natural requirement imposed on kinetic rate functions in Chapter 3: for each reaction $y \to y'$, $\mathscr{K}_{y\to y'}(c)$ is strictly positive when $\text{supp}\, y$ is contained in supp c and is zero otherwise. Thus, if reaction $y_1 \to y_2$ is "switched on" at the equilibrium c^*, then so must every other reaction along the chain (7.13) be switched on as well.

In particular, if the network contains a directed reaction cycle

$$y_1 \to y_2 \to \cdots \to y_n \to y_1, \tag{7.14}$$

then the support of *each* complex in the cycle is contained in supp c^* or the support of *none* is. Similarly, either *all* reactions in the cycle are switched on at the equilibrium c^* or else *none* are. From this it is easy to see that similar considerations apply to complexes residing in a particular strong-linkage class (Section 6.6): If $\Lambda \subset \mathscr{C}$ is a strong-linkage class, then either *all* complexes within Λ have their supports entirely within supp c^* or *none* do. And either *all* reactions $y \to y'$ with $y \in \Lambda$ are switched on at c^* or *none* are.

What we've said so far applies to kinetic systems derived from networks of arbitrary deficiency. When the deficiency is zero, we can say much more. For *deficiency zero networks*, the following theorem tells us that, at any equilibrium c^*, a reaction $y \to y'$ *must* be turned *off* whenever its reactant complex y is not a member of a *terminal* strong-linkage class. In fact, when y is not a terminal complex, it must be the case that $\text{supp}\, y \not\subset \text{supp}\, c^*$. In informal terms, this is to say that *when the deficiency is zero, each nonterminal complex must contain at least one species that is absent at the equilibrium c^*. At c^* only reactions originating in a terminal strong-linkage class can be switched on.* The proof is given in Part III, Section 16.5.6.

Theorem 7.4.1 (Deficiency Zero Boundary Equilibria Theorem). *Suppose that $\{\mathscr{S},\mathscr{C},\mathscr{R}\}$ is a deficiency zero reaction network, that $\mathscr{K}$ is a kinetics for the network (not necessarily mass action), and that $c^* \in \overline{\mathbb{R}}_+^{\mathscr{S}}$ is an equilibrium of the kinetic system $\{\mathscr{S},\mathscr{C},\mathscr{R},\mathscr{K}\}$.*

(i) If $y \in \mathscr{C}$ is not a terminal complex, then $\text{supp}\, y \not\subset \text{supp}\, c^*$. *In particular, there is at least one species in* $\text{supp}\, y$ *that is absent at composition c^*, and, for*

any reaction $y \to y'$ having y as its reactant complex, it must be the case that $\mathscr{K}_{y\to y'}(c^*) = 0$.

(ii) If $y^* \in \mathscr{C}$ is a member of terminal strong-linkage class Λ and $\text{supp}\, y^* \subset \text{supp}\, c^*$, *then* $\text{supp}\, y \subset \text{supp}\, c^*$ *for every complex $y \in \Lambda$. Moreover, $\mathscr{K}_{y\to y'}(c^*) > 0$ for every reaction $y \to y'$ such that y is in Λ.*

The Lotka mass action system (7.8) demonstrates that a kinetic system derived from a deficiency *one* network can admit an equilibrium at which reactions originating in nonterminal complexes are switched on. Recall that the Lotka system admits a positive equilibrium. At that equilibrium all three reactions proceed at positive rate even though each of the three has a nonterminal reactant complex.

Theorem 7.4.1 can, in some instances, have implications that go beyond those in the theorem statement itself. Consider, for example, a kinetic system that derives from the deficiency zero network (7.15).

$$\begin{array}{ccc} A & \to & 2B \\ A+C & \rightleftarrows & D \\ & \nwarrow \quad \swarrow & \\ & B+E & \end{array} \tag{7.15}$$

Suppose that c^* is an equilibrium. Because complex A is nonterminal, Theorem 7.4.1 tells us that its support—i.e., species A—is not contained in the support of c^*, in which case $c_A^* = 0$. But then the support of complex $A+C$ cannot be contained entirely in the support of c^* either. In this case, Theorem 7.4.1 tells us that the supports of complexes D and $B+E$ cannot be contained entirely in supp c^*, so we must have $c_D^* = 0$ and either $c_B^* = 0$ or $c_E^* = 0$ (or both). Theorem 7.4.1 is telling us indirectly that, at the equilibrium c^*, *all* reactions must be switched off, even those with reactant complexes in the *terminal* strong-linkage class $\{A+C, D, B+E\}$.

Remark 7.4.2 (For a deficiency-zero kinetic system, which species can coexist along a composition cycle?). The ideas behind Theorem 7.4.1 extend to cyclic composition trajectories. Suppose that $c : [t_1, t_2] \to \overline{\mathbb{R}}_+^{\mathscr{S}}$ is a solution of the differential equation for a kinetic system $\{\mathscr{S}, \mathscr{C}, \mathscr{R}, \mathscr{K}\}$, with $c(t_1) = c(t_2)$. Recall from Remark 3.7.7 that the corresponding cyclic composition trajectory *has support* $\mathscr{S}^* \subset \mathscr{S}$ if, for all $t \in [t_1, t_2]$, supp $c(t) = \mathscr{S}^*$.

If the putative cyclic composition trajectory has support $\mathscr{S}^*$ and the underlying reaction network $\{\mathscr{S}, \mathscr{C}, \mathscr{R}\}$ has a deficiency of zero, then conclusions very much like those of Theorem 7.4.1 obtain: *If $y \in \mathscr{C}$ is not terminal, then there is at least one species in* supp y *that is not present in $\mathscr{S}^*$. On the other hand, if y^* is a terminal complex with* supp y^* *contained in $\mathscr{S}^*$, then, for every other complex y in the terminal strong-linkage class containing y^*, $\mathscr{S}^*$ contains* supp y. See Part III, Section 16.5.6.

7.5 Multicell Systems with Deficiency Zero Intracellular Chemistry

In Sections 4.2.4 and 5.2.2 we considered two different examples of multicell mass action systems. In particular, we showed how the multicell system described in Section 5.2.2 gives rise to exotic behavior (e.g., pattern formation), at least for certain parameter values. In each example there was an intracellular chemistry, identical in all cells, and there was also intercellular transport of various species from one cell to another. In these examples there were only two cells, but it is easy to imagine multicell systems comprised of many more cells and in which the intracellular chemistry is far more complex.

We described in Section 4.2.4 ways in which multicell systems can be described in reaction network terms, with reactions such as $A1 \rightleftarrows A2$ invoked to model transport of species A between cell 1 and cell 2. The very simple example considered there, involving only the trivial *intracellular* deficiency zero chemistry $A \rightleftarrows 2B$, destroys any hope that reaction networks invoked to encode *multicell* models will often have deficiencies of zero. The deficiency of the simple multicell network (4.24) is one.

Our purpose in this section is to describe a theorem of Shapiro and Horn [152, 153]. It indicates how, when the intracellular kinetics is mass action and when intercellular transport of material satisfies weak requirements, a multicell system, however complex, *must* behave in a certain dull way so long as the *intracellular* reaction network is weakly reversible and has a deficiency of zero. In particular, pattern formation of the kind exhibited in Section 5.2.2 is precluded.

It will be understood hereafter that, in our consideration of multicell systems, the intracellular chemistry—reaction network and kinetics—is identical in all cells and that the cells have identical volume. Moreover, we shall presume that the intercellular transport is of a certain general nature:

First, we shall suppose that, as in the examples, the molar rate of receipt of a species, say A, in cell j due to transport of A from cell i is proportional to the concentration of A in cell i. That is, we shall assume that, when there is transport of A from cell i to cell j, the molar rate of receipt of A in cell j (per unit cell volume) is of the form $k_{Ai \to Aj} c_{Ai}$. In this case, its contribution to the differential equations governing the multicell system can, as in the examples, be taken into account by inclusion of a "reaction" such as $A_i \to A_j$ with associated mass action rate constant $k_{Ai \to Aj}$.

Second, we shall suppose that, *were there no intracellular reactions* and were the cells initially charged—not necessarily uniformly—with mixture in which all species are present, then the intercellular transport would invariably cause the multicell system to come to an equilibrium in which, for each species, the concentration of that species is the same in every cell and is positive. Note that if species A is transported directly from cell i to cell j, we do not insist that there also be direct transport of A from cell j to cell i. If not, then there must be some indirect transport of A from cell j to cell i, via other cells, for otherwise A would eventually be depleted from cell i. Note also that there can be no entrapment of a particular species within a proper subset of the cells.

In the following theorem statement, when we refer to a stoichiometric compatibility class, we mean a stoichiometric compatibility class for the reaction network built, as in Sections 4.2.4 and 5.2.2, to encode the full multicell system under consideration. When we refer to the differential equations for a multicell system, we mean the differential equations constructed as in Sections 4.2.4 and 5.2.2.

The Shapiro-Horn theorem [152, 153] tells us that, *despite what might be a high deficiency for the multicell network*, weak reversibility and deficiency zero in the *intracellular* (mass action) chemistry impart to the full multicell system the same behavioral dullness expected of a weakly reversible deficiency zero single-cell mass action system. And there is more: *At equilibria of the multicell system, pattern formation is precluded. There can be no cell-to-cell composition variations.* Ideas underlying the Shapiro-Horn Theorem are discussed in Sections 15.3 and 16.8.3.

Theorem 7.5.1 (Deficiency Zero Multicell Theorem). *Consider a multicell system in which the intracellular kinetics is mass action and the underlying intracellular reaction network is weakly reversible and has a deficiency of zero. Then, regardless of the deficiency of the full multicell network and regardless of parameter values, the corresponding differential equations for the multicell system have the following properties: In each positive stoichiometric compatibility class, there is precisely one muticell equilibrium; that equilibrium is asymptotically stable; and at that equilibrium, the concentration of each species is the same in every cell. Moreover, the differential equations cannot admit a cyclic composition trajectory along which all species concentrations are positive.*

It should be clearly understood that Theorem 7.5.1, like the Deficiency Zero Theorem, gives immediate qualitative information about systems of differential equations that can go on for pages. Recall that the multicell system described in Section 5.2.2 *did* give rise to a positive equilibrium in which there was pattern formation—that is, in which there were cell-to-cell composition variations. However, the deficiency of the intracellular reaction network (5.6) was *one*.

7.6 A Striking Counterpoint to the Deficiency Zero Theorem: Star-Like Networks

To the extent that a reaction network is identified with its standard reaction diagram, the network is in part a directed graph, with the complexes playing the role of vertices and the reaction arrows playing the role of directed edges. However, it is more than just a directed graph, for the vertices—that is, the complexes—carry algebraic information in the form of network's stoichiometric coefficients. In this sense, a reaction network has *both* graphical aspects and algebraic aspects.

At least within the class of weakly reversible networks, the Deficiency Zero Theorem is almost entirely about consequences of a network's *algebraic* aspects. Recall from Remark 6.3.3 that a network's deficiency depends not at all on the fine details of how the complexes are joined by reaction arrows within the various linkage classes.

The insensitivity of the deficiency to graphical structure becomes clearest for networks in which there is just one linkage class. In this case, the deficiency of the network is zero precisely when the complexes, say $\{y_1, y_2, \ldots, y_n\}$, are *affinely independent*—that is, precisely when the only solution of the equations

$$\sum_{\theta=1}^{n} \xi_\theta y_\theta = 0 \qquad \sum_{\theta=1}^{n} \xi_\theta = 0 \tag{7.16}$$

is $\xi_\theta = 0,\ \theta = 1,\ldots,n$.[1] Affine independence is a purely *algebraic* condition, having nothing at all to do with how the complexes in the sole linkage class are joined by reaction arrows. When this *algebraic* condition is satisfied, and when the kinetics is mass action, *any* arrangement of arrows that results in weak reversibility will suffice to give the behavior described in part (ii) of the Deficiency Zero Theorem.

As a striking counterpoint to the largely algebraic character of the Deficiency Zero Theorem, we present next a single-linkage-class theorem that is purely *graphical. The nature of the complexes will not matter at all!* By a *reversible star-like network* we mean one for which the standard reaction diagram has the shape depicted in Figure 7.2: there is a central complex and an arbitrary number of other complexes, each connected only to the central complex by two-way reaction arrows.

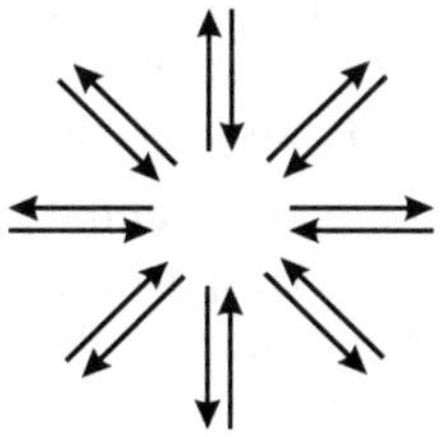

Fig. 7.2. The shape of a reversible star-like network

The following theorem, which is of more mathematical than practical interest, is proved Section 13.5 of Part III. (It was stated without proof in [72].) What makes the theorem compelling is its purely graphical hypothesis, *a hypothesis that is completely indifferent to the nature of the complexes.*

Theorem 7.6.1 (Star-like Network Theorem). *Consider a reversible star-like network taken with mass action kinetics. Regardless of the rate constants* ***and regardless of the complexes****, the resulting differential equations have the following properties: There exists within each positive stoichiometric compatibility class precisely one equilibrium; that equilibrium is asymptotically stable; and there cannot exist a nontrivial cyclic composition trajectory along which all species concentrations are positive.*

---
[1] This follows from ideas in Sections 6.2 and 6.3.

It is not difficult to see that there are reversible star-like networks of arbitrarily high deficiency. In fact, one can create high-deficiency examples quite simply, as in the single-species network shown in Figure 7.3.

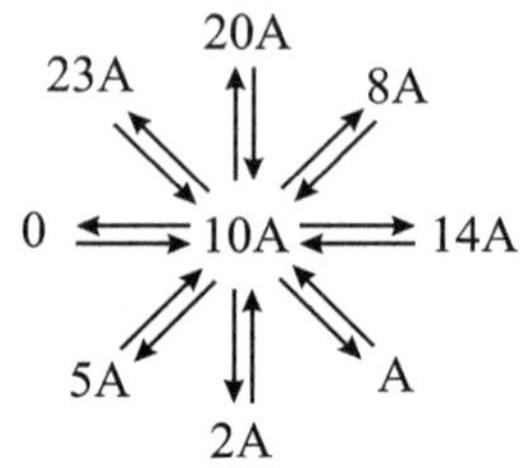

Fig. 7.3. A single-species reversible star-like network

There the deficiency is seven ($n = 9, \ell = 1, s = 1$). In this case, Theorem 7.6.1 tells us the not-so-obvious fact that the equation of equilibrium for the resulting mass action system, which amounts to a single polynomial equation in c_A, has precisely one positive root, no matter what the rate constants are. Moreover, one gets the same result *no matter how one alters the coefficients of A in the various complexes*. Our focus on the single-species network in Figure 7.3 derived solely from the fact that it has a high deficiency. Far more interesting is what Theorem 7.6.1 tells us about star-like networks having richer multi-species collections of complexes.

Remark 7.6.2. Theorem 7.6.1 admits a generalization to networks having several linkage classes: If the subnetworks corresponding to the various linkage classes are all reversible star-like and if the deficiencies of the individual linkage classes sum to the deficiency of the network itself, then the conclusion of Theorem 7.6.1 obtains. In fact, when the deficiencies sum in that way, the conclusion obtains when the subnetworks corresponding to the various linkage classes are either reversible star-like or are weakly reversible and have a deficiency of zero.

7.7 Complex Balancing and Origins of the Deficiency Zero Theorem

As we indicated earlier, the Deficiency Zero Theorem had its origins in three papers that appeared in 1972 [67, 107, 109]. Nominally, all were about *complex balancing*, an idea in chemical kinetics originated by F. J. M. Horn and Roy Jackson [109] . Complex balancing, in turn, amounted to a generalization of an older idea, *detailed balancing*. One has detailed balancing or complex balancing at an equilibrium if, at that equilibrium, rates of the various reactions together satisfy certain conditions which, in some instances, go beyond those required for the vanishing of the species-formation-rate function.

For a kinetic system $\{\mathscr{S},\mathscr{C},\mathscr{R},\mathscr{K}\}$ in which the underlying reaction network is reversible, there is detailed balancing at an equilibrium c^* if, at c^*, the rate of each reaction is precisely equal to the rate of the reverse reaction—that is, if for each $y \to y' \in \mathscr{R}$,

$$\mathscr{K}_{y\to y'}(c^*) = \mathscr{K}_{y'\to y}(c^*). \tag{7.17}$$

There are mass action systems for which one has detailed balancing at some positive equilibrium only when the rate constants are orchestrated in a precise way. Necessary and sufficient conditions for detailed balancing in mass action systems of arbitrary complexity are given in [75] and will be discussed further in Part III of this book.

On the other hand, for a kinetic system $\{\mathscr{S}, \mathscr{C}, \mathscr{R}, \mathscr{K}\}$ (in which the underlying reaction network is not necessarily reversible), there is complex balancing at an equilibrium c^* if, for each complex y, the sum of the rates at c^* of all reactions having y as the product complex is exactly equal to the sum of the rates at c^* of all reactions having y as the reactant complex—that is, if for each $y \in \mathscr{C}$,

$$\sum_{y'\to y \,\in\, \mathscr{R}_{\to y}} \mathscr{K}_{y'\to y}(c^*) = \sum_{y\to y' \,\in\, \mathscr{R}_{y\to}} \mathscr{K}_{y\to y'}(c^*). \tag{7.18}$$

Here the set of all reactions having y as the product (respectively, reactant) complex is denoted $\mathscr{R}_{\to y}$ (respectively, $\mathscr{R}_{y\to}$). Detailed balancing at an equilibrium implies complex balancing, but for networks that are not forest-like, the converse is not generally true.

Horn and Jackson showed in [109] that *if a mass action system admits even one positive equilibrium at which complex balancing obtains, then, remarkably, the differential equations for that mass action system can only give rise to behavior of a very dull kind*—in particular, the behavior described in the second part of Theorem 7.1.1. In [68] I showed that *for any kinetic system, not necessarily mass action, in which the underlying reaction network has a deficiency of zero, complex balancing must obtain at all equilibria*. Horn showed in [107] that, for a mass action system in which the underlying network is weakly reversible and has a deficiency of zero, there will indeed be a positive equilibrium. In that case, because the deficiency is zero, the extant positive equilibrium *must* be complex balanced, whereupon the dull behavior resulting from complex balancing is ensured.[2]

The idea of complex balancing provided a scaffold upon which the Deficiency Zero Theorem was built. That scaffold, once utilized, could then be pulled away, as Horn and I did in [79], to state a theorem solely about the behavior of the large class of deficiency zero reaction networks, a theorem in which complex balancing need not be defined or even mentioned. Nevertheless, we shall see in Part III, where we prove the Deficiency Zero Theorem, how this important idea of Horn and Jackson lurks powerfully in the background. The following remark contains some further elaboration.

[2] In [107] Horn showed that weak reversibility is necessary for the existence of a complex balanced positive equilibrium, and he also went beyond the zero deficiency case to work out necessary and sufficient conditions the rate constants must satisfy in order that a weakly reversible positive-deficiency mass action system admit a complex balanced positive equilibrium. For any weakly reversible positive-deficiency network, there invariably exists an assignment of rate constants such that the resulting mass action differential equations admit no complex balanced positive equilibrium.

Remark 7.7.1 (Quasi-thermostatic and quasi-thermodynamic kinetic systems). Here we provide a little more detail about the connection drawn by Horn and Jackson [109] between the existence of a complex balanced positive equilibrium in a mass action system and the severely circumscribed behavior the system might possibly admit. Again, they showed—and we will show in a different way—that for *any* mass action system $\{\mathscr{S},\mathscr{C},\mathscr{R},k\}$, regardless of the network's deficiency, the existence of a positive equilibrium at which complex balancing obtains has, in itself, striking consequences.

One of these is that the system is *quasi-thermostatic*. We say that a kinetic system $\{\mathscr{S},\mathscr{C},\mathscr{R},\mathscr{K}\}$ is quasi-thermostatic if it admits a positive equilibrium, say c^*, and if the set E of *all* positive equilibria (spread over the various stoichiometric compatibility classes) is given by

$$E := \{c \in \mathbb{R}_+^{\mathscr{S}} : \ln c - \ln c^* \in S^\perp\}, \tag{7.19}$$

where S is the underlying network's stoichiometric subspace. As we shall see (Part III), in this case each positive stoichiometric compatibility class contains *precisely one* member of E.

A stronger consequence of the existence of a complex balanced positive equilibrium for a mass action system is that the system is not only quasi-thermostatic but also *quasi-thermodynamic*. We say that a kinetic system $\{\mathscr{S},\mathscr{C},\mathscr{R},\mathscr{K}\}$ is quasi-thermodynamic if the system is quasi-thermostatic and if, in addition, it has the following property: If $c^* \in \mathbb{R}_+^{\mathscr{S}}$ is an equilibrium, then the scalar function $h(\cdot)$ given, for each $c \in \mathbb{R}_+^{\mathscr{S}}$, by

$$h(c) := \sum_{s \in \mathscr{S}} c_s (\ln c_s - ln c_s^* - 1) \tag{7.20}$$

is, in the positive stoichiometric compatibility class containing c^*, strictly decreasing along composition trajectories of the differential equations for $\{\mathscr{S},\mathscr{C},\mathscr{R},\mathscr{K}\}$, except at the equilibrium c^*. In this case, $h(\cdot)$ serves as a Lyapunov function [30, 34, 104, 164], one that guarantees the asymptotic stability of the equilibrium c^* and the absence of cyclic composition trajectories in the positive stoichiometric compatibility class containing it. (The function $h(\cdot)$ was inspired by the Helmholtz free energy function for ideal gas mixtures. See Chapter 15.)

As we indicated earlier, the connection to the class of deficiency zero networks resides in the fact that at all equilibria of a kinetic system (mass action or otherwise), in which the underlying network has a deficiency of zero, complex balancing *must* obtain [68]. Thus, a deficiency zero mass action system that admits a positive equilibrium inherits all of the dull static and dynamical properties of quasi-thermostatic and quasi-thermodynamic kinetic systems.

It should be noted, however, that the existence of a complex balanced positive equilibrium is not necessary in order that a mass action system be either quasi-thermostatic or quasi-thermodynamic. In fact, all mass action systems in which the underlying network is reversible and star-like (Section 7.6) are quasi-thermodynamic (and therefore quasi-thermostatic) regardless of rate constant values, even in the absence of any complex balanced equilibria. This is proved in Part III. Moreover, we shall see in Chapter 8 that there is a broad class of mass action systems, again described in terms of deficiencies but not limited to the deficiency zero class, that are also quasi-thermostatic (but not necessarily quasi-thermodynamic) regardless of rate constant values. Within that class there are mass action systems for which there are no complex balanced positive equilibria at all. This too is proved in Part III.

7.8 The Global Attractor Conjecture and the Persistence Conjecture

As it is stated in Section 7.1, the Deficiency Zero Theorem asserts among other things that, for a mass action system in which the underlying network is weakly reversible and has a deficiency of zero, the unique equilibrium in each positive stoichiometric compatibility class is asymptotically stable. This is a *local* statement, asserting that for initial conditions sufficiently close to (and stoichiometrically compatible with) a positive equilibrium, the resulting composition trajectory will approach that equilibrium in the limit as time gets arbitrarily large.

In part because we also know that there cannot be cyclic composition trajectories, it is reasonable to conjecture that *all* composition trajectories originating within a positive stoichiometric compatibility class will, in the limit with increasing time, approach its unique positive equilibrium. Indeed, Horn and Jackson [109] claimed something even broader:

Assertion A: *For any complex balanced mass action system—not necessarily one derived from a deficiency zero network—the unique equilibrium in each positive stoichiometric class is approached in the limit with increasing time by any trajectory originating in that positive stoichiometric compatibility class.*

However, shortly after its publication, Arnold Shapiro, a graduate student of Horn, noticed that the proof in [109], while adequate to support the (local) asymptotic stability of positive equilibria, was not sufficiently cognizant of a certain possibility: that a trajectory originating within a positive stoichiometric compatibility class might approach a nonpositive equilibrium on its boundary, even while values of the Lyapunov function given by (7.20) are decreasing.

Horn acknowledged that possibility (and the gap) in [108] but conjectured that Assertion A is nevertheless true. Proof of this *global attractor conjecture* has been the subject of a great deal of work—see, for example, [2, 3, 6, 50, 135, 161, 162]—with success in important special cases.[3] *In particular, David Anderson [3] proved*

[3] From results in Chapter 10, it will follow that the conjecture cannot be violated by any network in the very large class of *concordant* networks: Weak reversibility is a necessary condition for complex balancing [107]. For any weakly reversible concordant network, no positive stoichiometric compatibility class can have on its boundary a reaction-transitive composition, in particular an equilibrium. Siegel and MacLean [161] showed that any violation of Assertion A requires such a boundary equilibrium.

the conjecture for networks with a single linkage class. At the time of this writing, it appears that the conjecture has been proved in full by Gheorghe Craciun [45].

In connection with Assertion A, there is another conjecture worth mentioning. Soon after the gap in the global stability argument became apparent to us, Horn, Shapiro, and I became increasingly convinced that it was not complex balancing itself but, rather, *weak reversibility* (a necessary condition for complex balancing) that, for mass action systems, thwarts a longtime approach to the boundary of a positive stoichiometric compatibility class from its interior. This second conjecture would support the first, but its proof eluded us.

The second conjecture, along with a discussion of issues connected with Assertion A, was stated in [73] in the following form:

Assertion B: *For any weakly reversible network taken with mass action kinetics, the corresponding differential equations have the property that no composition trajectory beginning with all species concentrations positive evolves in increasing time toward a limit in which one or more species concentrations are zero.*

Assertion B, like Assertion A, has been the subject of considerable research. It has come to be called the *persistence conjecture*, largely because it precludes the longtime extinction of one or more species in a population in which all species are initially present. Much of the literature cited above in connection with Assertion A also addresses Assertion B.[4]

[4] In more recent research, the technical formulation of Assertion B usually requires that, for weakly reversible mass action systems, the ω-limit set associated with any bounded trajectory originating in $\mathbb{R}_+^{\mathscr{S}}$ lies entirely in $\mathbb{R}_+^{\mathscr{S}}$. A discussion of various technical formulations of Assertion B can be found in [3].

8

Deficiency One Theory

The Deficiency Zero Theorem tells us, among other things, that for all weakly reversible deficiency zero networks taken with mass action kinetics, the induced differential equations admit precisely one equilibrium in each positive stoichiometric compatibility class. This holds true regardless of values the rate constants take and regardless of how intricate the differential equations might be. It turns out that there is an easily described but even broader class of networks for which the same statement can be made. This is the subject of the Deficiency One Theorem.

8.1 Motivation for the Deficiency One Theorem

In order to motivate the Deficiency One Theorem, it will be helpful to recall some ideas from Chapter 6, in particular ideas connected with the deficiencies of individual linkage classes. Recall that, for a network under study, we can associate with each linkage class its own deficiency and that these individual linkage class deficiencies might or might not sum to the deficiency of the parent network. Nevertheless, if $\delta_1, \delta_2, \ldots, \delta_\ell$ are the deficiencies of the individual linkage classes and δ is the network's deficiency, we always have

$$\sum_{\theta=1}^{\ell} \delta_\theta \leq \delta. \tag{8.1}$$

From this, it is easy to see that the deficiency *zero* networks are precisely those that satisfy both of the following conditions: First, the deficiency of each linkage class is zero and, second, the deficiency of the entire network is equal to the sum of the deficiencies of the individual linkage classes.

With this in mind, we can state part of the Deficiency Zero Theorem in the following seemingly awkward way:

M. Feinberg, *Foundations of Chemical Reaction Network Theory*,
Applied Mathematical Sciences 202,
https://doi.org/10.1007/978-3-030-03858-8_8

Theorem 8.1.1 (A seemingly awkward rephrasing of part of the Deficiency Zero Theorem). *Consider a mass action system for which the underlying reaction network is weakly reversible and has ℓ linkage classes. Moreover, suppose that the deficiency of the entire network and the deficiencies of the individual linkage classes satisfy the following conditions:*

(i) $\delta_\theta = 0, \quad \theta = 1,2,\ldots,\ell$

(ii) $\sum_{\theta=1}^{\ell} \delta_\theta = \delta.$

Then, no matter what positive values the rate constants take, the corresponding differential equations admit precisely one equilibrium in each positive stoichiometric compatibility class.

The Deficiency *One* Theorem replaces condition (i) with the far weaker requirement that the deficiencies of the individual linkage classes not exceed *one*! (hence the theorem's name). And it goes further.

Recall that for deficiency zero networks, there is no possibility of a positive equilibrium unless the network is weakly reversible. For networks of nonzero deficiency, weak reversibility is no longer a precondition for the existence of positive equilibria, so for them the possibility of *multiple* stoichiometrically compatible positive equilibria becomes viable even in the absence of weak reversibility.

With respect to preclusion of multiple positive equilibria, the Deficiency One Theorem replaces the weak reversibility requirement in Theorem 8.1.1 by the far milder requirement that *each linkage class contain just one terminal strong-linkage class*—i.e., $t = \ell$ (Remark 6.6.5). Weakly reversible networks are just very special cases of $t = \ell$ networks: for weakly reversible networks, the terminal strong-linkage classes and the linkage classes are one and the same.

8.2 The Deficiency One Theorem

The Deficiency One Theorem appeared in [71], [72], and [73], and a full proof was given in [76]. Part III of this book contains a sketch of how the proof goes. See also [25].

Theorem 8.2.1 (The Deficiency One Theorem). *Consider a reaction network* $\{\mathscr{S},\mathscr{C},\mathscr{R}\}$ *having ℓ linkage classes, each containing just one terminal strong-linkage class. Moreover, suppose that the deficiency of the entire network and the deficiencies of the individual linkage classes satisfy the following conditions:*

(i) $\delta_\theta \leq 1, \quad \theta = 1,2,\ldots,\ell$

(ii) $\sum_{\theta=1}^{\ell} \delta_\theta = \delta$

If, for a particular $k \in \mathbb{R}_+^{\mathscr{R}}$*, the mass action system* $\{\mathscr{S},\mathscr{C},\mathscr{R},k\}$ *admits a positive equilibrium, then each positive stoichiometric compatibility class contains precisely one equilibrium, and at every positive equilibrium, the derivative of the species-formation-rate function is nonsingular. If the network is weakly reversible, then, for every* $k \in \mathbb{R}_+^{\mathscr{R}}$*, the mass action system* $\{\mathscr{S},\mathscr{C},\mathscr{R},k\}$ *admits a positive equilibrium.*

Remark 8.2.2. When in the theorem we refer to the derivative of the species-formation-rate function, we mean, as in Section 3.6, the derivative restricted to the stoichiometric subspace. When its hypothesis is satisfied, the theorem asserts, in the language of Section 3.6, that all of the positive equilibria are nondegenerate.

Remark 8.2.3. The network-structural conditions of Theorem 8.2.1 are clearly satisfied for any deficiency one reaction network containing just one linkage class, so long as there is just one terminal strong-linkage class. In this context, see [26] for results on the existence of positive equilibria when weak reversibility does not obtain.

Remark 8.2.4 (About the proof). In Remark 7.7.1 we discussed, in a cursory way, ideas behind the proof of the Deficiency Zero Theorem. A key idea was that, for any mass action system $\{\mathscr{S},\mathscr{C},\mathscr{R},k\}$ having a positive equilibrium c^* *at which complex balancing obtains*, the full set of positive equilibria $E \subset \mathbb{R}_+^{\mathscr{S}}$ has what Horn and Jackson [109] called the *quasi-thermostatic* property. This to say that, with S denoting the stoichiometric subspace,

$$E := \{c \in \mathbb{R}_+^{\mathscr{S}} : \ln c - \ln c^* \in S^{\perp}\}. \tag{8.2}$$

As we will show in Part III, the quasi-thermostatic property in turn implies that there is precisely one equilibrium in each positive stoichiometric compatibility class. For mass action systems in which the underlying reaction network is weakly reversible and has a deficiency of *zero*, a complex balanced positive equilibrium invariably exists.

To prove the Deficiency One Theorem in Part III, we will show that, for any mass action system in which the underlying network satisfies the structural requirements of the theorem and for which there exists *some* positive equilibrium c^*, the set of positive equilibria has the quasi-thermostatic property *whether or not there is a complex balanced positive equilibrium*. From this it follows that there is precisely one equilibrium in each positive stoichiometric compatibility class.

Example 8.2.5. Occasionally it will be useful to write out, in full, the differential equations for a fairly complex reaction network so that the power of one or another theorem might become more tangible. Consider network (8.3), in which there are 8 species and 20 reactions.

$$\begin{gathered} 2A \rightleftarrows D \rightleftarrows A+B \rightleftarrows C \rightleftarrows 2B \\ C+D \rightleftarrows E+F \rightleftarrows H \rightleftarrows C+E \rightleftarrows D+F \\ A+D \rightleftarrows G \rightleftarrows B+H \end{gathered} \tag{8.3}$$

Taken with mass action kinetics, the network gives rise to the system of eight coupled nonlinear differential equations displayed as (8.4). The rate constants are numbered according to the order in which the corresponding reactions appear in (8.3).

$$
\begin{aligned}
\dot{c}_A &= -2k_1{c_A}^2 + k_6 c_C + k_{18} c_G + (k_3 + 2k_2)\, c_D - (k_5 + k_4)\, c_A c_B - k_{17} c_A c_D \\
\dot{c}_B &= -2k_8{c_B}^2 + (2k_7 + k_6)\, c_C + k_{19} c_G + k_3 c_D - (k_5 + k_4)\, c_A c_B - k_{20} c_B c_H \\
\dot{c}_C &= k_8{c_B}^2 - (k_6 + k_7)\, c_C + k_5 c_A c_B + k_{10} c_E c_F - k_9 c_C c_D + k_{16} c_D c_F \\
&\quad - (k_{14} + k_{15})\, c_C c_E + k_{13} c_H \\
\dot{c}_D &= k_1{c_A}^2 + k_{18} c_G - (k_3 + k_2)\, c_D + k_4 c_A c_B + k_{10} c_E c_F - k_9 c_C c_D \\
&\quad - k_{17} c_A c_D - k_{16} c_D c_F + k_{15} c_C c_E \\
\dot{c}_E &= -(k_{10} + k_{11})\, c_E c_F + k_9 c_C c_D + k_{16} c_D c_F - (k_{14} + k_{15})\, c_C c_E + (k_{13} + k_{12})\, c_H \\
\dot{c}_F &= -(k_{10} + k_{11})\, c_E c_F + k_9 c_C c_D - k_{16} c_D c_F + k_{15} c_C c_E + k_{12} c_H \\
\dot{c}_G &= -(k_{18} + k_{19})\, c_G + k_{20} c_B c_H + k_{17} c_A c_D \\
\dot{c}_H &= k_{19} c_G + k_{11} c_E c_F - k_{20} c_B c_H + k_{14} c_C c_E - (k_{13} + k_{12})\, c_H
\end{aligned}
\tag{8.4}
$$

We can ask whether there is *some* combination of positive values for the 20 rate constants that is consistent with the existence of two distinct stoichiometrically compatible positive equilibria. Taken on an *ad hoc* basis, this is an extremely difficult question in polynomial algebra.

Nevertheless, the Deficiency One Theorem gives the answer almost immediately. From top to bottom, the three linkage classes in (8.3) have deficiencies of *one*, *one*, and *zero*, so condition (i) is satisfied. Moreover, the deficiency of the network as a whole is *two*, so condition (ii) is satisfied. Finally, as with all reversible (and weakly reversible) networks, the terminal strong-linkage classes coincide with the linkage classes, so the $t = \ell$ requirement is also satisfied. The Deficiency One Theorem tells us, then, that every positive stoichiometric compatibility class contains *precisely one* equilibrium, *no matter what positive values the 20 rate constants take.* Moreover, *at any positive equilibrium, the derivative of the species-formation-rate function is invariably nonsingular.*

8.3 What the Deficiency One Theorem Does Not Preclude

It is natural to conjecture that, for every weakly reversible network satisfying conditions (i) and (ii) of the Deficiency One Theorem, we get *all* of the corresponding weakly reversible mass action results of the Deficiency Zero Theorem, not just the preclusion of multiple positive equilibria or degenerate positive equilibria. That is, it is reasonable to suppose that, for every assignment of rate constants, not only are we guaranteed uniqueness and nondegeneracy of equilibria within positive stoichiometric compatibility classes *but also that there cannot be an unstable positive equilibrium or a cyclic composition trajectory.*

In fact, the conditions of the Deficiency One Theorem preclude neither of these. A counterexample is provided by mass action system (8.5)[1].

---

[1] This variant of the mass action system (5.2) was discussed earlier in Appendix 6.A of Chapter 6.

$$
\begin{array}{ccc}
 & A \underset{1}{\overset{8.5}{\rightleftarrows}} 2A & \\
 & 0.01 \downharpoonleft\upharpoonright 0.08 & \\
A + B \underset{1}{\overset{1}{\rightleftarrows}} & C & \underset{0.2}{\overset{1}{\rightleftarrows}} B
\end{array}
\tag{8.5}
$$

Note that the network shown in (8.5) satisfies all the conditions of the Deficiency One Theorem: its deficiency is one, and there is only one linkage class, which coincides with the sole terminal strong-linkage class. Note too that the dimension of the stoichiometric subspace for this three-species network is three, so there is but one positive stoichiometric compatibility class, consisting of the entire positive orthant of $\mathbb{R}^{\mathscr{S}}$. In this case, the Deficiency One Theorem tells us that there is precisely one positive equilibrium in all of $\mathbb{R}^{\mathscr{S}}_{+}$ and that the sole positive equilibrium is nondegenerate.

However, for the rate constants indicated, *the only equilibrium is unstable, and there is in* $\mathbb{R}^{\mathscr{S}}_{+}$ *a stable limit cycle.*

8.4 All of the Network-Structural Conditions of the Deficiency One Theorem Are Essential

There are three network-structural conditions in the Deficiency One Theorem's hypotheses: conditions (i) and (ii) and also the requirement that each linkage class contain no more than one terminal strong-linkage class (i.e., that $t = \ell$). By means of three examples, we show in this section that if any of the three conditions is omitted, the theorem becomes false.

Example 8.4.1 (Omission of the $t = \ell$ condition). Consider network (8.6).

$$2A \overset{\alpha}{\leftarrow} A + B \overset{\beta}{\rightarrow} 2B \tag{8.6}$$

With rate constants as indicated in (8.6), the corresponding mass action differential equations are those shown in (8.7). Note that when α and β are unequal, there are no positive equilibria at all. When α and β are equal, however, *every positive composition is an equilibrium*! In this case, every positive stoichiometric compatibility class contains an *infinite* number of equilibria.[2]

$$
\begin{aligned}
\dot{c}_A &= (\alpha - \beta) c_A c_B \\
\dot{c}_B &= (\beta - \alpha) c_A c_B
\end{aligned}
\tag{8.7}
$$

The deficiency of network (8.6) is easily seen to be one ($n = 3$, $\ell = 1$, $s = 1$), as is the deficiency of its sole linkage class. Thus, conditions (i) and (ii) of the Deficiency One Theorem are satisfied, but the sole linkage class contains *two* terminal

[2] The stoichiometric subspace is one-dimensional and is spanned by the reaction vector $B - A$. The stoichiometric compatibility classes are line segments parallel to the stoichiometric subspace.

strong-linkage classes, $\{2A\}$ and $\{2B\}$. The example tells us that the Deficiency One Theorem would be false in the absence of the requirement that each linkage class contain no more than one terminal strong-linkage class.

Remark 8.4.2. The sudden explosion of positive equilibria—from none when $\alpha \neq \beta$ to an infinite number for the exceptional case $\alpha = \beta$—is, of course, a bizarre phenomenon that we would not expect to be exhibited in real reaction networks. In fact, the "equilibria explosion" disappears completely so long as one or both of the reactions in (8.6) are made even slightly reversible (in which case $t = \ell$), no matter how tiny might be the rate constants for the reactions added. Not only, then, does the mass action model (8.6) give rise to strange behavior; the model itself is not robust. A striking phenomenon it exhibits vanishes following an ever-so-tiny perturbation of the model.

More generally, mass action systems such as (8.6) that admit positive equilibria and in which there is an excess of terminal strong-linkage classes are often poor models for real chemistry. The aberrant behavior of such systems has roots in issues discussed earlier in Appendix 3.A. That discussion is resumed in Appendix 8.A at the end of this chapter, where we examine properties of $t > \ell$ mass action systems more fully. For reasons considered there, it is not surprising that the hypothesis of the Deficiency One Theorem should preclude them from consideration at the outset.

Example 8.4.3 (Omission of condition (i)). Recall network (8.8), studied by Horn and Jackson [109]. Recall too that, with rate constants as indicated in (8.8), the corresponding mass action differential equations admit *three* equilibria in each positive stoichiometric compatibility class whenever ε is less than $\frac{1}{6}$. (See Figure 5.2.)

$$\begin{array}{ccc} 3A & \overset{\varepsilon}{\rightarrow} & A+2B \\ 1\uparrow & & \downarrow 1 \\ 2A+B & \underset{\varepsilon}{\leftarrow} & 3B \end{array} \tag{8.8}$$

There is only one linkage class, so condition (ii) of the Deficiency One Theorem is satisfied. So too is the requirement that each linkage class contain just one terminal strong-linkage class. However, the deficiency of the network's sole linkage class is *two* ($n = 4,\ ,\ell = 1,\ s = 1$). The example demonstrates not only that condition (i) cannot be dropped but also that the deficiency bound in condition (i) cannot be increased to two.

Example 8.4.4 (Omission of condition (ii)). Recall the mass action system studied in Section 5.1.2 and repeated here as (8.9). Recall too that, for the rate constants indicated, there are certain positive stoichiometric compatibility classes containing *three* equilibria. See Figure 5.3. (There are also two positive stoichiometric compatibility classes containing a [unique] equilibrium at which the derivative of the species-formation-rate function is singular.)

$$\begin{array}{c} A \underset{1}{\overset{8.5}{\rightleftarrows}} 2A \\[1em] A+B \underset{1}{\overset{1}{\rightleftarrows}} C \underset{0.2}{\overset{1}{\rightleftarrows}} B \end{array} \tag{8.9}$$

In this case the underlying network is reversible so that the linkage classes coincide with the terminal strong-linkage classes; the $t = \ell$ condition is satisfied. Moreover, each of the two linkage classes has a deficiency of zero, so condition (i) is satisfied. On the other hand, the full network's deficiency is one, so condition (ii) is not satisfied. The example indicates that the Deficiency One Theorem would be false were condition (ii) omitted.

8.5 The Deficiency One Algorithm: A First Glimpse

For a *deficiency one* network, condition (ii) fails precisely when each linkage class has a deficiency of zero. Among such networks there are some—as we have just seen—for which there are rate constants such that the resulting mass action differential equations admit multiple equilibria within a positive stoichiometric compatibility class. However, there are also some that have all properties that the Deficiency One Theorem guarantees: For *every* assignment of positive rate constants, the resulting mass action differential equations admit no more than one equilibrium in any positive stoichiometric compatibility class, nor do they admit a degenerate positive equilibrium. We would like to be able to distinguish between these deficiency one networks in a decisive way.

By way of preparation, it will be useful to first recognize clearly the surprising boldness of statements that the Deficiency Zero and Deficiency One Theorems make. Because these theorems encourage the expectation that additional theory might speak in similarly bold terms, we will want to examine the extent to which questions of the kind we have been asking can invariably be answered with the same broad strokes.

8.5.1 Describing Reaction Network Behavior in Broad, Bold Terms: When Do the Fine Details of Network Structure Matter?

At least for weakly reversible[3] networks, there is a magic in the Deficiency Zero and Deficiency One Theorems that is easily overlooked: *their hypotheses are largely indifferent to the fine details of reaction network structure*. For example, if the hypotheses of one or another of the two theorems are satisfied for a given network, then they will also be satisfied for any other network having the same complexes, no matter how reaction arrows might be added or omitted, so long as the partition of complexes into linkage classes remains the same and so long as weak reversibility is maintained. This is because fine details of the way complexes are linked affect neither the deficiency of the network nor the deficiencies of the individual linkage classes. (Recall Remark 6.3.3.)

For networks that violate condition (ii) of the Deficiency One Theorem (and that therefore violate the hypothesis of the Deficiency Zero Theorem), there is no hope

[3] The focus on weakly reversible networks here is meant only to make the discussion less cumbersome.

that we can invariably speak in the same bold terms—that is, in terms indifferent to the fine details of reaction network structure. This is made clear by consideration of the three toy networks displayed in Table 8.1. The first and second networks describe continuous-flow stirred-tank reactors in which the operative reactions are, respectively, $A + B \rightleftarrows 2A$ and $2A + B \rightleftarrows 3A$. The third network is a reversible version of the so-called Brusselator, discussed in Section 5.3.2. Note that each of the three networks has a deficiency of one, and in each case the deficiencies of the various linkage classes are zero. Thus, condition (ii) of the Deficiency One Theorem is violated for all three networks.

In each case we indicate in the table whether there are *some* rate constant values such that the resulting mass action system admits two or more distinct positive equilibria. Note that it is only the second network for which such rate constants exist. Yet the second network has precisely the same set of reaction vectors as the first network; the second differs from the first only in the addition of species A to both sides of a single reversible reaction. And the second network has precisely the same complexes and linkage classes as the third network; the two networks differ only to the extent that the complexes 0 and A are permuted.

Table 8.1. Three simple deficiency one networks and their capacity for multiple equilibria

Network	Multiple Equilibria?
1. $A + B \rightleftarrows 2A$ $A \rightleftarrows 0 \rightleftarrows B$	No
2. $2A + B \rightleftarrows 3A$ $A \rightleftarrows 0 \rightleftarrows B$	Yes
3. $2A + B \rightleftarrows 3A$ $0 \rightleftarrows A \rightleftarrows B$	No

The table tells us that any complete theory that purports to distinguish decisively between reaction networks on the basis of their capacity to admit multiple positive equilibria cannot speak solely in terms of deficiencies. Fine details matter. We must expect instead that such a theory would have an algorithmic character, relying on procedures that discriminate between networks on the basis of highly subtle differences between them.

8.5.2 What the Deficiency One Algorithm Does: Translating Difficult Questions About *Nonlinear Equations* into Easy Questions About *Linear Inequalities*

For a regular deficiency one network that fails to satisfy condition (ii), the *Deficiency One Algorithm* [74, 77] provides means to determine *decisively* whether, for the network, there are rate constants such that the resulting mass action differential equations admit multiple positive equilibria or a degenerate positive equilibrium. When no such rate constants exist, the algorithm will say so. When they do exist, the algorithm will again say so, and, moreover, it will provide an example.

A *regular* reaction network is one that satisfies three mild conditions: First, its reaction vectors are positively dependent. (Recall that this is a necessary condition for a positive equilibrium.) Second, each linkage class contains no more than one terminal strong-linkage class (i.e., $t = \ell$). (In the appendix to this chapter, we argue that a mass action system that derives from a deficiency one network for which $t > \ell$ will typically be pathological: it will admit an *infinite* number of equilibria in a positive stoichiometric compatibility class if it admits even one.) Third, each direct link of two *terminal* complexes is a cut-link. (Recall Section 6.7.)

This last condition is satisfied by network (8.10) but not by network (8.11). In (8.11) the direct link connecting the terminal complexes E and F is not a cut-link. (Note that in both networks the direct link connecting complexes C and D is not a cut-link, but this does not constitute a violation of the third condition because neither C nor D is terminal.)

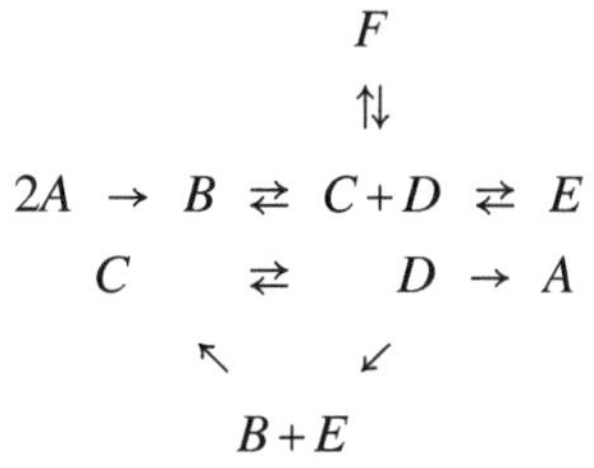

(8.10)

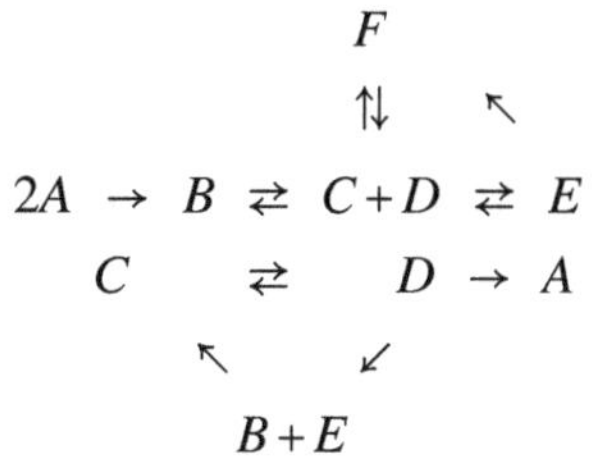

(8.11)

Our aim in this section is to indicate, in a very impressionistic way, how the deficiency one algorithm works. We will have much more to say in Part III. For a network $\{\mathscr{S}, \mathscr{C}, \mathscr{R}\}$ under study, the algorithm begins by associating with each species $s \in \mathscr{S}$

a real variable μ_s. This is to say, equivalently, that we associate with the network a vector variable $\mu \in \mathbb{R}^{\mathscr{S}}$. Moreover, with each complex $y \in \mathscr{C}$, we associate the linear form $y \cdot \mu$. Thus, for the five complexes in network (2) of Table 8.1, the set of associated linear forms is

$$\{2\mu_A + \mu_B, 3\mu_A, \mu_A, 0, \mu_B\}. \tag{8.12}$$

The algorithm then indicates (in Part III) how the linear forms for the network are joined by inequality or equality signs to produce one or more systems of linear inequalities. In the case of the networks shown in Table 8.1, the algorithm produces, for each network, *two* linear inequality systems. These are displayed in Table 8.2.[4]

Table 8.2. Three simple deficiency one networks and their inequality systems

Network	Inequality Systems	Comment
1. $A + B \rightleftarrows 2A$	(a) $\mu_A > 0 > \mu_B > 2\mu_A > \mu_A + \mu_B$	No solution for (a) or (b)
$A \rightleftarrows 0 \rightleftarrows B$	(b) $\mu_B > 0 > \mu_A > \mu_A + \mu_B > 2\mu_A$	Multiple equilibria impossible
2. $2A + B \rightleftarrows 3A$	(a) $\mu_A > 0 > \mu_B > 3\mu_A > 2\mu_A + \mu_B$	No solution for (a)
$A \rightleftarrows 0 \rightleftarrows B$	(b) $\mu_B > 0 > \mu_A > 2\mu_A + \mu_B > 3\mu_A$	Solution for (b): $\mu_A = -2$, $\mu_B = 1$
3. $2A + B \rightleftarrows 3A$	(a) $0 = \mu_A > \mu_B > 3\mu_A > 2\mu_A + \mu_B$	No solution for (a) or (b)
$0 \rightleftarrows A \rightleftarrows B$	(b) $\mu_B > \mu_A = 0 > 2\mu_A + \mu_B > 3\mu_A$	Multiple equilibria impossible

The key question then becomes: For the network under study, does there exist a vector $\mu^* \in \mathbb{R}^{\mathscr{S}}$ that satisfies at least one of the network's inequality systems and is *sign-compatible with the network's stoichiometric subspace*? This last requirement is given meaning in the following definition:

Definition 8.5.1. *For a reaction network* $\{\mathscr{S}, \mathscr{C}, \mathscr{R}\}$ *with stoichiometric subspace* S*, we say that* $\mu \in \mathbb{R}^{\mathscr{S}}$ *is* **sign-compatible with** S *if there exists* $\sigma \in S$ *such that* $\operatorname{sign} \mu_s = \operatorname{sign} \sigma_s$ *for all* $s \in \mathscr{S}$.

Here is the central result of the Deficiency One Algorithm: *If, for a regular deficiency one network, none of the inequality systems produced by the algorithm admits a solution that is sign-compatible with the network's stoichiometric subspace, then*

[4] In the Comments, by a solution we mean one sign-compatible with the stoichiometric subspace.

for no set of rate constants can the corresponding mass action differential equations give rise to multiple equilibria within a positive stoichiometric compatibility class, nor can they give rise to a degenerate positive equilibrium. If even one of the inequality systems admits such a solution, then there exist for the network rate constants such that the resulting mass action differential equations give rise to multiple equilibria within a positive stoichiometric compatibility class. Moreover, there exist rate constants such that the resulting mass action differential equations give rise to a degenerate positive equilibrium.

Example 8.5.2. For all of the networks in Table 8.2, the stoichiometric subspace coincides with $\mathbb{R}^{\mathscr{S}}$, so for them the issue of stoichiometric sign compatibility becomes vacuous. Note that for neither network (1) nor network (3) do the any of the corresponding linear inequality systems shown in the table admit a solution. Thus, neither network can, when taken with mass action kinetics, admit multiple equilibria within a positive stoichiometric compatibility class or a degenerate positive equilibrium. On the other hand, one of the inequality systems for network (2) does admit a solution. For network (2), then, the mass action differential equations admit multiple stoichiometrically compatible positive equilibria and a degenerate positive equilibrium for certain (perhaps different) rate constant values. Despite the great similarities between networks (1), (2), and (3), the Deficiency One Algorithm draws sharp behavioral distinctions between them.

8.6 About Higher Deficiency Networks and Irregular Deficiency One Networks

For a regular deficiency one network, the Deficiency One Algorithm will *invariably* decide in a conclusive way—either yes or no—whether there exist for the network rate constants such that the resulting mass action system admits either a degenerate positive equilibrium or multiple equilibria within a positive stoichiometric compatibility class. Again, these determinations are made by translating questions about nonlinear differential equations into questions about systems of linear inequalities.

When the deficiency is higher than one or when a network is irregular, the same questions can *often*, but not always, be answered using similar ideas, again by translating these questions into ones about systems of linear inequalities. Major advances along these lines came in the Ph.D. thesis of Phillipp Ellison [59] and then, subsequently, in the Ph.D. thesis of Haixia Ji [115].[5] See also [60, 61] and work by Conradi and Flockerzi[42]. Although the Deficiency One Algorithm will be explored in considerable depth in Part III of this book, higher deficiency mass action theory, which is more complicated, is beyond the book's intended scope.

[5] The work of Ellison and Ji is implemented in the Chemical Reaction Network Toolbox [62]. In fact, work by Ellison was implemented in the much earlier DOS version of the Toolbox [78].

In Chapter 11, centered on the *Species-Reaction Graph*, there are theorems, like those considered here, that attempt to answer questions about the capacity of a reaction network to admit a degenerate positive equilibrium or to admit multiple equilibria within a stoichiometric compatibility class. There, however, the questions are not specifically about mass action systems. In some cases, Chapter 11's theorems will deny the capacity for multiple stoichiometrically compatible equilibria for networks of high deficiency, and not just for mass action kinetics but for more general kinetics as well. On the other hand, there are questions specific to mass action systems that are answered by the deficiency-oriented theory discussed here but for which the Species-Reaction Graph theory stands silent.

8.7 Two Remarks About the Existence of Positive Equilibria for Mass Action Systems

The main interest that deficiency one theory holds for us is in the guarantee of *uniqueness* of positive equilibria. We should not, however, overlook the considerable mathematical difficulties connected with the *existence* of *positive* equilibria for mass action systems generally. In this connection we offer two related remarks, the first a precursor to the second.

Remark 8.7.1 (For networks that are not weakly reversible the existence of positive equilibria can depend on rate constant values). Recall that for a kinetic system to admit a positive equilibrium, it is necessary that the underlying reaction network have positively dependent reaction vectors. When this is the case, it is a simple exercise to show that there is *some* assignment of positive rate constants to the various reactions such that the resulting mass action system admits a positive equilibrium. But this is not to say that, when the reaction vectors are positively dependent, the mass action differential equations will admit a positive equilibrium for *every* assignment of positive rate constants.

In fact, even among reaction networks that are not weakly reversible but which otherwise satisfy all conditions of the Deficiency One Theorem, there are networks that admit a positive equilibrium for some rate constants but not for others. An example is provided by the mass action system displayed as (8.13). For $k > 1$ there is precisely one equilibrium in each positive stoichiometric compatibility class. For $k \leq 1$ there are no positive equilibria at all.

$$2A \underset{k}{\overset{1}{\rightleftarrows}} A + B \xrightarrow{1} 2B \tag{8.13}$$

Moreover, even for a reaction network that admits a positive equilibrium for all assignments of rate constants, it is not necessarily true that, for a given rate constant assignment, there is an equilibrium in every positive stoichiometric class. As network (5.3) and Figure 5.4 demonstrate, there might be equilibria in some positive stoichiometric compatibility classes but none in others.

Remark 8.7.2 (The existence of positive equilibria for all *weakly reversible mass action systems, regardless of rate constant values and regardless of deficiencies).* It is safe to say that most chemists would, on intuitive grounds, take for granted the *existence* of a positive equilibrium for *every* mass action system in which the underlying reaction network is weakly reversible, *regardless of rate constant values and regardless of deficiencies.* And, were the question put to them in a comprehensible way, they might also expect the existence of at least one equilibrium in each positive stoichiometric compatibility class.

Yet, a glance at (8.4)—the differential equations for *just one* weakly reversible mass action system—makes evident the difficulty in proving the existence of a positive equilibrium *for all weakly reversible networks*, again, *regardless of rate constant values and regardless of deficiencies.* Proof for the subclass of weakly reversible *deficiency zero* networks [107] was already nontrivial. More difficult still was proof [76] for the larger class of weakly reversible networks satisfying conditions (i) and (ii) of the Deficiency One Theorem. However, conditions (i) and (ii) clearly narrow the class of weakly reversible networks considered. The difficulty of their removal represents a significant barrier to proof of what chemists presumably believe more generally: *Taken with mass action kinetics, every weakly reversible network invariably admits positive equilibria, regardless rate constant values—in fact, at least one equilibrium in each positive stoichiometric compatibility class.*

Proof for *all* weakly reversible networks, *regardless of deficiencies and regardless of whether condition (ii) is satisfied*, appear in [55] and [28], almost entirely as a result of work by Adrian Nachman, Jian Deng, and Balázs Boros. Here is some history: In 1983, subsequent to lectures I gave on the Deficiency One Theorem, Nachman sent some notes in which he proved the existence of a positive equilibrium for any mass action system in which the underlying network is weakly reversible, is of *arbitrary* deficiency, *but has just one linkage class.* The extension—including the existence of an equilibrium in each positive stoichiometric compatibility class—to weakly reversible networks having several linkage classes *but which satisfy condition (ii)* was an almost immediate consequence of arguments already given in the proof of the Deficiency One Theorem. The very difficult removal of the condition (ii) requirement was largely the work of Deng and Boros.

The results in [55] were unknown to Balázs Boros when he nevertheless proved in [27] a more limited, but still difficult, result: the existence of an equilibrium in each positive stoichiometric compatibility class for every mass action system in which the underlying reaction network is weakly reversible and has a deficiency of one, *regardless of whether condition (ii) is satisfied.*

Appendix 8.A Why Mass Action Models with an Excess of Terminal Strong-Linkage Classes Are Problematic

The Deficiency One Theorem and the Deficiency One Algorithm preclude from consideration $t > \ell$ mass action systems—that is, mass action systems in which the underlying reaction network has more terminal strong-linkage classes than linkage classes. In fact, this is no great loss, for such systems often exhibit phenomena (e.g., the existence of multiple equilibria within a positive stoichiometric compatibility class) that disappear when the system is perturbed in the slightest way. With respect to their precise reflection in real chemistry, ephemeral mathematical phenomena of this kind should be viewed with skepticism.

That $t > \ell$ mass action systems should behave in this way has roots in Appendix 3.A, where we considered distinctions between the kinetic and stoichiometric subspaces. There we indicated how, when the kinetic subspace for a particular kinetic system is smaller than the stoichiometric subspace for the underlying reaction network, strange non-robust behavior can result. Here we extend that discussion. In particular, for mass action systems, we relate the lack of coincidence of the kinetic and stoichiometric subspaces directly to an excess of terminal strong-linkage classes.

8.A.1 The Kinetic Subspace Revisited

We will begin by reviewing and elaborating somewhat on ideas from Appendix 3.A. For a kinetic system $\{\mathscr{S},\mathscr{C},\mathscr{R},\mathscr{K}\}$, the *kinetic subspace*, denoted K, is the smallest linear subspace of $\mathbb{R}^{\mathscr{S}}$ that contains the image of the species-formation-rate function. Because the species-formation-rate function takes values in the stoichiometric subspace S for the underlying network $\{\mathscr{S},\mathscr{C},\mathscr{R}\}$, we always have the inclusion $K \subset S$. For a particular kinetics $\mathscr{K}$, the kinetic subspace might be smaller than the stoichiometric subspace.

Recall that the differential equation for a kinetic system $\{\mathscr{S},\mathscr{C},\mathscr{R},\mathscr{K}\}$ is

$$\dot{c} = f(c), \tag{8.A.1}$$

where $f(\cdot)$ is the system's species-formation-rate function. Because the "velocity vector" $\dot{c}$ takes values in the image of the species-formation-rate function, it invariably points not only along S but, indeed, along K. As we pointed out in Appendix 3.A, the fact that K might be smaller than S gives rise to a sharpening of ideas in Chapter 3.

In particular, using ideas very much like those in Chapter 3, we can see without difficulty that a composition c' might follow a composition c along a solution of (8.A.1) only if $c' - c$ is a member not only of the stoichiometric subspace S but also of the possibly smaller kinetic subspace $K \subset S$. This provides motivation for the following definitions, all related to *kinetic compatibility*, which parallel our earlier definitions related to *stoichiometric compatibility*.

Definition 8.A.1. *Let $\{\mathscr{S},\mathscr{C},\mathscr{R},\mathscr{K}\}$ be a kinetic system, and let $K \subset \mathbb{R}^{\mathscr{S}}$ be its kinetic subspace. Two vectors c' and $c \in \overline{\mathbb{R}}_+^{\mathscr{S}}$ are* **kinetically compatible** *if $c' - c$ lies in K. Kinetic compatibility is an equivalence relation that induces a partition of $\overline{\mathbb{R}}_+^{\mathscr{S}}$ into equivalence classes called the* **kinetic compatibility classes** *for the kinetic system. In particular, the kinetic compatibility class containing $c \in \overline{\mathbb{R}}_+^{\mathscr{S}}$ is the set $(c+K) \cap \overline{\mathbb{R}}_+^{\mathscr{S}}$.*

When, for a particular kinetic system, K and S coincide, there is no distinction between kinetic and stoichiometric compatibility classes. However, when K is smaller than S, then each stoichiometric compatibility class is stratified into a collection of finer kinetic compatibility classes. This will be illustrated later on in Figure 8.A.2 when we consider a concrete example. *A trajectory of* (6.1.1) *that begins in a particular kinetic compatibility class resides entirely within that same kinetic compatibility class.*

Hereafter in this appendix, our focus will mostly be on mass action systems. The discussion is close to one given in [73].

8.A.2 The Kinetic Subspace for a Mass Action System

Consider a mass action system $\{\mathscr{S},\mathscr{C},\mathscr{R},k\}$. For each $y \in \mathscr{C}$ we denote by $\mathscr{R}_{y\to}$ the set of all reactions having y as their reactant complex. That is,

$$\mathscr{R}_{y\to} := \{\bar{y} \to y' \in \mathscr{R} : \bar{y} = y\}. \tag{8.A.2}$$

The species-formation-rate function (3.8) for the system can be rearranged:

$$f(c) := \sum_{y\in\mathscr{C}} c^y \sum_{\mathscr{R}_{y\to}} k_{y\to y'}(y'-y). \tag{8.A.3}$$

Thus, we can write

$$f(c) := \sum_{y\in\mathscr{C}} c^y d_y, \tag{8.A.4}$$

where

$$d_y := \sum_{\mathscr{R}_{y\to}} k_{y\to y'}(y'-y). \tag{8.A.5}$$

When $\mathscr{R}_{y\to}$ is empty—that is, when y is not a reactant complex for any reaction—it is understood that $d_y = 0$.

From (8.A.4) it follows that, for the mass action system under consideration, the kinetic subspace K—again, the smallest linear subspace of $\mathbb{R}^{\mathscr{S}}$ containing the image of $f(\cdot)$—is contained in the span of the set $\{d_y : y \in \mathscr{C}\}$. In fact, we actually have equality[6]:

$$K = \text{span}\{d_y : y \in \mathscr{C}\}. \tag{8.A.6}$$

Because each d_y is a linear combination of reaction vectors, it becomes evident in another way that K resides in S, the stoichiometric subspace for the reaction network $\{\mathscr{S},\mathscr{C},\mathscr{R}\}$. *Whether K coincides with S or is smaller than S might depend on the particular values that the rate constants take, for these values determine the vectors* $\{d_y : y \in \mathscr{C}\}$.

We will eventually see that there is a surprisingly large class of reaction networks having the property that, *for every assignment of rate constants*, the kinetic and stoichiometric subspaces are identical. First, we will examine the kinds of anomalous behavior that can result for networks outside this class when, for a particular rate constant assignment, the kinetic subspace is smaller than the stoichiometric subspace.

[6] This follows from arguments similar to those used in [80], in particular from Lemma 1. See also Lemma 16.B.3 and equation (16.B.13) in Part III.

8.A.3 An Instructive Mass Action Example Revisited

It will be helpful to revisit the mass action system (8.A.7) introduced earlier in Appendix 3.A, this time with a more detailed discussion of the behavior the system admits:

$$B \overset{\alpha}{\leftarrow} A \overset{1}{\rightarrow} C \qquad (8.A.7)$$
$$C + B \overset{1}{\rightarrow} 2A.$$

The differential equations for the system in component form are

$$\begin{aligned} \dot{c}_A &= -(\alpha+1)c_A + 2c_B c_C \\ \dot{c}_B &= \alpha c_A - c_B c_C \\ \dot{c}_C &= c_A - c_B c_C. \end{aligned} \qquad (8.A.8)$$

Recall that the stoichiometric subspace for the underlying network is given by

$$S = \text{span}\{B - A, C - A, 2A - C - B\}; \qquad (8.A.9)$$

its dimension is two. Stoichiometric compatibility classes take the form of triangles, as shown in Figure 8.A.1.

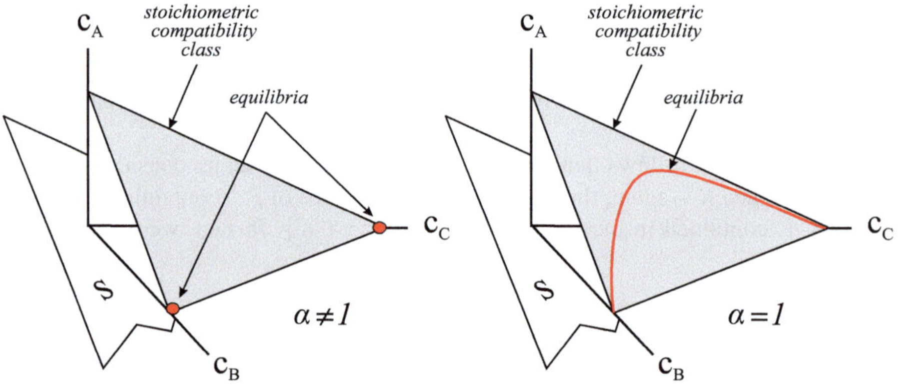

Fig. 8.A.1. Stoichiometric compatibility class and equilibria for the mass action system (8.A.7)

From (8.A.8) it is easy to see that for $\alpha \neq 1$ there are no positive equilibria at all: as indicated in Figure 8.A.1, the equilibria are those composition states in which only B or C is present. On the other hand, when $\alpha = 1$ there is an explosion of equilibria. In fact, within each positive stoichiometric compatibility class, there is an *infinite* number of equilibria, depicted schematically in the right panel of Figure 8.A.1.

To a great extent, this singular behavior is connected to the fact that the stoichiometric and kinetic subspaces coincide *except in the exceptional case* $\alpha = 1$. It is helpful to examine the nature of the kinetic subspace in light shed by the relation (8.A.6). For the mass action system under consideration, we have

$$d_A = \alpha(B-A)+(C-A) \quad d_{A+B} = 2A-C-B \quad d_B = 0 \quad d_C = 0. \tag{8.A.10}$$

Clearly, then

$$K = \text{span}\{d_A, d_{A+B}\}. \tag{8.A.11}$$

Note that d_A and d_{A+B} constitute an independent set whenever $\alpha \neq 1$, in which case $K \subset S$ is two-dimensional and therefore coincides with S. However, when $\alpha = 1$ d_A and d_{A+B} are colinear, in which case K is one-dimensional and is smaller than S.

Composition trajectories for both the typical $\alpha = 2$ and the exceptional $\alpha = 1$ cases are sketched in Figure 8.A.2. For $\alpha = 2$ the kinetic and stoichiometric compatibility classes are identical, and trajectories are free to meander freely within such a class. The situation for $\alpha = 1$ is very different. As depicted in the right panel of Figure 8.A.2, the kinetic subspace K is a line residing in the two-dimensional stoichiometric subspace S. The triangular stoichiometric compatibility class depicted previously is now shown striated into kinetic compatibility classes that appear as line segments. For $\alpha = 1$ composition trajectories are no longer free to wander: instead, they must reside entirely within the straight line segments corresponding to the kinetic compatibility classes.

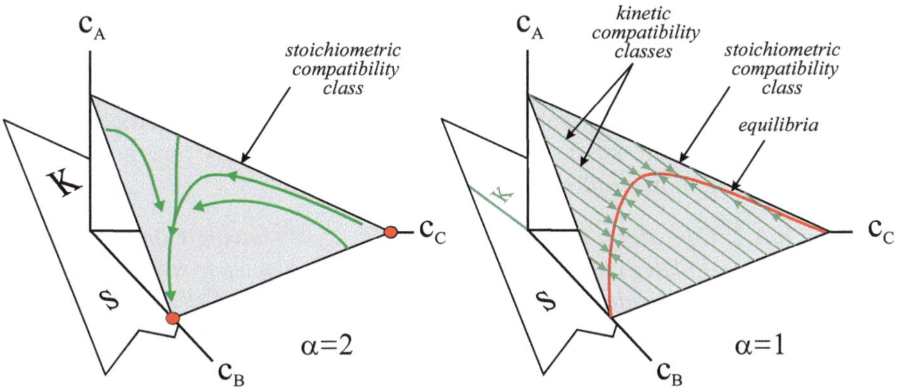

Fig. 8.A.2. Phase portraits for the mass action system (8.A.7), $\alpha = 2$ and $\alpha = 1$

For a kinetic system in which the kinetic subspace is smaller than the stoichiometric subspace, Figure 8.A.2 provides an indication of why it is that *a stoichiometric compatibility class that contains a positive equilibrium will usually contain an infinite number of them*: Consider a positive equilibrium c^* that resides within a particular stoichiometric compatibility class for the underlying network. When K is smaller than S, that stoichiometric compatibility class is stratified into an infinite number of finer kinetic compatibility classes, one of which contains c^*. As the right panel of Figure 8.A.2 suggests, the dynamics in the kinetic compatibility class containing

c^* can be expected to be replicated very closely in "nearby" kinetic compatibility classes. In particular, it is natural to expect that every nearby kinetic compatibility class will contain an equilibrium close to c^*. (This can be made precise using an implicit function theorem argument that applies to smooth kinetics, not just mass action kinetics.)

These considerations tell us something important about theorems or algorithms (such as the Deficiency One Theorem or the Deficiency One Algorithm) that purport to ensure that, for a particular reaction network taken with kinetics constrained only to lie within a large family (e.g., mass action), no positive stoichiometric compatibility class can ever contain more than one equilibrium: *Their hypotheses must, perhaps implicitly, preclude the possibility that the kinetic subspace can be smaller than the stoichiometric subspace for any member of the indicated kinetics family.*

As we shall see in the remainder of this appendix, when the kinetics is mass action, that preclusion is tied in a surprisingly simple and powerful way to the *number of terminal strong-linkage classes* possessed by the particular network under study. We shall also see why mass action models in which there is an excess of terminal strong-linkage classes make for problematic descriptions of real chemistry.

8.A.4 The Importance of the Number of Terminal Strong-Linkage Classes

Each linkage class in a reaction network contains at least one terminal strong-linkage class (Section 6.6). Thus, we always have $t \geq \ell$, and we might or might not have $t = \ell$. The following theorem [73, 80] tells us that if $t = \ell$, then, so long as the kinetics is mass action, the kinetic subspace *must* coincide with the stoichiometric subspace, *no matter what values the rate constants take*. It also tells us circumstances in which we can be sure that the kinetic subspace is smaller than the stoichiometric subspace. The roots of the theorem are discussed in Chapter 16 of Part III. See, in particular, the proof provided in Appendix 16.B.

Theorem 8.A.2. *Consider a reaction network $\{\mathscr{S},\mathscr{C},\mathscr{R}\}$ having ℓ linkage classes, t terminal strong-linkage classes, deficiency δ, and stoichiometric subspace $S \subset \mathbb{R}^{\mathscr{S}}$. Suppose that for a particular $k \in \mathbb{R}^{\mathscr{R}}_{+}$, the mass action system $\{\mathscr{S},\mathscr{C},\mathscr{R},k\}$ gives rise to the kinetic subspace $K \subset \mathbb{R}^{\mathscr{S}}$. Then the following statements hold true:*

(i) If $t = \ell$, then, regardless of rate constant values, $K = S$. That is, if each linkage class in the underlying network contains no more than one terminal strong-linkage class, there is no distinction between the kinetic and stoichiometric subspaces.

(ii) If $t - \ell > \delta$, then, regardless of rate constant values, K is smaller than S. In fact,

$$\dim S - \dim K \geq t - \ell - \delta. \tag{8.A.12}$$

(iii) If $t > \ell$, if $t - \ell \geq \delta$, and if the rate constants are such that the mass action system $\{\mathscr{S},\mathscr{C},\mathscr{R},k\}$ admits a positive equilibrium, then K is smaller than S. In fact,

$$\dim S - \dim K > t - \ell - \delta. \tag{8.A.13}$$

Remark 8.A.3. For a proof see Appendix 16.B. Parts *(i)* and *(ii)* were proved in [80]. Part *(iii)* was discussed in [73] and is an easy extension of arguments in [80]. As we shall see in Appendix 16.B, the requirement in *(iii)* that there be a positive equilibrium is stronger than it need be. It is enough that there be an equilibrium $c^* \in \overline{\mathbb{R}}_+^{\mathscr{S}}$ such that for some nonterminal complex y, it is true that $\text{supp}\, y \subset \text{supp}\, c^*$. In Appendix 16.B and in [80], it is shown that for mass action systems in which the underlying network has a deficiency of *zero*, we always have the *equality* $\dim S - \dim K = t - \ell$.

Because for weakly reversible networks the linkage classes are identical to the terminal strong-linkage classes, we have the following important corollary of Theorem 8.A.2:

Corollary 8.A.4. *For any mass action system in which the underlying reaction network is weakly reversible, the kinetic and stoichiometric subspaces are identical. In particular the kinetic and stoichiometric subspaces are identical if the underlying reaction network is reversible.*

We should keep in mind that this chapter is about *deficiency one* theory. In this case, Theorem 8.A.2 and, in particular, the inequality (8.A.13) provide considerable insight about what our expectations should be. The essential fact is summarized in the following corollary.

Corollary 8.A.5. *Consider a deficiency one reaction network having more terminal strong-linkage classes than linkage classes. For any assignment of rate constants such that the resulting mass action system admits a positive equilibrium, the kinetic subspace is smaller than the stoichiometric subspace.*

The corollary tells us why it was inevitable that $t > \ell$ networks be excluded in the formulation of the deficiency one results described earlier, the aim of which was to describe mass action systems that could *not* admit multiple positive equilibria within a stoichiometric compatibility class. For reasons discussed earlier, when the kinetic subspace is smaller than the stoichiometric subspace, a positive equilibrium in a stoichiometric compatibility class will usually be accompanied by an infinite number of them.

Example 8.A.6. We return to the network shown in (8.A.7), this time in light shed by Theorem 8.A.2. For the network $\delta = 1, \ell = 2$, and $t = 3$. Thus, we have $t > \ell$ and $t - \ell = \delta$. When the rate constants for the network are such that the corresponding mass action system admits no positive equilibrium, Theorem 8.A.2 is silent; indeed, despite the fact that $t > \ell$, the kinetic and stoichiometric subspaces nevertheless coincide. On the other hand, when there is a positive equilibrium—in particular, when $\alpha = 1$ in (8.A.7)—Theorem 8.A.2 tells us that the kinetic subspace is smaller than stoichiometric subspace, with $\dim S - \dim K = 1$. In this case there is an infinite number of equilibria in each positive stoichiometric compatibility class.

8.A.5 The Fragility of Phenomena Emerging from Mass Action Models Having an Excess of Terminal Strong-Linkage Classes

We have seen that there are certain dramatic phenomena exhibited by mass action systems—for example, the $\alpha = 1$ equilibria explosion exhibited in Figure 8.A.1—that ultimately derive from the fact that the kinetic subspace is smaller than the stoichiometric subspace. Even if we leave aside the fact that these phenomena might be highly singular, occurring for only very singular rate constant values, there is still something else to consider: *their lack of persistence against seemingly inconsequential perturbations of the reaction network itself.*

To make this apparent, we repeat below the special mass action system we've been studying,

$$B \overset{\alpha}{\leftarrow} A \overset{1}{\rightarrow} C \qquad C + B \overset{1}{\rightarrow} 2A, \tag{8.A.14}$$

in order that we might contrast its underlying network with a slightly different one:

$$B \leftarrow A \rightleftarrows C \qquad C + B \rightarrow 2A \tag{8.A.15}$$

Note that this second network differs from the first only to the extent that the reaction $A \rightarrow C$ has been made reversible. But now there are only *two* terminal strong-linkage classes, $\{B\}$ and $\{2A\}$, whereas before there were *three*, $\{B\}$, $\{C\}$, and $\{2A\}$. Before, we had $t > \ell$, whereupon Corollary 8.A.5 ensured that the kinetic and stoichiometric subspaces *do not* coincide, so long as the rate constants are such that there is a positive equilibrium. Now, for network (8.A.15), the same theorem ensures that the kinetic and stoichiometric subspaces *must* coincide, *no matter what rate constants are assigned to the various reactions.* Thus, dramatic phenomena that, for (8.A.14), rely on $K \neq S$ will be absent for any mass action system in which (8.A.15) is the underlying reaction network.

Clearly, any reaction network for which $t > \ell$ can, as in the example, be perturbed to a $t = \ell$ network by the addition of sufficiently many reverse reactions. It should be understood that chemists often insist that *all* reactions are reversible, if only to a very small extent (*i.e.*, perhaps with some reverse reactions having, in the mass action context, almost negligible rate constants). For this reason, it would not seem wise to place untempered credence in dramatic mass-action-model phenomena that derive from a lack of coincidence of the kinetic and stoichiometric subspaces (and, therefore, from the presence of an excess of terminal strong-linkage classes).

It is one thing to employ, for practical modeling purposes, networks in which there is a lack of reversibility in some or all reactions. It is quite a different thing to expect in nature precise replication of certain dramatic phenomena that result *only* from networks that offend chemists' sensibilities. Indeed, it is fitting that otherwise broad theorems should, in their hypotheses, exclude such networks from consideration at the outset. The exclusion of $t > \ell$ networks in the Deficiency One Theorem not only makes its conclusion possible, the exclusion also aligns with what chemists believe.

8.A.6 Summary: The Good News That Theorem 8.A.2 Contains

We first introduced the seemingly troublesome distinction between the kinetic and stoichiometric subspaces in an appendix to Chapter 3. There we emphasized that the stoichiometric subspace is an attribute of a *reaction network alone*, while the more fundamental kinetic subspace is an attribute of a *kinetic system* that might change from one kinetics to another, perhaps as kinetic parameters vary ever so slightly.

However, that appendix ended on a happy note (Section 3.A.3): we asserted that, at least for mass action systems, there is a very large and highly robust family of reaction networks for which there is no distinction between the kinetic and stoichiometric subspaces. To know the stoichiometric subspace for a network in that family is to know the kinetic subspace; the two subspaces coincide *no matter what values the mass action rate constants take*.

At the end of Chapter 3, we did not yet have the language available to describe that congenial reaction network family meaningfully, but now we do. The kinetic and stoichiometric subspaces coincide for any mass action system such that, in the underlying reaction network, no linkage class contains two or more terminal strong-linkage classes. The family contains all weakly reversible networks and, in particular, all reversible networks, but it is *far* broader.

To the extent that a mass action model might exhibit a lack of coincidence of the kinetic and stoichiometric subspaces, it cannot do so robustly: the lack of coincidence will vanish in a "nearby" mass action system in which one or more reactions are made reversible, perhaps with tiny rate constants for the reactions added. That modification is one that chemists would bless.

9

Concentration Robustness and Its Importance in Biology: Some More Deficiency-Oriented Theorems

This chapter is somewhat different from the surrounding ones, in which the interest is largely in network-structural conditions that confer a degree of stable behavior against composition perturbations *within* stoichiometric compatibility classes. The interest here is again in a kind of stability—in this case, precise maintenance of a desired concentration of a critical species—but now against composition perturbations, even very large ones, *across* stoichiometric compatibility classes. This chapter's placement just after the preceding one results from its close connection to deficiency one theory. The connection will become more transparent in Part III, where we consider the common mathematics underlying both chapters.

Here our primary purpose is to elaborate more fully on a robustness theorem reported in [155], which in turn resulted from interesting questions considered in the molecular cell biology laboratory of Professor Uri Alon at the Weizmann Institute of Science [159, 160]. Guy Shinar in particular was instrumental in work that both stimulated and resulted in the various theorems in this chapter.

9.1 A Motivating Thought Experiment

We begin this chapter with consideration of a thought experiment, involving a simple mass action system (9.1) that we had studied earlier. We presume at the outset that the chemistry is operating in a closed well-stirred reactor, the very reactor studied in Example 3.4.7. This time, however, we will think of B as a protein monomer that can combine with itself (reversibly) to form, via $2B \to A$, the protein dimer A. Thus, a molecule A contains two copies of a monomer unit.

$$A \underset{k'}{\overset{k}{\rightleftarrows}} 2B \tag{9.1}$$

The governing differential equations are those shown as (9.2). A phase portrait was depicted first in Figure 3.1 and then again in Figure 5.1. That same drawing appears below as Figure 9.2, along with some embellishments to be discussed shortly.

M. Feinberg, *Foundations of Chemical Reaction Network Theory*,
Applied Mathematical Sciences 202,
https://doi.org/10.1007/978-3-030-03858-8_9

$$\begin{aligned}\dot{c}_A &= -kc_A + k'c_B^2 \\ \dot{c}_B &= 2kc_A - 2k'c_B^2\end{aligned} \tag{9.2}$$

Note that at every mixture composition we have $2\dot{c}_A + \dot{c}_B = 0$, from which it follows that $2c_A(t) + c_B(t) = 2c_A(0) + c_B(0)$ for all t. This is just another way of saying something we said before: composition trajectories reside entirely in stoichiometric compatibility classes for network (9.1).

For the purposes of our thought experiment, however, it will be useful to retell that same story in more physical terms, with special reference to the monomer-dimer interpretation of species A and B. The various stages of the thought experiment we'll consider are depicted schematically in Figure 9.1.

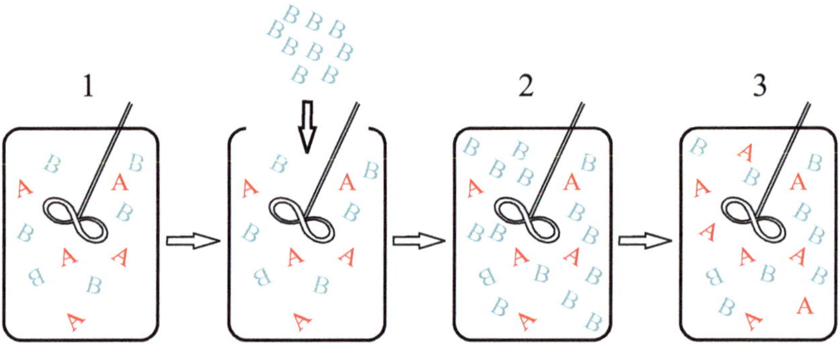

Fig. 9.1. Schematic picture of the thought experiment

When our closed reactor is initially charged, we presumably place within it a certain number of monomer units, some as freestanding molecules of B and some coupled to each other within the dimer A . If we denote by T^* the total concentration of monomer units in one form or another at time zero, simple accounting gives the equation $2c_A(0) + c_B(0) = T^*$. That $2c_A(t) + c_B(t) = T^*$ for all subsequent t results from the fact that the reactions in (9.1) neither create nor destroy monomer units.

In more geometric terms, we can say that all compositions along a particular trajectory (and residing in a particular stoichiometric compatibility class) are consistent with the same particular value of T—that is, the same particular value of total monomer concentration at time zero. In fact, there is precisely one positive stoichiometric compatibility class corresponding to each positive value of T.

Recall that, for the mass action system (9.1), in each positive stoichiometric compatibility class, there is exactly one equilibrium; trajectories beginning in that stoichiometric compatibility class approach its unique equilibrium as time becomes large. See Figure 9.2. It is not difficult to determine that equilibrium concentrations of A and B depend on T according to formulae given in (9.3).

$$c_A = \frac{4\frac{k'}{k}T + 1 - \sqrt{1 + 8\frac{k'}{k}T}}{8\frac{k'}{k}} \qquad c_B = \frac{-1 + \sqrt{1 + 8\frac{k'}{k}T}}{4\frac{k'}{k}} \tag{9.3}$$

Now suppose that, in our thought experiment, the mixture in the reactor chamber has come to the equilibrium labeled "1" in Figure 9.2, the equilibrium corresponding to the initial concentration of monomer units T^*. We now continue the experiment by first opening the reactor chamber and then inserting more *B*—in fact sufficiently much as to elevate the total concentration of monomer units to the substantially higher value T^{**}. For the purposes of the thought experiment, we presume that the newly added *B* is inserted into and mixes with the chamber solution instantaneously.[1]

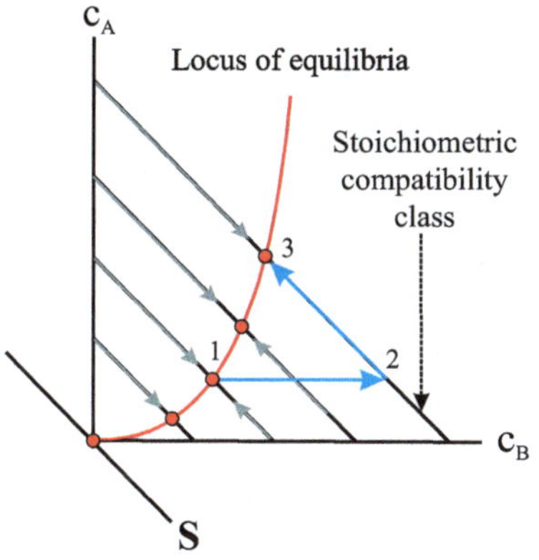

Fig. 9.2. A thought experiment with a mass action system derived from the network $A \rightleftarrows 2B$

In this case, the mixture within the reactor chamber is instantaneously brought to the new, more *B*-rich composition indicated in Figure 9.2 by the point labeled "2," which is not an equilibrium. Indeed, the occurrence of the reactions (9.1) will now cause the composition to move away from "2," along the stoichiometric compatibility class corresponding to total monomer unit concentration T^{**}, and ultimately approach the equilibrium labeled "3"—the sole equilibrium consistent with T^{**}.

Note that at equilibrium "3" the concentrations of both *A* and *B* might be substantially larger than they were at equilibrium "1," in particular if the total concentration of monomer unit is substantially different at the two equilibria.

9.2 A Problem for the Cell

Here we continue our discussion of the monomer-dimer system, but now we suppose that the dimer *A* is a biologically active form of a protein, while the monomer *B* is inactive. We shall also suppose that for a cell to function properly, it is important that the concentration of the active form be maintained within very narrow bounds. (We are considering a cartoon cell, in which concentrations of *A* and *B* are spatially

[1] This presumption is unnecessary to the point we shall make; it only makes for a cleaner picture.

uniform.) If there are periods of dramatic waxing and waning in the cell's total supply of monomer units—perhaps because of the transport of monomer units across the cell membrane or because of relatively slow synthesis of them by transcription and translation reactions—then one can imagine that the concentrations of both A and B would drift substantially:

In the sense of the caricature given in Figures 9.1 and 9.2, variations in the total supply of monomer units would give rise to drift from one stoichiometric compatibility class to another. If that drift is much slower than the occurrence of the reactions $A \rightleftarrows 2B$, one would expect the mixture composition to ride along the locus of equilibria in Figure 9.2, with perhaps considerable excursions upward and downward. In the presence of large, natural fluctuations in the monomer unit supply, then, the cell's critical concentration of A could not be maintained steady.

The fact is that network (9.1), like most networks, is poorly equipped to provide homeostasis in the concentration of a critical species against substantial fluctuations in the overall supply of reacting material. There are, however, reaction networks which, by virtue of their very special structure, can maintain concentration robustness very precisely. They possess architectural features that nature can—and probably does—exploit to do just that.

We turn now to consideration of a toy network that exhibits *absolute* concentration robustness—that is, invariance of the concentration of a crucial species at *all* positive equilibria, even against arbitrarily large fluctuations in the overall supply of reacting material.

9.3 A Toy Network Exhibiting Absolute Concentration Robustness

Here we consider the mass action system (9.4). Again we shall suppose that A and B are different variants of a protein, with A the sole form that plays a certain critical intracellular role. Production of A proceeds directly from B by means of the reaction $B \to A$, but B also serves as a catalyst to convert A back to B via the reaction $A + B \to 2B$. We presume once more that, for proper functioning of the cell, the concentration of A must be maintained at or near a certain fixed value, no matter what might be the *total* concentration of the protein in its two forms.

$$\begin{aligned} A + B &\overset{\alpha}{\to} 2B \\ B &\overset{\beta}{\to} A \end{aligned} \tag{9.4}$$

The differential equations for the (closed-reactor) mass action system (9.4) are shown in (9.5).

$$\begin{aligned} \dot{c}_A &= -\alpha c_A c_B + \beta c_B \\ \dot{c}_B &= \alpha c_A c_B - \beta c_B \end{aligned} \tag{9.5}$$

Clearly, we have $\dot{c}_A + \dot{c}_B = 0$, which implies that $c_A(t) + c_B(t) = c_A(0) + c_B(0)$ for all t. Thus, the total protein concentration, whatever its form, remains what it was initially. As in the preceding example, this is just another way of saying that composition trajectories reside entirely within stoichiometric compatibility classes, this time for network (9.4). There is precisely one stoichiometric compatibility class associated with each value Θ of the total protein concentration, described by the equation

$$c_A + c_B = \Theta. \tag{9.6}$$

From (9.5) and (9.6), we can determine that, at the positive equilibria, the concentrations of A and B are given by

$$c_A = \frac{\beta}{\alpha} \qquad c_B = \Theta - \frac{\beta}{\alpha}. \tag{9.7}$$

There are also nonpositive equilibria characterized by $c_B = 0$.

It is important to note the striking contrast between the positive equilibria given by (9.7) and those given by (9.3) for the monomer-dimer example. There, the equilibrium concentrations of *both* A and B depended on T, the total monomer unit concentration, and, hence, *on the stoichiometric compatibility class in which the equilibrium resides.* Here, the equilibrium concentration of B depends on Θ, the total protein concentration, *but the concentration of the crucial species A does not.* In fact, *at all positive equilibria, the concentrations of A is the same, regardless of the stoichiometric compatibility class.*

A phase portrait is shown in Figure 9.3. Note that there are stoichiometric compatibility classes close to the origin that contain no positive equilibria. These stoichiometric compatibility classes correspond to values of the *total* protein concentration, Θ, that are less than or equal to $\frac{\beta}{\alpha}$, the obligatory concentration of protein A in every positive equilibrium. On the other hand, in *every* stoichiometric compatibility class for which the total protein concentration exceeds $\frac{\beta}{\alpha}$, the mixture composition will, if there is *some* B present, evolve toward an equilibrium in which the concentration of A is exactly $\frac{\beta}{\alpha}$, the presumably happy value the cell requires.

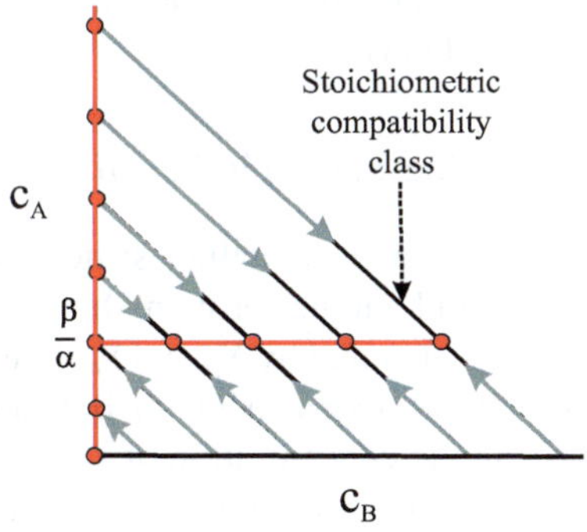

Fig. 9.3. Phase portrait for mass action system (9.4)

So long as the total concentration of protein remains above $\frac{\beta}{\alpha}$, then, even a large fluctuation in the total protein supply, resulting in a large movement from one stoichiometric compatibility class to another, will not result in a change in the positive-equilibrium concentration of A. If the reactions in network (9.5) are fast relative to the time scale of total protein supply fluctuations, the concentration of A will remain very close to $\frac{\beta}{\alpha}$, except perhaps for departures of brief duration, while equilibration is proceeding.

Motivated by these considerations, we say that a kinetic system exhibits *absolute concentration robustness relative to a particular species* if there is at least one positive equilibrium and if the concentration of that species is the same at all positive equilibria, regardless of stoichiometric compatibility class. Thus, the mass action system (9.4) exhibits absolute concentration robustness relative to species A.

Definition 9.3.1. *A kinetic system* $\{\mathscr{S},\mathscr{C},\mathscr{R},\mathscr{K}\}$ *has* **absolute concentration robustness relative to a species** $s \in \mathscr{S}$ *if the system has a positive equilibrium* c^* *and if, for every other positive equilibrium* c^{**}, $c^{**}_s = c^*_s$.

9.4 More Substantial Toy Networks: Biochemical Concentration Robustness Models Inspired by Experiments

Experiments suggest strong concentration robustness in two well-studied biochemical systems discussed below, each involving a bifunctional enzyme. By way of preparation, it should be understood that proteins are often available in multiple guises, corresponding to the number and location of attached *phosphoryl groups*. The function and activity of a protein can depend heavily on its "phosphorylation state."

Certain enzymes (kinases) catalyze the addition of phosphoryl groups to particular proteins, while other enzymes (phosphatases) catalyze removal of them. Kinases and phosphatases are themselves proteins, whose activity (or lack of it) can depend on *their* phosphorylation states. In the cell, adenosine triphosphate (ATP), *usually present in very large supply*, is the principle donor of phosphoryl groups to proteins. In the process, an ATP molecule becomes a molecule of ADP (adenosine diphosphate). ADP can also be a phosphoryl donor.

9.4.1 The EnvZ-OmpR Signaling System in *E. coli*

The EnvZ-OmpR is a two-component signaling system, consisting of the kinase EnvZ, denoted hereafter by X, and another protein OmpR, denoted by Y. Both proteins have phosphorylated forms, denoted by X_p and Y_p. Even against large variations in the supplies of EnvZ and OmpR, experiments of Batchelor and Goulian [17, 96] suggest a strong homeostasis in the concentration of Y_p.

In an attempt to capture the concentration robustness observed for Y_p, while also taking account of other observations about EnvZ-OmpR, Shinar et al. [159] proposed the mass action reaction network model displayed in (9.8). There T denotes ATP, and $[T]$ denotes the presumed time-invariant concentration of ATP. (Recall our discussion of the Edelstein model (Section 4.2.3), in which the concentrations of species available in large supply were subsumed within rate constants.) The first line of (9.8) represents a reversible binding of ATP to X, followed by phosphorylation of X.[2] The second line represents a binding of Y to phosphorylated X, followed by a transfer of the phosphoryl group from X to Y. The third line indicates that XT acts as a phosphatase that removes the phosphoryl group from Y_p.

$$
\begin{aligned}
&X \underset{k_{-1}}{\overset{k_1[T]}{\rightleftarrows}} XT \overset{k_2}{\rightarrow} X_p \\
&X_p + Y \underset{k_{-3}}{\overset{k_3}{\rightleftarrows}} X_pY \overset{k_4}{\rightarrow} X + Y_p \\
&XT + Y_p \underset{k_{-5}}{\overset{k_5}{\rightleftarrows}} XTY_p \overset{k_6}{\rightarrow} XT + Y
\end{aligned}
\tag{9.8}
$$

Note that these reactions preserve both the number of X-containing molecules and the number of Y-containing molecules. Consequently, along a composition trajectory for the mass action system (9.8), the total concentrations of X- and Y-containing molecules, denoted Θ_X and Θ_Y, are invariant. Specification of Θ_X and Θ_Y serves to characterize a unique stoichiometric compatibility class containing all trajectories consistent with that Θ_X, Θ_Y pair. Indeed, there is a bijective correspondence between the positive stoichiometric compatibility classes and the pairs of positive values for Θ_X and Θ_Y. (Each stoichiometric compatibility class resides in a parallel of the five-dimensional stoichiometric subspace for network (9.8).)

There is a wealth of positive equilibria for the mass action system (9.8), distributed across positive stoichiometric compatibility classes for which Θ_Y exceeds a certain critical value, discussed below. (In every stoichiometric compatibility class containing a positive equilibrium, there is precisely one such equilibrium.) Apart from one exceptional species, the individual species concentrations at the various positive equilibria depend upon the stoichiometric compatibility class in which each particular equilibrium resides, which is to say that there is a dependence on Θ_X and Θ_Y.

The exceptional species is Y_p. *At every positive equilibrium, the concentration of Y_p is the same, regardless of stoichiometric compatibility class [159].* That concentration is given by (9.9), in which there is no dependence at all on Θ_X or Θ_Y, nor is there a dependence on the presumed-constant value of $[T]$, the ATP concentration. *The mass action system* (9.8) *exhibits absolute concentration robustness relative to Y_p.*

$$
c^*_{Y_p} = \frac{k_2\,(k_{-5} + k_6)}{k_5 k_6}
\tag{9.9}
$$

Note that there can be no positive equilibrium in any stoichiometric compatibility class for which $\Theta_Y \le c^*_{Y_p}$: In any positive equilibrium, the concentration of Y_p is $c^*_{Y_p}$, which clearly cannot exceed or equal Θ_Y, the total concentration of all Y-containing species.

---

[2] The phosphorylation reaction produces not only X_p but also ADP, which is not displayed; like ATP it is considered to be present in very large supply.

9.4.2 The IDHKP-IDH Glyoxylate Bypass Regulation System

In *E. coli* the distribution of carbon flux between the tricarboxylic acid (TCA) cycle and the glyoxylate bypass is controlled by the so-called IDHKP-IDH glyoxylate bypass regulation system. Regulation is achieved by control of the phosphorylation state of the TCA cycle enzyme isocitrate dehydrogenase (IDH). We denote the unphosphorylated version of IDH by I and its phosphorylated version by I_p. The concentration of the unphosphorylated form, I, determines how much of the carbon flux flows through TCA cycle and how much through the glyoxylate bypass. Experiments by LaPorte et al. [121] suggest strong robustness in the concentration of I, the unphosphorylated form of IDH, against large variations in the total supply of IDH.

Central to both the phosphorylation and dephosphorylation of IDH is the enzyme isocitrate dehydrogenase kinase phosphatase (IDHKP), which we denote by E. At the heart of several models by Shinar et al. [160] for the phosphorylation-dephosphorylation interplay of IDH and IDHKP lies a core mass action system shown in (9.10). Here again, [T] indicates the presumed-constant concentration of ATP.

$$\begin{aligned} EI_p + I &\underset{k_{-1}}{\overset{k_1}{\rightleftarrows}} EI_pI \overset{k_2[T]}{\rightarrow} EI_p + I_p \\ E + I_p &\underset{k_{-3}}{\overset{k_3}{\rightleftarrows}} EI_p \overset{k_4}{\rightarrow} E + I \end{aligned} \tag{9.10}$$

There is an abundance of positive equilibria for (9.10), spread over a variety of stoichiometric compatibility classes, corresponding to different total concentrations of IDH and IDHKP proteins. In all positive equilibria, however, the concentration of I takes the same value, given by (9.11)—a value that is independent of the total IDH and IDHKP levels associated with the various stoichiometric compatibility classes. *That is, the mass action system* (9.10) *exhibits absolute concentration robustness relative to I.* As a model, then, (9.10) provides a caricature of the strong concentration robustness suggested by experiments described in [121].

$$c_I^* = \frac{k_4\,(k_{-1} + k_2[T])}{k_1 k_2[T]} \tag{9.11}$$

9.5 A Theorem: Structural Sources of Absolute Concentration Robustness

The literature mass action models for both the EnvZ-OmpR and IDHKP-IDH systems were guided by what was known empirically about each, not by an overarching theory about which reaction network architectures give rise to concentration robustness. The two networks have somewhat vague structural features in common, but they also have some clear differences, especially from a mechanistic biochemical standpoint. What exactly, then, unites the two models in their ability to give the precise concentration robustness that nature seems to require?

An answer is provided by a theorem that can be stated once we have a small amount of new vocabulary at our disposal. We say that *two complexes y' and y in a reaction network differ in only species $\mathscr{s}$ if there is a nonzero number α such that $y' - y = \alpha \mathscr{s}$.* For example, the complexes $A+B$ and A differ only in species B because $A+B-A = 1B$. The complexes $3A+B$ and $A+B$ differs only in species A because $3A+B-(A+B) = 2A$. The complexes $2A$ and A differ only in species A because $2A - A = 1A$. Finally, remember that a *nonterminal complex* is one that does not reside in a terminal strong-linkage class (Section 6.6).

The following theorem appeared in [155]:

Theorem 9.5.1 (Deficiency One Concentration Robustness Theorem). *Consider a mass action system that admits a positive equilibrium, and suppose that the deficiency of the underlying reaction network is one. If, in the network, there are two nonterminal complexes that differ only in species $\mathscr{s}$, then the system has absolute concentration robustness relative to $\mathscr{s}$.*

Despite the somewhat disparate appearances of the three robustness examples considered so far, Theorem 9.5.1 homes in on the subtle structural features that unite them—features that ensure in each example constancy in the concentration of a critical species over all positive equilibria, regardless of stoichiometric compatibility class. We revisit the examples in light shed by the theorem:

Example 9.5.2 (The toy example). Network (9.4) has a deficiency of one ($n = 4$, $\ell = 2$, $s = 1$). The two complexes $A+B$ and B are nonterminal, and they differ only in species A. As (9.7) indicates, the equilibrium concentration of A depends only on the two rate constants and—in contrast to the equilibrium concentration of B—not at all on the total monomer concentration, Θ. This is to say that the concentration of A is identical across all positive equilibria, regardless of which stoichiometric compatibility class a particular equilibrium might reside.

Example 9.5.3 (EnvZ-OmpR). The deficiency of network (9.8) is one ($n = 9$, $\ell = 3$, $s = 5$). The nonterminal complexes $XT + Y_p$ and XT differ only in species Y_p. The theorem tells us that, in all positive equilibria, regardless of stoichiometric compatibility class, the concentration of Y_p should be the same. Indeed, (9.9) tells us that the equilibrium concentration of Y_p depends only on *certain* rate constant values and not at all on Θ_X or Θ_Y, the total concentrations of the proteins X and Y, which in turn characterize the stoichiometric compatibility classes. Later in this chapter, we shall have interest in the fact that the equilibrium concentration of Y_p depends on some rate constants but not on others.

Example 9.5.4 (IDHKP-IDH). Network (9.8) has a deficiency of one ($n = 6$, $\ell = 2$, $s = 3$). The nonterminal complexes $EI_p + I$ and EI_p differ only in I, so Theorem 9.5.1 tells us that its concentration should be invariant over all positive equilibria. In fact, the concentration of I in positive equilibria, given by (9.11), depends only on rate constant values and not on the stoichiometric compatibility class in which a particular equilibrium might reside. As with EnvZ-OmpR we shall have interest in the fact that only *certain* rate constant values affect the result.

Remark 9.5.5 (Very strong but imperfect concentration robustness). Each of the examples considered made free use of irreversible reactions. If some of the reactions (and certainly if all of the reactions) were made reversible, Theorem 9.5.1 might stand silent, for then formerly nonterminal complexes could become terminal. In such cases, absolute concentration robustness might be lost. In fact, though, it is unreasonable to expect *absolute* concentration robustness in nature. The robustness inferred for IDHKP-IDH and EnvZ-OmpR from experiments was approximate, not absolute. *Strict* constancy in the equilibrium concentration of a certain species against *all* perturbations of stoichiometric compatibility class, *however large*, is far too much to ask. That certain important biochemical systems exhibit even rough constancy against moderately large perturbations is already striking.

Note that if irreversible reactions in the examples were made reversible, but with sufficiently small rate constants for the reactions added, we would expect the resulting mass action systems to exhibit pronounced but *approximate* concentration robustness over a broad swath of stoichiometric compatibility classes. That is, if rate constants for the added reactions are sufficiently small, we would expect that, over a fairly broad range of stoichiometric compatibility classes, the equilibrium concentration of the critical species would change only very slightly. (For an analysis of specific cases, see the supplementary material provided with [155].)

It is in this sense that mass action systems satisfying the conditions of Theorem 9.5.1 should be understood: *They serve as absolute-concentration-robustness structural skeletons upon which more fully fleshed-out mass action systems can be built, systems that support concentration robustness that is imperfect but nevertheless strong.*

Remark 9.5.6 (The deficiency one requirement cannot be dropped). Theorem 9.5.1 becomes false if the deficiency one requirement is omitted. Consider, for example, the two networks shown below. Both have a deficiency of two, and, in each case, the nonterminal complexes $A+B$ and B differ only in species A. Taken with mass action kinetics, network (9.12) exhibits absolute concentration robustness relative to A, but network (9.13) does not. (Both admit positive equilibria regardless of rate constant values.)

$$\begin{aligned} A+B &\to 2B \\ B &\to A \\ 2A+B &\to 3B \end{aligned} \tag{9.12}$$

$$\begin{aligned} A+B &\to 2B \\ B &\to A \\ A+2B &\to 3B \end{aligned} \tag{9.13}$$

We can also say the following: No mass action system in which the underlying network is conservative and has a deficiency of zero can exhibit absolute concentration robustness relative to any species. This is a consequence of Theorem 9.7.1, which we shall meet later in this chapter.

Remark 9.5.7 (Absolute concentration robustness for networks that do not satisfy the hypothesis of Theorem 9.5.1). There are mass action systems (not necessarily deficiency one) that exhibit concentration robustness in a particular species even when, in the underlying network, there do not exist two nonterminal complexes that differ only in that species. One such surprising deficiency one example, due to Germán Enciso, is a mass action system having as its underlying network the one shown in (9.14). In this case, there is absolute concentration robustness in species C! The supplementary material provided with [155] contained a proof of a theorem that is broader than Theorem 9.5.1 and that also accounts for the robustness exhibited by (9.14). The idea behind the theorem is discussed in Chapter 18.

$$\begin{aligned} &A + 2B \rightarrow 3B \\ &B \rightarrow A \\ &A + B \rightleftarrows C. \end{aligned} \tag{9.14}$$

9.6 How the Robust Concentration Depends on Rate Constants: Implications for Biology

In the preceding section, we took note of the fact that, in two of the examples, the robust concentration depended on certain mass action rate constants but not on others. In particular, for the EnvZ-OmpR mass action model (9.8), the equilibrium concentration of Y_p, given by (9.16), did not depend on the rate constant $k_1[T]$. Recall that $[T]$ denotes the cellular concentration of the phosphoryl-group donor ATP. Recall too that, relative to species appearing explicitly in the network, ATP is presumed available in sufficiently large supply that its concentration can, for practical purposes, be regarded fixed, at least on a moderately small time scale. Whatever [T] might be, $c^*_{Y_p}$, given by (9.16), is independent of it.

Now consider the also-plausible mass action description of EnvZ-OmpR given by (9.15). In this case two different phosphoryl donors, ATP (denoted T) and ADP (denoted D), bind reversibly to the kinase X. As before, it is XT that gives rise to X_p, the phosphorylated form of X. Now, however, it is XD that acts as the phosphatase that removes a phosphoryl group from Y_p. In (9.15), $[D]$ and $[T]$ denote the presumed-constant concentrations of ADP and ATP, which are deemed to be available in large supply.

$$
\begin{aligned}
&XD \underset{k_9[D]}{\overset{k_{-9}}{\rightleftarrows}} X \underset{k_{-1}}{\overset{k_1[T]}{\rightleftarrows}} XT \xrightarrow{k_2} X_p \\
&X_p + Y \underset{k_{-3}}{\overset{k_3}{\rightleftarrows}} X_pY \xrightarrow{k_4} X + Y_p \\
&XD + Y_p \underset{k_{-7}}{\overset{k_7}{\rightleftarrows}} XDY_p \xrightarrow{k_8} XD + Y
\end{aligned}
\tag{9.15}
$$

The deficiency of the network is one ($n = 10, \ell = 3, s = 6$), and there are two nonterminal complexes, $XD + Y_p$ and XD, that differ only in Y_p. Thus, from Theorem 9.5.1 we again have invariance of the concentration of Y_p across all positive equilibria, regardless of stoichiometric compatibility class—that is, regardless of the total concentrations of X- and Y-containing proteins. This time the equilibrium concentration of Y_p is given [155] by (9.16).

$$
c^*_{Y_p} = \frac{k_2}{k_{-1} + k_2} \frac{k_{-7} + k_8}{k_7} \frac{k_{-9}}{k_8} \frac{k_1}{k_9} \frac{[T]}{[D]}. \tag{9.16}
$$

Again there is a dependence on some rate constants and not on others, but this time there is an important difference: Although $c^*_{Y_p}$ given in (9.9) for the mass action system (9.8) was independent of $[T]$ (and of course $[D]$), the equilibrium concentration of Y_p, given by (9.16) for the mass action system (9.15), now depends on both $[T]$ and $[D]$, through their ratio.

Were nature to select one mass action system over another to ensure that the equilibrium concentration of Y_p remain constant across a wide range of stoichiometric compatibility classes, it would seem that our earlier one, (9.8), is to be preferred. The architecture of (9.8) shields the equilibrium concentration of Y_p against long-term drifts (or cell-to-cell differences) in the relatively large but nevertheless variable concentrations of ATP and ADP.

The contrast between the two designs teaches a lesson: there can be important strategic differences between seemingly similar network architectures that implement the same concentration robustness objective, differences that hinge on the particular set of rate constants that actually affect the result. There is something more that should be said: It has always been our presumption that the systems under study are isothermal, that at least in the short run temperature can be regarded time-invariant. Temperatures do drift, however, even in biological systems, and rate "constants" can and do depend on the operating temperature, some with marked sensitivity. It would seem, then, that concentration robustness designs in nature would favor networks in which the robust concentration is largely independent of rate constants that are highly temperature-sensitive.

These considerations prompt a question: *If we restrict attention to mass action systems that satisfy the conditions of Theorem 9.5.1, how does reaction network structure dictate even further which particular rate constants do or do not influence the final robust species concentration.* We consider this question in the next two sections, with proofs provided in Part III. The answers are surprisingly subtle.

9.6.1 First Theorem on Rate Constant Independence

To begin, it will be helpful if we review a little vocabulary from Chapter 6, this time less formally: We say that the reaction(s) directly connecting complexes y and y' constitute a *cut-link* if, after removal of the arrow(s) joining y and y', there is no longer a (not necessarily directed) sequence of reaction arrows that connects y and y'. For example, in network (9.17), the reaction $B \to C$ is a cut-link, but the same reaction in network (9.18) is not a cut-link. In network (9.18), $B \rightleftarrows D$ is not a cut-link, but $C \to A$ is a cut-link.

$$\begin{array}{c} B \;\to\; C \;\rightleftarrows\; A \\ A + B \to 2B \end{array} \tag{9.17}$$

$$\begin{array}{c} \begin{array}{ccc} B & \to & C \to A \\ & \nwarrow\!\!\searrow \quad \swarrow & \\ & D & \end{array} \\ \\ A + B \to 2B \end{array} \tag{9.18}$$

When we refer to the reaction network obtained by removing reaction(s) of a cut-link, we mean the reaction network obtained by deleting the reaction(s) of the cut-link and any complex which, after that deletion, is not a participant in any reaction. Thus, for example, if we remove the cut-link $C \rightleftarrows A$ from network (9.17), we obtain network (9.19).

$$\begin{array}{c} B \to C \\ A + B \to 2B \end{array} \tag{9.19}$$

Theorem 9.6.1. *Consider a mass action system that satisfies all the conditions of Theorem 9.5.1, in particular the presence of nonterminal complexes y and y' that differ only in species $\mathscr{s}$. Suppose that removal of reaction(s) in a cut-link directly connecting complexes $\bar{y}$ and $\bar{y}'$ results in a network that again has y and y' as non-terminal complexes. Then, for the original mass action system, the concentration of $\mathscr{s}$ is not only identical in all positive equilibria, it is also independent of the rate constant(s) for reaction(s) directly connecting complexes $\bar{y}$ and $\bar{y}'$.*

Remark 9.6.2. In Theorem 9.6.1, proved in Section 18.2, independence with respect to rate constants is meant to be understood in the following sense: Consider a second mass action system that admits a positive equilibrium and that differs from the mass action system of Theorem 9.6.1 at most in the value(s) of the rate constant(s) for reaction(s) directly connecting complexes $\bar{y}$ and $\bar{y}'$ of the theorem statement. When the conditions of Theorem 9.6.1 are satisfied, the concentration of species $\mathscr{s}$ is the same in all positive equilibria of both systems.

Example 9.6.3. Consider the mass action system (9.20) with rate constants as indicated. Note that the underlying deficiency one network satisfies the conditions of

Theorem 9.5.1. In particular the two nonterminal complexes $A+B$ and B differ only in species A, so there is absolute concentration robustness in A.

$$B \overset{\alpha}{\rightarrow} C \underset{\xi}{\overset{\gamma}{\rightleftarrows}} A \tag{9.20}$$

$$A+B \overset{\beta}{\rightarrow} 2B$$

If we remove reactions of the cut-link $C \rightleftarrows A$, the resulting network is the one already displayed as (9.19), in which the complexes $A+B$ and B are again nonterminal and differ only in species A. In this case, Theorem 9.6.1 tells us that c_A^*, the equilibrium concentration of A for (9.20), cannot depend on the rate constants γ or ξ. In fact, it is easy to determine that $c_A^* = \frac{\alpha}{\beta}$.

Suppose, on the other hand, that we remove from network (9.20) the cut-link $B \rightarrow C$. The result is network (9.21). In this case, there are no longer two nonterminal complexes that differ only in species A, and the theorem stands silent. In particular, we cannot assert that c_A^* is independent of α, and, indeed, it is not.

$$C \rightleftarrows A \tag{9.21}$$

$$A+B \rightarrow 2B$$

Example 9.6.4. Consider the mass action system (9.22), which is a variant of the system considered in Example 9.6.3. Here again the underlying deficiency one network satisfies the conditions of Theorem 9.5.1. In particular, the nonterminal complexes $A+B$ and B differ only in species A, so there is absolute concentration robustness in A.

$$B \underset{\xi}{\overset{\alpha}{\rightleftarrows}} C \overset{\gamma}{\rightarrow} A \tag{9.22}$$

$$A+B \overset{\beta}{\rightarrow} 2B$$

This time, however, there is no cut-link that satisfies the requirements of Theorem 9.6.1. For example, removal of the cut-link $C \rightarrow A$ results in network (9.23). Although complexes $A+B$ and B again differ only in species A, in network (9.23) the complex B is not nonterminal. Thus, Theorem 9.6.1 does not permit us to assert that c_A^*, the (robust) concentration of A in every positive equilibrium of system (9.22), is independent of α and ξ. In fact, $c_A^* = \frac{\alpha\gamma}{\beta(\gamma+\xi)}$.

$$B \rightleftarrows C \tag{9.23}$$

$$A+B \rightarrow 2B$$

Example 9.6.5 (The EnvZ-OmpR models revisited). In light shed by Theorem 9.6.1, we can now re-examine the rather different rate constant dependencies in the two EnvZ-OmpR models considered earlier.

Note that if we remove the cut-link $X \rightleftarrows XT$ from network (9.8), we still have two nonterminal complexes, XT and $XT + Y_p$, that differ only in the critical species Y_p. Thus, Theorem 9.6.1 tells us that the robust equilibrium concentration of Y_p should be independent of the rate constants $k_1[T]$ and k_{-1}. This is confirmed by (9.9). Similarly, Theorem 9.6.1 predicts that, for the mass action system (9.8), $c^*_{Y_p}$ should not depend on k_3, k_{-3}, or k_4. It does not.

The situation is very different for the EnvZ-OmpR variant given by (9.15). Removal of the cut-link $X \rightleftarrows XT$ results in a network that still has two complexes, XD and $XD + Y_p$, that differ only in Y_p, *but in this case XD is terminal*. Theorem 9.6.1 is silent, and, indeed, $c^*_{Y_p}$ given by (9.16) *does* depend on both $k_1[T]$ and k_{-1}, the rate constants for $X \rightleftarrows XT$. Removal of the cut-link $X \rightleftarrows XD$ results in a network having no pair of complexes differing only in Y_p. Theorem 9.6.1 is again silent, this time for a different reason. Equation (9.16) indicates that $c^*_{Y_p}$ depends on both $k_9[D]$ and k_{-9}.

9.6.2 Second Theorem on Rate Constant Independence

In Theorem 9.6.1 we cannot drop the requirement that the reactions connecting complexes y and y' be a cut-link. The mass action system (9.24) not only provides a counterexample, it also indicates how subtle the situation for non-cut-link reactions can be. A discussion of that system (and a *seemingly* slight variant of it) will provide motivation for our second rate-constant-independence theorem.

$$\begin{array}{ccccc} B & & \overset{\alpha}{\rightarrow} & & C \overset{\gamma}{\rightarrow} A \\ & \rho \searrow\!\!\nwarrow \tau & & \swarrow \varepsilon & \\ & & D & & \\ & & & & \\ & A+B \overset{\beta}{\rightarrow} 2B & & & \end{array} \tag{9.24}$$

For the mass action system (9.24), Theorem 9.5.1 ensures absolute concentration robustness relative to species A. In particular, in the underlying deficiency one reaction network, there are two nonterminal complexes, $A+B$ and B, that differ only in species A. In fact, in all positive equilibria, the concentration of A has the same value, given by (9.25).

$$c^*_A = \frac{\alpha\gamma}{\beta(\gamma+\varepsilon)} \tag{9.25}$$

Note that c^*_A depends explicitly on ε, the rate constant for $C \rightarrow D$, despite the fact that removal of $C \rightarrow D$ results in network (9.26), which satisfies the removal requirement of Theorem 9.6.1; that is, the nonterminal complexes $A+B$ and B differ only in species A. In this case, however, $C \rightarrow D$ is not a cut-link, so the theorem does not preclude the dependence of c^*_A on ε. This tells us that Theorem 9.6.1 becomes false if, in its statement, we simply remove the requirement that reactions connecting y and y' be a cut-link.

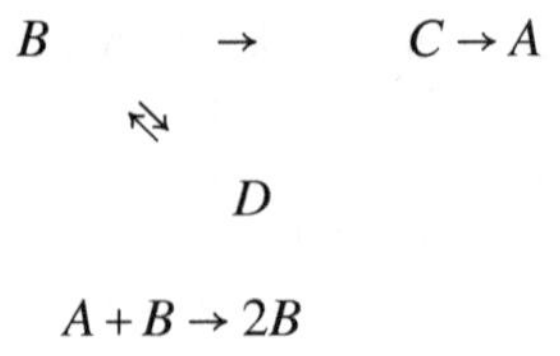

(9.26)

The situation for reactions $B \rightleftarrows D$ in (9.24) is surprisingly different. Removal of those reactions from the original network (9.24) results in a network (9.27), in which there are again the nonterminal complexes $A+B$ and B that differ only in species A. Despite the fact that network (9.27) meets the requirements of Theorem 9.6.1, the theorem itself again remains silent because, like $C \to D$, the reactions $D \rightleftarrows B$ do not constitute a cut-link. Yet, in this case the conclusion of the theorem nevertheless obtains: *values of the rate constants* ρ *and* τ *associated with* $B \rightleftarrows D$ *exert no influence at all on the value of* c_A^* *given by* (9.25).

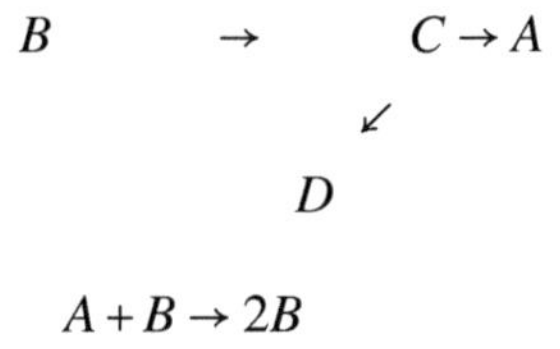

(9.27)

The qualitatively different outcomes for the reactions $C \to D$ and $B \rightleftarrows D$ indicate that there can be a considerable subtlety in the influence of rate constants for non-cut-link reactions on the equilibrium concentration of the robust species. Even more subtlety is indicated by consideration of the mass action system (9.28), which differs from (9.24) only in the addition of the reverse reaction $D \to C$. The conditions of Theorem 9.5.1 are satisfied, and we again have absolute concentration robustness relative to species A.

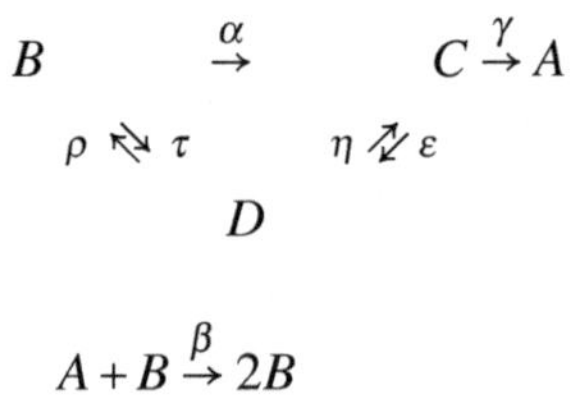

(9.28)

Recall that, for the original mass action system (9.24), c_A^* was independent of ρ and τ, the rate constants for reactions $B \rightleftarrows D$. Although we might expect this independence to persist after the seemingly innocuous addition of the *reverse* of a *different* reaction, now c_A^* depends explicitly on the rate constants of *all (!)* reactions, including ρ and τ, through the formula (9.29).

$$c_A^* = \frac{(\alpha\rho + \alpha\eta + \eta\tau)\gamma}{\beta(\gamma\rho + \gamma\eta + \varepsilon\rho)} \tag{9.29}$$

The next theorem, proved in Section 18.3, brings coherence to these vexing examples.

Theorem 9.6.6. *Consider a mass action system that satisfies all the conditions of Theorem 9.5.1, in particular the presence of nonterminal complexes y and y' that differ only in species $\mathscr{s}$. Suppose that removal of reaction(s) directly connecting complexes $\bar{y}$ and $\bar{y}'$ results in a network that (i) again has y and y' as nonterminal complexes and (ii) has more terminal strong-linkage classes than the original network. Then, for the original mass action system, the concentration of $\mathscr{s}$ is not only identical in all positive steady states, it is also independent of the rate constant(s) for reaction(s) directly connecting complexes $\bar{y}$ and $\bar{y}'$.*

In light shed by the theorem, we can now return to consideration of the mass action system (9.24), for which c_A^* was given by equation (9.25). Note that removal of the reactions $D \rightleftarrows B$ resulted in network (9.27), in which there is one more terminal strong-linkage class than there was in the original network (9.24). In this case Theorem 9.6.6 indicates correctly that c_A^* should be independent of ρ and τ. On the other hand, removal of $C \to D$ from network (9.24) resulted in network (9.26), in which the number of terminal strong-linkage classes is identical to the number of terminal strong-linkage classes in network (9.24). In this case, Theorem 9.6.6 does not preclude dependence of c_A^* on ε.

We can also reconsider the mass action system (9.28) in light of Theorem 9.6.6. Despite the close similarity of (9.28) to the mass action system (9.24), the situation for reactions $D \rightleftarrows B$ is now different. Removal of them from network (9.28) results in the network (9.30), in which the number of terminal strong-linkage classes is the same as in the original network. In this case Theorem 9.6.6 is silent, and dependence of c_A^* on ρ and τ is not precluded.

$$
\begin{array}{ccc}
B & \rightarrow & C \rightarrow A \\
 & & \nearrow\!\!\swarrow \\
 & D & \\
 & & \\
A + B \rightarrow 2B & &
\end{array}
\tag{9.30}
$$

9.6.3 A Useful but Less Interesting Observation About Rate Constant Independence

Consider the mass action system (9.31). Comparing the underlying reaction network here with the highly similar one in the mass action system (9.28), we can see that the only difference between the two lies in a permutation of complexes B and C.

$$
\begin{array}{ccc}
C & \overset{\gamma}{\rightarrow} & B \overset{\alpha}{\rightarrow} A \\
\rho \nwarrow\!\!\searrow \tau & & \eta \nearrow\!\!\swarrow \varepsilon \\
 & D & \\
 & & \\
A + B \overset{\beta}{\rightarrow} 2B & &
\end{array}
\tag{9.31}
$$

In both deficiency one networks, there are the same two nonterminal complexes, $A+B$ and B, that differ only in species A. Thus, in each mass action system, we have invariance of the concentration of A across all positive equilibria, regardless of stoichiometric compatibility class. While for mass action system (9.28) the dependence of that concentration on rate constants, given by (9.29), was complicated and involved *all* of the rate constants, the situation for the mass action system (9.31) is very different:

$$c_A^* = \frac{\alpha}{\beta}. \tag{9.32}$$

Neither Theorem 9.6.1 nor Theorem 9.6.6 accounts for the great simplicity of (9.32), in particular the dependence of c_A^* on α and β *alone*.

In this case, the explanation is a shallow one: The network in (9.31) can be decomposed into two *independent* subnetworks (Appendix 6.A):

$$\begin{array}{c} A+B \rightarrow 2B \\ B \rightarrow A \end{array} \tag{9.33}$$

and

$$\begin{array}{ccc} C & \rightarrow & B \\ & \searrow\!\!\!\nwarrow \quad \nearrow\!\!\!\swarrow & \\ & D. & \end{array} \tag{9.34}$$

Any equilibrium for the mass action system (9.31) must in fact be an equilibrium for each of the smaller mass action systems *considered separately*, those corresponding to the independent subnetworks (9.33) and (9.34). (See Appendix 6.A.) From this follows the simple equation (9.32) for the equilibrium value of c_A^*. The rate constants associated with the complementary subnetwork (9.34) have no bearing on the result. Simple considerations such as these, suitably generalized, are useful, but they do not rise to the level of a theorem.

It is worth noting that a slightly augmented version of the mass action system (9.31), containing the single addition

$$D \xrightarrow{\theta} A,$$

again satisfies the conditions of Theorem 9.5.1 and again exhibits absolute concentration robustness relative to species A. In this case, however, the robust concentration of A has a complicated dependence on *all* rate constant values. The seemingly minor addition of a single reaction prevents the formerly congenial decomposition into independent subnetworks.

9.7 Reaction Network Architectures That Thwart Concentration Robustness

Theorem 9.5.1 described a class of reaction networks that exhibit *absolute* concentration robustness for each choice of rate constants consistent with the existence of positive equilibria. This is to say that, for any such rate constant assignment, the concentration of a critical species is *exactly* the same in *all* positive equilibria across *all* stoichiometric compatibility classes.

In Remark 9.5.5 we pointed out that a mass action system that satisfies the hypothesis of Theorem 9.5.1 will inherently contain reactions that are not reversible. However, we also suggested that "nearby" reversible mass action systems, containing newly added reverse reactions with extremely small rate constants, would give rise to strong but imperfect concentration robustness relative to the critical species. That is, if the rate constants for the added reactions are sufficiently small, then, over some wide range of stoichiometric compatibility classes, the equilibrium concentration of the critical species should change very little from one stoichiometric compatibility class to another.

By way of contrast, we describe in this section large common classes of reaction networks which, when taken with mass action kinetics, actually *thwart* concentration robustness *no matter how rate constants are assigned.* For such networks, the positive-equilibrium concentration of each species *must* change with a *pronounced rapidity* with every shift from one stoichiometric compatibility class to another nearby one—a change quantifiable by means of lower bounds on certain derivatives. *These lower bounds depend only on reaction network structure; they do not depend on rate constants, nor do they depend on the particular stoichiometric compatibility classes under consideration.* For a network in this family, absolute concentration robustness is not only impossible, *it cannot even be approached arbitrarily closely.*

Here we provide only a glimpse at a far richer story, told in [154, 156]. We'll begin with a theorem (proved in Part III, Section 18.4) and then indicate one way in which the theorem can be sharpened considerably.

9.7.1 First Robustness-Thwarting Theorem

Theorem 9.7.1. *Consider a mass action system that admits a positive equilibrium. Moreover, suppose that the underlying reaction network is conservative and satisfies at least one of the following conditions:*

(i) The network has a deficiency of zero.
(ii) The network satisfies the conditions of the Deficiency One Theorem.
(iii) The network is reversible and star-like.

Then there is no species with respect to which the mass action system exhibits absolute concentration robustness.

Remark 9.7.2. Theorem 9.7.1 becomes false if the requirement that the underlying network be conservative is omitted. For the mass action system

$$A + B \underset{1}{\overset{1}{\rightleftarrows}} A \tag{9.35}$$

the set of positive equilibria consists of all compositions for which c_A is positive and $c_B = 1$. Thus, there is absolute concentration robustness relative to B, for its concentration is the same in all positive equilibria. Although the underlying reaction network has a deficiency of zero, it is not conservative.

Remark 9.7.3. As shown in [154] and in Chapter 18, the absence of absolute concentration robustness actually obtains for *any* kinetic system, not necessarily mass action, that is *quasi-thermostatic* (Sections 7.7.1 and 13.3), so long as the underlying reaction network is conservative. The hypothesis of Theorem 9.7.1 merely serves to ensure that the mass action system under consideration is quasi-thermostatic.

9.7.2 Second Robustness-Thwarting Theorem: Architectures That Deny Even Approximate Concentration Robustness

In [160] it was shown that, when the hypothesis of Theorem 9.7.1 is satisfied,[3] not only is absolute concentration robustness impossible, there is a concrete sense in which absolute concentration robustness cannot even be approximated arbitrarily closely. Although the story told in [160] is far more general, it will be useful to circumscribe the class of networks considered here so that the essence of that story can be told in simpler, more tangible terms.

To describe the class of networks on which we'll focus, it will be useful to begin with an example, network (9.36). Think of A, B, and C as monomers from which the larger molecules A_2, A_2B, and A_2BC are built. In this case, we can regard the full species set to be partitioned into a set of "elements" $\{A,\ B,\ C\}$ and a set of "compounds" $\{A_2,\ A_2B,\ A_2BC\}$.[4] In (9.36) each reversible reaction produces a new compound that is not present in any other reversible reaction pair.

$$2A \rightleftarrows A_2 \qquad A_2 + B \rightleftarrows A_2B \qquad A_2B + C \rightleftarrows A_2BC \tag{9.36}$$

Network (9.36) is an example of what we shall call a *constructive* network—roughly, a network in which the species set contains certain distinguished members (elements) and reactions that serve to construct the remaining species (compounds) from the elements. We do not insist that an element appear explicitly in every reaction. For example, were (9.36) supplemented with the additional reaction pair $A_2B + A_2BC \rightleftarrows A_4B_2C$, in which the new species A_4B_2C is formed, the augmented network would also be constructive.

Note that the stoichiometric subspace for network (9.36) has the obvious basis

$$\{A_2 - 2A,\ A_2B - A_2 - B,\ A_2BC - A_2B - C\},$$

but it also has the more suggestive basis

---

[3] The results in [160] obtain for a general kinetic system (not necessarily mass action) that satisfies two conditions: it is quasi-thermostatic (Sections 7.7.1 and 13.3) and the underlying reaction network is conservative. As we shall see in Part III, the hypothesis of Theorem 9.7.1 ensures that these conditions are met.

[4] The "elements" should not be construed to be elements of the periodic table. Rather, they are fundamental species of the network from which other compound species are built.

$$\{A_2 - 2A,\ A_2B - (2A + B),\ A_2BC - (2A + B + C)\}. \tag{9.37}$$

In this basis each member—precisely one for each compound—can be regarded as a "reaction vector" corresponding to a "genesis reaction" that builds a particular compound *directly* and solely from the elements of which it is composed. For example, the last member of (9.37) is the "reaction vector" corresponding to the "genesis reaction" for compound A_2BC:

$$2A + B + C \to A_2BC.$$

Note that a genesis reaction might or might not be a "true" reaction—that is, a member of the actual reaction set that defines the network. In the genesis reaction corresponding to a particular compound, the stoichiometric coefficient of element e on the reactant side is, of course, the number of units of element e in that compound.

These considerations set the stage for a formal statement of what we shall mean by a constructive network.

Definition 9.7.4. *A reaction network* $\{\mathscr{S}, \mathscr{C}, \mathscr{R}\}$ *is* **constructive** *if the species set* $\mathscr{S}$ *admits a partition into two nonempty subsets—a set* $\mathscr{E}$ *of* **elements** *and a set* $\mathscr{Q}$ *of* **compounds**—*such that there are nonnegative numbers* $\{M_e^q\}_{q \in \mathscr{Q}, e \in \mathscr{E}}$ *having the property that the vectors*

$$\{q - \sum_{e \in \mathscr{E}} M_e^q e\}_{q \in \mathscr{Q}} \subset \mathbb{R}^{\mathscr{S}} \tag{9.38}$$

constitute a basis for the network's stoichiometric subspace. The number M_e^q *is called the* e**-content of element** e **in compound** q.

In the set (9.38), the vector corresponding to a particular compound q can be regarded as a "reaction vector" for the "genesis reaction," whereby q is formed from the elements. The number M_e^q is the number of units of element e contained in a molecule of compound q. Thus, for our sample constructive network (9.36),

$$\begin{array}{lll} M_A^{A_2} = 2 & M_B^{A_2} = 0 & M_C^{A_2} = 0 \\ M_A^{A_2B} = 2 & M_B^{A_2B} = 1 & M_C^{A_2B} = 0 \\ M_A^{A_2BC} = 2 & M_B^{A_2BC} = 1 & M_C^{A_2BC} = 1. \end{array} \tag{9.39}$$

Remark 9.7.5. For a *conservative* constructive network, it must be the case that, in the basis (9.38), for each compound $q \in \mathscr{Q}$, there is at least one element $e \in \mathscr{E}$ such that M_e^q is positive. In fact, suppose on the contrary that for a particular compound q^* the corresponding vector in (9.38) reduces simply to q^*. In this case, q^* is a member of S, the stoichiometric subspace for the network. Because the network is conservative, there is a strictly positive vector p in $S^\perp$. Thus, we must have $p \cdot q^* = 0$. However, $p \cdot q^* = p_{q^*} > 0$, so we have a contradiction.

Remark 9.7.6. It is not difficult to see that, for a constructive network with stoichiometric subspace S, the vectors

$$M_e := e + \sum_{q \in \mathscr{Q}} M_e^q q, \quad \forall e \in \mathscr{E} \tag{9.40}$$

constitute a basis for $S^{\perp}$. For our example

$$\begin{aligned} M_A &= A + 2A_2 + 2A_2B + 2A_2BC \\ M_B &= B + A_2B + A_2BC \\ M_C &= C + A_2BC. \end{aligned} \tag{9.41}$$

Note that, for each element $e \in \mathscr{E}$, the vector M_e carries complete information about the e-content of the various species. Moreover, if $c \in \overline{\mathbb{R}}_+^{\mathscr{S}}$ is a mixture composition, then the number

$$M_e \cdot c = M_e c_e + \sum_{q \in \mathscr{Q}} M_e^q c_q$$

has a simple interpretation: It is the total concentration of element e when the mixture composition is c, taking account of all copies of e, be they overt copies or else covert copies contained latently within the compound species.

Remark 9.7.7. Consider a kinetic system $\{\mathscr{S}, \mathscr{C}, \mathscr{R}, \mathscr{K}\}$ in which the underlying network is constructive. Suppose that $c^* \in \overline{\mathbb{R}}_+^{\mathscr{S}}$ is a composition and that $c \in \overline{\mathbb{R}}_+^{\mathscr{S}}$ is any other composition along a composition trajectory passing through c^*. Then, from considerations discussed in Chapter 3, it must be the case that $c - c^*$ is a member of S, the stoichiometric subspace for the network $\{\mathscr{S}, \mathscr{C}, \mathscr{R}\}$. Thus, from Remark 9.7.6 we also have $M_e \cdot (c - c^*) = 0$ or, equivalently, $M_e \cdot c = M_e \cdot c^*$, for each element $e \in \mathscr{E}$. This merely asserts that everywhere along the composition trajectory containing c^*, the total concentration of each element $e \in \mathscr{E}$ is equal to the constant $T_e^* := M_e \cdot c^*$.

Now suppose that $\{T_e^*\}_{e \in \mathscr{E}}$ is a fixed set of nonnegative numbers. This we will regard to be a fixed specification of total concentrations of the various elements, manifesting themselves either overtly or covertly within the compound molecules. Then the set of all compositions $c \in \overline{\mathbb{R}}_+^{\mathscr{S}}$ consistent with this element-concentration specification is precisely the set of compositions that satisfies the system of equations

$$M_e \cdot c = T_e^*, \quad \forall e \in \mathscr{E}. \tag{9.42}$$

If $c^* \in \overline{\mathbb{R}}_+^{\mathscr{S}}$ is a particular solution of (9.42), then it is not difficult to see that the *set of all solutions in* $\overline{\mathbb{R}}_+^{\mathscr{S}}$ *coincides precisely with the stoichiometric compatibility class containing* c^*.

Indeed, *the various stoichiometric compatibility classes can be put into bijective correspondence with the totality of all distinct sets of nonnegative numbers* $\{T_e\}_{e \in \mathscr{E}}$. In this way, each stoichiometric compatibility class can be identified with precisely one such a numerical set, the numbers in the set being associated with a particular specification of the total concentrations of the elements. A shift from one stoichiometric compatibility class to a different one is synonymous with a change in the total concentration of at least one element.

This idea provides a way to discuss sensitivities of equilibrium concentrations against movements from one stoichiometric compatibility class to another. Our interest will be in equilibria that reside in positive stoichiometric compatibility classes. With this in mind, we can, in the sense of Remark 9.7.7, associate each positive stoichiometric compatibility class with a specification of (positive) total element concentrations $\{T_e\}_{e \in \mathscr{E}}$ or, equivalently, with a vector $T \in \mathbb{R}_+^{\mathscr{E}}$.

Hereafter in this section, it will be understood that we have in mind a mass action system $\{\mathscr{S}, \mathscr{C}, \mathscr{R}, k\}$ *that satisfies the hypothesis of Theorem 9.7.1 and in which the underlying network is constructive.* In fact, though, everything we will say holds true for any quasi-thermostatic kinetic system (Sections 7.7.1 and 13.3) in which the underlying network is constructive and conservative [154].

For the system under study, there is precisely one equilibrium in each positive stoichiometric compatibility class. Moreover, there is a smooth function that takes each $T \in \mathbb{R}_+^{\mathscr{E}}$, characterizing a particular positive stoichiometric compatibility class, into the unique equilibrium within that positive stoichiometric compatibility class. Because it is more convenient to work with logarithms, we introduce the function $\bar{c} : \mathbb{R}^{\mathscr{E}} \to \mathbb{R}_+^{\mathscr{S}}$ that takes the logarithm of T characterizing a positive stoichiometric compatibility class into its unique equilibrium composition. That is, $\bar{c}(\ln T)$ is the equilibrium in the positive stoichiometric compatibility class corresponding to element-concentration specification T.

We are now in position to study the sensitivity of a particular species's equilibrium concentration to small shifts from one stoichiometric compatibility class to another or, what amounts to the same thing, to small changes in the total concentrations of the elements. (It is useful to keep in mind an experiment of the kind depicted in Figure 9.1.) As before, the element set is denoted $\mathscr{E}$.

Definition 9.7.8. *Let* $T^* \in \mathbb{R}_+^{\mathscr{E}}$ *be a fixed specification of the total concentrations of the elements. The* **equilibrium sensitivity at** T^* **of species** $\mathscr{s}$, *denoted* $\Lambda^{\mathscr{s}}(T^*)$, *is given by*

$$\Lambda^{\mathscr{s}}(T^*) := \max_{e \in \mathscr{E}} \left\{ \left| \frac{\partial \ln \bar{c}_{\mathscr{s}}}{\partial \ln T_e} \right| \right\}_{T^*}. \tag{9.43}$$

Thus, at T^* the equilibrium sensitivity of species $\mathscr{s}$ captures the magnitude of the most pronounced of the variations in the equilibrium concentration of $\mathscr{s}$ against individual changes in the total concentration of each of the elements.[5]

[5] Some time ago, two Duke University Ph.D. students of Michael Reed—Karen Israel [114] and Jean Nailor [132]—wrote theses in which there was interest in the *signs* of derivatives related to those shown in (9.43), signs that might be determined from experiments. The idea was that such experimentally determined sign patterns might provide clues about the presumed-unknown operative reaction network.

Remark 9.7.9. Note that *if the kinetic system under study exhibits absolute concentration robustness relative to species $\mathcal{s}$, then the equilibrium sensitivity of $\mathcal{s}$ must be zero at every T^**. This is because absolute concentration robustness relative to $\mathcal{s}$ requires that the equilibrium concentration of $\mathcal{s}$ be the same in every positive stoichiometric compatibility class and, therefore, that it be independent of values of the total element concentrations.

Remark 9.7.9 tells us that the equilibrium sensitivity of a particular species $\mathcal{s}$, as it ranges over all positive stoichiometric compatibility classes (or, equivalently, over all values of $T^* \in \mathbb{R}_+^{\mathscr{E}}$), provides a measure of the departure from absolute concentration robustness. We shall claim that, for any constructive mass action system that satisfies the hypothesis Theorem 9.7.1, there is for each species a positive lower bound on the equilibrium sensitivity, *a bound that is an attribute of the reaction network alone*, independent of kinetics and independent of T^*. In this sense, constructive *reaction network architectures* consistent with the hypothesis of Theorem 9.7.1 not only deny the possibility of absolute concentration robustness, they also deny the possibility of an arbitrarily close approximation to it.

To describe the bound, we will first need some more vocabulary. By the *connectivity* of an element $e \in \mathscr{E}$, we mean the number of elements, including e itself, that appear with e in at least one species. Those "collaborators" of e are, in more precise terms, the members of the set

$$\Gamma(e) = \left\{e' \in \mathscr{E} : \operatorname{supp} M_{e'} \cap \operatorname{supp} M_e \neq \varnothing\right\}. \tag{9.44}$$

The connectivity of e is then the number of such collaborators:

$$\operatorname{conn}(e) := \#(\Gamma(e)). \tag{9.45}$$

For our sample network (9.36), element A has three collaborators, so $\operatorname{conn}(A) = 3$: it collaborates with itself in A (and also in A_2, A_2B, and A_2BC); it collaborates with B in A_2B (and A_2BC), and it collaborates with C in A_2BC. In fact, we also have $\operatorname{conn}(B) = 3$ and $\operatorname{conn}(C) = 3$. That all three elements have maximal connectivity results, in this case, from the fact that all collaborate in the compound A_2BC.

Finally, for each element e, we denote by M_e^{max} the maximal content of e in any species. That is,

$$M_e^{max} := \max_{\mathcal{s} \in \mathscr{S}} \{M_e^{\mathcal{s}}\}. \tag{9.46}$$

For our example $M_A^{max} = 2$, $M_B^{max} = 1$, and $M_C^{max} = 1$.

Theorem 9.7.10. *Consider a mass action system in which the underlying network is constructive and satisfies the hypothesis of Theorem 9.7.1. Then, for each species $\mathcal{s}$ and each $T^* \in \mathbb{R}_+^{\mathscr{E}}$,*

$$\Lambda^{\mathcal{s}}(T^*) \geq \max_{e \in \mathscr{E}} \left\{ \frac{M_e^{\mathcal{s}}}{\operatorname{conn}(e) M_e^{max}} \right\}. \tag{9.47}$$

The foundations of Theorem 9.7.10 are contained in [154]. There the argument is given for any quasi-thermostatic kinetic system in which the underlying reaction network is conservative. (Mass action systems that satisfy the hypothesis of Theorem 9.7.1 are just special cases. So too are conservative mass action systems that comply with detailed balancing or, more generally, complex balancing.) Sensitivity bounds are given in [154] even when the underlying network is not constructive, in which case a more generalized notion of "element" is required. Moreover, in the constructive case, the bounds given in [154] are sharper but slightly more complex than those given by (9.47).

Example 9.7.11. For our sample network (9.36), Theorem 9.7.10 provides a simple way to calculate a lower bound on the equilibrium sensitivity of, for example, species A_2BC: At each T^*

$$\begin{aligned} \Lambda^{A_2BC}(T^*) &\geq \max\left\{\frac{M_A^{A_2BC}}{\text{conn}(A)M_A^{max}}, \frac{M_B^{A_2BC}}{\text{conn}(B)M_B^{max}}, \frac{M_C^{A_2BC}}{\text{conn}(C)M_C^{max}}\right\} \\ &= \max\left\{\frac{2}{3\cdot 2}, \frac{1}{3\cdot 1}, \frac{1}{3\cdot 1}\right\} = \frac{1}{3}. \end{aligned} \tag{9.48}$$

Remark 9.7.12. Note that the right side of (9.47) depends *only* on the network; it depends not at all on kinetic parameters or even on T^*. Note too that for *every* species, the corresponding lower bound is *strictly positive*. (This is ensured by Remark 9.7.5 and the fact that the reaction network under consideration is conservative.) As claimed earlier, then, mass action systems that satisfy the conditions of Theorem 9.7.1 are not only denied absolute concentration robustness with respect to any species, it is also the case for no species can there be an arbitrarily close approach to absolute concentration robustness. Such systems are "bounded away from absolute concentration robustness."

9.7.3 Networks Involving Large Complex Molecules Made from Many Distinct Gregarious Subunits

Theorem 9.7.10 suggests interesting connections to biology. For the class of systems to which it applies, the theorem bounds the extent to which absolute concentration robustness might even be approached. The *value* of the bound, however, depends on the reaction network under consideration. This is to say that some networks might be far more congenial to concentration robustness than others.

For a particular species, a low value of the right side of (9.47) is more conducive to concentration robustness than a high one. For species $\mathscr{S}$ a low value will require for each element e appearing in $\mathscr{S}$ one of the following, or a mixture of both: (i) high connectivity of e ($\text{conn}(e) \gg 1$), which in turn means that e collaborates gregariously with many other elements, or (ii) relative to its content in $\mathscr{S}$, the content of e in some other species is substantially greater—that is, $M_e^{max}/M_e^{\mathscr{S}} \gg 1$.

Biochemistry would seem to provide fertile ground in which these conditions might be met. A biomolecule is often a very large one, built from multiple copies of several distinct elements that collaborate over and over again in the formation of many other biomolecules, some large and some small, which in turn react with each other to form still other biomolecules.

Remark 9.7.13. Note that the hypotheses of Theorems 9.7.1 and 9.7.10 invoke mass action kinetics, a kinetics that is less likely to be apt for reactions among large molecules in crowded cell-like environments. It should be remembered, however, that the conclusions of these theorems obtain for all quasi-thermostatic kinetic systems, not necessarily mass action, in which the underlying network is conservative [154].[6] The hypotheses of the two theorems, as stated here, are merely meant to provide certain concrete circumstances under which the quasi-thermostatic property is sure to be realized.

[6] Even the requirement that the network be constructive can be substantially generalized [154].

10

Concordant Reaction Networks: Architectures That Promote Dull, Reliable Behavior Across Broad Kinetic Classes

In this chapter we'll resume a narrative that ran through Chapters 7 and 8. There our interest was largely in the relationship between the structure of a reaction network and its capacity (or lack of it) to give rise to instabilities or multiple equilibria within a positive stoichiometric compatibility class. The principal theorems centered mostly on a reaction network's deficiency or, in some cases, on the deficiencies of the network's linkage classes. In this chapter, however, deficiencies will play almost no role at all.

The key idea in this chapter, based on work with Guy Shinar[157], will be *concordance*, a purely network-structural property that enforces a degree of dull, reliable behavior even against great varieties of kinetics within a very expansive class; mass action kinetics is only a special case. For kinetics in this broad class, there is a sense in which the absence or presence of concordance is actually *synonymous* with a network's capacity or incapacity to admit certain kinds of behavior. Whether or not a network is concordant can be determined computationally (and definitively) with freely available easy-to-use software [62].

As we shall see in the next chapter, concordance is also related in subtle ways to properties of a network's *Species-Reaction Graph*, a reaction network depiction similar to those often invoked in biochemistry. Although the powerful behavioral information that the Species-Reaction Graph gives can be stated and understood without reference to concordance, it is concordance that for us will provide the underlying theoretical foundation. Moreover, we shall see in the next chapter how connections between concordance and the Species-Reaction Graph suggest strongly that *only certain exceptional networks lack concordance*. To a great extent, *this tells us why dull, stable dynamics is prevalent so widely across the reaction network landscape*.

The original version of this chapter was revised: An equation reference in Example 10.6.18 and text below Eq. 10.39 have been corrected. The correction to this chapter is available at https://doi.org/10.1007/978-3-030-03858-8_20

M. Feinberg, *Foundations of Chemical Reaction Network Theory*,
Applied Mathematical Sciences 202,
https://doi.org/10.1007/978-3-030-03858-8_10

10.1 Some Advice for the Reader

Something needs to be said about this chapter, culminating in some guidance about how the reader might proceed. This chapter is somewhat more technical in tone than those preceding it. Indeed, the definition of concordance is itself technical, involving the signs of the components of certain vectors. However, the results given in this chapter are very, very powerful. If one knows that a network is concordant—and, again, computer programs (e.g.,[62]) are available to make that determination—then one knows *a lot* about how kinetic systems based on that network behave, even when the fine details of the kinetics are unknown.

The next chapter, about the Species-Reaction Graph, is far less technical. Although the results there derive from concordance theory, the present chapter need not be read in detail in order to understand reasonably well the principal results of the next one.

Here, then, is some advice: Readers who wish to skip over this chapter (or, preferably, to read it lightly) can do so and move to the next chapter. In this case, there will be a sacrifice of power and of the deeper meaning that this chapter carries, but there will be no irredeemable loss with respect to the understanding of the next one. In particular, the most important theorem of the next chapter, Theorem 11.6.1, makes no mention of concordance at all.[1]

10.2 A Puzzle: The Seemingly Unreasonable Stability of Classical Isothermal Continuous-Flow Stirred-Tank Reactors

The theorems of this chapter have broad application, impinging, for example, on cell biology. However, they actually had their origins in practical questions of chemical engineering, in particular questions about the classical continuous-flow stirred-tank reactor (CFSTR). The CFSTR origins of these theorems explain in part the occasional role played by what we shall call "fully open reaction networks." We'll begin the story there.

The continuous-flow stirred-tank reactor is a mainstay of chemical engineering. Recall from Section 4.2.1 that it consists of a stirred tank to which reactants are added continuously in a feed stream and from which tank mixture is withdrawn continuously in an effluent stream. The reactions occur in the presumed-homogeneous mixture that the tank contains. In the classical picture, it is generally supposed that the effluent stream faithfully samples the mixture in the tank so that all species present in the tank are also present in the effluent stream. The CFSTR is said to be *isothermal* when the presumed-uniform temperature of the tank mixture is kept constant over time, perhaps by means of a thermostatic control device.

[1] The theorem does, however, refer to "nondegenerate networks," defined and discussed in Section 10.6 of this chapter. The important thing to know is that degenerate networks, especially those that can admit a positive equilibrium, are highly pathological and rare. For example, no weakly reversible network is degenerate. Moreover, any degenerate network becomes nondegenerate when certain irreversible reactions are made reversible.

In Section 4.2.1 we drew the classical picture and provided, in the isothermal case, the governing differential equations for a simple chemistry consisting of a single reversible reaction. Keep in mind that the chemistry might be (and often is) far more complex, involving many species and many reactions. In those instances the governing differential equations will be enormously intricate, even in the isothermal case.[2] When the kinetics is mass action, the differential equations will involve polynomials in the species concentrations, and there will be many parameters (e.g., rate constants). Moreover, each new network of chemical reactions induces a new system of differential equations, almost always highly complicated. Stability questions would, therefore, seem to be extremely compelling—in particular very practical questions about the possibility of two distinct steady states having different production rates of a prized species, one high and one low.

Yet, judging from standard chemical engineering textbooks, stability properties of isothermal CFSTRs appear to hold little interest. Although there is much attention paid to the stability properties of *nonisothermal* CFSTRs, with an abundance of well-worked bifurcation diagrams provided for simple chemistries, the general attitude toward *isothermal* CFSTRs seems to be this: *They are boring, regardless of parameter values, even when the chemistry is complex. There is exactly one steady state; that steady state is stable; and, moreover, initial compositions different from the unique steady state will evolve toward it with the passage of time.*

This is a prejudice, not a fact derived from mathematical analysis, and it is not universally true.[3] The prejudice is nevertheless a useful one, held by intelligent people, that for the most part serves them well. It is a prejudice based, like much of engineering, on a long record of accumulated experience. For isothermal CFSTRs, even ones with intricate chemistry, the possibility of complex unstable behavior is largely ignored.

What, then, might be the mathematical source of the observed dullness of CFSTR dynamics, a dullness that prevails across a wide region of the reaction network landscape, regardless of values that parameters might take? Can the prevalence of dull behavior be explained, for example, by the Deficiency Zero Theorem of Chapter 7?

Although deficiency zero reaction networks are indeed common, even when there are many reactions involving many species, the Deficiency Zero Theorem provides little guidance when it comes to classical CFSTRs, for we must keep in mind how CFSTRs are described in reaction network terms. Recall from Section 4.2.1 that for each species $\mathscr{S}$ in the network, one augments the true chemical reactions with a reaction $\mathscr{S} \to 0$ to account for the removal of $\mathscr{S}$ in the effluent stream, and for each species $\mathscr{S}$ in the feed stream, one adds a reaction of the form $0 \to \mathscr{S}$ to account for the influx of new material.

---

2 In the nonisothermal case, one considers kinetic parameters to be functions of temperature. The differential equations for the species concentrations are supplemented by an additional one, involving the derivative of the temperature and coming from an energy balance[9, 138].

3 For experimental studies of complex CFSTR behavior involving the Belousov-Zhabotinsky chemistry (Section 1.2), see, for example, [53, 89, 97, 111, 149].

Consider, for example, a CFSTR in which the true chemistry is given by the very simple network

$$A + B \rightarrow C \qquad A + C \rightarrow D, \tag{10.1}$$

and suppose that only A and B are in the feed stream. (In the classical picture, all species are in the effluent stream.) In the sense of Section 4.2.1, the CFSTR is described not by the simple true-chemistry deficiency zero network (10.1) but, instead, by the more elaborate network (10.2) for which the deficiency is *two* $(n = 7, \ell = 1, s = 4)$.

$$\begin{array}{ccccc} A + B \rightarrow C & & & & A \\ & \searrow & & \nearrow\!\!\swarrow & \\ & & 0 & & \\ & \nearrow & & \nwarrow\!\!\searrow & \\ A + C \rightarrow D & & & & B \end{array} \tag{10.2}$$

In this case the Deficiency Zero and the Deficiency One Theorems are both silent. The example teaches a lesson: Even very modest complexity in a true chemistry can—and almost always does—give rise to a deficiency of two or more in the corresponding CFSTR network description [123]. The low-deficiency theorems cannot, then, explain broadly the prevalence of dull behavior in classical isothermal CFSTRs.

There is, however, good news in this. The long-acquired experience of chemical engineers hints at the existence of bold sweeping theorems that, at least for CFSTRs, give qualitative information when low-deficiency theory says nothing. Indeed, there are such theorems, specific to CFSTR-like kinetic systems in which, for every species $\mathscr{S}$ in the underlying network, there is an efflux reaction of the form $\mathscr{S} \rightarrow 0$—see, for example, very important contributions by D. Rumschitzki, P. Schlosser, G. Craciun, M. Banaji, P. Donnell, and S. Baigent [14–16, 44, 46, 48, 144, 145, 147, 148]. In particular, surprising work reported in [14–16] indicated what could be said in the CFSTR-like setting even for kinetic classes significantly broader than mass action.

As indicated in the introduction to this chapter, our route to understanding will instead be through the theory of *concordant networks*, developed in collaboration with Guy Shinar [118, 157, 158] and, later, with Daniel Knight [117, 118]. The theory is not specific to CFSTR-like kinetic systems and relies not at all on the presumption of a rich supply of $\mathscr{S} \rightarrow 0$ reactions. Nevertheless, there are certain features of CFSTR-like "fully open systems"—some transparent, others quite subtle—that make them easier to study and, as we shall see, give them a pivotal theoretical role that transcends considerably the modeling of archetypal chemical engineering reactors. Before we launch into an examination of concordance, it will be useful to say a few words about special properties of fully open networks.

10.3 Fully Open Networks and a Network's Fully Open Extension

In some ways reaction networks that describe classical CFSTRs are easier to study than networks that describe closed reactors or systems such as the enzymatic reactor in Section 4.2.2 that are semi-open. There are several reasons for this, and the most transparent is related to the idea of stoichiometric compatibility. In earlier chapters when we asked, for example, about a network's capacity for multiple equilibria, the concern was with the possibility of multiple equilibria *that are stoichiometrically compatible*. This was a restriction that made the question more nuanced and, to some extent, more difficult.

For a network $\{\mathscr{S},\mathscr{C},\mathscr{R}\}$ that describes a classical CFSTR, *questions of stoichiometric compatibility become largely moot*, for *any* two compositions are stoichiometrically compatible.[4] This is because $\mathscr{R}$ contains a full set of effluent reactions of the form $s \to 0$, one for *every* species $s \in \mathscr{S}$. Therefore, the set of reaction vectors for the network will contain the subset $\{-s\}_{s \in \mathscr{S}} \subset \mathbb{R}^{\mathscr{S}}$. This is clearly a basis for $\mathbb{R}^{\mathscr{S}}$, in which case the stoichiometric subspace coincides with $\mathbb{R}^{\mathscr{S}}$. For *any* choice of two compositions, their difference lies in the stoichiometric subspace, so the two compositions are stoichiometrically compatible. There is just one stoichiometric compatibility class, consisting of all of $\overline{\mathbb{R}}_{+}^{\mathscr{S}}$, and there is just one positive stoichiometric compatibility class, consisting of all of $\mathbb{R}_{+}^{\mathscr{S}}$.

These considerations, and others yet to be discussed, serve as motivation for the following definition.

Definition 10.3.1. *A reaction network with species set $\mathscr{S}$ is* **fully open** *if, for each $s \in \mathscr{S}$, there is a reaction of the form $s \to 0$.*

Remark 10.3.2. Although our consideration of fully open networks was motivated by the CFSTR picture, there are other situations in which such networks arise. For example, in certain contexts each species s might be subject to a degradation whereby it "dies" at a rate proportional to its current concentration, perhaps by conversion to an inert species not considered in the network's species set. In such a case, a reaction of the form $s \to 0$ would be included in the network to account for the degradation. In fact, we will sometimes refer to reactions of the form $s \to 0$ as *degradation reactions*.

Remark 10.3.3. Clearly all fully open networks, regardless of their origin, share the pleasant "omnipresent stoichiometric compatibility" property discussed earlier. However, there is a far deeper—and more technical—property shared by all fully open networks. *They are never degenerate.* Degeneracy and nondegeneracy of a *network* will be defined soon, in Section 10.6, but for now we will just say this: The impossibility of degeneracy appears, in retrospect, to have made considerably easier the attainment of certain powerful earlier results specific to fully open networks—results that do not hold generally in any wider class of networks in which network degeneracy is a possibility.

---

[4] In the classical situation, in which the rate constants for the effluent reactions are deemed to be the same positive number—see Section 4.2.1—any two equilibrium compositions will in fact be stoichiometrically compatible *with respect to the stoichiometric subspace for the true chemistry, before the addition of inflow-outflow "reactions."* Unlike constraints considered in earlier chapters, this is not a stricture that needs to be imposed separately. Rather, it is a *result* that follows *automatically* from the governing differential equations.

When a network is not fully open, we will sometimes find it useful to consider a larger network obtained by adding all reactions of the form $\mathcal{s} \to 0$ that were not already present in the original network. Not only will the reaction set be larger than that of the original network, so too might the complex set be larger. When, for example, $A \to 0$ is added as a reaction, the original network might not have contained as complexes either A or 0. These augmentations are implied in the following definition.

Definition 10.3.4. *By the* **fully open extension** *of a reaction network with species set $\mathscr{S}$ we mean the reaction network obtained by adding all reactions of the form $\mathcal{s} \to 0$, $\mathcal{s} \in \mathscr{S}$, that were not present in the original network.*

As we shall see, properties of a network's fully open extension often provide surprisingly powerful information about the original network's behavior. In Appendix 10.A we discuss some aspects of that relationship in very general terms that transcend (but are related to) the concordance ideas of this chapter.

10.4 Reaction Network Concordance: Definition and Examples

Here we begin our discussion of reaction network concordance. In this section we will say what it means for a network to be concordant or discordant; every reaction network is one or the other. We do *not* restrict our attention to fully open networks. In the next section, we will begin to describe the wealth of information that flows from knowledge of a network's concordance properties.

The definition of concordance is slightly technical. The important thing to know, however, is that determination of whether a network is concordant—and is therefore guaranteed to have certain behavioral properties—can be decided definitively by a freely available, easily used computer package (Remark 10.4.3). Moreover, we will also indicate in the next chapter how one can affirm—often very quickly—that a network is concordant merely by inspecting its Species-Reaction Graph. As we shall see, theorems connecting concordance to the nature of the Species-Reaction Graph are highly delicate. For this reason, these theorems provide considerable insight into the great subtlety that can exist in the relationship between reaction network structure and mixture behavior.

Throughout this section we consider a reaction network $\{\mathscr{S},\mathscr{C},\mathscr{R}\}$ with stoichiometric subspace $S \subset \mathbb{R}^{\mathscr{S}}$. Recall that the reaction vectors for the network are the members of the set $\{y'-y\}_{y\to y'\in\mathscr{R}} \subset \mathbb{R}^{\mathscr{S}}$. Recall also that the stoichiometric subspace is just the set of all linear combinations of the reaction vectors—that is, the set of all vectors in $\mathbb{R}^{\mathscr{S}}$ representable in the form

$$\sum_{y\to y'\in\mathscr{R}} \alpha_{y\to y'}(y'-y). \tag{10.3}$$

We shall also have interest in the set of all possible solutions to the linear equation

$$\sum_{y\to y'\in\mathscr{R}} \alpha_{y\to y'}(y'-y) = 0. \tag{10.4}$$

With these ideas in mind, we will find it convenient to let $L : \mathbb{R}^{\mathscr{R}} \to S$ be the linear map defined by

$$L\alpha = \sum_{y \to y' \in \mathscr{R}} \alpha_{y \to y'}(y' - y). \tag{10.5}$$

In standard terminology, the *kernel* of L (denoted $\ker L$) is the set of all vectors $\alpha \in \mathbb{R}^{\mathscr{R}}$ such that $L\alpha = 0$. This is just the set of all vectors in $\mathbb{R}^{\mathscr{R}}$ whose components satisfy (10.4). Recall that the *support* of complex y ($\mathrm{supp}\, y$) is the set of species "appearing in" that complex. For example, the support of complex $A + B$ is $\{A, B\}$, and the support of complex $2B$ consists only of species B. The support of the zero complex is empty. In what follows $\mathrm{sgn}(\cdot)$ is the sign function, taking values of $+,-$, or 0.

Definition 10.4.1. *The reaction network* $\{\mathscr{S}, \mathscr{C}, \mathscr{R}\}$ *is* **discordant** *if there exist an* $\alpha \in \ker L$ *and a nonzero* $\sigma \in S$ *having the following properties:*

(i) For each $y \to y'$ *such that* $\alpha_{y \to y'} \neq 0$, $\mathrm{supp}\, y$ *contains a species* $s \in \mathscr{S}$ *for which* $\mathrm{sgn}\, \sigma_s = \mathrm{sgn}\, \alpha_{y \to y'}$.

(ii) For each $y \to y'$ *such that* $\alpha_{y \to y'} = 0$, *either* $\sigma_s = 0$ *for all* $s \in \mathrm{supp}\, y$ *or else* $\mathrm{supp}\, y$ *contains species* s *and* s' *for which* $\mathrm{sgn}\, \sigma_s = -\mathrm{sgn}\, \sigma_{s'}$, *both not zero.*

The network is **concordant** *if it is not discordant.*

It will be convenient to have available the following terminology:

Definition 10.4.2. *A* **discordance** *for a network* $\{\mathscr{S}, \mathscr{C}, \mathscr{R}\}$ *is a pair* $\{\alpha, \sigma\}$, *with* $\alpha \in \ker L$ *and* σ *a nonzero member of the stoichiometric subspace, that satisfies conditions (i) and (ii) in Definition 10.4.1.*

Clearly, a discordance exists for a network if and only if the network is discordant. A search for discordances amounts, in effect, to a query of sign patterns in the linear subspaces S and $\ker L$ generated by the network at hand.

Remark 10.4.3. For a user-specified reaction network, The Chemical Reaction Network Toolbox [62] will test definitively for concordance.[5] When the network is not concordant, it will provide an example of a discordance.

Example 10.4.4 (Tiny Example 1). The network $A \rightleftarrows B$ is concordant. For the network

$$S := \{\sigma \in \mathbb{R}^{\mathscr{S}} : \sigma_A = -\sigma_B\} \tag{10.6}$$

and

$$\ker L := \{\alpha \in \mathbb{R}^{\mathscr{R}} : \alpha_{A \to B} = \alpha_{B \to A}\}. \tag{10.7}$$

From this it is easy to see that there is no discordance.

[5] Inclusion of concordance tests began with the 2011 version of the *Toolbox*. For descriptions of concordance algorithms used, see [115, 117]. Definitive determinant tests for concordance [118] can be found in Remark 10.6.11. See also [177]. These are very similar to determinant tests for injectivity in mass action systems given earlier in [46, 49, 84].

Example 10.4.5 (Tiny Example 2). The network $A+B \rightleftarrows 2B$ is discordant. For this network, the stoichiometric subspace and $\ker L$ are very much like they were in the preceding example:

$$S := \{\sigma \in \mathbb{R}^{\mathscr{S}} : \sigma_A = -\sigma_B\} \tag{10.8}$$

and

$$\ker L := \{\alpha \in \mathbb{R}^{\mathscr{R}} : \alpha_{A+B\to 2B} = \alpha_{2B\to A+B}\}. \tag{10.9}$$

Now, however, there is a discordance: $\{\alpha^*, \sigma^*\}$ with $\alpha^*_{A+B\to 2B} = 1, \alpha^*_{2B\to A+B} = 1$ and $\sigma^*_A = -1, \sigma^*_B = 1$.

Example 10.4.6 (Concordance is very subtle). Consider the following highly similar networks that first appeared, with surprising consequences, in [147, 148]:

$$\begin{aligned} A+B &\rightleftarrows P \\ B &\rightleftarrows 2A \end{aligned} \tag{10.10}$$

$$\begin{aligned} A+B &\rightleftarrows P \\ B+C &\rightleftarrows Q \\ C &\rightleftarrows 2A \end{aligned} \tag{10.11}$$

$$\begin{aligned} A+B &\rightleftarrows P \\ B+C &\rightleftarrows Q \\ C+D &\rightleftarrows R \\ D &\rightleftarrows 2A \end{aligned} \tag{10.12}$$

Of these, the first and third are concordant, *but the second is discordant.*[6] These examples indicate just how subtle and discerning the concordance property can be.

In fact, the second chemistry behaves very differently from the others in a CFSTR context (say with all species on the left in the feed stream): When the kinetics is mass action, there are parameter values such that, for (10.11), the corresponding CFSTR differential equations admit multiple positive equilibria. That is not the case for (10.10) or (10.12). For them, multiple equilibria are impossible even when the kinetics is not mass action, so long as the kinetics resides in the very broad *weakly monotonic* class we'll discuss in the next section.

10.5 Behavioral Consequences of Concordance and Discordance, Part I

Here we will begin to describe the wealth of behavioral information that flows from knowledge that a particular reaction network is concordant or discordant. There is

[6] Determination of the concordance status of the three networks can be made with the Chemical Reaction Network Toolbox [62], which will also indicate that the fully open extensions of the three networks have the same concordance status as the originals.

even more that can be said when one *also* knows that the network's fully open extension is concordant or discordant. In this section we will focus on consequences of concordance and discordance of the network itself. In the section immediately following this one, we will digress to discuss networks that are so pathological that they deserve to be called *degenerate*. They are not at all robust and make for poor models of real chemistry. Nevertheless, degenerate networks are mathematically annoying and need to be confronted before moving forward. Then we will be in a better position to consider what can be said when one knows not only the concordance properties of a network but also the concordance properties of the network's fully open extension.

As we've already indicated, some of the central results about reaction network concordance hold true when the kinetics resides in the very broad *weakly monotonic* class. This is the subject of the next subsection.

10.5.1 Weakly Monotonic Kinetics

In words, a kinetics for a reaction network is weakly monotonic if it satisfies two quite weak conditions: First, for each reaction in the network, its rate at composition c^{**} is *greater than* it is at composition c^* only if, at c^{**}, the concentration of at least one of the reaction's *reactant* species is greater than it is at c^*. Second, for each reaction in the network, its rate at c^{**} is *identical to* its rate at c^* only if the concentrations of all of its *reactant* species are identical at the two compositions or else there are two reactant species whose concentration levels at c^{**} and c^* are qualitatively offsetting.

This is made more precise in the following definition.

Definition 10.5.1. *A kinetics $\mathscr{K}$ for reaction network $\{\mathscr{S},\mathscr{C},\mathscr{R}\}$ is* **weakly monotonic** *if, for each pair of compositions c^* and c^{**}, the following implications hold for each reaction $y \to y' \in \mathscr{R}$ such that* $\operatorname{supp} y \subset \operatorname{supp} c^*$ *and* $\operatorname{supp} y \subset \operatorname{supp} c^{**}$:

(i) $\mathscr{K}_{y\to y'}(c^{**}) > \mathscr{K}_{y\to y'}(c^*) \quad \Rightarrow \quad$ *there is a species $s \in \operatorname{supp} y$ with $c_s^{**} > c_s^*$.*

(ii) $\mathscr{K}_{y\to y'}(c^{**}) = \mathscr{K}_{y\to y'}(c^*) \quad \Rightarrow \quad c_s^{**} = c_s^*$ *for all $s \in \operatorname{supp} y$ or else there are species s, $s' \in \operatorname{supp} y$ with $c_s^{**} > c_s^*$ and $c_{s'}^{**} < c_{s'}^*$.*

We say that the kinetic system $\{\mathscr{S},\mathscr{C},\mathscr{R},\mathscr{K}\}$ is weakly monotonic when its kinetics $\mathscr{K}$ is weakly monotonic.

Example 10.5.2. It is not difficult to see that every mass action system is weakly monotonic, but the class of weakly monotonic kinetic systems is substantially broader than the mass action class. For example, each of the following reaction-rate functions, where the k_i are positive numbers, complies with the requirements of weakly monotonic kinetics:

$$\mathscr{K}_{A+B\to C+F}(c) := \frac{k_1 c_A c_B}{k_2 + k_3 c_A + k_4 c_B},$$
$$\mathscr{K}_{C+F\to A+B}(c) := k_5 c_C^3 c_F^2, \tag{10.13}$$
$$\mathscr{K}_{C\to E}(c) := \frac{k_6 c_C}{k_7 + k_8 c_C}.$$

10.5.2 Concordance, Injectivity, and Multiple Equilibria

Very soon will assert, among other things, that for a concordant reaction network, no assignment of a weakly monotonic kinetics can give rise to multiple positive equilibria within a stoichiometric compatibility class. In fact, this is a consequence of an even stronger property of such kinetic systems: *For no two distinct stoichiometrically compatible compositions, at least one of which is positive, can the species-formation-rate function take precisely the same value.* (In particular, two such compositions could not both be equilibria, for then the species-formation-rate function would take the same value—zero—at each.)

There is much more that can be said, but first it will be helpful to introduce a little more vocabulary, beginning with codification of the stronger property we just discussed.

Definition 10.5.3. *A kinetic system* $\{\mathscr{S},\mathscr{C},\mathscr{R},\mathscr{K}\}$ *with species-formation-rate function* $f(\cdot)$ *is* **injective** *if, for each pair of distinct stoichiometrically compatible compositions* $c^* \in \overline{\mathbb{R}}_+^{\mathscr{S}}$ *and* $c^{**} \in \overline{\mathbb{R}}_+^{\mathscr{S}}$*, at least one of which is positive,* $f(c^*) \neq f(c^{**})$*; that is,*

$$\sum_{y\to y'\in\mathscr{R}} \mathscr{K}_{y\to y'}(c^{**})(y'-y) \neq \sum_{y\to y'\in\mathscr{R}} \mathscr{K}_{y\to y'}(c^*)(y'-y). \tag{10.14}$$

Remark 10.5.4. In language we now have available, it is worth repeating what was already said informally: *An injective kinetic system* $\{\mathscr{S},\mathscr{C},\mathscr{R},\mathscr{K}\}$ *cannot admit two distinct stoichiometrically compatible equilibria* $c^* \in \overline{\mathbb{R}}_+^{\mathscr{S}}$ *and* $c^{**} \in \overline{\mathbb{R}}_+^{\mathscr{S}}$*, at least one of which is positive.* Otherwise we would have $f(c^*) = f(c^{**}) = 0$.

The theorem we are about to state is one of the most important in this chapter. It tells us that a concordant reaction network taken with *any* weakly monotonic kinetics invariably gives rise to an injective kinetic system. *Consequently, reaction network concordance precludes multiple stoichiometrically compatible positive equilibria, against* every possible *assignment of a weakly monotonic kinetics.* The theorem tells us even more: The class of networks that is injective against all assignments of weakly monotonic kinetics is *synonymous* with the class of concordant chemical reaction networks. In this sense, concordance is *precisely* the structural attribute a reaction network must possess if injectivity is to be ensured no matter which weakly monotonic kinetics might adorn the network.

Theorem 10.5.5 ([157]). *The following are **equivalent**:*

(i) The reaction network $\{\mathscr{S},\mathscr{C},\mathscr{R}\}$ is concordant.
(ii) For every weakly monotonic kinetics $\mathscr{K}$ the kinetic system $\{\mathscr{S},\mathscr{C},\mathscr{R},\mathscr{K}\}$ is injective.

Corollary 10.5.6. *For a concordant reaction network, there is no weakly monotonic kinetics for which the resulting kinetic system admits two or more distinct stoichiometrically compatible equilibria, at least one of them positive.*

Remark 10.5.7. In [157] the proof that (ii) implies (i) amounts to showing that if a network is discordant, then there exists for it a power-law kinetics with respect to which the network is not injective. A kinetics $\mathscr{K}$ for a network $\{\mathscr{S},\mathscr{C},\mathscr{R}\}$ is a *power-law kinetics* if for each reaction $y \to y' \in \mathscr{R}$, there is a positive number $k_{y\to y'}$ and a vector $p_{y\to y'} \in \overline{\mathbb{R}}_+^{\mathscr{S}}$ with $\operatorname{supp} p_{y\to y'} = \operatorname{supp} y$ such that, for all $c \in \overline{\mathbb{R}}_+^{\mathscr{S}}$,

$$\mathscr{K}_{y\to y'}(c) = k_{y\to y'} c^{p_{y\to y'}}. \tag{10.15}$$

Every power-law kinetics is weakly monotonic. Because of the nature of the proof in [157], we also have

Proposition 10.5.8. *The following is equivalent to (i) and (ii) in Theorem 10.5.5:*

(iii) For every power-law kinetics $\mathscr{K}$ the kinetic system $\{\mathscr{S},\mathscr{C},\mathscr{R},\mathscr{K}\}$ is injective.

Remark 10.5.9 (Continuity is not essential.). In Chapter 3 we built into the definition of a kinetics the requirement that reaction-rate functions be continuous and, in fact, continuously differentiable. This was merely meant to avoid tedious stipulations of continuity whenever continuity (or something stronger) might be required in one theorem statement or another. Note that the definition of a weakly monotonic kinetics makes sense in the absence of continuity. In fact, *Theorem 10.5.5 holds true even when reaction-rate functions are not continuous.*

As we already indicated in Corollary 10.5.6, a consequence of Theorem 10.5.5 is that if a reaction network is concordant, then for no choice of a weakly monotonic kinetics can there be two distinct stoichiometrically compatible positive equilibria. However, we cannot assert the converse: Although we can say that for every discordant network there is weakly monotonic kinetics for which the resulting kinetic system is not injective, we cannot also say that for every discordant network there is a weakly monotonic kinetics that gives rise to two distinct stoichiometrically compatible positive equilibria.

Such an assertion would be false for the simple reason that not every discordant network has the capacity to admit a positive equilibrium. In particular, there are discordant networks—for example, $A + B \to 2A$—for which the reaction vectors are not positively dependent, and for such networks no choice of a kinetics, weakly monotonic or otherwise, can give rise to *even one* positive equilibrium (Section 3.5).

However, for a reaction network with positively dependent reaction vectors, discordance does indeed imply the existence of a weakly monotonic kinetics that gives rise to two distinct stoichiometrically compatible positive equilibria. (For a construction see Appendix B of [118].) Therefore, if we restrict our attention to those networks having positively dependent reaction vectors (a class that includes all weakly reversible networks), then we can state a theorem very much resembling Theorem 10.5.5, this one with a focus on multiple equilibria rather than on injectivity.

Theorem 10.5.10. *For a reaction network $\{\mathscr{S},\mathscr{C},\mathscr{R}\}$ having positively dependent reaction vectors, the following are* ***equivalent****:*

(i) The network $\{\mathscr{S},\mathscr{C},\mathscr{R}\}$ is concordant.
(ii) For no weakly monotonic kinetics $\mathscr{K}$ does the kinetic system $\{\mathscr{S},\mathscr{C},\mathscr{R},\mathscr{K}\}$ admit two or more distinct stoichiometrically compatible positive equilibria.

Remark 10.5.11 (The pivotal role of the concordance property: an instructive bio-fable). It is worth highlighting once more the canonical network-structural role played by the concordance property, this time within the class of networks having positively dependent reaction vectors.

A not-entirely-implausible biological fable might help make the point more tangible: Imagine that, in a cell biology context, there is a certain fixed network architecture that evolution has repeated over and over again in different guises, having different species sets in various instances. (For example, in different instances enzymes, substrates, and products might be different.) Suppose too that the architecture has evolved to produce, in each instance, a *unique* positive equilibrium that is—for reasons that might vary from one instance to another—biologically advantageous. Put differently, suppose that the architecture has evolved to deny in each instance the possibility of *multiple* stoichiometrically compatible positive equilibria, some of which might not only be unfavorable but also disastrous. Because the various instances feature different species sets, and therefore different kinetics, their common underlying network architecture faces the severe problem of thwarting multiple equilibria against what might be widely varying kinetics in the different instances.

How can this happen? Which network structures might do the job? So long as in each instance the kinetics satisfies the very mild requirements of weak monotonicity, *concordance in the underlying architecture ensures success.* Indeed, Theorem 10.5.10 tells us that, within the class of networks having positively dependent reaction vectors, concordance is *precisely* the network attribute that denies the possibility of multiple stoichiometrically compatible positive equilibria against *every* weakly monotonic kinetics.

Remark 10.5.12 (Questions restricted to mass action kinetics). For a given network, the preclusion of multiple stoichiometrically compatible positive equilibria against *all possible* weakly monotonic kinetics is a lot to ask. We now know that, within the class of reaction networks with positively dependent reaction vectors, concordance is precisely the network-design feature that ensures such a broad-spectrum preclusion.

We should point out, however, that there are *discordant* networks for which the preclusion obtains, not against *all* weakly monotonic kinetics but against kinetics residing within the smaller but iconic mass action class. For example, networks (10.10)–(10.12) are all reversible and have deficiencies of zero. Thus, for them the Deficiency Zero Theorem tells us that, when the kinetics is mass action, multiple stoichiometrically compatible positive equilibria are precluded *no matter what values the rate constants take*. Network (10.11) is discordant, so there exists for it a weakly monotonic kinetics for which such multiple equilibria *will* be exhibited; however, that kinetics cannot be mass action. On the other hand, networks (10.10) and (10.12) are concordant, so for them the preclusion obtains against *all* choices of weakly monotonic kinetics, mass action kinetics being only one special variety.

See Appendix 10.B for a more extensive discussion of what can be said when the kinetics is constrained to be mass action.

10.5.3 Concordance and the Nonsingularity of Derivatives

As we indicated in Remark 10.5.9, little in what we've said so far about concordance and weakly monotonic kinetics required constraints on the continuity or differentiability of reaction-rate functions. Especially when we want to discuss properties of the derivative of the species-formation-rate function, however, there will be a need to say more about differentiation. We begin with a strengthened version of weak monotonicity.

Definition 10.5.13. *A kinetics $\mathscr{K}$ for a reaction network $\{\mathscr{S},\mathscr{C},\mathscr{R}\}$ is* **differentiably monotonic** *at $c^* \in \mathbb{R}_+^{\mathscr{S}}$ if, for every reaction $y \to y' \in \mathscr{R}$, $\mathscr{K}_{y\to y'}(\cdot)$ is differentiable at c^* and, moreover, for each species $s \in \mathscr{S}$,*

$$\frac{\partial \mathscr{K}_{y\to y'}}{\partial c_s}(c^*) \geq 0, \tag{10.16}$$

with inequality holding if and only if $s \in \operatorname{supp} y$. A **differentiably monotonic kinetics** *is one that is differentiably monotonic at every positive composition.*

A consequence of the following theorem is that for a differentiably monotonic kinetic system in which the underlying reaction network is concordant, *the derivative of the species-formation-rate function is nonsingular at* every *positive composition, not just at equilibria.*

Theorem 10.5.14 ([157]). *Let $\{\mathscr{S},\mathscr{C},\mathscr{R},\mathscr{K}\}$ be a kinetic system with stoichiometric subspace S and species-formation-rate function $f : \overline{\mathbb{R}}_+^{\mathscr{S}} \to S$. Moreover, suppose that the kinetics is differentiably monotonic at $c^* \in \mathbb{R}_+^{\mathscr{S}}$. If the underlying network $\{\mathscr{S},\mathscr{C},\mathscr{R}\}$ is concordant, then the derivative $df(c^*) : S \to S$ is nonsingular (whereupon 0 is not one of its eigenvalues); in particular, if c^* is an equilibrium, then it is a nondegenerate equilibrium.*

Theorem 10.5.14, as formulated, is intended as a precursor of more powerful observations about eigenvalues that we shall meet later in the form of Theorems 10.7.2 and 10.7.7. In fact, though, Theorem 10.5.14 has something like a converse, stated below and discussed in Appendix 10.B.

Theorem 10.5.15. *For reaction network $\{\mathscr{S},\mathscr{C},\mathscr{R}\}$ the following are equivalent:*

(i) $\{\mathscr{S},\mathscr{C},\mathscr{R}\}$ is concordant

(ii) For every choice of a differentiably monotonic kinetics $\mathscr{K}$ the derivative of the species-formation-rate function for the kinetic system $\{\mathscr{S},\mathscr{C},\mathscr{R},\mathscr{K}\}$ is nonsingular at every positive composition.

10.5.4 A Substantial Biological Example: The Wnt Pathway

Every once in while it will be useful to display the full differential equations for a kinetic system illuminated by one or another of our theorems; it is easy to forget how complex these equations can actually be. In (10.17) we show, at a detailed level, a schematic reconstruction[7] of almost all of the reactions considered by Lee et al. [122] in their study of the Wnt signaling pathway—a cellular pathway believed to play a significant role in the development of cancer. The Wnt system will be instructive in this chapter, and it will also have echoes in the next one, where we shall see how the Species-Reaction Graph very quickly provides considerable and subtle insight.

$$
\begin{array}{c}
A_{10} \rightarrow 0 \rightleftarrows A_{11} \\
\upharpoonleft\downharpoonright \\
A_{12} \\
A_1 \rightleftarrows A_2 \\
A_2 + A_4 \rightarrow A_{16} \rightarrow A_2 + A_6 \\
A_3 \rightleftarrows A_4 \rightleftarrows A_6 \rightleftarrows A_7 + A_{12} \\
A_3 + A_{11} \rightleftarrows A_8 \rightarrow A_9 \rightarrow A_3 + A_{10} \\
A_{11} + A_{13} \rightleftarrows A_{14}
\end{array}
\tag{10.17}
$$

The system of differential equations corresponding to network (10.17) is shown as (10.18), in which the individual reaction-rate functions are assumed only to be weakly monotonic. (Although Lee et al. eventually made several rapid equilibrium approximations, they began by assuming mass action kinetics for the fundamental mechanistic reactions such as those depicted in (10.17).)

We can now ask a question: *Is there* any *choice of weakly monotonic kinetics (not necessarily mass action) such that the system* (10.18) *admits multiple stoichiometrically compatible positive equilibria?* The question is daunting, but we can answer it almost immediately. *The answer is no.* In fact, *network* (10.17) *is concordant*, as can be determined quickly, for example, by means of [62]. Moreover, so long as the kinetics is differentiably monotonic, the derivative of the species-formation-rate

[7] The species are numbered as in [122]. A species A_5 is not shown because it is presumed to be present in excess with largely time-invariant concentration. In the network displayed here, the reactions $A_2 + A_4 \rightarrow A_{16} \rightarrow A_2 + A_6$ are intended to model the important stimulatory effect of A_2 (Dsh_a) in an alternative route for the conversion of A_4 to A_6.

function is nonsingular at every positive composition. In particular, no positive equilibrium can be degenerate. And there's more: *We shall see in Section 10.7 that at a positive equilibrium, every real eigenvalue is negative.*

$$
\begin{aligned}
\dot{c}_1 &= -\mathscr{K}_{A_1\to A_2}(c_1) + \mathscr{K}_{A_2\to A_1}(c_2)\\
\dot{c}_2 &= \mathscr{K}_{A_1\to A_2}(c_1) - \mathscr{K}_{A_2\to A_1}(c_2) - \mathscr{K}_{A_2+A_4\to A_{16}}(c_2,c_4) + \mathscr{K}_{A_{16}\to A_2+A_6}(c_{16})\\
\dot{c}_3 &= \mathscr{K}_{A_4\to A_3}(c_4) - \mathscr{K}_{A_3\to A_4}(c_3) - \mathscr{K}_{A_3+A_{11}\to A_8}(c_3,c_{11}) + \mathscr{K}_{A_8\to A_3+A_{11}}(c_8)\\
&\quad + \mathscr{K}_{A_9\to A_3+A_{10}}(c_9)\\
\dot{c}_4 &= \mathscr{K}_{A_6\to A_4}(c_6) - \mathscr{K}_{A_4\to A_3}(c_4) + \mathscr{K}_{A_4\to A_6}(c_4) + \mathscr{K}_{A_3\to A_4}(c_3) - \mathscr{K}_{A_2+A_4\to A_{16}}(c_2,c_4)\\
\dot{c}_6 &= -\mathscr{K}_{A_6\to A_7+A_{12}}(c_6) - \mathscr{K}_{A_6\to A_4}(c_6) + \mathscr{K}_{A_4\to A_6}(c_4) + \mathscr{K}_{A_{16}\to A_2+A_6}(c_{16})\\
&\quad + \mathscr{K}_{A_{12}+A_7\to A_6}(c_7,c_{12})\\
\dot{c}_7 &= \mathscr{K}_{A_6\to A_7+A_{12}}(c_6) - \mathscr{K}_{A_7+A_{12}\to A_6}(c_7,c_{12}) \qquad (10.18)\\
\dot{c}_8 &= \mathscr{K}_{A_3+A_{11}\to A_8}(c_3,c_{11}) - \mathscr{K}_{A_8\to A_3+A_{11}}(c_8) - \mathscr{K}_{A_8\to A_9}(c_8)\\
\dot{c}_9 &= \mathscr{K}_{A_8\to A_9}(c_8) - \mathscr{K}_{A_9\to A_3+A_{10}}(c_9)\\
\dot{c}_{10} &= -\mathscr{K}_{A_{10}\to 0}(c_{10}) + \mathscr{K}_{A_9\to A_3+A_{10}}(c_9)\\
\dot{c}_{11} &= \mathscr{K}_{0\to A_{11}} - \mathscr{K}_{A_{11}\to 0}(c_{11}) + \mathscr{K}_{A_{14}\to A_{11}+A_{13}}(c_{14}) - \mathscr{K}_{A_{11}+A_{13}\to A_{14}}(c_{11},c_{13})\\
&\quad - \mathscr{K}_{A_3+A_{11}\to A_8}(c_3,c_{11}) + \mathscr{K}_{A_8\to A_3+A_{11}}(c_8)\\
\dot{c}_{12} &= \mathscr{K}_{0\to A_{12}} - \mathscr{K}_{A_{12}\to 0}(c_{12}) + \mathscr{K}_{A_6\to A_7+A_{12}}(c_6) - \mathscr{K}_{A_7+A_{12}\to A_6}(c_7,c_{12})\\
\dot{c}_{13} &= \mathscr{K}_{A_{14}\to A_{11}+A_{13}}(c_{14}) - \mathscr{K}_{A_{11}+A_{13}\to A_{14}}(c_{11},c_{13})\\
\dot{c}_{14} &= -\mathscr{K}_{A_{14}\to A_{11}+A_{13}}(c_{14}) + \mathscr{K}_{A_{11}+A_{13}\to A_{14}}(c_{11},c_{13})\\
\dot{c}_{16} &= \mathscr{K}_{A_2+A_4\to A_{16}}(c_2,c_4) - \mathscr{K}_{A_{16}\to A_2+A_6}(c_{16})
\end{aligned}
$$

Remark 10.5.16 (Seemingly innocent but behavior-altering reactions). We said at the outset that our depiction in (10.17) of the Wnt pathway is *almost* a reconstruction of the full chemistry described in [122]. Indeed, we omitted a seemingly innocuous (reversible) reaction,

$$A_7 + A_{11} \rightleftarrows A_{15}, \qquad (10.19)$$

that does little more than produce a new species A_{15}, a species that plays no role at all anywhere else in the network. True, the production of A_{15} consumes A_{11} and A_7, but those species are then regenerated by the reverse reaction in (10.19).

In deriving their differential equations, Lee et al. assumed that the reactions (10.19) equilibrate rapidly, but that was not our primary reason for omitting them from (10.17). Instead, the omission was for a mathematically instructive purpose—to indicate once again how subtle the concordance attribute can be in capturing the delicate structure-behavior relationship that reaction networks exhibit.

The fact is that, when the seemingly harmless reactions (10.19) are added to (10.17), *the slightly augmented network becomes discordant*. This can be determined by means of [62], which in its higher deficiency module will also indicate that *there exist for the augmented network rate constants such that the resulting mass action differential equations admit multiple stoichiometrically compatible positive equilibria.* In the next chapter, we shall be able to see more clearly, by means of the Species-Reaction Graph, just how addition of the apparently innocent reactions (10.19) serves to poison the concordance of network (10.17).

10.5.5 Weakly Reversible Concordant Networks: Absence of Boundary Equilibria or Cycles; Existence of a Unique Nondegenerate Equilibrium in Each Positive Stoichiometric Compatibility Class

It is worth remembering every once in a while that chemists sometimes insist that all "true" reaction networks should be deemed reversible, even when some "reverse reactions" can, in practical circumstances, be expected to have rates so small that they contribute little to mixture dynamics. As we indicated in Chapter 4, our view of reaction networks is more expansive, with allowance for not-necessarily-reversible reactions, including those such as $A \to 0$ or $0 \to A$ that can be used to model phenomena that are not always truly chemical. Still, the chemists' sensibilities provide good reason to pay special attention to reversible networks.

It happens that for the class of weakly reversible *concordant* networks, of which reversible concordant networks is a subclass, there are some interesting things that can be said, in particular about what might or might not be found within each nontrivial stoichiometric compatibility class. By a *nontrivial stoichiometric compatibility class*, we simply mean one that contains a positive composition—that is, a composition in which all species concentrations are strictly positive. That there can be trivial stoichiometric compatibility classes for a certain networks is demonstrated in the following elementary example.

Example 10.5.17 (A trivial stoichiometric compatibility class). For the reaction network $A + B \rightleftarrows C$, the composition $c^\dagger$ such that $c^\dagger_A = 0$, $c^\dagger_B = 1$, and $c^\dagger_C = 0$ is not stoichiometrically compatible with any other composition. In particular, $c^\dagger$ is not stoichiometrically compatible with any composition at which all species concentrations are positive. The set consisting only of $c^\dagger$ constitutes a trivial stoichiometric compatibility class. (Of course, for the network there are many nontrivial stoichiometric compatibility classes; every positive composition belongs to one.)

The first theorem we'll state about concordant weakly reversible networks has nothing at all to do with kinetics. Instead, *it constrains severely the geometric character of the network's nontrivial stoichiometric compatibility classes.* By way of background, we'll recall some ideas from Section 3.7. For a network $\{\mathscr{S}, \mathscr{C}, \mathscr{R}\}$ a (not-necessarily-positive) composition $c \in \overline{\mathbb{R}}^{\mathscr{S}}_{+}$ is *reaction-transitive* if, for each reaction $y \to y' \in \mathscr{R}$, the presence of all its reactant species (i.e., $\operatorname{supp} y$) at composition c implies the presence of all its product species (i.e., $\operatorname{supp} y'$) at composition c. That is, c is reaction-transitive if, for each $y \to y' \in \mathscr{R}$,

$$\operatorname{supp} y \subset \operatorname{supp} c \;\Rightarrow\; \operatorname{supp} y' \subset \operatorname{supp} c. \tag{10.20}$$

The following theorem tells us that for a weakly reversible concordant network, the geometry of nontrivial stoichiometric classes must be such as to deny the presence of reaction-transitive compositions on their boundaries. In turn, the theorem's corollaries will tell us that those boundaries can contain neither equilibria nor cyclic composition trajectories.

Theorem 10.5.18 ([157]). *For a weakly reversible concordant reaction network, no nontrivial stoichiometric compatibility class can have on its boundary a composition that is reaction-transitive.*

Recall from from the discussion in Section 3.7, in particular Proposition 3.7.5, that *every equilibrium composition of a kinetic system must be reaction-transitive.* Taken with that discussion, Theorem 10.5.18 has an immediate and important corollary. The corollary places no restrictions at all on kinetics.

Corollary 10.5.19. *For a kinetic system that derives from a weakly reversible concordant reaction network, no composition on the boundary of a nontrivial stoichiometric compatibility class is an equilibrium. In fact, at each boundary composition, the species-formation-rate vector points into the stoichiometric compatibility class in the sense that there is an absent species produced at strictly positive rate.*

Recall too from the discussion in Section 3.7—in particular Remark 3.7.7—that if a cyclic composition trajectory has species set $\mathscr{S}^*$ as its support, then $\mathscr{S}^*$ must be a reaction-transitive. This is to say that all compositions along the trajectory must be reaction-transitive. Thus, Theorem 10.5.18 has another corollary.

Corollary 10.5.20. *For a kinetic system that derives from a weakly reversible concordant reaction network, the support of a cyclic composition trajectory cannot reside in the boundary of a nontrivial stoichiometric compatibility class.*

In none of the results so far in this chapter have we asserted the *existence* of positive equilibria.[8] When the network is conservative, there is much that can be said. In the following theorem, we assume very little about kinetics, weak reversibility, or concordance at the outset. Then we add more about the network and more about the kinetics to make increasingly sharper assertions, derived largely from results we've already stated. It should be kept in mind that, in contrast to similar assertions made in the Deficiency Zero and Deficiency One Theorems, *there is no insistence here on mass action kinetics.*

Theorem 10.5.21 ([157]). *If $\mathscr{K}$ is a continuous kinetics for a conservative reaction network $\{\mathscr{S},\mathscr{C},\mathscr{R}\}$, then the kinetic system $\{\mathscr{S},\mathscr{C},\mathscr{R},\mathscr{K}\}$ has an equilibrium within each stoichiometric compatibility class. If the network is also weakly reversible and concordant, then within each nontrivial stoichiometric compatibility class, there is a positive equilibrium. If, in addition, the kinetics is weakly monotonic, then that positive equilibrium is the only equilibrium in the stoichiometric compatibility class containing it; if the kinetics is differentiably monotonic, then the sole equilibrium is nondegenerate.*

---

[8] Keep in mind that results about the existence of positive equilibria discussed in Remark 8.7.2 were restricted to mass action kinetics.

Remark 10.5.22 (Some tiny counterexamples). Network (10.21) is conservative and reversible but not concordant. There exists for it a differentiably monotonic kinetics—$\mathscr{K}_{2A+B\to A+2B}(c) = c_A c_B$, $\mathscr{K}_{A+2B\to 2A+B}(c) = c_A^2 c_B$—such that there are nontrivial stoichiometric compatibility classes without positive equilibria.

$$2A + B \rightleftarrows A + 2B \tag{10.21}$$

The network $0 \rightleftarrows A$ is reversible and concordant but not conservative. There exists for it a differentiably monotonic kinetics—$\mathscr{K}_{0\to A}(c) = 1$, $\mathscr{K}_{A\to 0}(c) = \frac{c_A}{1+c_A}$—such that there is no positive equilibrium.

Remark 10.5.23. Before closing this section, we will make one more observation that is related in flavor to Theorem 10.5.18 but for which the supporting argument is far simpler: *For a concordant network with positively dependent reaction vectors, no nontrivial stoichiometric compatibility class can have on its boundary a composition at which* every *reactant complex in the network contains at least one of the species absent at that composition.*

In fact, suppose there is a nontrivial stoichiometric compatibility class having on its boundary a composition c^* such that each reactant complex contains at least one species absent at c^*. Let c^{**} be a positive composition within that same nontrivial stoichiometric compatibility class. Now choose rate constants for the network such that c^{**} is an equilibrium of the resulting mass action system (Lemma 3.5.3). Note that c^* is a different equilibrium of the same mass action system, for at c^* all reactions proceed at zero rate (Definition 3.2.1). Because c^* and c^{**} are distinct stoichiometrically compatible equilibria of the (weakly monotonic) mass action system so constructed, there is a contradiction of Corollary 10.5.6.

If, on the boundary of that nontrivial stoichiometric compatibility class, there is a composition c^{**} such that each reactant complex contains at least one of the species absent at c^{**}, then c^{**} is a different equilibrium of the same mass action system, an equilibrium that is stoichiometrically compatible with c^*. This constitutes a contradiction of Corollary 10.5.6.

There is also a different argument that can be made, one that appeals directly to the definition of concordance and that makes no mention of kinetics.

10.6 Digression: Degenerate and Nondegenerate Networks

In the next section, we will resume our discussion of behavioral properties of concordant and discordant networks, but first we will digress to describe a class of networks so degenerate that we will sometimes want to exclude them from consideration. Failure to do so would make otherwise important and wide-ranging theorems false. This will not represent a great loss at all, for these bad actors are somewhat rare, and, because the exceptional phenomena they admit are never robust, they make for poor models of real chemistry. As we shall see, the degeneracy always disappears if sufficiently many irreversible reactions are made reversible, in which case the augmented network would actually be better attuned to the sensibilities of chemists, who often urge that all true reactions be considered reversible to some extent.

This section is, in parts, somewhat more technical than the others in this chapter. *It can be skipped without substantial loss of continuity, provided that the reader is willing to accept a few ideas:* First, there are networks, called *degenerate networks*, that are severely pathological but rare, especially among networks for which positive equilibria are admissible (*i.e.*, networks having positively dependent reaction vectors). Second, every degenerate network will become nondegenerate if sufficiently many (and often very few) irreversible reactions are made reversible; again, degeneracy is a price paid for failure to heed sufficiently the admonitions of chemists. Moreover, *all* weakly reversible networks are nondegenerate. Third, *all* fully open networks are nondegenerate. And, finally, there are readily accessible [62] computational means to determine with certainty whether or not a network is nondegenerate; some underlying methods are discussed in this section.

Readers who wish to do so should proceed to Section 10.7.

10.6.1 Nondegenerate (and Degenerate) Networks Defined

There are several equivalent ways to describe the pathology we have in mind—see Proposition 10.6.7—but perhaps the most tangible one is this: We shall say that a reaction network is *degenerate* if, for *every possible* assignment of a differentiably monotonic kinetics, the derivative of the species-formation-rate function is singular at *every* positive composition. (Yes, this can happen!) Note that in this context *degeneracy is an attribute of a network.*[9] Indeed, Proposition 10.6.7 relates network degeneracy directly to aspects of reaction network structure, divorced from any mention of kinetics.

In the next section and in the next chapter, our focus will be on *nondegenerate networks*—that is, on "reasonable" networks free of the pathology we just described. For this reason our preference is to frame the working definition not in terms of network degeneracy but, rather, in terms of nondegeneracy. This formulation has the benefit of bringing to light the great likelihood that a network of practical interest will be nondegenerate.

Definition 10.6.1. *A reaction network is* **nondegenerate** *if there exists for it* even one *differentiably monotonic kinetics such that at* some *positive composition* c^* *(not necessarily an equilibrium), the derivative of the species-formation-rate function* $df(c^*) : S \to S$ *is nonsingular. Otherwise, the network is* **degenerate**.

Remark 10.6.2. It follows immediately from Definition 10.6.1 and Theorem 10.5.14 that *every degenerate network is discordant.*

---

[9] Degeneracy of a *network* should not be confused with degeneracy of an *equilibrium* (Section 3.6). The latter is an attribute of a particular equilibrium composition for a given kinetic system. There is, however, a connection: for a degenerate network taken with *any* differentiably monotonic kinetics, *every* equilibrium is degenerate.

Example 10.6.3. We shall be primarily interested in networks that have the capacity to admit a positive equilibrium—that is, in networks that have positively dependent reaction vectors. Among these, degeneracy is rare but possible. Network (10.22) is an example, as can be confirmed by means of [62]. By way of contrast, network (10.23), obtained from (10.22) by making a single reaction reversible, is nondegenerate.

$$\begin{aligned} A &\to B \to A + C \\ D &\to E \to D + F \\ C + F &\to 0 \end{aligned} \tag{10.22}$$

$$\begin{aligned} A &\to B \rightleftarrows A + C \\ D &\to E \to D + F \\ C + F &\to 0 \end{aligned} \tag{10.23}$$

Remark 10.6.4 (A nondegenerate network can give rise to a degenerate equilibrium). For a degenerate network taken with a differentiably monotonic kinetics, every positive equilibrium is degenerate in the sense of Section 3.6. However, it should be clearly understood that, for a *nondegenerate* network $\{\mathscr{S},\mathscr{C},\mathscr{R}\}$, there can be a differentiably monotonic kinetics $\mathscr{K}$ such that the kinetic system $\{\mathscr{S},\mathscr{C},\mathscr{R},\mathscr{K}\}$ gives rise to a positive degenerate equilibrium. For example, the mass action system (5.2) does just that. The underlying Edelstein network is nondegenerate. (We shall see in the next section that nondegeneracy is attribute of all networks which, like the Edelstein network, are reversible.)

10.6.2 Reversibility and Nondegeneracy: Every Weakly Reversible Network Is Nondegenerate

As we indicated earlier, a degenerate network can always be made nondegenerate by making sufficiently many reactions reversible. This is the subject of Proposition 10.6.5. The proposition points to the great fragility of dynamic model phenomena rooted in network degeneracy; addition of reverse reactions—reactions normally deemed present by chemists—will cause such phenomena to disappear. See Remark 10.6.10.

Proposition 10.6.5 ([118]). *Suppose that a network $\{\mathscr{S},\mathscr{C},\mathscr{R}\}$ of rank s is degenerate. Then there is a nondegenerate network $\{\mathscr{S},\mathscr{C},\mathscr{R}^\dagger\}$, with $\mathscr{R} \subset \mathscr{R}^\dagger$, having no more than s additional reactions, each being the reverse of a reaction in the original network. (In fact, $\{\mathscr{S},\mathscr{C},\mathscr{R}^\dagger\}$ will be nondegenerate if it contains a reversible subnetwork of rank s.)*

The following important (and not-so-obvious) theorem is related in spirit and substance to Proposition 10.6.5. Proof of the theorem will be a simple consequence of material gathered in Part III. See Section 16.5.5 in Chapter 16.

Theorem 10.6.6 ([49, 118, 158]). *Every weakly reversible reaction network is nondegenerate.*[10]

10.6.3 Mild Network-Structural Conditions Equivalent to Nondegeneracy

Even when a network is not weakly reversible, it is nevertheless very likely to be nondegenerate. The following proposition provides very mild conditions on network structure that serve to ensure nondegeneracy. Each is, in fact, equivalent to nondegeneracy. (See also Proposition 10.6.23 for still another such condition.)

Proposition 10.6.7 ([118, 158]). *For a reaction network $\{\mathscr{S},\mathscr{C},\mathscr{R}\}$ of rank s, the following are equivalent:*

(i) The network is nondegenerate.

(ii) There is a choice of s reactions $\{y_i \to y'_i\}_{i=1...s}$ and s distinct species $\{s_i\}_{i=1...s}$ with $s_i \in \operatorname{supp} y_i$, $i = 1\ldots s$, such that the matrix

$$\left[s_i \cdot (y'_j - y_j)\right]_{i,j=1...s} \tag{10.24}$$

has nonzero determinant.

(iii) There is a choice of s reactions $\{y_i \to y'_i\}_{i=1...s}$ and s distinct species $\mathscr{S}^* = \{s_i\}_{i=1...s}$ with $s_i \in \operatorname{supp} y_i$, $i = 1\ldots s$, such that the $\mathscr{S}^*$-projected reaction vectors $\{\bar{y}'_i - \bar{y}_i\}_{i=1...s}$ are linearly independent.

In item *(iii)* when we refer to the $\mathscr{S}^*$- projection of a vector $x \in \mathbb{R}^{\mathscr{S}}$, we mean the vector $\bar{x} \in \mathbb{R}^{\mathscr{S}}$ such that $\bar{x}_s = x_s$ for all $s \in \mathscr{S}^*$ and $\bar{x}_s = 0$ for all $s \notin \mathscr{S}^*$.

Remark 10.6.8. Note that in items *(ii)* and *(iii)*, the *s* distinct species choices are required to reside in *reactant* complexes of the corresponding reactions. Thus, network nondegeneracy requires in part that there be at least *s* different *reactant* species, suitably distributed over *s* different reactions. There cannot, then, be too few *reactant* species in the network. Note, however, that for networks having positively dependent reaction vectors—a condition necessary for the existence of a positive equilibrium—*every* species is a reactant species for some reaction (and also a product species for a perhaps different reaction).

Example 10.6.9. For the rank 4 network (10.23), we can, in item *(iii)*, choose the reaction set to be

$$\{A \to B,\ A+C \to B,\ D \to E,\ C+F \to 0\}$$

and the corresponding species set to be $\mathscr{S}^* = \{A,C,D,F\}$. The $\mathscr{S}^*$-projected reaction vectors

$$\{-A, -(A+C), -D, -(C+F)\}$$

are linearly independent. Thus, (10.23) is nondegenerate. No such choices, satisfying the requirements of *(iii)*, can be made for the degenerate network (10.22).

[10] In [49] it was argued that every weakly reversible network is *normal*, a network attribute that in turn implies network nondegeneracy.

Remark 10.6.10. There are rate constants for the *degenerate* network (10.22) such that the resulting mass action differential equations admit multiple stoichiometrically compatible positive equilibria [62]. For the slightly augmented *nondegenerate* network (10.23)—differing from (10.22) in only one reverse reaction—no choice of rate constants will give rise to the same phenomenon. These examples demonstrate how degenerate networks can exhibit highly fragile phenomena that result simply from *omission* of certain reverse reactions—reactions that chemists might argue should be retained if nature is to be mirrored faithfully.

Remark 10.6.11 (Determinant conditions for concordance and discordance). Proposition 10.6.7 tells us, in condition *(ii)*, that a network of rank s is nondegenerate if and only if even one of certain $s \times s$ matrices has a nonzero determinant. In such a case, the variety of signs of the determinants of those matrices tells us unequivocally whether the network is concordant or discordant. That is the content of the following proposition.[11]

Proposition 10.6.12 ([118]). *For a reaction network of rank s, the following condition is sufficient for discordance: There is a choice of s reactions* $\{y_i^+ \to y_i'^+\}_{i=1\dots s}$ *and s distinct species* $\{\mathcal{s}_i^+\}_{i=1\dots s}$ *with* $\mathcal{s}_i^+ \in \operatorname{supp} y_i^+$, $i = 1\dots s$, *such that*

$$\operatorname{Det}\left[\mathcal{s}_i^+ \cdot (y_j'^+ - y_j^+)\right]_{i,j=1\dots s} > 0 \tag{10.25}$$

and another choice of s reactions $\{y_i^- \to y_i'^-\}_{i=1\dots s}$ *and s distinct species* $\{\mathcal{s}_i^-\}_{i=1\dots s}$ *with* $\mathcal{s}_i^- \in \operatorname{supp} y_i^-$, $i = 1\dots s$, *such that*

$$\operatorname{Det}\left[\mathcal{s}_i^- \cdot (y_j'^- - y_j^-)\right]_{i,j=1\dots s} < 0. \tag{10.26}$$

If the network is nondegenerate—in particular, if condition (ii) in Proposition 10.6.7 is satisfied—then the condition above is also necessary for discordance.

Recall that any degenerate network is discordant.

10.6.4 Every Fully Open Network Is Nondegenerate

The following corollary of Proposition 10.6.7 tells us that *fully open networks are never degenerate*. The corollary is more important than it might seem. It does much to explain, in retrospect, why success in moving beyond deficiency-oriented theory came first for studies aimed at understanding classical CFSTRs. Reaction network descriptions of classical CFSTRs are always fully open and, therefore, always nondegenerate. Development of CFSTR-motivated theory benefited greatly from the fact that network degeneracy could not have acted as a latent obstacle, complicating analysis and preventing proofs from going forward.

Corollary 10.6.13. *Every fully open network is nondegenerate.*

Proof. Suppose that $\{\mathscr{S},\mathscr{C},\mathscr{R}\}$ is fully open, in which case its rank is the number of species. Then, in item *(ii)* or item *(iii)*, we can choose the reaction set to be $\{\mathcal{s} \to 0\}_{\mathcal{s} \in \mathscr{S}}$ and the species set to be all of $\mathscr{S}$. □

[11] See Lemma C.16 in [118]. Proposition 10.6.12 is similar in spirit to Proposition 10.2 in [49], which gives determinant conditions for injectivity in mass action systems.

10.6.5 Nondegeneracy of Semi-open Reaction Networks: An Enzyme Example and the Wnt Pathway

Recall that a fully open reaction network is one for which there is an "effluent" reaction $\mathscr{S} \to 0$ for each species $\mathscr{S}$ that the network contains. There are networks of considerable practical interest for which there are effluent reactions for some species but not for others. By an *entrapped-species network*, we mean a network that is not fully open and that has the additional property that a reaction of the form $0 \to \mathscr{S}$ is always accompanied by the reaction $\mathscr{S} \to 0$. (However, a reaction of the form $\mathscr{S} \to 0$ need not be accompanied by its reverse.)

An entrapped-species network appeared in Section 4.2.2, where we considered an enzyme-driven reactor in which certain species, $S1$, $S2$, and P, were free to exit or (in the case of $S1$ and $S2$) to enter the reactor chamber. On the other hand, the enzyme-containing species $E, ES1, ES2$ and $ES1S2$ were deemed to remain forever entrapped within the chamber. The reactor was modeled by a network we repeat below as (10.27).

$$
\begin{array}{ccccc}
S1+E \rightleftarrows ES1 & & & & S2+E \rightleftarrows ES2 \\
& S2+ES1 & \rightleftarrows ES1S2 \leftrightarrows & S1+ES2 & \\
& & \downarrow & & \\
& & P+E & & \\
& & & & \\
& S1 \rightleftarrows & 0 & \leftrightarrows S2 & \\
& & \uparrow & & \\
& & P & &
\end{array}
\tag{10.27}
$$

In an entrapped-species network, an *entrapped species* is a species $\mathscr{S}$ for which there is in the network no reaction of the form $\mathscr{S} \to 0$ (or of the form $0 \to \mathscr{S}$). The remaining species are the *mobile species*. We denote by $\mathscr{E}$ the set of entrapped species and by $\mathscr{M}$ the set of mobile species. In our example $\mathscr{E} = \{E, ES1, ES2, ES1S2\}$ and $\mathscr{M} = \{S1, S2, P\}$. Recall from Section 4.2.2 that the "true" chemical reactions in (10.27) were those shown in (4.14); the remaining ones were those added to model the transport of $S1$, $S2$, and P into and out of the reactor chamber. With this in mind, we say that the *true chemistry* associated with an entrapped-species network is the subnetwork obtained by removing all reactions of the form $\mathscr{S} \to 0$ and $0 \to \mathscr{S}$. Thus, (10.28) is the true chemistry associated with the entrapped-species network (10.27).

$$
\begin{array}{ccccc}
S1+E \rightleftarrows ES1 & & & & S2+E \rightleftarrows ES2 \\
& S2+ES1 & \rightleftarrows ES1S2 \leftrightarrows & S1+ES2 & \\
& & \downarrow & & \\
& & P+E & &
\end{array}
\tag{10.28}
$$

In Corollary 10.6.4 we established that every fully open network is nondegenerate. We will be interested in the nondegeneracy of networks such as (10.27) that are "semi-open" in the sense that some but not all species are mobile. We are headed toward a theorem that gives such information, but first we'll need just a little more vocabulary. For an entrapped-species network, the *entrapped-species projection* of the network's true chemistry is the network obtained by stripping from the true chemistry all mobile species.[12] The entrapped-species projection of the true chemistry for (10.27) is shown in (10.29).

$$\begin{array}{ccccc} ES1 & \rightleftarrows & E & \rightleftarrows & ES2 \\ & \nwarrow\!\!\searrow & \uparrow & \swarrow\!\!\nearrow & \\ & & ES1S2 & & \end{array} \tag{10.29}$$

We are now in a position to connect the nondegeneracy of an entrapped-species network to the nondegeneracy of its true-chemistry entrapped-species projection. The following theorem has much in common with Proposition 11.3 in [49], a paper focused on injectivity in semi-open mass action systems.

Theorem 10.6.14. *An entrapped-species network is nondegenerate if the entrapped-species projection of its true chemistry is nondegenerate. In particular, an entrapped-species network is nondegenerate if the entrapped-species projection of its true chemistry is weakly reversible.*

Because network (10.29) is weakly reversible, Theorem 10.6.14 tells us very quickly that the entrapped-species network (10.27), which is neither fully open nor weakly reversible, is nevertheless nondegenerate.

Remark 10.6.15. That the true-chemistry entrapped-species projection in our example is weakly reversible is, in a sense, not surprising. We can expect the same result whenever the entrapped-species set consists of all species composed wholly or in part from one or more catalysts (that are not degraded)—in our example the enzyme E. In such instances, the true chemistry will exhibit a "recycling" of the catalyst to work its magic over and over again. As in our example, the catalyst recycling encourages weak reversibility in the true chemistry's entrapped-species projection.

Sketch of Proof of Theorem 10.6.14. For an entrapped-species network, we denote by s its rank, by r the rank of its true-chemistry entrapped-species projection, and by $\#(\mathscr{M})$ the number of its mobile species. It is not difficult to show that $s = r + \#(\mathscr{M})$.

Because the entrapped-species projection of the true chemistry is, by hypothesis, nondegenerate, it satisfies condition *(iii)* of Proposition 10.6.7. Thus, there is a set of distinct entrapped species $\mathscr{E}^* := \{e_i\}_{i=1\ldots r}$ and, in the true-chemistry entrapped-species projection, a set of distinct reactions $\{y_i^\dagger \to y_i'^\dagger\}_{i=1\ldots r}$ with $e_i \in \operatorname{supp} y_i^\dagger$, $i = 1\ldots r$, such that the $\mathscr{E}^*$-projected reaction vectors $\{\bar{y}_i'^\dagger - \bar{y}_i^\dagger\}_{i=1\ldots r}$ are linearly independent. From this it follows that in the original entrapped-species network, there are

[12] When the stripping results in "reactions" such as $0 \to 0$ or, more generally, "reactions" of the form $y \to y$, having the same reactant and product complexes, such "reactions" are simply omitted.

reactions $\mathscr{R}^\circ = \{y_i \to y'_i\}_{i=1\ldots r}$ with $e_i \in \operatorname{supp} y_i$, $i = 1\ldots r$, such that the corresponding $\mathscr{E}^*$- projected reaction vectors are linearly independent.

Now, for the original entrapped-species network, choose

$$\mathscr{S}^* := \mathscr{E}^* \cup \mathscr{M} \quad \text{and} \quad \mathscr{R}^* := \mathscr{R}^\circ \cup \{m \to 0, m \in \mathscr{M}\}.$$

It is not difficult to see that this choice, sequenced in the obvious way, will satisfy the requirements of condition *(iii)* of Proposition 10.6.7; in particular, the $\mathscr{S}^*$-projected reaction vectors corresponding to the members of $\mathscr{R}^*$ are linearly independent. By virtue of Proposition 10.6.7, this gives the nondegeneracy of the original entrapped-species network. □

Example 10.6.16 (Nondegeneracy of the Wnt pathway). The partial Wnt pathway displayed in (10.17) can be viewed as an entrapped-species network with mobile species A_{10}, A_{11}, and A_{12}. The entrapped-species projection of the true chemistry is shown as (10.30). Note that the projection this time is not weakly reversible, so we cannot *immediately* conclude that it is nondegenerate.

$$\begin{array}{l} A_2 + A_4 \to A_{16} \to A_2 + A_6 \\ A_1 \rightleftarrows A_2 \quad A_{13} \rightleftarrows A_{14} \\ A_3 \rightleftarrows A_4 \rightleftarrows A_6 \rightleftarrows A_7 \\ \quad \swarrow\!\!\nearrow \quad \nwarrow \\ A_8 \quad \to \quad A_9 \end{array} \tag{10.30}$$

However, (10.30) is *almost* weakly reversible, the failure coming only in the first line. This makes it easy to deduce the nondegeneracy of (10.30) and, consequently, of (10.17): First consider the subnetwork of (10.30) obtained by removing the troublesome first line. It is easy to see that this subnetwork has a rank of seven. Because the subnetwork is weakly reversible it is nondegenerate, and so for it there is a choice of seven reactions (and companion reactant species) that satisfy the requirements of (iii) in Proposition 10.6.7. (Indeed, such a choice is fairly obvious.) The rank of network (10.30) is easily seen to be eight. To satisfy the requirements of (iii) for (10.30), we need only augment the previous choice with the additional reaction $A_{16} \to A_2 + A_4$.

The fuller Wnt network obtained by adjoining to (10.17) the additional reactions $A_7 + A_{11} \rightleftarrows A_{15}$ is also nondegenerate; the argument proceeds in the same way. The nondegeneracy of both networks will play a role in the next chapter, where we will examine the networks from the viewpoint of the Species-Reaction Graph.

10.6.6 A Nondegenerate Network with a Concordant Fully Open Extension Is Itself Concordant

The following theorem connects nondegeneracy of a network, concordance of the network, and concordance of the network's fully open extension. The theorem will exert itself in important and surprising ways.

Theorem 10.6.17 ([118, 157]). *A nondegenerate network with a concordant fully open extension is itself concordant.*

Example 10.6.18 (A counterexample). Network (10.22) has a concordant fully open extension, but the network itself is discordant. There is no contradiction of the theorem because network (10.22) is degenerate. The example shows that the nondegeneracy requirement cannot be omitted.

Because every degenerate reaction network is discordant (Remark 10.6.2), Theorem 10.6.17 actually gives us an equivalence:

Corollary 10.6.19. *If a network* $\{\mathscr{S},\mathscr{C},\mathscr{R}\}$ *has a concordant fully open extension, the following are equivalent:*

(i) Network $\{\mathscr{S},\mathscr{C},\mathscr{R}\}$ *is nondegenerate.*
(ii) Network $\{\mathscr{S},\mathscr{C},\mathscr{R}\}$ *is concordant.*

Put differently, the nondegenerate networks are *precisely* the ones for which concordance of the fully open extension ensures concordance of the network itself.

Because every weakly reversible network is nondegenerate, Theorem 10.6.17 has an immediate corollary.

Corollary 10.6.20. *A weakly reversible network with a concordant fully open extension is itself concordant. Equivalently, a discordant weakly reversible reaction network has a discordant fully open extension.*

Remark 10.6.21 (About proofs of Theorem 10.6.17). Two very different proofs of Theorem 10.6.17 are available, one in [157, 158] and another in [118]. The first of these stays close to the definitions of nondegeneracy and concordance. The second, however, is heavily determinant-based. In this, the second proof is very much in the spirit of (but different from) earlier work focused on injectivity for mass action systems [46, 49].

In [46] determinant conditions were given for injectivity of *fully open* mass action systems, and these led to conditions represented in terms of the *Species-Reaction Graph* [48], again for fully open mass action systems. The problem, then, was to relax the fully open requirement. This was the subject of [49]. There it was argued, among other things, that when injectivity obtains for a network's fully open extension, then injectivity also obtains for the network itself, provided that the network is *normal*, a network property closely related to but somewhat stronger than nondegeneracy. In fact, if a network's fully open extension is concordant—the case of greatest interest—then for the original network, there is no distinction between nondegeneracy and normalcy [118].

10.6.7 Surprising "All-or-Nothing" Properties of Networks with Concordant Fully Open Extensions

In the next chapter, we shall see that a network's *Species-Reaction Graph* often tells us that the network's fully open extension is concordant. In this case there are surprising additional things that can be said. We'll begin with an example, contained in the following remark.

Remark 10.6.22. Suppose that we know, perhaps by means of the Species-Reaction Graph, that a network of interest, say $\{\mathscr{S},\mathscr{C},\mathscr{R}\}$, has a concordant fully open extension. How, then, can we ascertain that the network itself is concordant?

Imagine that we choose *some* concrete differentiably monotonic kinetics, $\mathscr{K}$, perhaps mass action with a fixed set of rate constants, and *some* positive composition c^*. For the resulting kinetic system $\{\mathscr{S},\mathscr{C},\mathscr{R},\mathscr{K}\}$, we can determine whether $df(c^*)$, the derivative at c^* of the resulting species-formation-rate function, is singular; calculation of a determinant is all that is needed.

There are two possibilities: the derivative is singular or it is not. *Either way, the concordance of the network $\{\mathscr{S},\mathscr{C},\mathscr{R}\}$ is decided.* If the derivative is singular, then the network is *not* concordant. This follows from Theorem 10.5.15. Suppose, on the other hand, that $df(c^*)$ is nonsingular. In this case the network $\{\mathscr{S},\mathscr{C},\mathscr{R}\}$ *is* concordant: Because the derivative is nonsingular, the network is nondegenerate (Definition 10.6.1). Then, because the fully open extension of $\{\mathscr{S},\mathscr{C},\mathscr{R}\}$ is concordant, the network itself is concordant (Theorem 10.6.17).

Determination of the concordance or discordance of network $\{\mathscr{S},\mathscr{C},\mathscr{R}\}$ began with two choices—first, a concrete choice of the differentiably monotonic kinetics $\mathscr{K}$ and, second, a choice of the fixed positive composition c^*. Both contributed in very tangible ways to the particular value of the derivative $df(c^*)$, whose singularity or nonsingularity determined the discordance or concordance of network $\{\mathscr{S},\mathscr{C},\mathscr{R}\}$. It is interesting to consider whether the choice of $\mathscr{K}$ or of c^* matters to the outcome.

Clearly, they don't, for the network is either concordant or it is not, and the *particular* path to the determination of one or the other result should make no difference. In fact, when the fully open extension of network $\{\mathscr{S},\mathscr{C},\mathscr{R}\}$ is concordant, it is a consequence of Theorems 10.5.15 and 10.6.17 that there are only two possibilities, these being "all-or-nothing" alternatives: *(i) for* every *choice of a differentiably monotonic kinetics for network $\{\mathscr{S},\mathscr{C},\mathscr{R}\}$, the derivative of the resulting species-formation-rate function is nonsingular at* all *positive compositions or (ii) for* no *choice of differentiably monotonic kinetics is the derivative nonsingular at* any *positive composition.*

This is to be contrasted with the behavior of network (10.31), which was a small animal exhibited in our toy zoo (Section 5.1.3):

$$
\begin{gathered}
B \to A \\
A + 2B \to 3B.
\end{gathered}
\tag{10.31}
$$

When the kinetics is mass action—the rate constants don't matter—the derivative of the species-formation-rate function is singular at one positive equilibrium and nonsingular at others. (A phase portrait was shown in Figure 5.4.) Although network (10.31) is nondegenerate, it has a discordant fully open extension.

When a network's fully open extension is known to be concordant, Remark 10.6.22, with its choose-any-kinetics, choose-any-composition, all-or-nothing methodology, provides a seemingly magical route to the determination of the concordance or discordance of the original network. Upon reflection, however, it is also a strange route, *for neither kinetics nor compositions enter at all into the definition of concordance*. It would seem, then, that lurking beneath the veneer of Remark 10.6.22 must be something more fundamental, something closer to the way in which concordance was defined at the outset.

There are already intimations of this in the combination of two results already stated: (i) Corollary 10.6.19, which, when a network's fully open extension is concordant, equates concordance of the original network to its nondegeneracy and (ii) Proposition 10.6.7, which gives tests for nondegeneracy of a network based solely on the network's structure, divorced from any mention of kinetics or compositions.

However, Proposition 10.6.7 does not exploit, as Remark 10.6.22 does, the "all-or-nothing" magic that derives from the stipulation that the network's fully open extension is concordant. We will examine that magic more closely, from a more fundamental perspective having nothing to do with kinetics or compositions. The result will be the "all-or-nothing" Proposition 10.6.27, which in turn will give rise to two decisive and practical tests for the determination of concordance of a network when it is known that the network has a concordant fully open extension.

First, we'll begin with two propositions [118, 157, 158] that are very similar in appearance; neither makes mention of a network's fully open extension. The first of our preliminary propositions gives an additional condition for network nondegeneracy beyond the two already given in Proposition 10.6.7.

Proposition 10.6.23. *For a reaction network* $\{\mathscr{S},\mathscr{C},\mathscr{R}\}$ *with stoichiometric subspace S, the following are equivalent:*

(i) The network is nondegenerate.

*(ii) For **some** choice of* $\{p_{y\to y'}\}_{y\to y'\in\mathscr{R}} \subset \overline{\mathbb{R}}_+^{\mathscr{S}}$, *with* $\operatorname{supp} p_{y\to y'} = \operatorname{supp} y$, $\forall y \to y' \in \mathscr{R}$, *the linear transformation* $T : S \to S$ *given by* (10.32) *is nonsingular.*

$$T\sigma = \sum_{y\to y'\in\mathscr{R}} p_{y\to y'} \cdot \sigma(y'-y) \tag{10.32}$$

A connection between concordance and the linear transformation (10.32) in Proposition 10.6.23 derives from a simple idea used in various ways in [157, 158]; in one variant, the idea is expressed in the following lemma:

Lemma 10.6.24. *Suppose that* $\{\alpha^*,\sigma^*\}$ *is a discordance for the reaction network* $\{\mathscr{S},\mathscr{C},\mathscr{R}\}$. *Then, for each reaction* $y \to y' \in \mathscr{R}$, *there is a vector* $p_{y\to y'} \in \overline{\mathbb{R}}_+^{\mathscr{S}}$ *with* $\operatorname{supp} p_{y\to y'} = \operatorname{supp} y$, *such that* $\alpha^* = p_{y\to y'} \cdot \sigma^*$.

The lemma provides the basis for our second proposition:

Proposition 10.6.25. *For a reaction network* $\{\mathscr{S},\mathscr{C},\mathscr{R}\}$ *with stoichiometric subspace S, the following are equivalent:*

(i) The network is concordant.

*(ii) For **every** choice of* $\{p_{y\to y'}\}_{y\to y'\in\mathscr{R}} \subset \overline{\mathbb{R}}_+^{\mathscr{S}}$, *with* $\operatorname{supp} p_{y\to y'} = \operatorname{supp} y$, $\forall y \to y' \in \mathscr{R}$, *the linear transformation* $T : S \to S$ *given by* (10.32) *is nonsingular.*

Remark 10.6.26. In our early discussions of what we now call concordant reaction networks, Guy Shinar and I began by studying networks having property *(ii)* in Proposition 10.6.25. Motivated by earlier ideas for mass action systems given in [46] and [49] (and discussed in Appendix 10.B), the aim was to study networks for which the derivative of the species-formation-rate function is, for every differentiably monotonic kinetics, nonsingular at every positive composition. (For a putative differentiably monotonic kinetic system, each $p_{y\to y'}$ was associated with the gradient of $\mathscr{K}_{y\to y'}(\cdot)$ at some particular positive composition. See, for example, Appendix D in [157].) Very early, though, it became clear, through the observation in Lemma 10.6.24, that property *(ii)* was equivalent to the concordance property as given in Definition 10.4.1, a network property that connected far more naturally with the injectivity idea given in Definition 10.5.3.

Now consider a reaction network $\{\mathscr{S},\mathscr{C},\mathscr{R}\}$ with a concordant fully open extension. From Corollary 10.6.19 nondegeneracy of $\{\mathscr{S},\mathscr{C},\mathscr{R}\}$ is *equivalent* to its concordance. In this case, *all* of the items in Propositions 10.6.23 and 10.6.25 are equivalent. In particular we have the following "all-or-nothing" proposition:

Proposition 10.6.27 (An "all or nothing" proposition). *Suppose that the network $\{\mathscr{S},\mathscr{C},\mathscr{R}\}$, with stoichiometric subspace S, has a concordant fully open extension. The following are equivalent:*

*(i) For **some** choice of $\{p_{y\to y'}\}_{y\to y'\in\mathscr{R}} \subset \overline{\mathbb{R}}_+^{\mathscr{S}}$, with $\operatorname{supp} p_{y\to y'} = \operatorname{supp} y$, $\forall y \to y' \in \mathscr{R}$, the linear transformation $T : S \to S$ given by (10.32) is nonsingular.*

*(ii) For **every** choice of $\{p_{y\to y'}\}_{y\to y'\in\mathscr{R}} \subset \overline{\mathbb{R}}_+^{\mathscr{S}}$, with $\operatorname{supp} p_{y\to y'} = \operatorname{supp} y$, $\forall y \to y' \in \mathscr{R}$, the linear transformation $T : S \to S$ given by (10.32) is nonsingular.*

Example 10.6.28 (A network for which "all or nothing" fails). Consider the network

$$2A \leftarrow A + B \rightarrow 2B. \tag{10.33}$$

In this case, S is one-dimensional and consists of all vectors of the form $\alpha(B-A)$. For the choice $p_{A+B\to 2A} = p_{A+B\to 2B} = A+B$, the linear transformation $T : S \to S$ given by (10.32) is singular. On the other hand, nonsingularity results for the different choice $p_{A+B\to 2A} = A+B$, $p_{A+B\to 2B} = A+2B$. Because the fully open extension of network (10.33) is discordant, the hypothesis of Proposition 10.6.27 is not satisfied.

As a prelude to Propositions 10.6.29 and 10.6.30, it will be useful to repeat what was said earlier: In Chapter 11 we shall see that if the Species-Reaction Graph for a network $\{\mathscr{S},\mathscr{C},\mathscr{R}\}$ satisfies certain mild conditions, then we will be able to assert that the network's fully open extension is concordant. We will want to know if network $\{\mathscr{S},\mathscr{C},\mathscr{R}\}$ is itself concordant. This amounts to determining whether the network is nondegenerate.

For this purpose, Proposition 10.6.27 has dramatic consequences: It tells us that, in applying the nondegeneracy test in Proposition 10.6.23, one needn't be concerned with searching for one of perhaps very few choices of

$$\{p_{y\to y'}\}_{y\to y'\in\mathscr{R}} \subset \overline{\mathbb{R}}_+^{\mathscr{S}} \text{ with } \operatorname{supp} p_{y\to y'} = \operatorname{supp} y,\ \forall y \to y' \in \mathscr{R}$$

that will make the linear transformation (10.32) nonsingular. Because the fully open extension of $\{\mathscr{S},\mathscr{C},\mathscr{R}\}$ is known to be concordant, Proposition 10.6.27 tells us that *any choice will do!* Each choice will give the same result.

In particular, we can choose $p_{y\to y'} = y$ for all $y \to y' \in \mathscr{R}$ to give the following proposition:

Proposition 10.6.29. *Suppose that network $\{\mathscr{S},\mathscr{C},\mathscr{R}\}$, with stoichiometric subspace S, has a concordant fully open extension. The following are equivalent:*

(i) Network $\{\mathscr{S},\mathscr{C},\mathscr{R}\}$ is itself concordant.
(ii) The linear transformation $T: S \to S$ given by (10.34) *is nonsingular or, equivalently,* $\det T \neq 0$.

$$T\sigma = \sum_{y\to y'\in\mathscr{R}} y\cdot\sigma(y'-y) \tag{10.34}$$

The following proposition is closely related to Proposition 10.6.29.[13]

Proposition 10.6.30. *Consider a reaction network $\{\mathscr{S},\mathscr{C},\mathscr{R}\}$ of rank s. The network is nondegenerate if it satisfies the following condition: There is a set of s reactions $\{y_i \to y'_i\}_{i=1...s}$ such that the matrix*

$$\left[y_i\cdot(y'_j - y_j)\right]_{i,j=1...s} \tag{10.35}$$

has nonzero determinant. If $\{\mathscr{S},\mathscr{C},\mathscr{R}\}$ has a concordant fully open extension, then this same nonzero determinant condition is not only sufficient but also necessary for the nondegeneracy of $\{\mathscr{S},\mathscr{C},\mathscr{R}\}$.

10.7 Behavioral Consequences of Concordance and Discordance, Part II

Here we resume our discussion of connections between concordance and behavior when the kinetics is weakly monotonic and, in particular, when the kinetics is differentiably monotonic. This time, however, our emphasis will be on things that can be said when it is known that a network's fully open extension is concordant. As we shall see in the next chapter, information of this kind very often comes quickly from study of the network's Species-Reaction Graph. It will be useful to keep in

[13] The first part of Proposition 10.6.30 is argued in [158]. The "necessary" part derives from Proposition 10.6.29 and Proposition C.10 in [118].

mind something we know from the preceding section: when a network's fully open extension is concordant, the network itself is concordant, so long as the network is not degenerate.

Remark 10.7.1. Especially in consideration of classical continuous-flow stirred-tank reactors, the following is worth noting: Suppose that a network $\{\mathscr{S},\mathscr{C},\mathscr{R}\}$ has a concordant fully open extension. It is not difficult to argue that also concordant will be any augmentation of the fully open extension with a combination of feed reactions of the form $0 \rightarrow s$, $s \in \mathscr{S}$.

10.7.1 Concordance and Eigenvalues

Stated informally, one of the principal and most useful ideas in the theory of ordinary differential equations is this: If $\dot{x} = g(x)$ indicates the vector form of a system of ordinary differential equations, then an equilibrium of that system x^* is asymptotically stable if all eigenvalues[14] of the derivative of $g(\cdot)$ at x^* have negative real part. If even one eigenvalue has a positive real part, then x^* is unstable [12, 104].

Here we shall be mainly concerned with what can be said about eigenvalues associated with the derivative of the species-formation-rate function, not just at positive equilibria but also at all positive compositions. From Theorem 10.5.14 we know that, for a differentiably monotonic kinetic system in which the underlying reaction network is concordant, the derivative of the species-formation rate is nonsingular at every positive composition, whereupon zero is never an eigenvalue.

When a network's fully open extension is concordant, there is something different and more dramatic that can be said. In the following theorem, which subsumes Theorem 10.5.14, we do not preclude the possibility that a reaction network's fully open extension is concordant while the network itself is discordant (and therefore degenerate). In the theorem statement, the composition c^* is not necessarily an equilibrium.

Theorem 10.7.2 ([157]). *Let $\{\mathscr{S},\mathscr{C},\mathscr{R},\mathscr{K}\}$ be a kinetic system with stoichiometric subspace S and species-formation-rate function $f : \overline{\mathbb{R}}_+^{\mathscr{S}} \rightarrow S$. Moreover, suppose that the kinetics is differentiably monotonic at $c^* \in \mathbb{R}_+^{\mathscr{S}}$. If the underlying network $\{\mathscr{S},\mathscr{C},\mathscr{R}\}$ is concordant, then the derivative $df(c^*) : S \rightarrow S$ is nonsingular (whereupon 0 is not one of its eigenvalues). If the network's fully open extension is concordant, then no real eigenvalue of $df(c^*)$ is positive.*

In contrast to Theorem 10.7.2, we stipulate in the following corollary that the reaction network under consideration is nondegenerate. In this case, concordance of the network's fully open extension gives concordance of the network itself.

[14] Strictly speaking, a map on a real vector space has only real eigenvalues. We are using "eigenvalue" here as a synonym for a root of the characteristic polynomial.

Corollary 10.7.3 ([157]). *Let $\{\mathscr{S},\mathscr{C},\mathscr{R},\mathscr{K}\}$ be a kinetic system with stoichiometric subspace S and species-formation-rate function $f : \overline{\mathbb{R}}_+^{\mathscr{S}} \to S$. Moreover, suppose that the kinetics is differentiably monotonic at $c^* \in \mathbb{R}_+^{\mathscr{S}}$. If the underlying network $\{\mathscr{S},\mathscr{C},\mathscr{R}\}$ is nondegenerate—in particular, if it is weakly reversible—and if the network's fully open extension is concordant, then every real eigenvalue of $df(c^*)$ is negative.*

Example 10.7.4 (The Wnt pathway again). As can be confirmed by means of [62], network (10.17) is not only concordant, it is also nondegenerate and has a concordant fully open extension. So long as it is taken with a differentiably monotonic kinetics, at each positive composition—in particular, at a positive equilibrium—every real eigenvalue of the derivative of the species-formation rate function is negative.

Remark 10.7.5 (The extent of dull behavior enforced by concordance). For a differentiably monotonic kinetic system in which the underlying network and its fully open extension are both concordant, we now have a considerable amount of substantive information about equilibria: We know from Corollary 10.5.6 that a positive equilibrium is the *only* equilibrium in the stoichiometric compatibility class containing it. Corollary 10.7.3 gives us *some* information about the stability of that equilibrium: all *real* eigenvalues associated with the equilibrium are *negative*.

Corollary 10.7.3 does not, however, preclude the possibility that there are complex eigenvalues with positive (or merely nonnegative) real part. We might hope that the corollary tells only part of a larger, more wonderful story, that concordance of the network and of its fully open extension actually suffice to ensure that the *real parts* of all eigenvalues are negative, for then we could assert the equilibrium's asymptotic stability.

That hope, unmoderated, can lead only to heartbreak. Even when both a network and its fully open extension are concordant, and even when the kinetics is differentiably monotonic, instability of a unique positive equilibrium cannot be excluded: Such an equilibrium, associated with complex eigenvalues having positive real part, exists for a kinetic system studied by Reidl et al. [141], in which network (10.36), taken with a differentiably monotonic (almost mass action) kinetics, served as a model for calcium oscillations in olfactory cilia. As determined by [62], network (10.36) and its fully open extension are both concordant.

$$
\begin{gathered}
A \rightleftarrows B \to D + B \\
C + 4D \rightleftarrows E \\
B + E \to F \rightleftarrows A + E \\
D \to 0
\end{gathered}
\tag{10.36}
$$

The example provides a cautionary lesson: When the kinetics is differentiably monotonic, concordance of a network (and of its fully open extension) enforce a *degree* of dull behavior, to the extent that they preclude multiple stoichiometrically compatible equilibria (and positive real eigenvalues). In particular instances, however, the capacity for other forms of interesting behavior might nevertheless remain.

We shall see in the next chapter that there are certain cases in which we can assert that eigenvalues are *all* real. In those cases, and when the conditions of Corollary 10.7.3 obtain, asymptotic stability of equilibria is ensured.

10.7.2 A Summary of Some Results for Nondegenerate Networks with Concordant Fully Open Extensions

In the following theorem, we summarize some of what can be said for a nondegenerate network—in particular, any weakly reversible network—having a concordant fully open extension. The summary will be especially useful in the next chapter, for there we will describe graphical means for establishing that a reaction network's fully open extension is concordant.

Theorem 10.7.6 (Properties of a nondegenerate network having a concordant fully open extension). *For any nondegenerate network with a concordant fully open extension, the following statements hold true:*

(i) For any choice of a weakly monotonic kinetics, each positive equilibrium is the only equilibrium within its stoichiometric compatibility class. That is, no positive equilibrium is stoichiometrically compatible with a different equilibrium, positive or otherwise. If the kinetics is differentiably monotonic, then every real eigenvalue[15] *associated with a positive equilibrium is strictly negative.*

(ii) If the network is weakly reversible, then, for each choice of a kinetics (not necessarily weakly monotonic), no nontrivial stoichiometric compatibility class has an equilibrium on its boundary. In fact, at each boundary composition in any nontrivial stoichiometric compatibility class, the species-formation-rate vector points into the stoichiometric compatibility class in the sense that there is an absent species produced at strictly positive rate. If, in addition, the network is conservative, then, for any[16] *choice of a weakly monotonic kinetics, there is precisely one equilibrium in each nontrivial stoichiometric compatibility class, and it is positive.*

10.7.3 Discordance and Eigenvalues: The Certainty of a Kinetics for Which There Is an Unstable Positive Equilibrium

Because in the preceding subsection our assertions were about the negativity of *real* eigenvalues, we were somewhat equivocal about implications of concordance for the stability of equilibria against all possible differentiably monotonic kinetics. Here we

[15] As usual, we are referring here to eigenvalues associated with eigenvectors in the network's stoichiometric subspace.

[16] Recall that, for reasons explained in Remark 3.2.3, continuity is built into the requirements of a general kinetics. In Remark 10.5.9 we pointed out that continuity is not essential to some of the assertions that can be made for concordant networks, but it is essential here.

will be able to make a statement that is not equivocal at all: *For every nondegenerate discordant network with positively dependent reaction vectors*[17]*—in particular, for every weakly reversible discordant network—there is invariably a differentiably monotonic kinetics (in fact, a power-law kinetics) such that the resulting kinetic system admits an unstable positive equilibrium.*

This tells us the essential role that concordance *must* play in the design of networks that are to ensure dull, stable behavior even against extremely broad assignments of reaction-rate functions. Among networks with positively dependent reaction vectors, *concordance is necessary in any network for which an unstable positive equilibrium is to be thwarted no matter which differentiably monotonic kinetics the network might carry.*

We begin with a theorem.

Theorem 10.7.7 ([157]). *Consider a reaction network with positively dependent reaction vectors. If the network is discordant, then there exists for it a differentiably monotonic kinetics such that the resulting kinetic system admits a positive* degenerate *equilibrium. If the network's fully open extension is discordant, then there exists for the original network a differentiably monotonic kinetics such that the resulting kinetic system admits an unstable positive equilibrium—in fact, a positive equilibrium associated with a positive real eigenvalue.*

Remark 10.7.8. In [157] the proof of each part of Theorem 10.7.7 proceeds by construction of a power-law kinetics (Remark 10.5.7). Thus, the theorem can also be stated in stronger terms, asserting in each part the existence of a power-law kinetics that fulfills the theorem's conclusions.

In each of the following corollaries, the unstable positive equilibrium mentioned can be taken to have associated with it a positive real eigenvalue.

Corollary 10.7.9. *For every discordant nondegenerate network with positively dependent reaction vectors, there is a differentiably monotonic kinetics such that the resulting kinetic system admits a positive unstable equilibrium.*

Proof. It follows from Theorem 10.6.17 that every nondegenerate discordant network has a discordant fully open extension. Corollary 10.7.9 is then a direct consequence of Theorem 10.7.7. □

Corollary 10.7.10. *For every weakly reversible discordant network, there is a differentiably monotonic kinetics such that the resulting kinetic system admits a positive unstable equilibrium.*

Proof. A weakly reversible network is nondegenerate (Theorem 10.6.6) and, moreover, has positively dependent reaction vectors. Corollary 10.7.10 is then a consequence of Corollary 10.7.9. □

---

[17] Recall from Lemma 3.5.2 that positively dependent reaction vectors are necessary for the existence of positive equilibrium.

Remark 10.7.11. Although concordance of a weakly reversible network is necessary if an unstable positive equilibrium is to be precluded against all possible differentiably monotonic kinetics, a similar preclusion can obtain for a *discordant* weakly reversible network, but for a narrower class of kinetics. For example, the discordant reversible deficiency zero network (10.11) can admit no unstable positive equilibrium, provided the kinetics is mass action. In fact, Theorem 7.3.1—a variant of the Deficiency Zero Theorem—ensures for (10.11) that, when the kinetics is mass action, all eigenvalues associated with a positive equilibrium are invariably real and negative. (The same information is given for the concordant reversible deficiency zero networks (10.10) and (10.12).) Again, see Appendix 10.B for a discussion of what can be said when the kinetics is mass action.

10.7.4 Concordant Weakly Reversible Networks with Discordant Fully Open Extensions: Sustained Periodic Composition Oscillations

Although a weakly reversible network with a concordant fully open extension must itself be concordant, it remains possible for a weakly reversible concordant network to have a discordant fully open extension. Networks (10.37) and (10.38) provide two such examples (as confirmed by [62]). The first was studied for a different purpose in [148]. The second is a weakly reversible version (resulting from the addition of $3Y \to X + 2Y$) of a network that, taken with mass action kinetics, induces differential equations studied by Tyson and Kauffman [166] and by Golubitsky and Langford [94].

$$\begin{array}{c} A + B \rightleftarrows E \qquad C + D \rightleftarrows F \\ C \rightleftarrows 2A \rightleftarrows D \rightleftarrows B \end{array} \tag{10.37}$$

$$\begin{array}{ccc} 0 & \rightarrow & X \\ & \nwarrow \quad \swarrow & \\ & Y & \\ \\ & X + 2Y \rightleftarrows 3Y & \end{array} \tag{10.38}$$

The following theorem is a compendium of results we've already stated, this time applied specifically to concordant weakly reversible networks with discordant fully open extensions.

Theorem 10.7.12 ([157]). *For every weakly reversible concordant reaction network with a discordant fully open extension, there exists for the original network a differentiably monotonic kinetics such that the resulting kinetic system (for the original network) has the following properties: There is an equilibrium that is positive, unique within its stoichiometric compatibility class, and* unstable. *Moreover, at each point on the boundary of that stoichiometric compatibility class, the species-formation-rate vector points inward in the sense that there is an absent species produced at strictly positive rate.*

For a kinetic system of the kind described in Theorem 10.7.12, a provocative picture emerges, especially when the underlying reaction network is conservative. (In that case stoichiometric compatibility classes are closed and bounded (Proposition 4.1.2).) Instability of the equilibrium in the theorem statement implies the existence of a trajectory, lying within that equilibrium's stoichiometric compatibility class, that begins near the equilibrium and then moves away from it.

But what is that trajectory's fate? It cannot proceed toward a different stoichiometrically compatible equilibrium because no such equilibria exist. And, if the network is conservative, the trajectory cannot "go off to infinity" because the stoichiometric class in which the trajectory resides is bounded. A strong possibility is that the trajectory is itself cyclic or else that it approaches a cyclic composition trajectory with increasing time. In any case, the presence of a cyclic composition trajectory for at least *some* (if not most) choices of a differentiably monotonic kinetics seems likely.

Indeed, *for every weakly reversible (not necessarily conservative) concordant reaction network with a discordant fully open extension, there are very good mathematical reasons to expect the existence of a differentiably monotonic kinetics such that the resulting kinetic system admits a nonconstant cyclic composition trajectory*—reasons that presage a Hopf bifurcation [94, 95, 99, 151, 164] in the presence of kinetic variations:

Consider a weakly reversible concordant network with discordant fully open extension. Because the network is weakly reversible, we can assign to it a complex balanced mass action kinetics [71, 107, 109]. In this case, the resulting mass action system will have the property that, at the unique equilibrium in each nontrivial stoichiometric compatibility class, all eigenvalues of the derivative of the species-formation-rate function have strictly negative real parts.[18] Theorem 10.7.7 ensures that, for the same network, there exists a different differentiably monotonic kinetics (in fact, a power-law kinetics) for which there is a nontrivial stoichiometric compatibility class SCC^* containing a unique (positive) equilibrium associated with a positive real eigenvalue. Along a suitably parameterized transition from the first differentiably monotonic kinetics to the second, we expect that a pair of complex-conjugate eigenvalues associated with the unique positive (parameterized) equilibrium in SCC^* will cross the imaginary axis. In fact, so long as the parameterized kinetics remains differentiably monotonic, Theorem 10.7.2 and concordance of the original network ensure that the crossing cannot be through the origin of the complex plane. The most substantive requirement of the Hopf Bifurcation Theorem is thereby satisfied. Apart perhaps from failure of highly technical and almost-always-satisfied other requirements, a cyclic composition trajectory is to be expected for a kinetics along the parameterized family.

---

[18] The construction of a complex balanced mass action kinetics and its consequences are related to the discussion in Section 7.7. We will have more to say about these things in Part III of this book.

10.8 Discordance, Degeneracy, and an Excess of Terminal Strong-Linkage Classes

In the next section, we will move on to consideration of *strong concordance*, a strengthened version of concordance that has more potent consequences. Before we do that, however, we'll connect network discordance and network degeneracy to network pathologies considered in earlier chapters, in particular to pathologies resulting from an excess of terminal strong-linkage classes.

We'll list these connections [118] and then provide some discussion of why they are indeed true. In what follows δ, ℓ, and t denote, respectively, a network's deficiency, the number of its linkage classes, and the number of its terminal strong-linkage classes.

(i) A network for which $t - \ell - \delta > 0$ is invariably discordant.
(ii) Among networks that have positively dependent reaction vectors, those that satisfy $t - \ell > 0$ and $t - \ell - \delta \geq 0$ are all discordant.
(iii) A network for which $t - \ell - \delta > 0$ can nevertheless be nondegenerate, but only if its fully open extension is discordant.
(iv) A network for which $t - \ell - \delta \leq 0$ can be degenerate (and therefore discordant).

Although these various items make sense on their own, their theoretical underpinnings derive from material discussed earlier in Appendices 3.A and 8.A, appendices devoted to discussion of the *kinetic subspace* for a kinetic system $\{\mathscr{S},\mathscr{C},\mathscr{R},\mathscr{K}\}$ and its relation to the stoichiometric subspace of the underlying network $\{\mathscr{S},\mathscr{C},\mathscr{R}\}$. (See also Appendix 16.B in Part III.) Recall that the kinetic subspace for a particular kinetic system is the smallest linear subspace of $\mathbb{R}^{\mathscr{S}}$ that contains the image of the species-formation-rate function. The kinetic subspace is always contained in the stoichiometric subspace, and it might or might not be smaller. We state here, as a lemma, a fact proved earlier in Remark 3.A.4 of Appendix 3.A

Lemma 10.8.1. *Consider a kinetic system $\{\mathscr{S},\mathscr{C},\mathscr{R},\mathscr{K}\}$ with kinetic subspace K, stoichiometric subspace S, and species-formation-rate function $f : \overline{\mathbb{R}}_+^{\mathscr{S}} \to S$. If K is smaller than S, then, at* every *composition $c \in \mathbb{R}_+^{\mathscr{S}}$, the derivative $df(c) : S \to S$ is singular.*

Placing this lemma alongside Theorems 10.5.14 and 10.6.17, we get as a result the following proposition:

Proposition 10.8.2. *Suppose that for reaction network $\{\mathscr{S},\mathscr{C},\mathscr{R}\}$ there is a differentiably monotonic kinetics $\mathscr{K}$ such that, for the kinetic system $\{\mathscr{S},\mathscr{C},\mathscr{R},\mathscr{K}\}$, the kinetic subspace is smaller than the stoichiometric subspace. Then the network $\{\mathscr{S},\mathscr{C},\mathscr{R}\}$ is discordant. If the network's fully open extension is concordant, then the network is also degenerate.*

We are now in a position to provide detailed justifications for items (i)–(iv) in our list. We begin with items (i) and (ii), which are both consequences of Proposition 10.8.2 and Theorem 8.A.2 in Appendix 8.A:

For a reaction network that satisfies the condition of (i), simply assign to it any mass action kinetics; the rate constants are of no consequence. By virtue of Theorem 8.A.2, the resulting mass action system will have a kinetic subspace that is smaller than the stoichiometric subspace. Discordance of the network then follows from Proposition 10.8.2.

For a reaction network that satisfies the conditions of (ii), we can, by virtue of Lemma 3.5.3, choose for it an assignment of rate constants such that the resulting mass action system admits a positive equilibrium. Theorem 8.A.2 then tells us that the resulting kinetic subspace will be smaller than the stoichiometric subspace. Discordance of the network again follows from Proposition 10.8.2.

With regard to (iii), consider network (10.39). For the network $\delta = 0$, $t = 2$, and $\ell = 1$. Thus, we have $t - \ell - \delta > 0$. By means of any of several methods given in Section 10.6 or by means of [62], it is easy to confirm that the network is nondegenerate. By virtue of (i) the network is discordant, so Theorem 10.6.17 requires that the network's fully open extension be discordant. It is, as can be confirmed by [62].

$$C \leftarrow A + B \rightarrow D \rightarrow 2A \tag{10.39}$$

In the case of item (iv), the degenerate network (10.22) serves as an example. In fact, $\delta = 1$, $t = 3$, and $\ell = 3$, so we actually have the strict inequality $t - \ell - \delta < 0$.

10.9 Strong Concordance and Product Inhibition

Especially in coarse-grained models of biochemical systems, it is not uncommon for an occurrence-rate function governing a particular reaction to reflect "product inhibition," which is to say the rate of the reaction might decrease when the concentration of a product species increases. Consider, for example, a reaction network containing the reaction $A + B \rightarrow C$. The rate function for the reaction might take the form

$$\mathscr{K}_{A+B\rightarrow C}(c) = \frac{\alpha c_A c_B}{\beta + \gamma c_C}, \tag{10.40}$$

a form intended to capture the idea that the product species C is inhibiting the reaction producing it. In this case, the kinetics for the network could not be weakly monotonic: We might, for example, have $\mathscr{K}_{A+B\rightarrow C}(c^{**}) < \mathscr{K}_{A+B\rightarrow C}(c^{*})$ even when at composition c^{**} the concentrations of A and B are both greater than they are at composition c^{*}.

When the kinetics is weakly monotonic, the rate of reaction $y \rightarrow y'$ is affected at most by the concentrations of the *reactant* species—that is, by members of the set $\text{supp}\, y$; the concentrations of the product species, residing in $\text{supp}\, y'$, have no effect. To accommodate models that embrace product inhibition, we will want to consider a broader class of kinetic that includes all instances of weakly monotonic kinetics.

So that there is no ambiguity in what is meant by a product or a reactant species for a particular reaction, *we will, for the remainder of this subsection, add the stipulation that for any reaction considered, the same species cannot be both a reactant and a product—i.e., that, for any reaction* $y \to y'$, $\operatorname{supp} y \cap \operatorname{supp} y'$ *is empty*. Note that, although we temporarily preclude from consideration a reaction such as $A + B \to 2A$, we do not preclude the more chemically plausible reaction sequence $A + B \to C \to 2A$.

By a *two-way weakly monotonic* kinetics, we mean one in which an increase in the rate of a given reaction requires that there be either an increase in the concentration of a reactant species or else a decrease in the concentration of a product species. Moreover, should the rate remain unchanged for a particular change in composition, we require that, in that composition change, there be no change in the concentration of any reactant species or else that there be appropriately opposing changes in concentrations of the reactant and/or product species. We make this precise in the following definition.

Definition 10.9.1 ([157]). *A kinetics* $\mathscr{K}$ *for a reaction network* $\{\mathscr{S}, \mathscr{C}, \mathscr{R}\}$ *is* **two-way weakly monotonic** *if, for each pair of compositions* c^* *and* c^{**}, *the following implications hold for each reaction* $y \to y' \in \mathscr{R}$ *such that* $\operatorname{supp} y \subset \operatorname{supp} c^*$ *and* $\operatorname{supp} y \subset \operatorname{supp} c^{**}$*:*

(i) $\mathscr{K}_{y\to y'}(c^{**}) > \mathscr{K}_{y\to y'}(c^*) \Rightarrow$ *there is a species* $\mathscr{s}$ *such that* $\operatorname{sgn}(c^{**}_{\mathscr{s}} - c^*_{\mathscr{s}}) = \operatorname{sgn}(y - y')_{\mathscr{s}} \neq 0$.

(ii) $\mathscr{K}_{y\to y'}(c^{**}) = \mathscr{K}_{y\to y'}(c^*) \Rightarrow c^{**}_{\mathscr{s}} = c^*_{\mathscr{s}}$ *for all* $\mathscr{s} \in \operatorname{supp} y$, *or else there are species* $\mathscr{s}, \mathscr{s}'$ *with* $\operatorname{sgn}(c^{**}_{\mathscr{s}} - c^*_{\mathscr{s}}) = \operatorname{sgn}(y - y')_{\mathscr{s}} \neq 0$ *and* $\operatorname{sgn}(c^{**}_{\mathscr{s}'} - c^*_{\mathscr{s}'}) = -\operatorname{sgn}(y - y')_{\mathscr{s}'} \neq 0$.

Remark 10.9.2. The class of two-way weakly monotonic kinetics *subsumes* the weakly monotonic kinetic class. Definition 10.9.1 is intended to permit, *but not require*, the concentration of a reaction's product species to influence its rate. Note that, in the definition, the reactant and product species are not on an equal footing: As with the special case of mass action kinetics, the rate of a reaction can remain unchanged even when the concentrations of all of the reaction's *product* species increase, the concentrations of all other species remaining fixed. This is not true of the reaction's *reactant* species. An increase in the concentrations of all of its reactant species, the others remaining fixed, *must* result in an increase in the reaction rate. It should be kept in mind that *any kinetics that is weakly monotonic is also two-way weakly monotonic*.

To the extent possible, we will want to extend what we've been able to say about the behavior of concordant weakly monotonic kinetic systems to the wider two-way weakly monotonic kinetic class. Because the two-way weakly monotonic kinetic class is far broader than the weakly monotonic class, analogous behavior assertions will not hold true for the full family of concordant reaction networks. Instead, they will hold true for the somewhat narrower class of *strongly concordant* networks. In the definition of strong concordance, the linear map $L : \mathbb{R}^{\mathscr{R}} \to S$ is defined just as it was earlier:

$$L\alpha = \sum_{y\to y' \in \mathscr{R}} \alpha_{y\to y'}(y' - y). \tag{10.41}$$

Definition 10.9.3 ([157]). *A reaction network $\{\mathscr{S},\mathscr{C},\mathscr{R}\}$ with stoichiometric subspace S is* **strongly concordant** *if there do not exist $\alpha \in \ker L$ and a nonzero $\sigma \in S$ having the following properties:*

(i) For each $y \to y'$ such that $\alpha_{y\to y'} > 0$, there exists a species s for which $\operatorname{sgn}\sigma_s = \operatorname{sgn}(y-y')_s \neq 0$.
(ii) For each $y \to y'$ such that $\alpha_{y\to y'} < 0$, there exists a species s for which $\operatorname{sgn}\sigma_s = -\operatorname{sgn}(y-y')_s \neq 0$.
(iii) For each $y \to y'$ such that $\alpha_{y\to y'} = 0$, either (a) $\sigma_s = 0$ for all $s \in \operatorname{supp} y$ or (b) there exist species s, s' for which $\operatorname{sgn}\sigma_s = \operatorname{sgn}(y-y')_s \neq 0$ and $\operatorname{sgn}\sigma_{s'} = -\operatorname{sgn}(y-y')_{s'} \neq 0$.

A network that is strongly concordant is also concordant. Network (10.42) is concordant but not strongly concordant:

$$\begin{aligned} A+B &\to P \\ B+C &\to Q \\ C &\to 2A. \end{aligned} \tag{10.42}$$

Network (10.43) is another example of a network that is concordant but not strongly concordant [62], this one having positively dependent reaction vectors. It is a variant of the calcium oscillation network (10.36), modified to comply with the stricture introduced in this section that no species appear on both sides of the same reaction.

$$\begin{gathered} A \rightleftarrows B \to G \to D+B \\ C+4D \rightleftarrows E \\ B+E \to F \rightleftarrows A+E \\ D \to 0 \end{gathered} \tag{10.43}$$

Remark 10.9.4 (Determining strong concordance). Strong concordance of a network can be determined, either positively or negatively, by a sign-checking procedure implemented in [62]. Algorithmic underpinnings are available in [115]. We shall see in the next chapter that strong concordance of a nondegenerate network can often be ascertained quickly by inspection of a network's Species-Reaction Graph.

The following proposition and its corollary tell us, among other things, that *strong concordance of a reaction network serves to preclude multiple positive stoichiometrically compatible equilibria not only for all kinetics in the already-large weakly monotonic class but also for all kinetics in the substantially larger two-way weakly monotonic class.*

Proposition 10.9.5 ([157]). *If a reaction network is strongly concordant, then, for any choice of a two-way weakly monotonic kinetics, the resulting kinetic system is injective.*

Proposition 10.9.5 and Remark 10.5.4 give us the following corollary:

Corollary 10.9.6 ([157]). *No two-way weakly monotonic kinetic system in which the underlying network is strongly concordant can admit two distinct stoichiometrically compatible equilibria, at least one of which is positive.*

The following theorem is a strong-concordance version of Theorem 10.6.17.[19] It is useful in several important ways. In particular, when we know that a nondegenerate network's fully open extension is strongly concordant, perhaps by inspection of the Species-Reaction Graph, then we will know that the original network is also strongly concordant.

Theorem 10.9.7 ([157]). *A nondegenerate network with a strongly concordant fully open extension is itself strongly concordant. In particular, a weakly reversible network is strongly concordant if its fully open extension is strongly concordant.*

Theorem 10.7.2 also has a strong-concordance generalization. In preparation for it, we need the following definition:

Definition 10.9.8 ([157]). *A kinetics $\mathscr{K}$ for a reaction network $\{\mathscr{S},\mathscr{C},\mathscr{R}\}$ is* **differentiably two-way monotonic** *at $c^* \in \mathbb{R}_+^{\mathscr{S}}$ if, for every reaction $y \to y' \in \mathscr{R}$, $\mathscr{K}_{y\to y'}(\cdot)$ is differentiable at c^* and, moreover,*

$$(i)\ \frac{\partial \mathscr{K}_{y\to y'}}{\partial c_s}(c^*) > 0 \qquad \forall s \in \operatorname{supp} y,$$

$$(ii)\ \frac{\partial \mathscr{K}_{y\to y'}}{\partial c_s}(c^*) \le 0 \qquad \forall s \in \operatorname{supp} y',$$

$$(iii)\ \frac{\partial \mathscr{K}_{y\to y'}}{\partial c_s}(c^*) = 0 \qquad \forall s \in \mathscr{S} \setminus (\operatorname{supp} y \cup \operatorname{supp} y').$$

The following theorem and its corollary can be proved very much along the lines of proofs in [157] of Theorem 10.7.2 and its corollary.

Theorem 10.9.9 ([157]). *Let $\{\mathscr{S},\mathscr{C},\mathscr{R},\mathscr{K}\}$ be a kinetic system with stoichiometric subspace S and species-formation-rate function $f : \overline{\mathbb{R}}_+^{\mathscr{S}} \to S$. Moreover, suppose that the kinetics is differentiably two-way monotonic at $c^* \in \mathbb{R}_+^{\mathscr{S}}$. If the underlying network $\{\mathscr{S},\mathscr{C},\mathscr{R}\}$ is strongly concordant, then the derivative $df(c^*) : S \to S$ is nonsingular (whereupon 0 is not one of its eigenvalues). If the network's fully open extension is strongly concordant, then no real eigenvalue of $df(c^*)$ is positive.*

Corollary 10.9.10 ([157]). *Let $\{\mathscr{S},\mathscr{C},\mathscr{R},\mathscr{K}\}$ be a kinetic system with stoichiometric subspace S and species-formation-rate function $f : \overline{\mathbb{R}}_+^{\mathscr{S}} \to S$. Moreover, suppose that the kinetics is differentiably two-way monotonic at $c^* \in \mathbb{R}_+^{\mathscr{S}}$. If the underlying network $\{\mathscr{S},\mathscr{C},\mathscr{R}\}$ is nondegenerate—in particular, if it is weakly reversible—and if the network's fully open extension is strongly concordant, then every real eigenvalue of $df(c^*)$ is negative.*

[19] The corresponding Theorem 9.8 in [157] invoked not network nondegeneracy but instead the stronger condition of network normalcy. In fact, though, for a network with a concordant fully open extension, the two conditions are the same. This follows from Proposition 10.6.27.

10.10 More General Species Influences on Reaction Rates

For a weakly monotonic kinetics, a change in the rate of a particular reaction can occur only if there is a change in the concentration of one of its *reactant* species. For a two-way monotonic kinetics, the situation is different: there can be a change in the rate of the reaction only if there is a change in concentration of a *reactant* species or of a *product* species. In neither kinetic class can there be a change in a particular reaction's rate because of a change in concentration of a species that is *not* a reactant or a product (while the concentrations of all reactant and product species remain fixed).

However broad the weakly monotonic or two-way monotonic kinetic classes might be, it is useful to allow for consideration of more general species influences on reaction rates. This is especially true in coarse-grained models of chemistry that, at a finer level of description, might be driven by enzyme catalysis. In such instances, a particular species I produced by an enzyme-catalyzed sequence of reactions at the fine-description level might inhibit the action of a different enzyme by binding to it. That second enzyme might, for example, serve as catalyst in another sequence of reactions that ultimately has the effect of converting species A to species B. In a coarse-grained kinetic system model, in which the details of enzyme chemistry are suppressed, the "overall" reaction $A \to B$ might appear in the coarse-grained reaction network, taken with a corresponding coarse-grained kinetic rate function such as

$$\mathscr{K}_{A\to B}(c) := \frac{\alpha c_A}{\beta + \gamma c_I}. \tag{10.44}$$

In such a reaction-rate function, the presence of c_I—the concentration of a species that is neither a reactant nor a product in $A \to B$—is intended to model the inhibitory effect of I.

Such models can be accommodated within a broadened concordance-theory framework very much like what we've seen so far. It is the purpose of this section to indicate how this might be done and to give one example, Proposition 10.10.6, which is a generalization of results already stated in narrower settings. First we will say what we mean by a specification of species influences for a reaction network [157].

Definition 10.10.1. *An* **influence specification** $\mathscr{I}$ *for a reaction network* $\{\mathscr{S},\mathscr{C},\mathscr{R}\}$ *is an assignment to each reaction* $y \to y'$ *of a function* $\mathscr{I}_{y\to y'} : \mathscr{S} \to \{1,0,-1\}$ *such that*

$$\mathscr{I}_{y\to y'}(s) = 1, \quad \forall s \in \operatorname{supp} y. \tag{10.45}$$

If $\mathscr{I}_{y\to y'}(s) = 1$ *[respectively, -1] then species* s *is an* **inducer** [**inhibitor**] *of reaction* $y \to y'$.

For the purpose of constructing a coarse-grained kinetic system model based on a putative reaction network, a chemist will sometimes have a qualitative sense of which species *might* inhibit or induce the various reactions. That qualitative sense can be viewed as being expressed in an influence specification of the kind given by Definition 10.10.1, with the understanding that the inhibitors and inducers specified are to be regarded as the ones *permitted* when qualitative judgments are finally expressed quantitatively in a concrete kinetics.

Within the class of two-way weakly monotonic kinetics, we permitted, but did not *require*, the product species of a particular reaction to be rate inhibitors. In the following definition, our intent is to provide broad variants of the two-way weakly monotonic class, each variant corresponding to a different choice of influence specification. A kinetics within the corresponding kinetic class is supposed to respect the specification of which species are *permitted* to be rate inhibitors or rate inducers for the various reactions—without *requiring* them to be one or the other. Note that each variant includes all instances of ordinary weakly monotonic kinetics as a subclass.

Definition 10.10.2 ([157]). *A kinetics $\mathscr{K}$ for a reaction network $\{\mathscr{S},\mathscr{C},\mathscr{R}\}$ is* **weakly monotonic with respect to influence specification** $\mathscr{I}$ *if, for every pair of compositions c^* and c^{**}, the following implications hold for each reaction $y \to y' \in \mathscr{R}$ such that* $\operatorname{supp} y \subset \operatorname{supp} c^*$ *and* $\operatorname{supp} y \subset \operatorname{supp} c^{**}$*:*

(i) $\mathscr{K}_{y\to y'}(c^{**}) > \mathscr{K}_{y\to y'}(c^*) \Rightarrow$ *there is a species s such that* $\operatorname{sgn}(c^{**}_s - c^*_s) = \mathscr{I}_{y\to y'}(s) \neq 0$.
(ii) $\mathscr{K}_{y\to y'}(c^{**}) = \mathscr{K}_{y\to y'}(c^*) \Rightarrow$ *either (a) $c^{**}_s = c^*_s$ for all $s \in \operatorname{supp} y$ or (b) there are species s, s' with* $\operatorname{sgn}(c^{**}_s - c^*_s) = \mathscr{I}_{y\to y'}(s) \neq 0$ *and* $\operatorname{sgn}(c^{**}_{s'} - c^*_{s'}) = -\mathscr{I}_{y\to y'}(s') \neq 0$.

There is, for each influence specification, a companion notion of concordance. The linear map $L : \mathbb{R}^{\mathscr{R}} \to S$ is, as before, given by (10.41).

Definition 10.10.3 ([157]). *A reaction network $\{\mathscr{S},\mathscr{C},\mathscr{R}\}$ with stoichiometric subspace S is* **concordant with respect to the influence specification** $\mathscr{I}$ *if there do not exist $\alpha \in \ker L$ and a nonzero $\sigma \in S$ having the following properties:*

(i) For each $y \to y'$ such that $\alpha_{y\to y'} > 0$, there exists a species s for which $\operatorname{sgn}\sigma_s = \mathscr{I}_{y\to y'}(s) \neq 0$.
(ii) For each $y \to y'$ such that $\alpha_{y\to y'} < 0$, there exists a species s for which $\operatorname{sgn}\sigma_s = -\mathscr{I}_{y\to y'}(s) \neq 0$.
(iii) For each $y \to y'$ such that $\alpha_{y\to y'} = 0$, either (a) $\sigma_s = 0$ for all $s \in \operatorname{supp} y$ or (b) there are species s, s' for which $\operatorname{sgn}\sigma_s = \mathscr{I}_{y\to y'}(s) \neq 0$ *and* $\operatorname{sgn}\sigma_{s'} = -\mathscr{I}_{y\to y'}(s') \neq 0$.

Note that if a network is concordant with respect to a particular influence specification $\mathscr{I}$ then it is also concordant in the original sense of Definition 10.4.1. Just as strong concordance is more demanding than the original concordance, so too is concordance with respect to an influence specification stronger, except when the influence specification is such that the two notions of concordance coincide. See Remark 10.10.5.

Remark 10.10.4. Determining whether a given network is or is not concordant with respect to a given influence function is a matter of sign-checking, which is best done by computer. In fact, [62] provides an example of a freely available computer application that does such sign-checking in response to an easily supplied input of both the network and the influence specification. Algorithmic considerations are discussed in [117].

Remark 10.10.5. As we've already suggested, some of our earlier definitions are in fact special cases of definitions given in this section, applied for certain canonical influence specifications. For a reaction network $\{\mathscr{S},\mathscr{C},\mathscr{R}\}$ the definitions of weakly monotonic kinetic class (Definition 10.5.1) and of ordinary network concordance (Definition 10.4.1) are just special cases of Definitions 10.10.2 and 10.10.3, each implemented for the particular influence specification given by $\mathscr{I}_{y\to y'} = \operatorname{sgn} y, \forall y \to y' \in \mathscr{R}$. The definitions of the two-way weakly monotonic kinetics and of strong concordance emerge from the influence specification given by $\mathscr{I}_{y\to y'} = \operatorname{sgn}(y - y'), \forall y \to y' \in \mathscr{R}$.[20]

The following proposition generalizes Proposition 10.9.5.

Proposition 10.10.6 ([157]). *A kinetic system $\{\mathscr{S},\mathscr{C},\mathscr{R},\mathscr{K}\}$ is injective whenever there exists an influence specification $\mathscr{I}$ such that:*

(i) The kinetics $\mathscr{K}$ is weakly monotonic with respect to $\mathscr{I}$.
(ii) The underlying network $\{\mathscr{S},\mathscr{C},\mathscr{R}\}$ is concordant with respect to $\mathscr{I}$.

In this case, the kinetic system cannot admit multiple stoichiometrically compatible equilibria, at least one of which is positive.

10.11 A Concluding Remark About Concordance and Mathematical Aesthetics

Before the last quarter of the 20th century, a theorem, if it were to be deemed useful in applications, would have a hypothesis that could be checked, in a practical way, for compliance in particular instances. Such checks would normally be "by eye" or perhaps "by hand" when some computations might be involved. The Deficiency Zero Theorem provides an example, at least in applications to networks having just several species and several reactions.

The theorems of this chapter provide powerful information, but they require means to determine whether a particular network is or is not concordant. In an earlier time, such determinations could not be made easily, either by eye or by hand, except in the most trivial cases. Accustomed as we are to a classical aesthetic, the concordance theorems seem less appealing than the more flashy Deficiency Zero Theorem.

But we are not in that earlier time. The concordance or discordance of a network can now be determined readily and quickly, for example, by means of [62], at least when the network is not very large. The easy availability of computational power

[20] For a vector $x \in \mathbb{R}^{\mathscr{S}}$, we mean by $\operatorname{sgn}(x)$ the function from $\mathscr{S}$ to $\{-1,0,1\}$ that assigns to each species s the sign of the number x_s.

does that for us, and so our aesthetic sense probably needs revision. It is, after all, a thing of beauty that we can know even this: When a network is concordant—a condition stated in terms of *linear* transformations and *linear* subspaces—we *also* have injectivity for all *nonlinear* kinetics within a very broad class.

Still, there is something lacking. It is one thing to be able to determine on a case-by-case basis the concordance or discordance of this or that reaction network. It is quite another thing to have a sense of structure, of particular features of a reaction network that make for concordance or discordance. To a significant extent, this lack will be mitigated in the next chapter.

Appendix 10.A Deducing Behavior of a Reaction Network from What Is Known About Its Behavior in a Fully Open Context

The body of work alluded to in Section 10.2 was aimed at understanding the surprisingly dull behavior exhibited, over a very wide range of different chemistries, by (isothermal) classical continuous-flow stirred-tank reactors (CFSTRs). In their underlying arguments, these results relied heavily on the mathematical features of the fully open setting—features that are apt for classical CFSTRs but not broadly. In biological applications, for example, we should not expect that, for *every* species $\mathcal{S}$, there will be a degradation reaction of the form $\mathcal{S} \to 0$.

When results derived in the fully open context are available, it is natural to wonder about the extent to which the fully open requirement can be relaxed. Stated differently, we would like to know when a statement that holds true for a network's fully open extension, modified to account for stoichiometric compatibility, extends to the network itself. In this chapter there was considerable discussion of that very issue, but always in consideration of what can be said about a network's behavior when it is known that the network's fully open extension is concordant.

It is the purpose of this appendix to discuss, briefly and informally, certain results along these lines, results that transcend concordance considerations or even restrictions to particular classes of kinetics. For the sake of concretion, we will restrict our attention to information that derives from knowledge of the impossibility of multiple positive equilibria in a fully open setting.

It is reasonable to conjecture that, for a given chemistry, the impossibility of multiple positive equilibria in a fully open CFSTR context (including a feed stream) implies the impossibility of multiple positive stoichiometrically compatible equilibria for the chemistry itself, removed from the fully open setting. That conjecture, unmodified, would be false. Consider, for example, the pathological mass action system (10.A.1) considered earlier in Section 8.A.3. Recall that in each positive stoichiometric compatibility class there are multiple equilibria, in fact an *infinite* number of them. (See Figure 8.A.2.) Yet, it can be confirmed that, in a fully open CFSTR context, there is precisely *one* positive equilibrium, so long as there is some A in the feed stream.

$$B \overset{1}{\leftarrow} A \overset{1}{\rightarrow} C \tag{10.A.1}$$
$$C + B \overset{1}{\rightarrow} 2A$$

The conjecture is, however, *almost* true. Implicit-function-theorem arguments in [47] indicate that if a reaction network, taken with a smooth kinetics, admits two distinct stoichiometrically compatible *nondegenerate*[21] positive equilibria, then there are parameter values for a classical CFSTR, within which the same kinetic system is operative, such that there are again two distinct positive equilibria. Therefore, a theorem that serves to deny the capacity for multiple positive equilibria in the fully open CFSTR context also gives information about the capacity for multiple stoichiometrically compatible positive equilibria for the original chemistry: There can be multiple stoichiometrically compatible positive equilibria for the original chemistry *only if all but perhaps one of those equilibria are degenerate.* Note that all positive equilibria of the mass action system (10.A.1) are degenerate.

Appendix 10.B Mass Action Injectivity

The principal theorems of this chapter circumscribe behavior for the very large class of concordant reaction networks when the kinetics is subject to very weak constraints (e.g., when the kinetics is differentiably monotonic). To a great extent, these theorems have significant roots in prior work restricted to mass action kinetics. In fact, some of the earlier theory tailored *specifically* to mass action systems permit similar theorems for a still broader class of networks, including some discordant ones. It is the purpose of this appendix to illuminate, briefly, some relationships between results in this chapter and results that derive from earlier mass-action-specific theory.

10.B.1 Two Similar Theorems, One Broad and One Mass-Action-Specific

By way of example, we will state two theorems—Theorems 10.B.2 and 10.B.3—one very general in the kinetics it invokes and one specific to mass action kinetics. Once these are in place, we will be in a position to consider an example that indicates how mass-action-restricted theorems can give information when broader theorems are silent.

As a precursor to the statement of Theorem 10.B.2, we recall Theorem 10.5.15, repeated below as Theorem 10.B.1. (See Section 10.B.3.1 for remarks about the proof.)

Theorem 10.B.1. *For reaction network* $\{\mathscr{S},\mathscr{C},\mathscr{R}\}$ *the following are equivalent:*

(i) $\{\mathscr{S},\mathscr{C},\mathscr{R}\}$ *is concordant*

(ii) For every choice of a differentiably monotonic kinetics $\mathscr{K}$ *the derivative of the species-formation-rate function for the kinetic system* $\{\mathscr{S},\mathscr{C},\mathscr{R},\mathscr{K}\}$ *is nonsingular at every positive composition.*

[21] The reference here is to a nondegenerate *equilibrium*. (Recall Section 3.6.) This should not be confused with a nondegenerate *reaction network*, which was the subject of Section 10.6.

Items (i) and (ii) in Theorem 10.B.1 are then equivalent to item (ii) in Theorem 10.5.5 and item (iii) in Proposition 10.5.8. For the purposes of this appendix, the important equivalence is the one expressed in the following theorem, which is merely a corollary of Theorems 10.B.1 and 10.5.5:

Theorem 10.B.2. *For reaction network $\{\mathscr{S},\mathscr{C},\mathscr{R}\}$ the following are equivalent:*

(i) For every weakly monotonic kinetics $\mathscr{K}$ the kinetic system $\{\mathscr{S},\mathscr{C},\mathscr{R},\mathscr{K}\}$ is injective.

(ii) For every choice of a differentiably monotonic kinetics $\mathscr{K}$ the derivative of the species-formation-rate function for the kinetic system $\{\mathscr{S},\mathscr{C},\mathscr{R},\mathscr{K}\}$ is nonsingular at every positive composition.

In preparation for the statement of a mass action version of Theorem 10.B.2, we will want to qualify the meaning of kinetic-system injectivity just slightly. Recall that a kinetic system is injective if its species-formation-rate function $f(\cdot)$ has the property that $f(c^*) \neq f(c^{**})$ whenever c^* and c^{**} are distinct stoichiometrically compatible compositions, *at least one of which is positive*. We shall say that a kinetic system is *positive-composition injective* if $f(c^*) \neq f(c^{**})$ whenever c^* and c^{**} are distinct stoichiometrically compatible compositions, *both positive*.

Proof of the following theorem is given in Section 10.B.3.2. See also Remark 10.B.4.

Theorem 10.B.3. *For reaction network $\{\mathscr{S},\mathscr{C},\mathscr{R}\}$ the following are equivalent:*

(i) For every choice of rate constants $k \in \mathbb{R}_+^{\mathscr{R}}$ the mass action system $\{\mathscr{S},\mathscr{C},\mathscr{R},k\}$ is positive-composition injective.

(ii) For every choice of rate constants $k \in \mathbb{R}_+^{\mathscr{R}}$ the derivative of the species-formation-rate function for the mass action system $\{\mathscr{S},\mathscr{C},\mathscr{R},k\}$ is nonsingular at every positive composition.

If a network satisfies either of the two equivalent conditions in Theorem 10.B.3, we will say that the network is *mass action injective*. For a network of interest, condition (ii) provides an attractive route to a computational test for mass action injectivity. Such a test is available in [62]. Computational aspects are discussed later in Section 10.B.4 and more fully in [115].

In fact, the two equivalences given by Theorems 10.B.2 and 10.B.3 have computational virtues: in both cases they tie difficult-to-study *global* properties in (i) of the *nonlinear* species-formation-rate function $f : \overline{\mathbb{R}}_+^{\mathscr{S}} \to S$ directly to properties in (ii) of a more *local* object, the *linear* transformation $df(c) : S \to S$.

Remark 10.B.4. Theorem 10.B.3 has origins in the Ph.D. thesis of Gheorghe Craciun [44], which focused on fully open reaction networks. For the fully open case, proofs appeared there and in [46], along with determinant-based theorems for assessing injectivity of fully open mass action systems. These in turn provided foundations for Species-Reaction Graph mass action theorems given in [47] and [52].

For networks that are not fully open, stoichiometric considerations exert themselves, and the proof becomes slightly more complicated. For this more general case,

Theorem 10.B.3 was stated in Remark 6.4 of [49], where determinant conditions for mass action injectivity were again provided, this time for networks that are not presumed to be fully open.[22] For the purpose of documenting the mass-action-injectivity computational module of [62], a proof of Theorem 10.B.3—essentially the one given in Section 10.B.3.2—was included in the Ph.D. thesis of Haixia Ji [115]. A proof was also given [84], where a determinant-based Mathematica-like script for assessment of mass action injectivity was provided.

10.B.2 Two Instructive Examples

In this section we examine two networks that teach very different lessons.

Example 10.B.5. Consider first network (10.B.2).

$$\begin{array}{ccccc} A+B \rightleftarrows C & & & & A \\ & \searrow & & \swarrow\!\!\nearrow & \\ & & 0 & & \\ & \nearrow & & \nwarrow\!\!\searrow & \\ 2A+B \rightleftarrows D & & & & B \end{array} \tag{10.B.2}$$

The network is *discordant*, as can be determined by means of [62]. Theorem 10.5.5 then tells us that there is a weakly monotonic kinetics for the network such that the resulting kinetic system is *not* injective. In fact, Theorem 10.5.10 tells us more: there is for the network a weakly monotonic kinetics for which the resulting kinetic system *admits a pair of distinct positive stoichiometrically compatible equilibria*. Moreover, Theorem 10.B.1 ensures for the network the existence of a differentiably monotonic kinetics such that at some positive composition the derivative of the species-formation-rate function is singular. In particular, Theorem 10.7.7 tells us that the kinetics can be chosen in such a way as to engender *a degenerate positive equilibrium.*

On the other hand, *the network is mass action injective*; this too can be determined by means of [62]. Theorem 10.B.3 then tells us that, while there is indeed *some* kinetics for the network that give rise to the various phenomena just described, there is no *mass action* kinetics for which these phenomena will be admitted, no matter what rate constants are assigned to the various reactions.[23]

The example demonstrates concretely that, for at least certain networks, the equivalent behavior-restricting conditions in the mass action Theorem 10.B.3 might be satisfied, while those in the more kinetically expansive Theorem 10.B.2 are not. Indeed, theory focused exclusively on the archetypal mass action kinetics has a special richness all its own.

---

[22] See in particular Proposition 10.3 of [49].

[23] That this fully open network is mass action injective is also a consequence of the fact that its Species-Reaction Graph (Chapter 11) consists of a single *c-cycle*—that is, a cycle whose edge set is the union of c-pairs [44, 147, 148]. See in particular Section 11.11.

Example 10.B.6. There is a possible source of confusion that we would do well to mitigate by means of an example. Recall network (10.B.3), which we met earlier in Section 10.4.

$$\begin{aligned} A+B &\rightleftarrows P \\ B+C &\rightleftarrows Q \\ C &\rightleftarrows 2A \end{aligned} \tag{10.B.3}$$

The network is not concordant, nor is it mass action injective [62]. In particular, neither of the two equivalent conditions in Theorem 10.B.3 is satisfied. This tells us that there is a rate constant specification such that the resulting mass action system admits a pair of stoichiometrically compatible positive compositions, c^* and c^{**}, at which the species-formation-rate function takes identical values: $f(c^*) = f(c^{**})$. Moreover, there is a (perhaps different) rate constant specification such that at some positive composition, the derivative of the species-formation-rate function has a singular derivative.

Note, however, that *network* (10.B.3) *has a deficiency of zero* ($n = 6, \ell = 3, s = 3$).

The Deficiency Zero Theorem (Theorem 7.1.1) tells us that *for no assignment of rate constants* can the resulting mass action system admit a pair of stoichiometrically compatible positive *equilibria*. This is to say that, although for some rate constant assignment the species-formation-rate function can indeed take the same value at two different stoichiometrically compatible positive compositions, *that value cannot be zero.*

Moreover, the local version of the Deficiency Zero Theorem (Theorem 7.3.1) tells us that *for no assignment of rate constants* can there be a positive *equilibrium* that is degenerate. This is to say that, although for some rate constant assignment there can be a positive composition at which the derivative of species-formation-rate function is singular, *that composition cannot be an equilibrium.*

The lesson here is this: Although mass action injectivity is a network attribute that does indeed preclude multiple stoichiometrically compatible positive equilibria and degenerate positive equilibria, regardless of rate constant values, the same preclusions might result even for networks that are not mass action injective.

10.B.3 Some Proofs

In this section we provide proofs of Theorems 10.B.1 and 10.B.3.

10.B.3.1 Proof of Theorem 10.B.1

The implication (i) ⇒ (ii) is the subject of Theorem 10.5.14, which was proved in [157]. Proof that (ii) ⇒ (i) is essentially contained in the proof of Theorem 10.7.7, also given in [157]; we provide the relevant part of that argument below. (Theorem 10.7.7 more or less gives what we want here, but its emphasis is on the stability of positive equilibria.)

Proof ((ii)⇒ (i)). Suppose that (ii) holds but that the network is discordant. In this case, Proposition 10.6.25 tells us that we can choose $\{p_{y\to y'}\}_{y\to y'\in\mathscr{R}} \subset \overline{\mathbb{R}}^{\mathscr{S}}_{+}$, with $\operatorname{supp} p_{y\to y'} = \operatorname{supp} y$, $\forall y\to y' \in \mathscr{R}$, such that the linear transformation $T : S \to S$ given by

$$T\sigma := \sum_{y\to y'\in\mathscr{R}} p_{y\to y'} \cdot \sigma(y'-y) \tag{10.B.4}$$

is singular. Now let $c^* \in \overline{\mathbb{R}}^{\mathscr{S}}_{+}$ be some fixed positive composition, and for the network $\{\mathscr{S},\mathscr{C},\mathscr{R}\}$ let $\mathscr{K}$ be the kinetics defined, for each $y\to y' \in \mathscr{R}$, by

$$\mathscr{K}_{y\to y'}(c) := \eta_{y\to y'} c^{q_{y\to y'}}, \tag{10.B.5}$$

where

$$q_{y\to y'} := c^* p_{y\to y'} \quad \text{and} \quad \eta_{y\to y'} := [(c^*)^{q_{y\to y'}}]^{-1}. \tag{10.B.6}$$

For this choice, the derivative at c^* of the species-formation-rate function for the differentiably monotonic kinetic system $\{\mathscr{S},\mathscr{C},\mathscr{R},\mathscr{K}\}$ coincides with the singular linear transformation $T : S \to S$ given by (10.B.4). This contradicts (ii). □

10.B.3.2 Proof of Theorem 10.B.3

Recall from Remark 3.6.2 that, for a mass action system $\{\mathscr{S},\mathscr{C},\mathscr{R},k\}$ with stoichiometric subspace S, the derivative of the species-formation-rate function $f(\cdot)$ has a special form: At composition $a \in \mathbb{R}^{\mathscr{S}}_{+}$ the derivative $df(a) : S \to S$ is given by the requirement that, for each $\sigma \in S$,

$$df(a)\sigma = \sum_{y\to y'\in\mathscr{R}} \eta_{y\to y'}(y *_a \sigma)(y'-y), \tag{10.B.7}$$

where, for each $y\to y' \in \mathscr{R}$,

$$\eta_{y\to y'} := k_{y\to y'}(a)^y \tag{10.B.8}$$

and "$*_a$" indicates the scalar product in $\mathbb{R}^{\mathscr{S}}$ defined by

$$v *_a w := \sum_{s\in\mathscr{S}} \frac{v_s w_s}{a_s}, \quad \forall v,w \in \mathbb{R}^{\mathscr{S}}. \tag{10.B.9}$$

With this in mind, we can restate Theorem 10.B.3 in the following way, essentially asserting, in more detail, the equivalence of the negations of (i) and (ii) in the original statement.

Theorem 10.B.7 ([49, 115]). *For reaction network $\{\mathscr{S},\mathscr{C},\mathscr{R}\}$ with stoichiometric subspace S the following are equivalent:*

(i) There is a choice of rate constants $k \in \mathbb{R}^{\mathscr{R}}_{+}$ and two distinct positive compositions c^* and c^{**}, with $c^* - c^{**} \in S$, such that

$$\sum_{y\to y'\in\mathscr{R}} k_{y\to y'}(c^*)^y(y'-y) = \sum_{y\to y'\in\mathscr{R}} k_{y\to y'}(c^{**})^y(y'-y). \tag{10.B.10}$$

(ii) There is a choice of $\eta \in \mathbb{R}_+^{\mathscr{R}}$ *and* $a \in \mathbb{R}_+^{\mathscr{S}}$ *such that the linear transformation* $T_{\eta,a} : S \to S$ *given by*

$$T_{\eta,a}\sigma := \sum_{y \to y' \in \mathscr{R}} \eta_{y \to y'} (y *_a \sigma)(y' - y) \tag{10.B.11}$$

is singular.

Proof. $(i) \Rightarrow (ii)$: With k, c^*, and c^{**} as in (i), we can rewrite (10.B.10) in the following form:

$$\begin{aligned} 0 &= \sum_{y \to y' \in \mathscr{R}} k_{y \to y'} (c^*)^y \Big[\frac{(c^{**})^y}{(c^*)^y} - 1\Big](y' - y) \\ &= \sum_{y \to y' \in \mathscr{R}} k_{y \to y'} (c^*)^y [e^{y \cdot \mu} - 1](y' - y), \end{aligned} \tag{10.B.12}$$

where "$\cdot$" is the standard scalar product in $\mathbb{R}^{\mathscr{S}}$ and

$$\mu := \ln c^{**} - \ln c^*. \tag{10.B.13}$$

Now let

$$\sigma := c^{**} - c^* \in S. \tag{10.B.14}$$

Note that, componentwise, $\operatorname{sgn} \mu = \operatorname{sgn} \sigma$. Therefore, there is an $a \in \mathbb{R}_+^{\mathscr{S}}$ such that $\mu = \frac{1}{a}\sigma$. Then, in (10.B.12), we can replace $y \cdot \mu$ by $y *_a \sigma$, where "$*_a$" is the scalar product in $\mathbb{R}^{\mathscr{S}}$ given by (10.B.9).

Note also that $\operatorname{sgn}(e^{y *_a \sigma} - 1) = \operatorname{sgn}(y *_a \sigma)$. Thus, for each $y \in \mathscr{C}$, there is a positive number p_y such that $e^{y *_a \sigma} - 1 = p_y (y *_a \sigma)$. If, for each $y \to y' \in \mathscr{R}$, we set $\eta_{y \to y'} := k_{y \to y'} (c^*)^y p_y$, it follows from (10.B.12) that

$$\sum_{y \to y' \in \mathscr{R}} \eta_{y \to y'} (y *_a \sigma)(y' - y) = 0. \tag{10.B.15}$$

Because σ is not zero, it follows from (10.B.15) that the linear transformation $T_{\eta,a} : S \to S$ defined in (ii) is singular.

$(ii) \Rightarrow (i)$: Suppose that $\eta \in \mathbb{R}_+^{\mathscr{R}}$, $a \in \mathbb{R}_+^{\mathscr{S}}$, and nonzero $\sigma \in S$ are such as to make (10.B.15) true. Let

$$\mu := \frac{1}{a}\sigma, \tag{10.B.16}$$

in which case (10.B.15) becomes

$$\sum_{y \to y' \in \mathscr{R}} \eta_{y \to y'} (y \cdot \mu)(y' - y) = 0. \tag{10.B.17}$$

Because $\operatorname{sgn}(y \cdot \mu) = \operatorname{sgn}(e^{y \cdot \mu} - 1)$, it is not difficult to see that we can choose $\bar{\eta} \in \mathbb{R}_+^{\mathscr{R}}$ such that

$$\sum_{y \to y' \in \mathscr{R}} \bar{\eta}_{y \to y'} (e^{y \cdot \mu} - 1)(y' - y) = 0. \tag{10.B.18}$$

Because, componentwise, $\operatorname{sgn}(\mu) = \operatorname{sgn}(\sigma)$, we also have $\operatorname{sgn}(e^{\mu} - 1) = \operatorname{sgn}\sigma$. Thus, there is $c^* \in \mathbb{R}_+^{\mathscr{S}}$ such that $\sigma = c^*(e^{\mu} - 1)$. Now let $c^{**} := c^* e^{\mu}$. Note that $c^{**} \neq c^*$, $c^{**} - c^* = \sigma \in S$, and $\mu = \ln c^{**} - \ln c^*$. After choosing $k \in \mathbb{R}_+^{\mathscr{R}}$ to satisfy $\bar{\eta}_{y\to y'} = k_{y\to y'}(c^*)^y, \forall\, y \to y' \in \mathscr{R}$, we can begin with (10.B.18) and then reverse the steps in (10.B.12) to obtain (10.B.15). □

10.B.4 A Route to the Determination of Mass Action Injectivity

Condition (ii) of Theorem 10.B.7 suggests several routes to establishing that a given network is or is not mass action injective. In particular, it is easy to see that (ii) is equivalent to the following:

(iii) There is a choice of $\eta \in \mathbb{R}_+^{\mathscr{R}}$ such that the linear transformation $\bar{T}_\eta : \mathbb{R}^{\mathscr{S}} \to S$ given by

$$\bar{T}_\eta \gamma := \sum_{y\to y' \,\in\, \mathscr{R}} \eta_{y\to y'}(y \cdot \gamma)(y' - y) \tag{10.B.19}$$

has in its kernel a vector that is sign-compatible with S.

Here "·" denotes the standard scalar product in $\mathbb{R}^{\mathscr{S}}$. Recall that a vector $\gamma^* \in \mathbb{R}^{\mathscr{S}}$ is *sign-compatible with S* (Definition 8.5.1) if there exists $\sigma \in S$ such that, componentwise, $\operatorname{sgn}\gamma^* = \operatorname{sgn}\,\sigma$. This formulation permits a connection with the linear transformation $L : \mathbb{R}^{\mathscr{R}} \to S$, given by

$$L\alpha = \sum_{y\to y' \in \mathscr{R}} \alpha_{y\to y'}(y' - y), \tag{10.B.20}$$

the linear transformation that appears in the definition of concordance. Clearly, (iii) is equivalent to the following:

(iv) There exist $\alpha \in \ker L$ and a nonzero $\gamma \in \mathbb{R}^{\mathscr{S}}$, sign-compatible with S, such that $\operatorname{sgn}(y \cdot \gamma) = \operatorname{sgn}\alpha_{y\to y'}$ for all $y \to y' \in \mathscr{R}$.

Mass action injectivity (or lack of it) can be determined—as it is in [62]—by means of sign-checking along lines described in (iv). For algorithmic aspects see [115].

11

The Species-Reaction Graph

The *Species-Reaction Graph* is a diagrammatic representation of a reaction network that closely resembles those commonly used to depict biochemical pathways. We shall see that a network's Species-Reaction Graph often carries an extraordinary amount of far-from-obvious information about how the network might behave. In fact, the theorems in this chapter will tell us a great deal about behavior across the entire reaction network landscape, in particular about why dull, stable behavior is more prevalent than one might expect within a mathematical macrocosm so rife with nonlinearity.

11.1 About the Connection of This Chapter to the Preceding One

The principal theorems of this chapter—Theorems 11.6.1 and 11.9.1—are freestanding ones that can be read and understood without reference to the concordance results of Chapter 10. The theorems are, however, stated for *nondegenerate networks*, which are defined and discussed there. That is, certain highly pathological networks—*degenerate networks*—are precluded from consideration at the outset, much as $t > \ell$ networks were precluded in the statement of the Deficiency One Theorem. Degenerate networks are ones that are so pathological that, for *any* differentiably monotonic kinetics, the derivative of the species-formation-rate function is singular at *every* positive composition.

Readers who read only lightly in Chapter 10 should not be very concerned if they lack a detailed understanding of what it means for a *network* to be degenerate: Degeneracy is rare, especially among networks that have the capacity to admit a positive equilibrium. For this reason preclusion of degenerate networks in theorem statements casts very few networks aside. Moreover, degeneracy is never robust. A degenerate network can always be made nondegenerate by adding reverse reactions to certain (but not necessarily all) reactions deemed irreversible in the original network. Thus, exotic kinetic system phenomena resulting solely from network degeneracy will invariably disappear under tiny perturbations of the model, such as

M. Feinberg, *Foundations of Chemical Reaction Network Theory*,
Applied Mathematical Sciences 202,
https://doi.org/10.1007/978-3-030-03858-8_11

adding certain reverse reactions that proceed only at vanishingly small rates. *All reversible networks are nondegenerate, as are all weakly reversible networks. So too are all fully open networks.* In any case, affirmation that a given network is indeed nondegenerate can be made computationally by means of [62].

Although this chapter's *principal* theorems make no mention of concordance, they are strongly connected to the concordance theorems of the previous chapter. The connection proceeds in the following way: A "preliminary theorem" will assert that *if a network's Species-Reaction Graph has Property X, then the network is concordant.* In this case, all of the behavioral consequences of concordance described in Chapter 10 accrue to the network. Such an accrual then permits a second and far more tangible theorem—actually a corollary of the first—stating that *Property X in a network's Species-Reaction Graph ensures behavior of a very circumscribed kind, so long as the kinetics satisfies certain very mild constraints.* The *principal* theorems of this chapter are of the second kind, in which the concordance-based connective tissue is invisible. Again, they can be read on their own terms, with no need for an understanding of concordance-related material in Chapter 10.

Still, we have included in this chapter the preliminary theorems that connect concordance (or strong concordance) to features of a network's Species-Reaction Graph. There are two reasons for this. First, these preliminary theorems are the difficult-to-prove ones that we will examine in Part III. Second, these theorems will give a sense of the structural features in a reaction network that ensure concordance. More to the point, they will reveal certain very exceptional features in a network's Species-Reaction Graph that are *necessary* in order that the network be *discordant.* In this way we will see just why concordance in chemical reaction networks is so common.

11.2 A Little History

The main results in this chapter derive from work with Guy Shinar and Daniel Knight [117, 118, 157, 158]. These Species-Reaction Graph theorems are sharper than earlier ones; they are not at all restricted to fully open networks, as almost all of the prior Species-Reaction Graph theorems were; and their theoretical underpinnings are located differently, in the concordance ideas of Chapter 10. Nevertheless, the debt to earlier work is immense, especially work by Paul Schlosser, Gheorghe Craciun, and Murad Banaji:

Motivated by the inability of deficiency-oriented theory to account for the widely prevalent behavioral dullness of isothermal continuous-flow stirred-tank reactors, early work took some meaningful but tentative steps toward an explanation—see, for example, [19, 20, 144, 145]. But it was the Ph.D. research of Paul Schlosser, through his invention of the *Species-Complex-Linkage Graph (SCL Graph)*—a precursor of the Species-Reaction Graph—that first brought great range and subtlety to the broad understanding of CFSTRs governed by mass action kinetics [147, 148]. Even the decades-later theorems in this chapter do not entirely subsume all the fruits of the Schlosser work, which sometimes gives information that more recent theorems do not.

The Ph.D. research of Gheorghe Craciun—focused also on mass action CFSTRs—took a different turn. It was based on different methods [44, 46], essentially rooted in an early fully open version of Theorem 10.B.3. These gave rise to a test for preclusion of multiple equilibria involving consistency of signs of various determinants, each determinant associated with a particular subnetwork of the reaction network under study. Although the test could be implemented computationally, it was shown that its conclusion could sometimes be deduced instead from simple inspection of the network's Species-Reaction Graph—a graph similar to but more fine-grained than the SCL Graph. Early Species-Reaction Graph theorems appeared in [44, 47, 48, 52].

Based on still other methods [16], very surprising work by Banaji and Craciun [14, 15] showed—again for fully open networks—that extant mass action Species-Reaction Graph results actually extend to a far wider kinetic class, a class resembling (but not identical to) the differentiably two-way weakly monotonic class described in Chapter 10.

11.3 How the Species-Reaction Graph Is Drawn

We are about to explain how the Species-Reaction Graph for a given network is drawn. By way of preparation, however, we will, in the following remark, place a very minor restriction on the networks admitted for consideration.

Remark 11.3.1. Throughout this chapter it will be understood that, for technical reasons, we preclude from consideration networks containing any reaction in which the same species resides within both the reactant and product complexes. That is, we preclude networks containing reactions such as $A+B \to 2A$ or $A+B \to A+C$. Note that we do *not* preclude far more realistic reaction sequences, such as $A+B \to D \to 2A$ or $A+B \to D \to A+C$, that proceed by way of an intermediate.

The Species-Reaction Graph for network (11.1), exhibited in Figure 11.1, will serve to illustrate how the graph is constructed [118].

$$\begin{aligned} A+B &\rightleftarrows C+D \\ B &\rightleftarrows C \to E \\ A+E &\to F+G \end{aligned} \tag{11.1}$$

11.3.1 Species Vertices and Reaction Vertices

The Species-Reaction Graph for a reaction network is a graph having two kinds of vertices, *species vertices* and *reaction vertices*, along with *edges* that connect species vertices to reaction vertices according to a certain prescription. There is one species vertex for each species in the network, that vertex carrying the species's name. Similarly, the reaction vertices are associated with the various reactions in the network, with the understanding that, in the case of a reversible reaction pair, a single vertex represents both members of the pair. However, in forming the Species-Reaction Graph, certain reactions are omitted altogether, as we explain in the following remark.

Remark 11.3.2. If a network under study contains degradation reactions of the form $\mathscr{S} \to 0$ or synthesis reactions of the form $0 \to \mathscr{S}$, it will be understood that the Species-Reaction Graph is drawn for the network with those reactions removed. If there are reactions such as $2A \to 0$ or $0 \to 3A$, containing a single species on one side of the reaction, those reactions are also removed. A reaction such as $A+B \to 0$ or $0 \to A+B$ is retained. Thus, the Species-Reaction Graph for network (11.1) is the same as the graph for the network's fully open extension; it is also the same as the graph for the fully open extension augmented by the reactions $0 \to A$ and $0 \to B$.

Remark 11.3.3 (The base subnetwork). By the *base subnetwork* of a reaction network, we will mean the subnetwork of the original network consisting of all reactions that are retained in the drawing of the Species-Reaction Graph. That is, the base subnetwork is the subnetwork consisting of all reactions of the original network apart from those of the form $0 \to n\mathscr{S}$ or $m\mathscr{S} \to 0$.

11.3.2 How Edges Are Drawn

An edge is drawn to connect a species vertex with a reaction vertex if the species appears in that reaction, either as a reactant or as a product. The connecting edge is then labeled with the complex in which the species appears. For example, in Figure 11.1 there is an edge, labeled $A+B$, connecting species A to the reaction(s) $A+B \rightleftarrows C+D$.

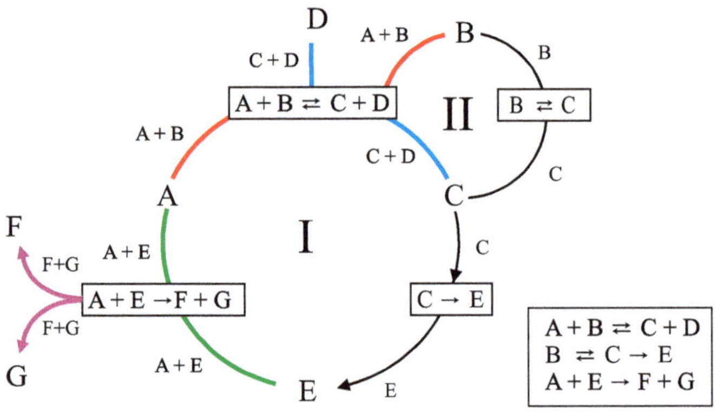

Fig. 11.1. An illustrative Species-Reaction Graph

A *fixed-direction edge* is an edge connecting an *irreversible* reaction to a species that is either (i) a product of the reaction or (ii) the *sole* reactant. For example, an edge joining an irreversible reaction $2A \to B+C$ with species B would be a fixed-direction edge because B is a product of the reaction. The edge joining that same reaction with species A would also be a fixed-direction edge because A is the *sole* reactant species. On the other hand, an edge joining an irreversible reaction $A+D \to E$ with species A would not be a fixed-direction edge because A is not the *sole* reactant species.

Each fixed-direction edge is given a *direction* according to the following prescription: The edge is directed from the species vertex to the reaction vertex if, in the reaction, the species appears as the (sole) reactant. The edge is directed from the reaction vertex to the species vertex if the species appears as a product of the reaction. Examples in Figure 11.1 are afforded by edges adjacent to the irreversible reactions $C \to E$ and $A + E \to F + G$. In the figure, directions assigned to the fixed-direction edges are indicated by arrows. Note the absence of arrows on the two edges connecting species A and E to reaction $A + E \to F + G$; these are not fixed-direction edges.

Once the vertices are connected by edges in the manner described, the drawing of the Species-Reaction Graph is complete.

11.4 Some Species-Reaction Graph Vocabulary

Very soon we will be in a position to state some powerful theorems. First, however, we will need a modest amount of vocabulary with which features of a particular Species-Reaction Graph might be described.

11.4.1 Complex Pairs, Odd Cycles, and Even Cycles

A *complex pair* (or *c-pair*) in a Species-Reaction Graph is a pair of edges adjacent to the same *reaction* vertex that carry the same complex label. For example, in Figure 11.1 the red edges adjacent to the reaction vertex $A + B \rightleftarrows C + D$ constitute a c-pair because they are both labeled by the complex $A + B$. In our display of Species-Reaction Graphs, c-pairs will be colored so that they can be recognized quickly. There are four c-pairs in Figure 11.1. Note that the two edges carrying the complex label C do not constitute a c-pair, for they are not adjacent to the same *reaction* vertex.

In the figure there are three cycles: cycle I, cycle II, and the large unlabeled outer cycle that traverses species A, B, C and E. An *odd cycle* is a cycle that contains among its edges an odd number of c-pairs. Similarly, an *even cycle* is one that contains an even number of c-pairs. Cycle I is odd because it contains only one c-pair, the green one. Cycle II is even because it contains no c-pairs. The large outer cycle is even because it contains two c-pairs, colored red and green.

11.4.2 Orientable Cycles

In a Species-Reaction Graph, a cycle $s_1 R_1 s_2 R_2 \ldots s_n R_n s_1$ is *orientable* if its edges can be assigned directions, either

$$s_1 \to R_1 \to s_2 \to R_2 \to \cdots \to s_n \to R_n \to s_1$$

or

$$s_1 \leftarrow R_1 \leftarrow s_2 \leftarrow R_2 \leftarrow \cdots \leftarrow s_n \leftarrow R_n \leftarrow s_1,$$

that is consistent with the fixed-direction edges in the cycle. A cycle having no fixed-direction edges is orientable in either direction.

In Figure 11.1 cycle I and the large outer cycle are both orientable, but only in the clockwise direction. Cycle II is orientable in both the clockwise and counterclockwise directions. Were reactions $A+B \rightleftarrows C+D$ replaced by the irreversible reaction $C+D \rightarrow A+B$, cycle I would no longer be orientable, nor would the large outer cycle. Cycle II would remain orientable, but now only in the clockwise direction.

A pair of cycles has a *consistent orientation* if each can be oriented in such a way that, in both of the given orientations, each edge common to the two cycles has the same direction. Each pair of cycles in Figure 11.1 admits a consistent orientation. For example, cycles I and II can be oriented consistently by giving cycle I a clockwise orientation and cycle II a counterclockwise orientation.

11.4.3 Stoichiometrically Expansive Orientations

In a Species-Reaction Graph, we can associate with each edge a *stoichiometric coefficient* in the following way: the stoichiometric coefficient of the edge connecting species s to reaction R is the (positive) stoichiometric coefficient of s in reaction R. For example, an edge connecting species A to reaction $2A \rightarrow B+C$ would have a stoichiometric coefficient of 2. An edge connecting species B to that same reaction would have a stoichiometric coefficient of 1.

Consider an oriented cycle, having orientation

$$R_1 \rightarrow s_1 \rightarrow R_2 \rightarrow s_2 \rightarrow \cdots \rightarrow R_n \rightarrow s_n \rightarrow R_1. \tag{11.2}$$

We denote the stoichiometric coefficient of a species-to-reaction edge $s \rightarrow R$ by $e_{s\rightarrow R}$ and the stoichiometric coefficient of a reaction-to-species edge $R \rightarrow s$ by $f_{R\rightarrow s}$. The cycle is *stoichiometrically expansive* relative to the given orientation if

$$\frac{f_{R_1\rightarrow s_1} f_{R_2\rightarrow s_2} \cdots f_{R_n\rightarrow s_n}}{e_{s_1\rightarrow R_2} e_{s_2\rightarrow R_3} \cdots e_{s_n\rightarrow R_1}} > 1. \tag{11.3}$$

In Figure 11.2 we show the Species-Reaction Graph for the simple network consisting of reactions $A+B \rightleftarrows P$ and $B \rightleftarrows 2A$. Because there are no fixed-direction edges,

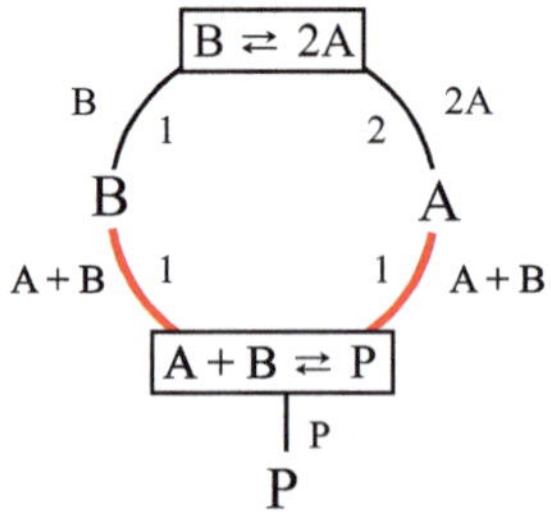

Fig. 11.2. A cycle with a stoichiometrically expansive orientation

the cycle in the figure can be oriented in either direction. Stoichiometric coefficients are displayed along the edges. The clockwise orientation is stoichiometrically expansive, but the counterclockwise orientation is not.

Remark 11.4.1 (Why stoichiometrically expansive cycles are not common). In the Species-Reaction Graph shown in Figure 11.1, the stoichiometric coefficient of every edge is 1, from which it follows that no cycle is stoichiometrically expansive relative to any possible orientation. It is fair to say that in most naturally occurring reaction networks, all stoichiometric coefficients will be 1. It is certainly true that two identical molecules can, in a single reaction, come together to form a product, but it is far more common for there to be a single reactant molecule or two different ones. Similarly, a single reaction can produce two molecules of the same species, but it is far more common for there to be one product molecule or two different ones. For this reason, we should not expect to find stoichiometrically expansive cycles very often in Species-Reaction Graphs.

The paucity of naturally occurring stoichiometrically expansive cycles will have considerable bearing on the range of the theorems in this chapter.

11.4.4 The Intersection of Two Cycles

By the *intersection* of two cycles in the Species-Reaction Graph, we mean the subgraph consisting of all vertices and edges common to the two cycles. Such a subgraph will consist of one or more connected components. Except in the trivial case in which a component consists of a single vertex, a component will be a path[1] common to the two cycles. For example, in Figure 11.1 the large outer cycle and cycle I intersect in the path that has species C at one end, the reactions $A+B \rightleftarrows C+D$ at the other end, and that traverses species E and A along the way. Cycles I and II intersect in the single-edge path that directly connects $A+B \rightleftarrows C+D$ with C.

To demonstrate that the intersection of two cycles can consist of multiple paths, we provide Figure 11.3, which is the Species-Reaction Graph drawn for network (11.1) augmented by the single reaction $F \to A$. The intersection of the large outer cycle with cycle I consists of two paths, one connecting A directly with $A+B \rightleftarrows C+D$ and the other connecting C with $A+E \to F+G$ via E.

Our primary interest will be in the intersection of two cycles that are oriented consistently. In such a case, a path in the intersection inherits a direction, pointing from one end vertex to the other. For example, when in Figure 11.3 cycle I and the large outer cycle are given clockwise orientations, the longer-path component of the intersection is directed from species C to reaction $A+E \to F+G$. The shorter single-edge path of the intersection is directed from species A to reactions $A+B \rightleftarrows C+D$. Each is an example of a *directed species-to-reaction path.*

When in the same figure cycle I is oriented clockwise and cycle II is oriented counterclockwise, their intersection consists of a single edge, directed from reactions $A+B \rightleftarrows C+D$ to species C. That intersection is an example of a *directed reaction-to-*

[1] When we say that a component of the cycle intersection is a path, we mean that it is a sequence of one or more edges that connect a sequence of distinct vertices.

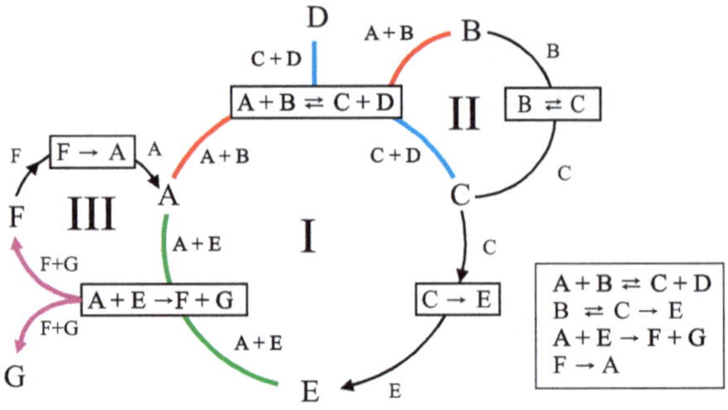

Fig. 11.3. Two cycles having a multiple-path intersection

species path. Various figures elsewhere in this chapter will demonstrate that, in the intersection of two consistently oriented cycles, there can also be components that are *directed reaction-to-reaction* and *directed species-to-species* paths.

11.5 First Preliminary Theorem: What the Species-Reaction Graph Tells Us About Network Concordance

We come now to the first of our preliminary theorems, one that connects concordance of a nondegenerate network to properties of its Species-Reaction Graph. This is a precursor of our first principal (and more tangible) theorem that will connect properties of a network's Species-Reaction Graph directly to kinetic system behavior—behavior that derives from concordance.

Theorem 11.5.1 is a consequence of a more nuanced but more technical theorem in [117]. Arguments underlying Theorem 11.5.1 and the more technical one will be discussed in Part III.

Theorem 11.5.1 ([117, 118, 158]). *Suppose that a reaction network's Species-Reaction Graph has the following properties:*

*(i) No **even** cycle admits a stoichiometrically expansive orientation.*
*(ii) No two consistently oriented **even** cycles have as their intersection a **single** directed path **originating at a species vertex and terminating at a reaction vertex**.*

Then the network's fully open extension is concordant. Moreover, so long as the network is nondegenerate, the network itself is concordant. In particular, the network is concordant if it is weakly reversible.

For the most part, we will defer examples—and counterexamples—to the next section, where we will be able to discuss more tangible consequences of a network's Species-Reaction Graph for behavior. Meanwhile, we can offer a few remarks in passing that pertain directly to concordance.

Remark 11.5.2 (Why concordance is common). Theorem 11.5.1 makes clear why concordance is common across the broad landscape of nondegenerate reaction networks:

For a reaction network to violate condition (i), there must be, somewhere in the network, a stoichiometric coefficient of 2 or more (assuming that all stoichiometric coefficients are integers). As we indicated in Remark 11.4.1, in most networks reflecting real chemistry, all nonzero stoichiometric coefficients are 1. Even when a 2 is present, violation of condition (i) requires that there be an *even* cycle that admits a stoichiometrically expansive orientation. An odd cycle won't do.

For a network to violate condition *(ii)*, the Species-Reaction Graph must contain two *even* cycles that admit a consistent orientation having a nontrivial intersection. An odd and an even cycle won't do, nor would two odd ones. And, *in addition*, the two consistently oriented cycles *must have an intersection of a very special kind.* The intersection must consist of *precisely one* path—it can't, for example, be two disconnected paths—and that lone path must, with respect to the given orientations, be directed *from an initial species vertex to a terminating reaction vertex.* The path can't be directed from a reaction vertex to a species vertex, from a species vertex to a species vertex, or from a reaction vertex to a reaction vertex.

Remark 11.5.3. Theorem 11.5.1 provides highly nuanced conditions that a nondegenerate network must satisfy if it is to be discordant: In its Species-Reaction Graph, either condition *(i)* or condition *(ii)* must fail. There is a far less nuanced but nevertheless useful necessary condition for discordance that derives directly from these, a condition identical to one first invoked by Badal Joshi and Anne Shiu [116] for different but related purposes. The focus for them was on injectivity in mass action systems, with arguments ultimately deriving from determinant-based theorems in [46, 49].

By the *total molecularity* of species $\mathscr{s}$ in a given network, Joshi and Shiu mean the sum of all stoichiometric coefficients of $\mathscr{s}$ (understood to be nonnegative integers) in the network, with summing effected over all reactions individually (with a reversible reaction pair counted only once). Excluded from their count are reactions such as $A \to 0$, $0 \to A$, and $2A \to 0$, having the zero complex on one side and only one species on the other. For example, in the reaction network consisting only of reactions $2A \to B$ and $A \rightleftarrows C$, the molecularity of A is three and the molecularity of C is one. The result is the same for the network's fully open extension. The following is a consequence of Theorem 11.5.1.

Corollary 11.5.4. *For a nondegenerate network to be discordant, it is necessary that it contain a species having a total molecularity greater than two.*

Proof. If, for the network's Species-Reaction Graph, condition *(i)* is violated, then, in the putative cycle with a stoichiometrically expansive orientation, not all edge stoichiometric coefficients can be 1. Thus, there must be an edge $s - R$ having a stoichiometric coefficient of at least 2 adjacent to a different edge $s - R'$ having a stoichiometric coefficient of at least 1. Thus, the total molecularity of species s is at least three. Suppose, on the other hand, that condition *(ii)* fails and that s is the initial species of the directed species-to-reaction intersection of the oriented-even-cycle pair. At species s three different edges meet, each having a stoichiometric coefficient of at least 1. From this it again follows that s has a total molecularity of at least three. □

Remark 11.5.5. To see that a network can be degenerate (and therefore discordant) even if its Species-Reaction Graph satisfies both conditions of Theorem 11.5.1, it is enough to consider the simple degenerate network

$$C \leftarrow A \rightarrow B. \tag{11.4}$$

Its Species-Reaction Graph has no cycles at all, so the conditions of the theorem are satisfied trivially.

Network (10.22) of the preceding chapter provides another example, this time one that has positively dependent reaction vectors and, consequently, the capacity to admit a positive equilibrium. Its Species-Reaction Graph is shown in Figure 11.4. All stoichiometric coefficients are 1, so condition *(i)* is satisfied. Moreover, condition *(ii)* is satisfied because the only two extant cycles do not intersect. Nevertheless, the network is degenerate and, therefore, discordant.

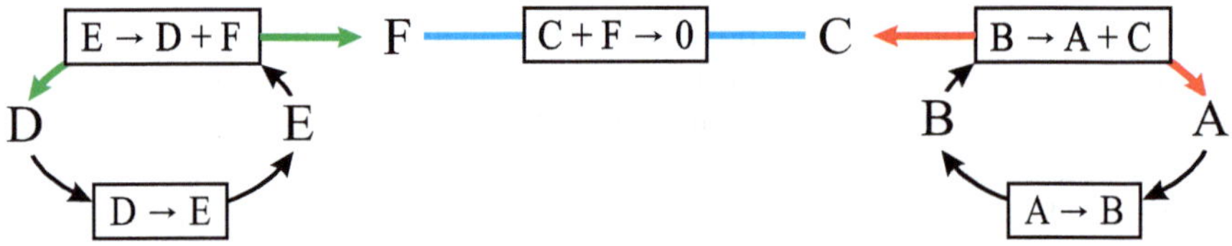

Fig. 11.4. A degenerate network's Species-Reaction Graph (edge labels omitted)

Remark 11.5.6 (A nondegenerate network can be concordant even if its Species-Reaction Graph violates condition (i) *or* (ii)*).* A reaction network for which the corresponding Species-Reaction Graph fails to satisfy condition *(i)* or *(ii)* will often be discordant. It should be clearly understood, however, that Theorem 11.5.1 provides *sufficient*, not necessary, conditions for concordance. This is to say that the class of concordant networks is even broader than the class described by Theorem 11.5.1.

An example [117, 118] is provided by network (11.5), in which an enzyme E acts as a catalyst for conversion of substrate S to product P. At the same time, S acts as a catalyst for conversion of E to an inactive form E^*. There is also a direct reversible conversion of E to E^*. The network's Species-Reaction Graph violates condition *(ii)* of Theorem 11.5.1. Yet the network is concordant.

$$\begin{array}{ccc} E+S \rightleftarrows ES \rightleftarrows E+P & & E \rightleftarrows E^* \\ \downarrow & & \\ E^*+S & & \end{array} \tag{11.5}$$

The concordance of network (11.5) is explained by another theorem, of which Theorem 11.5.1 is a corollary, that we will discuss in Part III, Chapter 19.

11.6 First Principal Theorem: What the Species-Reaction Graph Tells Us About Behavior (Monotonic Kinetics)

Theorem 11.5.1 of the preceding section posited very mild conditions on a network's Species-Reaction Graph that suffice for concordance. Here we reap the behavioral bounty that theorem affords. In our next theorem, concordance appears not at all but instead exerts its role in the background. Theorem 11.6.1 is a consequence of Theorem 11.5.1 and Theorem 10.7.6 taken together. In this way, behavior becomes tied directly to properties of a reaction network's Species-Reaction Graph. Recall that the very broad and natural weakly monotonic and differentiably monotonic kinetic classes were defined in Section 10.5.

Theorem 11.6.1 ([117, 118, 157, 158]). *Suppose that a nondegenerate reaction network's Species-Reaction Graph has the following properties:*

(i) No ***even*** *cycle admits a stoichiometrically expansive orientation.*
(ii) No two consistently oriented ***even*** *cycles have as their intersection a* ***single*** *directed path* ***originating at a species vertex and terminating at a reaction vertex****.*

Then the following statements hold true:

(A) For any choice of a weakly monotonic kinetics, each positive equilibrium is the only equilibrium within its stoichiometric compatibility class. That is, no positive equilibrium is stoichiometrically compatible with a different equilibrium, positive or otherwise. If the kinetics is differentiably monotonic, then every real eigenvalue associated with a positive equilibrium is strictly negative.

(B) If the network is weakly reversible, then, for each choice of a kinetics (not necessarily weakly monotonic), no nontrivial stoichiometric compatibility class has an equilibrium on its boundary. In fact, at each boundary composition in any nontrivial stoichiometric compatibility class, the species-formation-rate vector points into the stoichiometric compatibility class in the sense that there is an absent species produced at strictly positive rate. If, in addition, the network is conservative, then, for any choice of a weakly monotonic kinetics, there is precisely one equilibrium in each nontrivial stoichiometric compatibility class, and it is positive.

The hypothesis of Theorem 11.6.1 is the same as the hypothesis of the preliminary theorem of the preceding section. It is worth repeating here what was said there: the Species-Reaction Graph requirements posited by Theorem 11.6.1 are very, very mild, and they will be satisfied quite often by naturally occurring reaction networks. The theorem, then, tells us something remarkable: *Despite the presence of a great deal of nonlinearity, a degree of dull, stable behavior is enforced across a very wide region of the reaction network landscape, so long as the associated kinetic rate functions conform to very weak and highly natural constraints.*

11.7 Examples

We are about to examine a variety of reaction network examples viewed from the perspective of Theorem 11.6.1 (and, implicitly, of Theorem 11.5.1). In the examples, networks that *fail* to satisfy one or another of the conditions of these theorems will of course be represented. It should be kept in mind, however, that such examples appear precisely *because* they are aberrant, not because they are highly typical of what one finds among networks generally. (The situation for the narrower class of *biochemical* reaction networks is somewhat different; we shall have more to say about them in the next chapter.)

While discussing examples we will sometimes, but not always, consider dynamics in a fully open CFSTR setting. In such cases, remember from Section 4.2.1 that the appropriate network of interest is the *fully open extension* of the "true chemistry," taken together with feed (synthesis) reactions of the form $0 \to \mathcal{S}$, one for each species $\mathcal{S}$ deemed to be supplied to the reacting mixture. It should be kept in mind that *all fully open networks are nondegenerate* (Corollary 10.6.4). Remember too that the Species-Reaction Graph is drawn for the "true chemistry" base subnetwork, devoid of effluent and feed reactions.

Remark 11.7.1. When, in the consideration of a particular example, we indicate that for a certain weakly monotonic kinetics there are multiple stoichiometrically compatible positive equilibria, it will be because one or another of the conditions in Theorem 11.6.1 (and in Theorem 11.5.1) is violated. *When there are such multiple equilibria, we are also saying, in effect, that concordance is absent.*

11.7.1 Revisiting Some of Our First Species-Reaction Graphs

We will begin by revisiting some of the networks considered earlier in this chapter, in particular network (11.1) and variants of it. Their Species-Reaction Graphs are already available to us. In each case, we will ask about behavior in a fully open CFSTR setting, with species A and B in the feed stream. The network of interest then is the (nondegenerate) fully open extension of network (11.1)—or its particular variant—augmented by the reactions $0 \to A$ and $0 \to B$. Because in each case every stoichiometric coefficient is 1, condition *(i)* of Theorem 11.6.1 will be satisfied. Our focus will be on condition *(ii)*.

11.7.1.1 Network (11.1)

The appropriate Species-Reaction Graph was shown in Figure 11.1. In consideration of condition *(ii)*, we note that there are three cycles: cycle I, cycle II, and the large outer cycle. Of these, only cycle II and the large outer cycle are even. That pair admits only one consistent orientation, with both cycles oriented clockwise. With respect to that orientation, the intersection of the two cycles is a single directed path that passes through species vertex B and reaction vertex $B \rightleftarrows C$, but *it originates at a reaction vertex ($A + B \rightleftarrows C + D$) and terminates at a species vertex (C).* Thus, condition *(ii)* is satisfied. All of the conclusions given by part (A) of Theorem 11.6.1 obtain.

11.7.1.2 Network (11.1) *with $C \to E$ Made Reversible*

Now consider network (11.1) with reaction $C \to E$ replaced by $C \rightleftarrows E$. In this case the Species-Reaction Graph is very much like the one shown in Figure 11.1. However, the edges connecting species vertices C and E to the reaction vertex $C \rightleftarrows E$ are no longer fixed-direction edges, and the even cycle II can now be oriented either clockwise or counterclockwise. Moreover, cycle II and the large outer cycle, both even, now admit a consistent orientation (both counterclockwise) with respect to which their intersection is a single directed path, *this time originating at a species vertex (C) and terminating at a reaction vertex ($A+B \rightleftarrows C+D$).*

Condition *(ii)* of Theorems 11.5.1 and 11.6.1 is *not* satisfied, and both theorems are silent. In fact, the fully open network corresponding to the CFSTR under consideration—including reactions $0 \to A$ and $0 \to B$—is *discordant* [62]. *Moreover, there are mass action rate constants for the fully open network such that the resulting differential equations admit multiple positive equilibria.* Similarly, there are mass action rate constants for the fully open network such that the resulting differential equations admit a degenerate positive equilibrium [62].

11.7.1.3 Network (11.1) Augmented with Reaction $F \to A$

The appropriate Species-Reaction Graph was displayed in Figure 11.3. Relative to the graph displayed in Figure 11.1, any new cycles in Figure 11.3 result from the newly formed even cycle III. Motivated by our focus on condition *(ii)*, our interest will be restricted to interactions of *even* cycles: Cycle III has a nontrivial intersection with the even cycle formed by the perimeter of cycles I and II, but that perimeter and cycle III do not admit a consistent orientation. The even cycle II and the even cycle consisting of the perimeter of cycles I and III admit consistent orientations—the first counterclockwise and the second clockwise—but relative to these orientations, their intersection is a single directed path *beginning at a reaction vertex and terminating at a species vertex*. Condition *(ii)* is satisfied. All of the conclusions given by part (A) of Theorem 11.6.1 obtain.

11.7.2 Subtleties Present Even When There Is Only One Cycle

In light shed by the Species-Reaction Graph theorems, we can now revisit three highly similar networks, (11.6)–(11.8), having origins in [147, 148]. These were studied in Section 10.4, where we asserted, on the basis of computations [62], that the first and third of these are concordant while the second is discordant.

$$\begin{aligned} A+B &\rightleftarrows P \\ B &\rightleftarrows 2A \end{aligned} \tag{11.6}$$

$$\begin{aligned} A+B &\rightleftarrows P \\ B+C &\rightleftarrows Q \\ C &\rightleftarrows 2A \end{aligned} \tag{11.7}$$

$$\begin{aligned} A+B &\rightleftarrows P \\ B+C &\rightleftarrows Q \\ C+D &\rightleftarrows R \\ D &\rightleftarrows 2A \end{aligned} \tag{11.8}$$

We also noted that the second chemistry behaves very differently from the others in a fully open CFSTR context: When the kinetics is mass action, there are parameter values such that, for (11.7), the corresponding CFSTR differential equations admit multiple positive equilibria. That is not the case for (11.6) or (11.8). For them, multiple equilibria are impossible even when the kinetics is not mass action, so long as the kinetics is weakly monotonic.

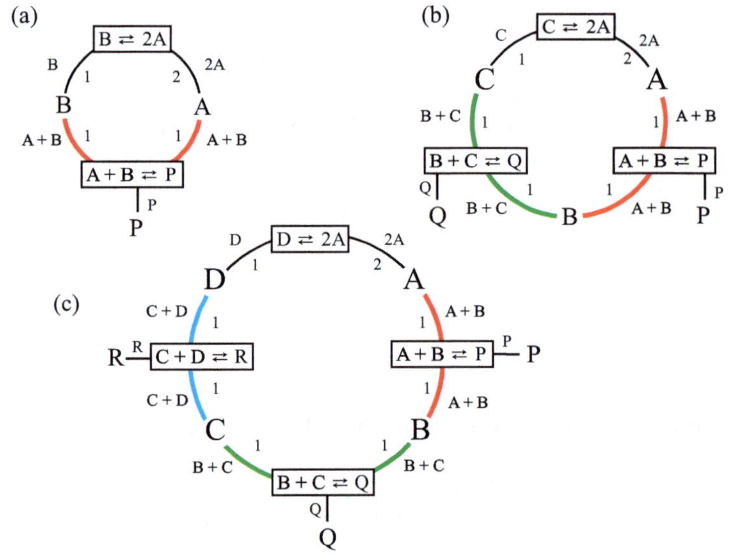

Fig. 11.5. Species-Reaction Graphs for networks (11.6)–(11.8)

The Species-Reaction Graphs for the three networks, displayed in Figure 11.5, offer a new perspective. (Recall that the Species-Reaction Graph for the fully open CFSTR network is the same as the graph for the true chemistry base subnetwork, devoid of feed or effluent reactions.) In each case there is just one cycle, so condition *(ii)* of Theorems 11.5.1 and 11.6.1 is satisfied trivially. On the other hand, in each case the lone cycle admits a stoichiometrically expansive orientation (clockwise). Nevertheless, condition *(i)* is *not* violated in the first and third cases because in each of these, the corresponding cycle is odd: there is an odd number of c-pairs. The conclusions of Theorems 11.5.1 and 11.6.1 thereby obtain for the fully open network appropriate to the CFSTR setting. Condition *(i) is* violated in Figure 11.5(b): there are two c-pairs, so the cycle is even. In this case Theorems 11.5.1 and 11.6.1 are silent. There is no preclusion of multiple positive CFSTR equilibria.

Remark 11.7.2. We have just considered the chemistry described by networks (11.6)–(11.8) operating in a CFSTR context. In each such case, the network of interest was the fully open extension of the corresponding network in (11.6)–(11.8), augmented by feed reactions of the form $0 \rightarrow \mathscr{S}$.

We can also consider the behavior of networks (11.6)–(11.8) on their own, removed from the CFSTR context. The corresponding Species-Reaction Graphs remain unchanged, and, because each of those reversible networks is nondegenerate, the conclusions of Theorems 11.5.1 and 11.6.1 for networks (11.6) and (11.8) remain what they were before. On the other hand, for network (11.7) both theorems are silent.

As we indicated in Remark 10.5.12, however, all three networks have a deficiency of zero. They are also forest-like. In this case highly detailed behavioral information is given for *all three* networks by the Deficiency Zero Theorem (Theorem 7.1.1) and by its derivative version (Theorem 7.3.1), *so long as the kinetics is mass action.* Note that Theorem 11.6.1 extends some of that same information to far larger classes of kinetics, but only for networks (11.6) and (11.8).

11.7.3 The Wnt Pathway Again

In Section 10.5.4 we discussed the Wnt signaling pathway, as modeled by Lee et al. [122]. In (11.10) we repeat the network displayed earlier as (10.17), a variant of the Lee et al. model that, in particular, omits the seemingly innocuous reversible reaction

$$A_7 + A_{11} \rightleftarrows A_{15}. \tag{11.9}$$

This reaction does little more than produce a new species A_{15} that appears nowhere else in the network. In (10.18) we displayed the imposing system of differential equations resulting from network (11.10) taken with a kinetics assumed only to be weakly monotonic.

$$
\begin{array}{c}
A_{10} \to 0 \rightleftarrows A_{11} \\
\upharpoonleft\downharpoonright \\
A_{12} \\
A_1 \rightleftarrows A_2 \\
A_2 + A_4 \to A_{16} \to A_2 + A_6 \\
A_3 \rightleftarrows A_4 \rightleftarrows A_6 \rightleftarrows A_7 + A_{12} \\
A_3 + A_{11} \rightleftarrows A_8 \to A_9 \to A_3 + A_{10} \\
A_{11} + A_{13} \rightleftarrows A_{14}
\end{array}
\tag{11.10}
$$

On the basis of computations [62], we asserted in Section 10.5.4 that network (11.10) is concordant. As a result, the dullness of behavior that concordance ensures is imposed on the highly complex and largely nonlinear system of differential equations (10.18), so long as the kinetics is weakly monotonic (or, for some purposes, differentiably monotonic).

We also asserted that, when the ostensibly benign reversible reaction (11.9) is adjoined to (11.10), the resulting network is *discordant.* Moreover, the higher deficiency module in [62] tells us that *there exist for the augmented network rate constants such that the resulting mass action differential equations admit multiple stoichiometrically compatible positive equilibria.*

It is remarkable that the addition of the seemingly innocent reactions (11.9) has such a behavior-altering effect. It is our purpose here to revisit the differences between network (11.10) and its slightly augmented version, this time with the benefit of the subtle discriminatory power of their Species-Reaction Graphs. In consideration of Theorems 11.5.1 and 11.6.1, we note that the nondegeneracy of the two networks was argued in Section 10.6.5.

Both graphs are shown in Figure 11.6. The Species-Reaction Graph for network (11.10) is represented by the figure with the shaded rectangular region excluded. The full figure, with the rectangular region included, is Species-Reaction Graph for network (11.10) with reactions $A_7 + A_{11} \rightleftarrows A_{15}$ adjoined. In both graphs all stoichiometric coefficients are 1, so condition *(i)* of Theorems 11.5.1 and 11.6.1 is satisfied. Our focus, then, will be on condition *(ii)*.

In the Species-Reaction Graph for network (11.10) (with the rectangular region excluded), there are only three orientable cycles, labeled I, II, and III. Cycles I and II are even and admit consistent orientations, shown in the figure. However, their intersection is a directed reaction-to-reaction path. Cycle III has no intersection with another cycle. Condition *(ii)* of Theorems 11.5.1 and 11.6.1 is satisfied. The conclusions of both theorems obtain. *It is worth a glance back at the differential equations* (10.18) *to remind ourselves, in a tangible way, of the extent to which Theorem 11.6.1 and the Species-Reaction Graph can quickly give powerful and far-from-obvious information about dauntingly complex systems.*

In the Species-Reaction Graph for network (11.10) (with the rectangular region included), the addition of a single reversible reaction makes for a very different situation. There are now additional cycles, in particular cycle IV, which can be oriented either clockwise or counterclockwise. Cycle IV contains two c-pairs, so it—like cycle III—is even. Those two *even* cycles can be oriented consistently, as shown in the figure, and, with respect to the given orientations, *their intersection is a single directed path originating at a species vertex* (A_3) *and terminating at a reaction vertex* ($A_3 + A_{11} \rightleftarrows A_8$). Condition *(ii)* is violated, so there is no preclusion of multiple stoichiometrically compatible positive equilibria.

The added reaction $A_7 + A_{11} \rightleftarrows A_{15}$ exerts its influence on the Species-Reaction Graph in two important ways: it adds new cycles,[2] and it adds a c-pair such that the newly created cycle IV is born even.

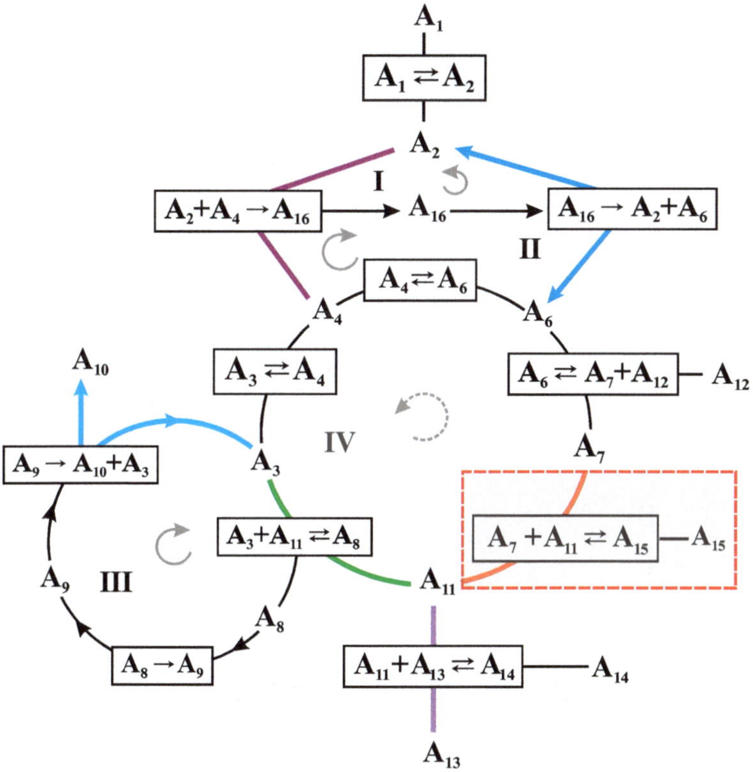

Fig. 11.6. Species-Reaction Graphs for the Wnt pathway (edge labels omitted)

11.7.4 Even Cycles Intersecting in Two Directed Species-to-Reaction Paths

The failure of condition *(ii)* in the full Wnt example of the preceding section resulted from the intersection of a pair of consistently oriented even cycles in a single path

[2] The cycle consisting of the outer perimeter of cycles III and IV is odd.

directed from a species vertex to a reaction vertex. That there was a *single* such path was critical. To make the point, we will consider the chemistry shown in (11.11) operating again in a fully open CFSTR context [117, 118].

$$
\begin{aligned}
A+B &\to C \to D+E \\
D+F &\to G \\
E &\to F \to A+H \\
C+H &\to J
\end{aligned}
\tag{11.11}
$$

The Species-Reaction Graph for network (11.11) is displayed in Figure 11.7. The same graph represents the larger fully open network (including feed and effluent reactions). Because so many of the reactions are not reversible, many of the edges have fixed directions. Although there are many cycles, there is a paucity of orientable cycles. We note that cycle II and the large outermost cycle have consistent orientations, both clockwise. These cycles are both *even*, and, with respect to the given orientations, they have an intersection consisting of *two* directed *species-to-reaction* paths, each consisting of a single edge: species vertex C to reaction vertex $C \to D+E$ and species vertex F to reaction vertex $F \to A+H$.

Because there are *two* such paths, there is no violation of condition *(ii)* in Theorems 11.5.1 and 11.6.1. In fact both conditions *(i)* and *(ii)* are satisfied. Moreover, the fully open CFSTR network under consideration is nondegenerate, as are all fully open networks. The network is indeed concordant, as can be confirmed computationally [62], and all of the behavioral consequences of Theorem 11.6.1 accrue to the CFSTR, so long as the kinetics has the stipulated monotonicity properties.

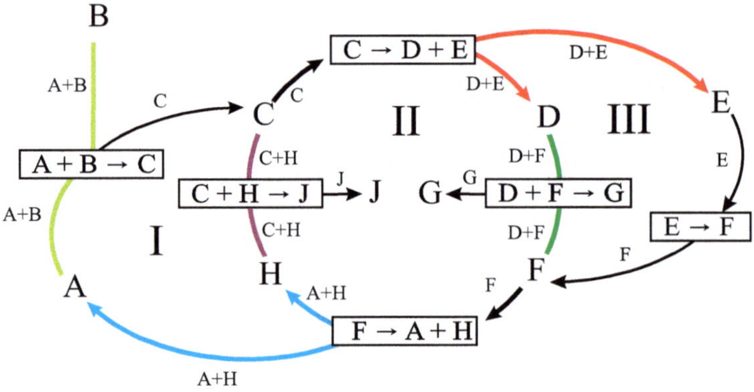

Fig. 11.7. Even cycles intersecting in two directed species-to-reaction paths. After [118]

Remark 11.7.3. If the reactions $C \to D+E$ and $F \to A+H$ are made reversible, cycle II can be oriented in both the clockwise and counterclockwise directions, and cycle III can be oriented in the clockwise direction. In this case cycles II and III admit a consistent orientation, with cycle II oriented counterclockwise and cycle III oriented clockwise. Cycles II and III are both even, and, with respect to the given orientations, their intersection is a *single* path directed from species vertex F to reaction vertex $C \rightleftarrows D+E$. Condition (ii) is no longer satisfied, whereupon Theorems 11.5.1 and 11.6.1 are silent. In fact, the new network is *discordant*, as can be confirmed computationally [62]. Moreover, it can also be confirmed by means of the higher deficiency module in [62] that there are parameter values such that the mass action CFSTR differential equations admit multiple positive equilibria.[3]

11.8 Second Preliminary Theorem: What the Species-Reaction Graph Tells Us About *Strong* Concordance

In Section 11.6 we stated Theorem 11.6.1, which, when its hypothesis is satisfied, gives considerable behavioral information so long as the network is taken with a kinetics that is weakly monotonic or, better still, differentiably monotonic. The route to that theorem was through the preliminary Theorem 11.5.1, which asserted that, when a nondegenerate network's Species-Reaction Graph satisfies certain conditions, the network is concordant. The information given in Theorem 11.6.1 then followed simply from the powerful behavioral constraints that concordance imposes.

In the next section, we will state another behavior-constraining theorem—this time one that embraces the much broader class of *two-way weakly monotonic kinetics*, a class that allows for product inhibition in the various kinetic rate functions. (See Section 10.9.) The route will be similar to the one we just traveled. The subject of this section will be another preliminary theorem. It will tell us that if a nondegenerate network's Species-Reaction Graph has certain attributes, then the network is *strongly concordant*. Strong concordance will then provide the underpinnings of our second principal theorem, stated in the next section, that connects properties of the Species-Reaction Graph directly to behavior when the kinetics is two-way monotonic.

In preparation for these theorems, we will need a new characterization of Species-Reaction Graph cycles. Because two-way monotonic kinetics is somewhat indifferent to reaction arrow directions, it is reasonable to expect that in theorems about such kinetics, there could be an indifference to *particular* cycle orientations. Indeed, it will turn out that in the hypotheses of these theorems, there will be a preclusion of cycles that are stoichiometrically expansive in *either* the clockwise or the counterclockwise direction. In the presence of such a preclusion, the only permissible cycles would be those for which the stoichiometric-coefficient ratio in (11.3) is *exactly one*. Such "good" cycles have, for a long time [147, 148], been called s-cycles:

An *s-cycle* in a Species-Reaction Graph is a cycle such that, with respect to some orientation,

$$R_1 \to s_1 \to R_2 \to s_2 \to \ldots R_n \to s_n \to R_1,$$

---

[3] In particular there are such parameters when all species apart from J and G are in the feed stream.

either clockwise or counterclockwise,

$$\frac{f_{R_1 \to s_1} f_{R_2 \to s_2} \cdots f_{R_n \to s_n}}{e_{s_1 \to R_2} e_{s_2 \to R_3} \cdots e_{s_n \to R_1}} = 1. \tag{11.12}$$

If (11.12) is satisfied for one such orientation, it will also be satisfied for the opposite orientation.

We are now in a position to state our second preliminary theorem. Conditions *(i)* and *(ii)* in Theorem 11.8.1 amount to orientation-free versions of their counterparts in Theorem 11.5.1. Note that they are required to be satisfied by *all* even cycles, not just orientable ones.

Theorem 11.8.1 ([117, 118, 158]). *Suppose that a reaction network's Species-Reaction Graph has the following properties:*

(i) Every ***even*** *cycle is an s-cycle.*
(ii) No two ***even*** *cycles have as their intersection a* ***single*** *path* ***with a species vertex at one end and a reaction vertex at the other****.*

Then the network's fully open extension is strongly concordant. Moreover, so long as the network is nondegenerate, the network itself is strongly concordant. In particular, the network is strongly concordant if it is weakly reversible.

11.9 Second Principal Theorem: What the Species-Reaction Graph Tells Us About Behavior (Two-Way Monotonic Kinetics)

The following theorem, in which strong concordance is not mentioned at all, is a direct consequence of Theorem 11.8.1 taken together with properties of strongly concordant networks discussed in the preceding chapter. Recall the definitions related to "two-way monotonic kinetics" in Section 10.9.

Theorem 11.9.1 ([117, 118, 157, 158]). *Suppose that a nondegenerate reaction network's Species-Reaction Graph has the following properties:*

(i) Every ***even*** *cycle is an s-cycle.*
(ii) No two ***even*** *cycles have as their intersection a* ***single*** *path* ***with a species vertex at one end and a reaction vertex at the other****.*

Then the conclusions of Theorem 11.6.1 obtain with "weakly monotonic" replaced by "two-way weakly monotonic" and "differentiably monotonic" replaced by "differentiably two-way monotonic."

Remark 11.9.2 (Conditions (i) and (ii) remain mild). Because Theorem 11.9.1 constrains behavior in the same way that Theorem 11.6.1 does, but for a far larger kinetic class, we should expect the family of networks to which it applies to be substantially narrower. Indeed, the two conditions in Theorem 11.9.1 are more demanding than their counterparts in Theorem 11.6.1.

Nevertheless, condition *(i)* remains very weak, at least when stoichiometric coefficients are integers: It will be satisfied for all networks in which every nonzero stoichiometric coefficient is 1, by far the most common situation.

Condition *(ii)* is more stringent than it was before, but it too remains relatively mild. A violation requires that in the Species-Reaction Graph, there be two *even* cycles that have a nontrivial intersection, and the intersection must consist of a *single* path *having a species vertex at one end and a reaction vertex at the other.* An intersection consisting of two disconnected components won't do, nor would a single path having species vertices at both ends or reaction vertices at both ends.

11.10 More Examples

Here we provide more examples, some visited before, this time with a focus on two-way monotonic kinetics (and, implicitly, on strong concordance).

11.10.1 Some of Our First Species-Reaction Graph Examples Revisited, Yet Again

Recall network (11.1), considered earlier in a CFSTR context. The appropriate Species-Reaction Graph was displayed in Figure 11.1. When the kinetics was weakly monotonic or, better still, differentiably monotonic, Theorem 11.6.1 gave powerful behavioral information. When the kinetics is two-way monotonic, however, Theorem 11.9.1 is silent: Although condition *(i)* is satisfied, condition *(ii)* is not. The large outer cycle, which is even, intersects the even cycle II in a single path having at its ends a reaction vertex and a species vertex. Although the nondegenerate network (11.1) is concordant, it is not strongly concordant, nor is the enlarged fully open network appropriate to the CFSTR setting [62].

Recall also the highly similar nondegenerate networks (11.6)–(11.8), for which the Species-Reaction Graphs were shown in Figure 11.5. When the kinetics is weakly monotonic (and, in particular, differentiably monotonic), Theorem 11.6.1 gave behavioral information for networks (11.6) and (11.8) but not for the "in-between" network (11.7) (which is, in fact, discordant). The same situation obtains when the kinetics is two-way monotonic: Networks (11.6) and (11.8) satisfy both conditions of Theorem 11.9.1 vacuously. Network (11.7) again fails to satisfy condition *(i)*.

11.10.2 The Wnt Pathway Again, This Time with Kinetics Admitting Product Inhibition

In the last chapter, we displayed in (10.18) the differential equations for network (11.10), an almost-complete variant of a Wnt pathway model studied by Lee et al. [122]. Those equations were written in terms of a weakly monotonic but otherwise arbitrary kinetics. Despite the great complexity of the system (10.18) and the wide kinetic class it embraces, Theorem 11.6.1 provided considerable information about the very limited varieties of behavior those equations can generate.

If for network (11.10) we expand the kinetic class considered to embrace product inhibition in the various reactions, we can ask if the resulting differential equations can admit new behavior that was precluded before. When the kinetics is two-way weakly monotonic, the differential equations become significantly more complicated than those shown in (10.18), for then the dependencies of rate functions on species concentrations become more expansive. For example, the rate of a reaction such as $A_6 \to A_7 + A_{12}$ could depend not only on the concentration of A_6 but also on the concentrations of A_7 and A_{12}. In this case, the rate $\mathscr{K}_{A_6 \to A_7 + A_{12}}(c_6)$ would be replaced in (10.18) by $\mathscr{K}_{A_6 \to A_7 + A_{12}}(c_6, c_7, c_{12})$.

Nevertheless, *when the kinetics is two-way weakly monotonic, our earlier conclusions remain unaltered.* For the nondegenerate network (11.10), the corresponding Species-Reaction Graph displayed in Figure 11.6—with the shaded rectangle excluded—satisfies the conditions of Theorem 11.9.1.

11.10.3 A More Incisive Example Illustrating Condition (ii)

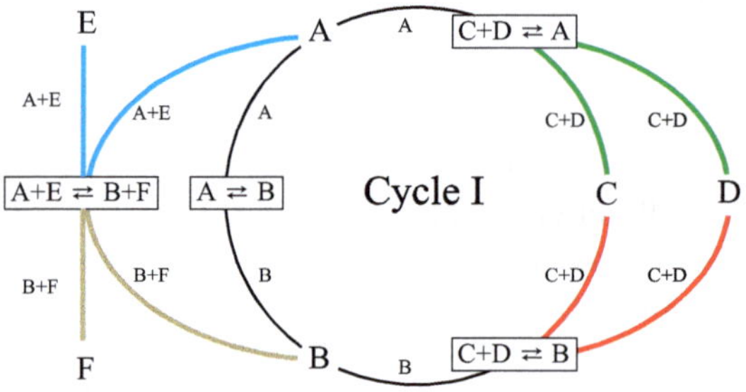

Fig. 11.8. Even cycles intersecting in two species-reaction paths. Figure adapted from [118]

In Figure 11.8 we show the Species-Reaction Graph for network (11.13).

$$\begin{array}{ccc} A & \rightleftarrows & B \\ & \nwarrow\!\searrow \quad \nearrow\!\swarrow & \\ & C+D & \end{array} \qquad A+E \rightleftarrows B+F \tag{11.13}$$

This example [118] will provide a useful illustration of the use of condition (ii) in Theorems 11.8.1 and 11.9.1.[4] (Because all stoichiometric coefficients are 1,

4 Network (11.13) was used also in [48] to make a different but related point about mass action systems.

condition (i) is satisfied.) There are several pairs of even cycles having nontrivial intersections, but most intersections consist of a path having species vertices at both ends or reaction vertices at both ends. Note, however, that the central cycle I and the large outer cycle, both even, have as their intersection *two* paths, each having a species vertex at one end and a reaction vertex at the other. Because there are *two* such paths in the intersection, this does not constitute a violation of condition *(ii)*. There is no pair of even cycles whose intersection consists of a *single* path with a species vertex at one end and a reaction vertex at the other.

Therefore, the Species-Reaction Graph displayed in Figure 11.8 complies with both conditions *(i)* and *(ii)*. Because network (11.13) is reversible, it is nondegenerate, as is any fully open augmentation of (11.13) constructed to model a CFSTR having that network as its true chemistry. Theorem 11.8.1 then tells us that (11.13) and CFSTR networks derived from it are strongly concordant.[5] So long as the kinetics is two-way weakly monotonic (and, in particular, differentiably two-way monotonic), the behavior described in Theorem 11.6.1 obtains.

11.11 Special Considerations for Mass Action Kinetics: The Knot Graph and the Species-Reaction Graph

When, for a nondegenerate network, the Species-Reaction Graph satisfies fairly mild conditions and when the kinetics is differentiably monotonic, Theorem 11.6.1 not only gives information about uniqueness of positive equilibria, it also gives *some* information about the stability of equilibria: all *real* eigenvalues associated with a positive equilibrium are negative. This is not quite enough to ensure asymptotic stability, for there might be complex eigenvalues with nonnegative real parts. For this reason, the theorem leaves the stability of positive equilibria unresolved.

When the kinetics is *mass action*, we can sometimes do better. For certain reaction networks, we can assert that, regardless of rate constants, *all* eigenvalues associated with positive equilibria are real and negative, in which case asymptotic stability of positive equilibria is ensured. Indeed, we have already seen a theorem that permits just such an assertion: Theorem 7.3.1 tells us that, for any mass action system in which the underlying reaction network is *forest-like and has a deficiency of zero*, all eigenvalues associated with a positive equilibrium are real and negative.

Theorem 7.3.1, however, has limited range, in particular for networks that derive from models of open systems. Consider for example the very simple mechanism for enzyme catalysis (11.14), in which a substrate S binds to an enzyme E, whereupon S gets converted to a product P that then unbinds from the enzyme.

$$E + S \rightleftarrows ES \rightleftarrows EP \rightleftarrows E + P \tag{11.14}$$

Network (11.14) is forest-like (in fact, tree-like) and has a deficiency of zero. When the kinetics is mass action, all the mass action properties of reversible defi-

[5] Recall that the Species-Reaction Graph for the fully open CFSTR network, including feed reactions of the form $0 \to \mathcal{s}$, is the same as the graph for (11.13).

ciency zero networks accrue to network (11.14), and, moreover, eigenvalues associated with positive equilibria are real and negative.

Now consider the same enzymatic chemistry, this time in the context of a semi-open flow reactor of the kind considered in Section 4.2.2. As in that example, we will suppose that the enzymatic species—in this case E, ES, and EP—are entrapped within the reactor chamber. Moreover, we will suppose that the substrate S is supplied to the reactor in the feed stream, while unconsumed substrate and product P are in the effluent stream. In this case, the governing differential equations derive, in the sense of Section 4.2.2, from the semi-open reaction network (11.15).

$$\begin{gathered} E + S \rightleftarrows ES \rightleftarrows EP \rightleftarrows E + P \\ P \rightarrow 0 \rightleftarrows S \end{gathered} \tag{11.15}$$

By virtue of the simple Species-Reaction Graph for the nondegenerate network (11.15), Theorem 11.6.1 tells us that, when the kinetics is mass action (and, therefore, differentiably monotonic), any positive equilibrium is the only equilibrium within its stoichiometric compatibility class. Moreover, the same theorem ensures that at every positive equilibrium, all *real* eigenvalues are negative. On the other hand, the deficiency of network (11.15) is now *one*, in which case Theorem 7.3.1 cannot preclude the existence of rate constants for which there is a positive equilibrium having complex eigenvalues. In fact, such rate constants do exist; they are exhibited in (11.16). For that mass action system, there is an equilibrium in which all species concentrations are 1, and at that equilibrium there is a pair of complex-conjugate eigenvalues (with negative real part).

$$\begin{gathered} E + S \underset{1}{\overset{5}{\rightleftarrows}} ES \underset{1}{\overset{5}{\rightleftarrows}} EP \underset{1}{\overset{5}{\rightleftarrows}} E + P \\ P \overset{4}{\rightarrow} 0 \underset{8}{\overset{4}{\leftrightarrows}} S \end{gathered} \tag{11.16}$$

The example tells us that even simple reversible forest-like deficiency zero networks with congenial Species-Reaction Graphs, when augmented by reactions of the type $\mathcal{S} \rightarrow 0$ or $0 \rightarrow \mathcal{S}$, can give rise to mass action systems having positive equilibria at which some eigenvalues are complex.

As we shall see, however, *there are networks which, when augmented by almost all choices of feed and effluent reactions, retain the invariably-real-negative-eigenvalues property, so long as the kinetics is mass action.* A class of such networks can be described by means of the *Knot Graph*, an idea originated and exploited in work by Delattre [54], by Hyver [112, 113], and by Beretta and co-workers [19, 20]. In the next two subsections, we will explain how the Knot Graph is constructed and what the graph can tell us. Then we will say something about connections between properties of the Knot Graph and the Species-Reaction Graph. In this way, we will be able to draw on prior Knot Graph work to expand the information that the Species-Reaction Graph provides.

11.11.1 The Knot Graph for a Reaction Network

Here, as in the rest of this chapter, it will be understood that we consider only networks in which there is no reaction having the same species as both a reactant and product. Although our interest is largely in networks for which there are reactions of the form $0 \to \mathcal{s}$ or $\mathcal{s} \to 0$, the Knot Graph is drawn for the base subnetwork (Remark 11.3.3) in which all such reactions have been removed (along with all reactions of the form $0 \to n\mathcal{s}$ or $m\mathcal{s} \to 0$). This is to say that all networks that have the same base subnetwork will have the same Knot Graph. Thus, the Knot Graph for the semi-open network (11.15) is drawn for the network

$$E+S \rightleftarrows ES \rightleftarrows EP \rightleftarrows E+P. \tag{11.17}$$

We will use (11.17) to illustrate how the Knot Graph is constructed.

Remark 11.11.1. Hereafter in Section 11.11 (and in Appendix 11.A), *we shall, to simplify the narrative, restrict our attention to networks for which the base subnetwork is reversible. Moreover, we shall suppose that for every (nonbasic) reaction of the form* $0 \to n\mathcal{s}$, *there is, for the same species* $\mathcal{s}$, *another reaction of the form* $m\mathcal{s} \to 0$, *with m and n not necessarily identical.* We do not insist that a reaction of the form $m\mathcal{s} \to 0$ be accompanied by a reaction of the form $0 \to n\mathcal{s}$. In almost all applications, the nonbasic reactions will be of the form $\mathcal{s} \to 0$ or $0 \rightleftarrows \mathcal{s}$.

To indicate how the Knot Graph is drawn, we will need to explain, informally at first, how the species set is partitioned into "knots." A knot is a maximal set of species such that each member interacts, either directly or transitively, with every other member through joint appearances in complexes. For example, in network (11.17), S, P, and E are members of the same knot: E and S interact directly through their joint appearance in the complex $E+S$. Similarly E and P interact directly through their joint appearance in the complex $E+P$. Finally S and P interact, transitively, through their shared relation with E. (As we shall see below, we permit such relations to be very remote ones, derived from a chain of direct interactions.) We regard every species to interact with itself. Thus, there are three knots in network (11.17): $\{E,S,P\}$, $\{ES\}$, and $\{EP\}$.

Remark 11.11.2 (A more precise definition of a knot). For chemical reaction network $\{\mathscr{S},\mathscr{C},\mathscr{R}\}$ we say that species $\mathcal{s}$ and $\mathcal{s}'$ are *knot-related*, denoted $\mathcal{s} \sim_k \mathcal{s}'$, if any of the following conditions are satisfied:

(i) $\mathcal{s} = \mathcal{s}'$
(ii) There is a complex $y \in \mathscr{C}$ with both $\mathcal{s}$ and $\mathcal{s}'$ in $\operatorname{supp} y$.
(iii) There is a sequence of species $\{\mathcal{s}_1, \mathcal{s}_2, \ldots, \mathcal{s}_m\} \subset \mathscr{S}$ and a sequence of complexes $\{y_0, y_1, y_2, \ldots, y_m\} \subset \mathscr{C}$ such that

$$\{\mathcal{s}, \mathcal{s}_1\} \subset \operatorname{supp} y_0,\ \{\mathcal{s}_1, \mathcal{s}_2\} \subset \operatorname{supp} y_1,\ \{\mathcal{s}_2, \mathcal{s}_3\} \subset \operatorname{supp} y_2,\ \ldots,\ \{\mathcal{s}_m, \mathcal{s}'\} \subset \operatorname{supp} y_m.$$

The relation "$\sim_k$" is an equivalence relation in the species set. The equivalence classes are the *knots* of the network $\{\mathscr{S},\mathscr{C},\mathscr{R}\}$.

The *Knot Graph* for a network has the network's knots as its vertices. Edges are drawn in the following way: An edge connects knots K and K' for each reversible reaction pair $y \rightleftarrows y'$ such that complex y contains a species in K and complex y' contains a species in K'—that is, such that $K \cap \text{supp}\, y \neq \emptyset$ and $K' \cap \text{supp}\, y' \neq \emptyset$. Note that there could be multiple edges directly connecting knots K and K', for there might be several different reversible-reaction pairs that serve to connect the two knots. Note also that a knot can be connected directly to itself by one or more edges.

In the following examples, we will always have in mind open reactors in which various species are being supplied and removed. Thus, in the full reaction network of interest, there will be reactions of the form $s \to 0$ or $0 \to s$. In each instance we show the Knot Graph and the Species-Reaction Graph, both of which, it will be recalled, are drawn for the base subnetwork.

Example 11.11.3. We show in Figure 11.9 both the Knot Graph and the Species-Reaction Graph for the semi-open network (11.15). The base subnetwork, for which the graphs are drawn, was given in (11.17). Recall that in (11.17) there are three knots: $\{E,S,P\}$, $\{ES\}$, and $\{EP\}$.

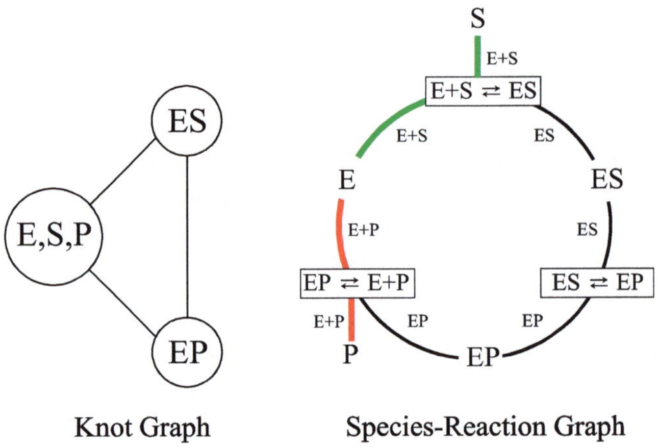

Fig. 11.9. The Knot Graph and the Species-Reaction Graph for network (11.15)

Example 11.11.4. Here we consider a classical CFSTR in which the true chemistry is shown in (11.18). The full CFSTR network, with feed and effluent reactions, was displayed as (10.B.2).

$$A + B \rightleftarrows C \quad 2A + B \rightleftarrows D \tag{11.18}$$

The Knot Graph and the Species-Reaction Graph, drawn for the base subnetwork (11.18), are displayed in Figure 11.10. The base subnetwork has three knots, $\{A,B\}$, $\{C\}$, and $\{D\}$.

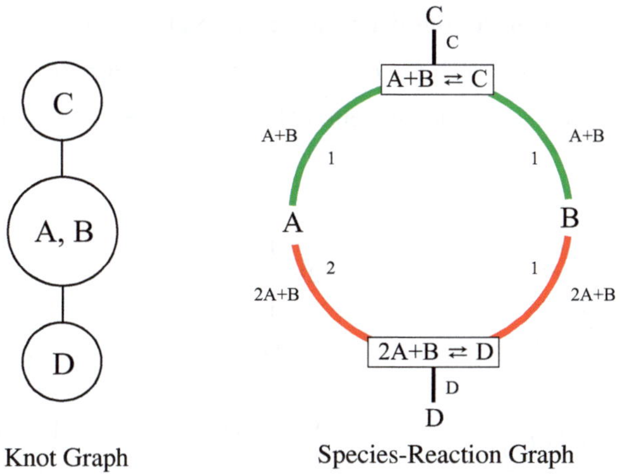

Fig. 11.10. The Knot Graph and the Species-Reaction Graph for network (11.18)

Example 11.11.5. Consider a classical CFSTR in which the chemistry is given by

$$A + B \rightleftarrows C \quad A \rightleftarrows B. \tag{11.19}$$

In this case the Knot Graph contains a loop—that is, a cycle having only a single vertex. See Figure 11.11.

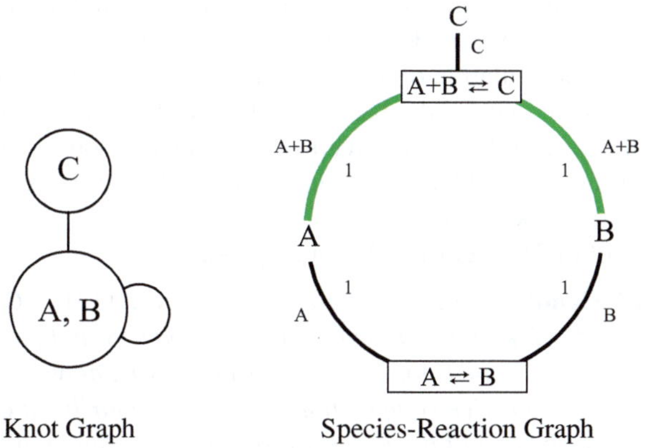

Fig. 11.11. The Knot Graph and the Species-Reaction Graph for network (11.19)

Example 11.11.6. Again we consider a classical CFSTR, this time with the chemistry

$$A+B \rightleftarrows 2C \quad C+D \rightleftarrows E \quad A+G \rightleftarrows F \quad 2B \rightleftarrows H. \tag{11.20}$$

The Knot Graph and the Species-Reaction Graph are displayed in Figure 11.12.

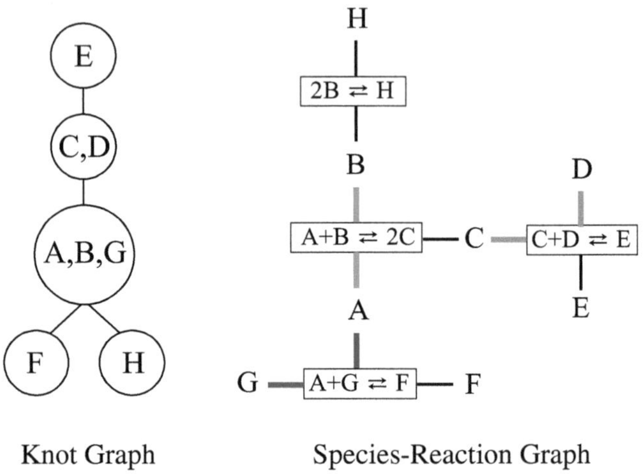

Fig. 11.12. The Knot Graph and the Species-Reaction Graph for network (11.20)

11.11.2 What the Knot Graph Can Tell Us

In the following theorem—due to Delattre, to Hyver, and to Beretta and co-workers—it should be clearly understood that the mass action system under consideration could be a representation of an open reactor. Although the Knot Graph in the theorem statement is drawn for the base subnetwork, the actual network of interest might, for example, be fully open. (Keep in mind that, in our exposition of Knot Graph theory, it is understood that attention is restricted reaction networks satisfying the constraints of Remark 11.11.1.)

We give a proof of Theorem 11.11.7 in Appendix 11.A.

Theorem 11.11.7 (Knot Graph Theorem [20, 54, 112, 113]**).** *Consider a mass action system* $\{\mathscr{S},\mathscr{C},\mathscr{R},k\}$ *with stoichiometric subspace S and species-formation-rate function* $f:\mathbb{R}^{\mathscr{S}} \to S$*. If the Knot Graph of the underlying network* $\{\mathscr{S},\mathscr{C},\mathscr{R}\}$ *has no cycles, then at each positive composition* c^* *(not necessarily an equilibrium), all eigenvalues of the derivative of the species-formation-rate function,* $df(c^*): S \to S$*, are real and negative. Moreover, there exists for S a basis consisting of eigenvectors of* $df(c^*)$*. If* c^* *is an equilibrium, it is asymptotically stable.*

Although a Knot Graph without cycles is, from a graphical standpoint, a relatively simple structure, it can nevertheless be the case that a mass action system from which the Knot Graph derives might have differential equations that are highly intricate. In (11.21) we show the differential equations for a CFSTR in which the chemistry is given by network (11.20); it is presumed that A, B, D, and G are in the feed stream. Recall that the Knot Graph and the Species-Reaction Graph were displayed in Figure 11.12. Neither has cycles.

By virtue of the acyclic Species-Reaction Graph, we know from Theorem 11.6.1 that there can be at most one positive equilibrium, regardless of parameter values. By virtue of the acyclic Knot Graph, Theorem 11.11.7 tells us more: *a positive equilibrium is asymptotically stable, again regardless of parameter values.*

$$
\begin{aligned}
\dot{c}_A &= k_{0\to A} - k_{A\to 0}c_A - k_{A+B\to C}c_Ac_B + k_{2C\to A+B}c_C^2 - k_{A+G\to F}c_Ac_G + k_{F\to A+G}c_F \\
\dot{c}_B &= k_{0\to B} - k_{B\to 0}c_B - k_{A+B\to C}c_Ac_B + k_{2C\to A+B}c_C^2 - 2k_{2B\to H}c_B^2 + 2k_{H\to 2B}c_H \\
\dot{c}_C &= -k_{C\to 0}c_C + 2k_{A+B\to C}c_Ac_B - 2k_{2C\to A+B}c_C^2 - k_{C+D\to E}c_Cc_D + k_{E\to C+D}c_E \\
\dot{c}_D &= k_{0\to D} - k_{D\to 0}c_D - k_{C+D\to E}c_Cc_D + k_{E\to C+D}c_E \\
\dot{c}_E &= -k_{E\to 0}c_E + k_{C+D\to E}c_Cc_D - k_{E\to C+D}c_E \\
\dot{c}_F &= -k_{F\to 0}c_F + k_{A+G\to F}c_Ac_G - k_{F\to A+G}c_F \\
\dot{c}_G &= k_{0\to G} - k_{G\to 0}c_G - k_{A+G\to F}c_Ac_G + k_{F\to A+G}c_F \\
\dot{c}_H &= -k_{H\to 0}c_H + k_{2B\to H}c_B^2 - k_{H\to 2B}c_H
\end{aligned}
\tag{11.21}
$$

11.11.3 Connections Between the Knot Graph and the Species-Reaction Graph

It is tempting to conjecture that the Knot Graph is without cycles precisely for those networks for which the Species-Reaction Graph is without cycles. However, that conjecture is belied by network (11.18) and Figure 11.10. A network that gives rise to a cycle-containing Species-Reaction Graph can indeed give rise to an acyclic Knot Graph, *but only when cycles in the Species-Reaction Graph are of a very special (and somewhat rare) kind.*

A *c-cycle* in a Species-Reaction Graph is a cycle in which the edge set is the union of c-pairs. The sole cycle in Figure 11.10 is an example of c-cycle: its four edges are comprised of two c-pairs. We leave proof of the following theorem to the reader.

Theorem 11.11.8. *The Knot Graph for a reaction network has no cycles if and only if every cycle in the network's Species-Reaction Graph is a c-cycle. In particular, the Knot Graph has no cycles if the Species-Reaction Graph has no cycles.*

Remark 11.11.9 (A little about c-cycles*).* In consideration of naturally occurring reaction networks, c-cycles are rare. Nevertheless, they do have a peculiar status in the mathematics of *mass action* systems [44, 147, 148]. Consider, for example, a fully open network, deriving perhaps from study of a classical CFSTR, and suppose that the Species-Reaction Graph contains just one cycle, that being a c-cycle. Then, for any assignment of positive rate constants, the resulting mass action system is injective (and there can be at most one positive equilibrium). *This is true regardless of* *whether the cycle is even and regardless of whether there exists for it a stoichiometrically expansive orientation.*

An example is provided by the CFSTR considered in Example 11.11.4. The full CFSTR network is shown in (11.22); A and B are presumed to be in the feed stream. The appropriate Species-Reaction and Knot Graphs were exhibited in Figure 11.10. Because the lone cycle in the Species-Reaction Graph is a c-cycle, network (11.22) is mass action injective [44, 62, 147, 148]. In particular, with mass action kinetics, multiple positive equilibria are impossible. Because the Knot Graph has no cycles, Theorem 11.11.7 tells us that the sole positive equilibrium is asymptotically stable; in fact, all eigenvalues associated with it are real and negative.

$$\begin{array}{ccccc} A+B \rightleftarrows C & & & & A \\ & \searrow & & \swarrow\!\!\nearrow & \\ & & 0 & & \\ & \nearrow & & \nwarrow\!\!\searrow & \\ 2A+B \rightleftarrows D & & & & B \end{array} \tag{11.22}$$

None of this need be true if the kinetics is not mass action. The lone cycle in the Species-Reaction Graph of Figure 11.10 fails to satisfy condition (i) in Theorems 11.5.1 and 11.6.1: the cycle is even, and it admits a stoichiometrically expansive orientation. In fact, network (11.22) is *discordant*, as can be determined by means of [62]. In this case, Theorem 10.5.5 tells us that there is for the network a weakly monotonic kinetics for which the resulting kinetic system is *not* injective. Because the reaction vectors are positively dependent, Theorem 10.5.10 ensures the existence of a weakly monotonic kinetics for which there are multiple positive equilibria. Moreover, Theorem 10.7.7 tells us that there exists for network (11.22) a differentiably monotonic kinetics such that the resulting kinetic system admits an *unstable* positive equilibrium, in fact one associated with a *real* eigenvalue that is *positive*.

Again, Species-Reaction Graph c-cycles are somewhat rare in the study of naturally occurring reaction networks, so, despite their benevolence when the kinetics is mass action, we will not try to say much more about them.

11.11.4 Mass Action Kinetics and Acyclic Species-Reaction Graphs

The following theorem is a consequence of Theorems 11.6.1, 11.11.7, and 11.11.8. In particular it improves part (A) in Theorem 11.6.1 to give stability information when certain additional conditions are satisfied. Very similar results were given in [147] and [148], with the asymptotic-stability assertion again based largely on the prior Knot Graph theory of Delattre, Hyver, and Beretta and co-workers.

Theorem 11.11.10 (Acyclic Species-Reaction Graph Theorem). *Suppose that a nondegenerate reaction network's Species-Reaction Graph has no cycles, and suppose also that the base subnetwork for which the Species-Reaction Graph is drawn is reversible. If the network is taken with mass action kinetics, then, regardless of rate* *constant values, the following statements hold true: Every positive equilibrium is unique within its stoichiometric compatibility class. That is, no positive equilibrium is stoichiometrically compatible with a different equilibrium, positive or otherwise. Moreover, every positive equilibrium is asymptotically stable. In fact, all eigenvalues associated with that equilibrium are real and negative.*

Remark 11.11.11. Theorem 11.11.10 is phrased to parallel Theorem 11.6.1. The non-degeneracy requirement in Theorem 11.11.10 is, however, redundant when the strictures of Remark 11.11.1 are in force. That this is so follows from arguments very much like those supporting Theorem 10.6.14.

Appendix 11.A Proof of Theorem 11.11.7

Here we give a proof of Theorem 11.11.7, which is repeated below as Theorem 11.A.1. The central idea of the proof is largely the one ("symmetrization") used in earlier Knot Graph work, but the style is different.

Theorem 11.A.1 ([20, 54, 112, 113]). *Consider a mass action system $\{\mathscr{S},\mathscr{C},\mathscr{R},k\}$ with stoichiometric subspace S and species-formation-rate function $f:\mathbb{R}^{\mathscr{S}} \to S$. If the Knot Graph of the underlying network $\{\mathscr{S},\mathscr{C},\mathscr{R}\}$ has no cycles, then at each positive composition c^* (not necessarily an equilibrium), all eigenvalues of the derivative of the species-formation-rate function, $df(c^*) : S \to S$, are real and negative. Moreover, there exists for S a basis consisting of eigenvectors of $df(c^*)$. If c^* is an equilibrium, it is asymptotically stable.*

Remark 11.A.2. Keep in mind that the base subnetwork of network $\{\mathscr{S},\mathscr{C},\mathscr{R}\}$ is presumed reversible. Keep in mind also the presumption that every reaction of the form $0 \to n\mathscr{s}$ is accompanied by another reaction of the form $m\mathscr{s} \to 0$. If, in the mass action system of Theorem 11.A.1, there is a reaction of the form $0 \to n\mathscr{s}$ unaccompanied by a reaction of the form $m\mathscr{s} \to 0$, we can still claim that all eigenvalues are real and nonpositive—see the proof—but an eigenvalue of 0 can occur. An example is provided by the simple network

$$0 \to A \rightleftarrows B. \tag{11.A.1}$$

This network is degenerate, so for every choice of mass action rate constants, we will have singularity of $df(c^*)$ (and an eigenvalue of 0) at every positive composition.

By way of preparation for the proof of Theorem 11.A.1, we review some fundamental ideas of linear algebra. Suppose that V is a real vector space with a scalar product, denoted "$\diamond$". A linear transformation $T : V \to V$ is *symmetric relative to the scalar product* $\diamond$ if

$$u \diamond Tw = w \diamond Tu \quad \forall u,w \in V. \tag{11.A.2}$$

Moreover, was say that T is *negative-semi-definite relative to the scalar product* $\diamond$ if, for all $v \in V$,

$$v \diamond Tv \leq 0 \tag{11.A.3}$$

and *negative-definite* relative to that scalar product if equality holds only when $v = 0$.

Remark 11.A.3. Suppose that V is finite-dimensional. To establish the symmetry of a linear transformation $T : V \to V$ relative to a scalar product $\diamond$, it is enough to test (11.A.2) against a basis for V. That is, if $\{v_1, v_2, \ldots, v_n\}$ is a basis for V, then T is symmetric if and only if

$$v_i \diamond Tv_j = v_j \diamond Tv_i, \quad i, j = 1, 2, \ldots, n. \tag{11.A.4}$$

The information contained in the following theorem is well known [98, 120].

Theorem 11.A.4 (Properties of a symmetric linear transformation). *Let V be a finite-dimensional real vector space and let $T : V \to V$ be a linear transformation. If there exists for V a scalar product "$\diamond$" with respect to which T is symmetric, then all roots of the characteristic polynomial of T are real. Moreover, there exists for V a basis that is orthonormal relative to $\diamond$ and consists entirely of eigenvectors of T. If T is negative-semi-definite [negative-definite] relative to $\diamond$, then all eigenvalues of T are nonpositive [negative].*

Proof (Theorem 11.A.1). To begin the proof of Theorem 11.A.1, we first note that, in the theorem statement, the derivative at c^* of the species-formation-rate function, $df(c^*) : S \to S$, is regarded as a map from the stoichiometric subspace into itself. For the purposes of the proof, it will be convenient to consider instead (temporarily) the extension of $df(c^*)$ to all of $\mathbb{R}^{\mathscr{S}}$, which we will denote by $Df(c^*) : \mathbb{R}^{\mathscr{S}} \to \mathbb{R}^{\mathscr{S}}$. This is just the linear map given by

$$Df(c^*)\sigma = \left. \frac{df(c^* + \theta\sigma)}{d\theta} \right|_{\theta=0}, \quad \forall \sigma \in \mathbb{R}^{\mathscr{S}}. \tag{11.A.5}$$

For the mass action system $\{\mathscr{S}, \mathscr{C}, \mathscr{R}, k\}$ of the theorem statement, the derivative $Df(c^*)$ takes the following form:

$$Df(c^*)\sigma = \sum_{y \to y' \in \mathscr{R}} \eta_{y \to y'} (y * \sigma)(y' - y), \quad \forall \sigma \in \mathbb{R}^{\mathscr{S}}, \tag{11.A.6}$$

where, for each $y \to y' \in \mathscr{R}$,

$$\eta_{y \to y'} := k_{y \to y'} (c^*)^y \tag{11.A.7}$$

and "$*$" indicates the scalar product in $\mathbb{R}^{\mathscr{S}}$ defined by

$$v * w := \sum_{s \in \mathscr{S}} \frac{v_s w_s}{c_s^*}, \quad \forall v, w \in \mathbb{R}^{\mathscr{S}}. \tag{11.A.8}$$

We denote by $\mathscr{B}$ the subset of $\mathscr{R}$ consisting of reactions of the base subnetwork. Because the base subnetwork is presumed reversible, we can choose for each reversible reaction pair a "forward" reaction, and we denote by $\mathscr{F}$ the subset of $\mathscr{B}$

consisting of all forward reactions. With this in mind, we can rewrite (11.A.6) in the following way: For all $\sigma \in \mathbb{R}^{\mathscr{S}}$

$$Df(c^*)\sigma = \sum_{y\to y'\in\mathscr{F}} [\eta_{y\to y'}(y*\sigma) - \eta_{y'\to y}(y'*\sigma)](y'-y) + \sum_{y\to y'\in\mathscr{R}\setminus\mathscr{B}} \eta_{y\to y'}(y*\sigma)(y'-y). \tag{11.A.9}$$

Our aim will be to show that, when the Knot Graph is acyclic, it is possible to choose $p \in \mathbb{R}^{\mathscr{S}}_+$ such that $Df(c^*)$ is symmetric relative to the scalar product "$\diamond$" defined by

$$v \diamond w := v * pw = \sum_{s\in\mathscr{S}} \frac{v_s p_s w_s}{c^*_s}, \quad \forall v, w \in \mathbb{R}^{\mathscr{S}}. \tag{11.A.10}$$

In fact, we will show that we can choose p to get symmetry term by term in (11.A.9). That is, we will show that we can choose p such that, for all σ and σ' in $\mathbb{R}^{\mathscr{S}}$, we have, for each $y \to y' \in \mathscr{F}$,

$$\begin{aligned}&[\eta_{y\to y'}(y*\sigma) - \eta_{y'\to y}(y'*\sigma)](y'-y)\diamond\sigma' \\ &= [\eta_{y\to y'}(y*\sigma') - \eta_{y'\to y}(y'*\sigma')](y'-y)\diamond\sigma\end{aligned} \tag{11.A.11}$$

and, for each $y \to y' \in \mathscr{R}\setminus\mathscr{B}$,

$$(y*\sigma)(y'-y)\diamond\sigma' = (y*\sigma')(y'-y)\diamond\sigma. \tag{11.A.12}$$

It is easy to see that the requirement (11.A.12) for reactions not in the base subnetwork is satisfied trivially no matter how $p \in \mathbb{R}^{\mathscr{S}}_+$ is chosen. Such reactions are either of the form $0 \to n s$ or $m s \to 0$. In the first case, both sides of (11.A.12) reduce to zero. In the second case, both sides reduce to

$$-\frac{p_s m^2 \sigma_s \sigma'_s}{(c^*_s)^2}. \tag{11.A.13}$$

We turn, then, to consideration of reactions in the base subnetwork. To begin, we will choose $p \in \mathbb{R}^{\mathscr{S}}_+$ to satisfy the requirement that when species s and s' are in the same knot, then $p_s = p_{s'}$. (There will eventually be additional constraints on p.) When K is a knot, we denote by p_K the value of p_s common to all $s \in K$. Moreover, when y is a complex, we denote by p_y the value of p_s common to all $s \in \operatorname{supp} y$. (All members of $\operatorname{supp} y$ are in the same knot.)

Now let $y \to y'$ be a particular reaction in $\mathscr{F}$. In light of Remark 11.A.3 and because we can regard the species set $\mathscr{S}$ as a basis for $\mathbb{R}^{\mathscr{S}}$, the symmetry condition (11.A.11) will be satisfied if it is satisfied for all choices of $\sigma = s$ and $\sigma' = s'$, where s and s' are distinct species.

For the particular reaction $y \to y' \in \mathscr{F}$ under study, we consider various cases:

(i) Suppose that neither s nor s' is in $\operatorname{supp} y'$. In this case (11.A.11) is satisfied, for both sides reduce to

$$-\frac{\eta_{y\to y'} p_y y_s y_{s'}}{c^*_s c^*_{s'}}. \tag{11.A.14}$$

(ii) Suppose that neither s nor s' is in supp y. Equation (11.A.11) is again satisfied, for both sides reduce to

$$-\frac{\eta_{y'\to y} p_{y'} y'_{s} y'_{s'}}{c^*_{s} c^*_{s'}}. \tag{11.A.15}$$

(iii) Now suppose that s is in supp y and that s' is in supp y'. In this case s and s' are in different knots, denoted K and K'. (Were the two species in the same knot, the Knot Graph would have a loop connecting the knot to itself, in contradiction to the hypothesis of Theorem 11.A.1.) Equation (11.A.11) reduces to

$$\frac{\eta_{y\to y'} p_{K'} y_{s} y'_{s'}}{c^*_{s} c^*_{s'}} = \frac{\eta_{y'\to y} p_{K} y_{s} y'_{s'}}{c^*_{s} c^*_{s'}}$$

This last equation will be satisfied if we choose

$$p_K = \frac{\eta_{y\to y'}}{\eta_{y'\to y}} p_{K'}. \tag{11.A.16}$$

There is precisely one equation of the form (11.A.16) for each edge in the Knot Graph. The question then becomes whether there is, for each knot K, a choice of $p_K > 0$ such that the resulting system of equations is satisfied, regardless of the values of $\{\eta_{y\to y'}\}_{y\to y' \in \mathscr{B}}$. Because the Knot Graph has no cycles, the answer is yes: The Knot Graph is a forest, so in each tree we can choose a leaf—that is, a knot K^* adjacent to precisely one edge—and then choose $p_{K^*} = 1$. The assignments of p_K for the remaining knots in the tree can be made by proceeding through the tree, applying (11.A.16) across each edge.

The resulting $p \in \mathbb{R}^{\mathscr{S}}_+$ then gives the scalar product $\diamond$ in $\mathbb{R}^{\mathscr{S}}$ with respect to which $Df(c^*) : \mathbb{R}^{\mathscr{S}} \to \mathbb{R}^{\mathscr{S}}$ is symmetric. Clearly, the stoichiometric subspace $S \subset \mathbb{R}^{\mathscr{S}}$ inherits the scalar product $\diamond$ from $\mathbb{R}^{\mathscr{S}}$, and, with respect to $\diamond$, the restriction of $Df(c^*)$ to S—*i.e.*, $df(c^*) : S \to S$—is also symmetric. From Theorem 11.A.4 it follows that all roots of the characteristic polynomial of $df(c^*)$ are real and that there exists for S a basis, orthonormal with respect to $\diamond$, consisting entirely of eigenvectors of $df(c^*)$.

It remains to be shown that the eigenvalues of $df(c^*)$ are all negative. When the Species-Reaction Graph of the network $\{\mathscr{S}, \mathscr{C}, \mathscr{R}\}$ has no cycles, the negativity of the eigenvalues already follows from Theorem 11.5.1 and Corollary 10.7.3, provided that $\{\mathscr{S}, \mathscr{C}, \mathscr{R}\}$ is nondegenerate. Nevertheless, we are required to prove the negativity of the eigenvalues under the hypothesis of Theorem 11.11.7 (and the constraints imposed by Remark 11.11.1).

This we shall do by arguing that $df(c^*)$ is negative-definite relative to the scalar product $\diamond$ just constructed. (See Theorem 11.A.4.) Thus, for $\sigma \in S$ we consider the sign of

$$\sigma \diamond df(c^*)\sigma = \tag{11.A.17}$$

$$\sum_{y\to y' \in \mathscr{F}} [\eta_{y\to y'}(y * \sigma) - \eta_{y'\to y}(y' * \sigma)](y' - y) \diamond \sigma + \sum_{y\to y' \in \mathscr{R}\setminus\mathscr{B}} \eta_{y\to y'}(y * \sigma)(y' - y) \diamond \sigma.$$

We will argue that the various individual terms in the two sums on the right side of (11.A.17) are all nonpositive with at least one negative so long as σ is not zero.

First consider a reaction $y \to y' \in \mathscr{F}$ with both y and y' nonzero. In this case, $\operatorname{supp} y$ and $\operatorname{supp} y'$ must reside in different knots K and K'. (Were they in the same knot, in the Knot Graph that knot would have a self-loop.) By virtue of the way $\diamond$ and p were constructed, the term in (11.A.17) corresponding to $y \to y'$ takes the form

$$\begin{aligned}
&[\eta_{y\to y'}(y*\sigma) - \eta_{y'\to y}(y'*\sigma)](y'-y)\diamond\sigma \\
&\quad = [\eta_{y\to y'}(y*\sigma) - \eta_{y'\to y}(y'*\sigma)][p_{K'}(y'*\sigma) - p_K(y*\sigma)] \\
&\quad = \frac{\eta_{y\to y'}}{p_K}[p_K(y*\sigma) - p_{K'}(y'*\sigma)][p_{K'}(y'*\sigma) - p_K(y*\sigma)] \\
&\quad = -\frac{\eta_{y\to y'}}{p_K}[(y'-y)\diamond\sigma]^2. \qquad (11.A.18)
\end{aligned}$$

The result is clearly nonpositive and is zero only when, for the particular $y \to y' \in \mathscr{F}$ under study, σ is orthogonal to the corresponding reaction vector relative to the $\diamond$ scalar product.

We do not preclude the possibility that there are in $\mathscr{F}$ reactions of the form $y \to 0$ (such as $A+B \to 0$) or $0 \to y$ (such as $0 \to A+B$). In such cases, it is not difficult to see that the corresponding terms in (11.A.17) reduce to, respectively,

$$-\frac{\eta_{y\to 0}}{p_y}(y\diamond\sigma)^2 \quad \text{or} \quad -\frac{\eta_{0\to y}}{p_y}(y\diamond\sigma)^2. \qquad (11.A.19)$$

Again, each is nonpositive and is zero only when σ is $\diamond$-orthogonal to the reaction vector for the corresponding reaction, either $y \to 0$ or $0 \to y$.

We turn now to terms in (11.A.17) that correspond to reactions that do not reside in the base subnetwork. These reactions are of the form $0 \to n\mathscr{s}$ or $m\mathscr{s} \to 0$. In the first case, the corresponding term in (11.A.17) reduces to zero. In the second case, the term reduces to

$$-\eta_{m\mathscr{s}\to 0}p_{\mathscr{s}}\left(\frac{m_{\mathscr{s}}\sigma_{\mathscr{s}}}{c^*_{\mathscr{s}}}\right)^2 = -\frac{\eta_{m\mathscr{s}\to 0}}{p_{\mathscr{s}}}(-m\mathscr{s}\diamond\sigma)^2. \qquad (11.A.20)$$

The result is again nonpositive and is zero only if $\sigma_{\mathscr{s}}$ is zero or, stated differently, only if σ is $\diamond$-orthogonal to the reaction vector $(-m\mathscr{s})$ corresponding the reaction $m\mathscr{s} \to 0$.

In summary, then, we have established that the right side of (11.A.17) is negative and is zero precisely when $\sigma \in S$ is $\diamond$-orthogonal to all reaction vectors corresponding to reactions in the set

$$\mathscr{R}^* := \mathscr{F} \cup \{y \to y' \in \mathscr{R} \setminus \mathscr{B} \text{ with } y' = 0\}. \qquad (11.A.21)$$

From the presumptions of Remark 11.11.1, it follows easily that

$$S := \operatorname{span}\{y'-y : y \to y' \in \mathscr{R}\} = \operatorname{span}\{y'-y : y \to y' \in \mathscr{R}^*\}. \qquad (11.A.22)$$

Thus, the right side of (11.A.17) is zero only when $\sigma \in S$ is in the orthogonal complement of S (relative to $\diamond$). This can happen only if $\sigma = 0$. We have established, then, that $df(c^*) : S \to S$ is negative-definite relative to $\diamond$, so all of its eigenvalues are negative.

When c^* is an equilibrium, it is asymptotically stable. This follows from the negativity of the eigenvalues associated with $df(c^*)$. [34, 104]. □

Remark 11.A.5. In Remark 11.11.1 we required that when, for a particular $\mathscr{s}$, there is a reaction of the form $0 \to n\mathscr{s}$, there is also one of the form $m\mathscr{s} \to 0$. In fact, this requirement is stronger than it needed to be. It is enough that $\mathscr{s}$, viewed as a vector of $\mathbb{R}^{\mathscr{S}}$, be a member of span$\{y' - y : y \to y' \in \mathscr{R}^*\}$.

12

The Big Picture Revisited

12.1 Back to the Beginning: Mysteries

We are now in a position to revisit the mysteries described in Chapter 1. Recall that these were not mysteries about the behavior of chemical reaction networks themselves but, rather, about the behavior of people. The questions posed were about styles and attitudes prevailing within different communities of chemists, engineers, biologists, and mathematicians—communities in which reaction network dynamics might be expected to be of compelling interest:

- Why, in view of the complexity of the mathematics governing chemical reactors, is the relationship between mathematics and chemistry so distant? Why, for mathematicians, doesn't chemistry—one of the great branches of physical science—have the rich mathematical tradition of the other branches or even of economics? Why is there so little importance attached to the basic mathematics of dynamical systems in the education of chemists?
- Why do chemical engineers believe, almost as a matter of doctrine, that isothermal continuous-flow stirred-tank reactors, even ones with highly intricate chemistry and governed by extremely complex nonlinear differential equations, behave in the most stable, mundane ways, regardless of the operative reaction network and regardless of parameter values?
- Why did the Belousov-Zhabotinsky chemistry, with its rich behavior, emerge so late and so singularly as a source of wonder among chemists and engineers? Or, put differently, why—given the vast landscape of chemical reaction networks—didn't many other dynamically rich chemistries emerge as examples long, long ago?

The original version of this chapter was revised: The caption of Fig. 12.13 has been updated. The correction to this chapter is available at https://doi.org/10.1007/978-3-030-03858-8_20

M. Feinberg, *Foundations of Chemical Reaction Network Theory*,
Applied Mathematical Sciences 202,
https://doi.org/10.1007/978-3-030-03858-8_12

- And what is it about *biological* chemistry that appears to be stimulating a now-burgeoning interest in reaction network dynamics, an interest that remains largely dormant among practitioners of "regular" chemistry?

Again, these were questions about the behavior of people, highly intelligent people with strong traditions based on long-acquired experience. In Chapter 1 we asserted that their attitudes toward the *anticipated* behavior of reaction networks—strange though those attitudes might seem from a mathematical standpoint—could provide clues not only about how reaction networks actually behave but also about the mathematical basis for that behavior.

After posing these very same mysteries in Chapter 1, we ended with a coda, worth repeating here, for it will provide the starting point for this chapter:

> *What is it, then, that imparts dull behavior across vast regions of the reaction network landscape but yet permits in that same landscape occasional eruptions of more exotic behavior? And what is it about biochemical networks that causes them to reside in that part of the landscape more prone to exotica? Can we somehow make sense of this landscape by drawing connections between the structure of a reaction network and the kinds of behavior its differential equations might admit?*

To a great extent, the subsequent chapters were devoted to answering these questions. We have already discussed in various places reasons for the great prevalence of dull behavior across vast regions of the reaction network landscape. A brief reprise is provided in the next two sections. What we have not addressed so far are mathematical reasons for the seemingly exceptional behavior to be found in *biological* chemistry. That discussion will occupy much of this chapter.

12.2 Reprise: Mathematical Roots of Enforced Dull, Stable Behavior in Complex Reaction Networks

As we suggested earlier, the prevalent behavior of chemical reaction networks is sufficiently dull as to make them uninteresting to mathematicians and advanced mathematics largely unnecessary to chemists. There is some irony, then, in the fact that a precise explanation for this largely dull behavior requires mathematics that is substantial. In different ways, the preceding chapters provide firm mathematical underpinnings for the (usually correct) prejudices of chemists and engineers about isothermal reaction network behavior, prejudices acquired not through analysis but, rather, through experience.

The Deficiency Zero Theorem and its companion theorems in Chapter 7 illuminate a significant part of the reaction network landscape, and they do so with a very broad beam. For *all* networks within the very large weakly reversible deficiency zero class—*including networks containing hundreds of species and hundreds of reactions*—dull behavior is ensured, so long as the kinetics is mass action, regardless of rate constant values: In each positive stoichiometric compatibility class, there is precisely one equilibrium; it is asymptotically stable; and there are no cyclic composition trajectories. Moreover, Chapter 8 tells us that, when the kinetics is mass action, the same uniqueness of equilibria obtains everywhere within the still larger class of networks that satisfy the hypothesis of the Deficiency One Theorem.

The theory of concordant networks, discussed in Chapter 10, sheds considerable light on very large parts of the reaction network landscape left unilluminated by the Deficiency Zero and Deficiency One Theorems, in particular on parts containing networks that incorporate degradation or synthesis reactions such as $\mathscr{S} \to 0$ or $0 \to \mathscr{S}$. Within the class of networks that have the capacity to admit a positive equilibrium—*i.e.*, those with positively dependent reaction vectors—the concordant networks are *precisely* those for which uniqueness of positive equilibria is ensured for *all* choices of weakly monotonic kinetics (not necessarily mass action). Concordance theory provides a degree of stability information as well: If a weakly reversible network's fully open extension is concordant and the kinetics is differentiably monotonic, then all real eigenvalues associated with the unique positive equilibrium are strictly negative.

That concordance theory illuminates a vast part of the reaction network landscape is made clear by connections drawn in Chapter 11 between concordance of a network and properties of the network's Species-Reaction Graph. Recall that within the class of nondegenerate networks, which includes all weakly reversible and all fully open networks, a network can *fail* to be concordant *only if its Species-Reaction Graph possesses at least one of two unusual features*:

(i) There is an *even* cycle that admits a stoichiometrically expansive orientation. For this to happen, there must be in the network a reaction in which a species has a stoichiometric coefficient of two or more,[1] which in turn requires that two or more *identical* molecules react directly with each other or are produced within a single reaction. This does indeed happen, but far more common are reactions in which all stoichiometric coefficients are one. Even when there is a stoichiometric coefficient greater than one, it must appear in a Species-Reaction Graph cycle that not only admits a stoichiometrically expansive orientation *but is also even*.

(ii) Within the network's Species-Reaction Graph, two *even* cycles admit consistent orientations such that, relative to the given cycle orientations, the intersection of the two cycles is a *single* directed path *that begins at a species vertex and terminates at a reaction vertex*. At the very least, this requires that there *exist* two cycles. (Many reaction networks give rise to Species-Reaction Graphs that have no cycles at all.) It is worth emphasizing again that both cycles must be *even*. An odd cycle and an odd cycle won't do, nor would an odd and an even. And, what's more, the two even cycles must have a nontrivial intersection of a very special kind: The intersection must be a *single* directed path that proceeds, relative to the given cycle orientations, *from a species vertex to a reaction vertex*. An intersection path having species or reaction vertices at both ends won't do, nor would a path directed from a reaction vertex to a species vertex.

---

[1] It is presumed in this discussion that stoichiometric coefficients are integers.

Because the presence of either of these features in a Species-Reaction Graph is a lot to ask, we should expect concordance to exert its powerful steadying influence very widely across the reaction network landscape.

Remark 12.2.1. The pervasiveness of dull behavior, at least with respect to the impossibility of multiple positive equilibria, is even broader than the discussion immediately preceding might suggest. When the kinetics is mass action, violation of the conditions of the Deficiency Zero or Deficiency One Theorems is only a *necessary* condition for the existence of rate constant values that give rise to multiple positive equilibria. Indeed, there are networks, *even discordant ones such as* (10.B.2), for which those conditions are violated but for which no such mass action rate constants exist. (The Deficiency One Algorithm and other tools in the Chemical Reaction Network Toolbox [62] will often identify such networks on a case-by-case basis.)

Moreover, even when a network's Species-Reaction Graph contains one of the features described in items *(i)* and *(ii)*, it might still happen that the network is concordant. (For a nondegenerate network, absence of these features is a *sufficient* condition for concordance, not a necessary one.) In such a case, multiple positive equilibria are again precluded, not just for mass action kinetics, regardless of rate constant values, but also for every choice of a weakly monotonic kinetics.

Remark 12.2.2. In this reprise, it is worth pointing out once again that, when the kinetics is mass action, results given by deficiency-oriented theory and those given by concordance theory are complementary. On one hand, it is easy to see that there are networks, especially those containing degradation and synthesis reactions, that are concordant but which do not satisfy conditions of the Deficiency Zero or Deficiency One Theorems. On the other hand, there are networks such as (11.7) that are discordant but which nevertheless have deficiencies of zero and thereby inherit all the stable behavior that the Deficiency Zero Theorem confers.

12.3 Reprise: Answers to Some Earlier Questions

The theorems we've met so far go a long way toward answering all but the last of the four questions posed at the beginning of this chapter. Because those theorems provide mathematical underpinnings for the great prevalence of dull dynamics in isothermal reactors, we can understand much more deeply the indifference of chemists and engineers to the possibility of rich behavior. And we can understand too the paucity of advanced mathematics in the education of chemists. Why study what dynamical systems theory teaches about complex phenomena that *might* arise when very few of them ever do? Of such ignorance chemistry is, in practice, highly forgiving.

As for the indifference of mathematicians to the intricate nonlinear differential equations of isothermal chemistry, that too has a basis—an ironic mathematical one—in those theorems that indicate the great prevalence of dull behavior. If, in the laboratory, dynamics is largely uninteresting and if isothermal reactors raise for engineers no stability concerns, what is there for mathematicians to study?

Against this backdrop the Belousov-Zhabotinsky chemistry was a disruptive anomaly. This was a platypus, to be studied because it was extraordinary, because it signaled that what might be taken for granted should not be. The rich dynamics of the Belousov-Zhabotinsky chemistry taught two things: First, *the very paucity of examples like it* pointed to the existence of deep, not entirely well understood, *mathematical* reasons for the broad prevalence of dull behavior across the reaction network landscape. Second, it taught that whatever impelled such dull behavior was not universal; there could be exceptions.

Remark 12.3.1 (The Belousov-Zhabotinsky chemistry). An article by Geiseler and Bar-Eli [89] provides experimental evidence that the Belousov-Zhabotinsky chemistry gives rise to multiple positive equilibria in a fully open CFSTR setting. That evidence is matched to a complex mass action model (for the internal chemistry) containing seven reversible reactions, only two of which are shown below in (12.1).

$$\begin{aligned} BrO_3^- + Br^- + 2H^+ &\rightleftarrows HBrO_2 + HOBr \\ 2HBrO_2 &\rightleftarrows BrO_3^- + HOBr + H^+ \end{aligned} \tag{12.1}$$

The reader can confirm that the Species-Reaction Graph for this small fragment of the full reaction network studied in [89] already contains even cycles that admit a stoichiometrically expansive orientations. The model thereby escapes the strictures against rich behavior provided by Theorem 11.6.1.

12.4 Useful Prejudices That Can Mislead in Fundamental Ways

The indifference of chemists and engineers to the possibility of exotic dynamics in isothermal reactors, even complex open ones, is largely based on a prejudice about how such reactors behave. The prejudice is a rational one, held by intelligent people, derived not from mathematical analysis but, rather, from long-standing experience. It is a highly useful prejudice that fosters practical progress, for without it each newly considered chemistry would give rise to fresh concerns, particular to that system, about the possibility of unreliable, perhaps wild, behavior.

But prejudices can mislead. When deviations from the expected occur in chemistries of no practical importance, such as the Belousov-Zhabotinsky chemistry, the lessons learned have no immediately compelling exigency. It is another matter, however, if prejudices can mislead in consideration of chemistries that lie at the very foundations of one or another sectors of scientific endeavor.

As we shall see in the remainder of this chapter, those theorems that impart dull behavior so broadly and powerfully across the reaction network landscape are circumvented—disconcertingly—by simple iconic reaction networks that are fundamental to commerce and, especially, to biology. It is biology that we turn to next.

12.5 Biochemistry: What Makes Life Interesting?

The manifestations of life are a wonderment, involving clocks, switches, computation, and all manner of vibrant dynamical behavior—behavior that is almost entirely absent in dead matter. Some of that vibrancy, such as the computational capacities of assemblages of neurons, resides at a level of description different from the one we have been considering. Nevertheless, some of the magic of biology results from rich behavior that can be captured by reaction-network-induced differential equations of the kind studied in this book.[2]

But if, as we have argued, dull behavior is prevalent across the reaction network landscape, what is it that gives *biochemistry* at the reaction network level its vitality? Yes, evolution might have made strategic selections of this or that network from scattered parts of reaction network landscape to perform specific functions. Still, we might ask if there something more pervasive in biochemistry that makes these selections less dependent upon improbable serendipity? Is there something that makes biochemistry more prone to be situated in the dynamically vibrant regions of the reaction network landscape?

A somewhat shallow answer is that biochemistry, as depicted in standard textbook pathway diagrams, is replete with cycles. Cycles, after all, provide feedback, which can be a source of instability. There is some truth in this, but it is a truth without much specificity, precision, or nuance. It is one we should explore critically from the perspective of theory accumulated so far. For this purpose, we will begin by considering, as a concrete example, a small pathway diagram adapted closely from a very well-regarded biochemistry book [168].

In Figure 12.1 we show a pathway diagram indicating three coupled reactions that serve to produce dTMP, a relative of thymine. Thymine is one of the four bases—adenine [A], thymine [T], guanine [G], and cytosine [C]—that lie at the heart of DNA encoding. Synthesis of dTMP is imperative if a cell is to generate enough DNA to replicate. The three small circles in the figure (and the arcs leading to and from them) indicate the three reactions. Thus, for example, the pathway segment near the filled red circle indicates the reaction (12.2) whereby NADPH acts on dihydrofolate (in the presence of an abundance of hydrogen ions) to form tetrahydrofolate and $NADP^+$. The diagram contains a cycle that is suggestive of feedback, which in turn suggests (vaguely) the *possibility* of rich behavior.

$$\mathrm{H_2Folate} + \mathrm{NADPH}\ \left[+\,\mathrm{H^+}\right] \rightarrow \mathrm{H_4Folate} + \mathrm{NADP^+} \tag{12.2}$$

We can, however, say much more and in a precise, detailed way. Note that pathway diagrams such as the small textbook example shown in Figure 12.1 are very

[2] For example, the attempt to understand rich dynamical behavior in the cell cycle has been a particularly strong source of differential equations derived from experimentally inspired reaction network models. See, in particular, work by Tyson, Novak, Chen, and collaborators [32, 33, 167]. Recall too the discussion in Chapter 1 of the Huang-Ferrell network, formulated to engender differential equations governing the mitogen-activated protein kinase cascade [110, 137].

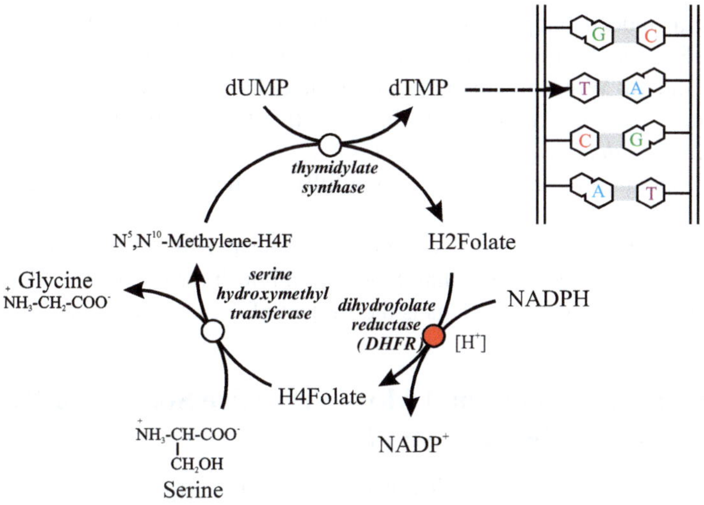

Fig. 12.1. A biochemical pathway resulting in the synthesis of thymine, after Voet and Voet [168]

close to being Species-Reaction Graphs. Replace each small circle in Figure 12.1 with a node containing a label that explicitly names the corresponding reaction—e.g., the label indicated in (12.2)—and a Species-Reaction Graph does indeed result.

The theorems in Chapters 10 and 11 can then give information that is precise and nuanced: If we want to know, for example, if the chemistry depicted in Figure 12.1 can give rise to multiple positive equilibria, those theorems provide an answer. There is but one cycle, and it is even. (There are c-pairs but none contained in the cycle.) All stoichiometric coefficients are 1, so there is no orientation with respect to which the cycle is stoichiometrically expansive; in fact, the cycle is an s-cycle. The network is strongly concordant (and nondegenerate), as is its fully open extension.

Even in a fully open setting, then, *there is no possibility of multiple stoichiometrically compatible positive equilibria resulting from the indicated three-reaction dTMP-synthesis cycle, so long as the kinetics for each of the three reactions is weakly monotonic (or even two-way weakly monotonic). If the kinetics is differentiably monotonic (or even differentiably two-way monotonic), then all real eigenvalues associated with a positive composition (in particular an equilibrium) are negative.* These same statements also obtain when the three reactions are deemed to be reversible.

The example we considered is a small one, but it serves as a reminder that, when the kinetics satisfies very weak constraints, the mere presence of cycles in a biochemical pathway diagram (or in the associated Species-Reaction Graph) is hardly enough to support, for example, multiple positive equilibria. Indeed, theorems in Chapters 10 and 11 will often preclude them. *Even when in a pathway diagram there are several interlaced cycles, multiple positive equilibria can result only when those cycles (as reflected in the corresponding Species-Reaction Graph) satisfy one of the highly demanding conditions (i) or (ii) in Section 12.2.*

This can indeed happen in model biochemical settings. Such an example appeared in Section 11.7.3, where we exhibited in Figure 11.6 the Species-Reaction Graph for the full Wnt signaling pathway, as modeled by Lee et al. [122]. Despite the Wnt example, one nevertheless gets the sense that the supply of networks which, *at the traditionally depicted pathway level*, satisfy conditions *(i)* or *(ii)* of Section 12.2 is meager—perhaps too meager to account fully for the great dynamical richness that evolution delivered.

Is there, then, a more pervasive (and perhaps deeper) instrumentality, unrevealed in standard pathway diagrams, that evolution might have exploited to give life the behavioral variety it requires? This is the subject of the next section.

12.6 Are There Deeper and More Pervasive Sources of Rich Behavior in Biochemistry?

This section is close in spirit and substance to [52], written in collaboration with Gheorghe Craciun and Yangzhong Tang. Here, however, we are relying on Species-Reaction Graph theorems somewhat different from (and considerably sharper than) earlier ones invoked in [52].

12.6.1 Pathway Diagrams Often Mask Crucial Mechanistic Detail

Although pathway diagrams such as the one in Figure 12.1 convey important information, they are often just summaries that hide a great deal of more fundamental detail. An individual "reaction" depicted frequently amounts to little more than a headline, a coarse statement that certain reactants can come together to form certain products. The statement suppresses details of the underlying mechanism, involving several true elementary reactions, by which the transformation is actually accomplished. The coarse-grained overall "reactions" appearing in the pathway are often made possible or are greatly hastened by the intervention of enzymes, which are fairly large proteins. In fact, each of the three overall reactions depicted in Figure 12.1 is catalyzed by its own enzyme, named alongside the corresponding circle.

At the fine mechanistic level, an enzyme's action proceeds by its own sequence of elementary reactions that involve binding of reactants (often called substrates) to the enzyme, conversion of the bound reactants to products, and unbinding of products from the enzyme. These elementary reactions are usually presumed to be governed by mass action kinetics. The detailed mechanism—that is, the elementary steps—might vary from one enzyme to the next.

In Section 4.2.2 we already met an example of one such elementary-step mechanism, for catalysis by enzyme E, of the simple overall reaction

$$S1 + S2 \to P. \tag{12.3}$$

That mechanism is repeated below as (12.4). Recall that in the first line, the substrates $S1$ and $S2$ bind to the enzyme to form $ES1$ and $ES2$. In the second line, $S2$ binds to $ES1$ and $S1$ binds to $ES2$ to form $ES1S2$. Finally, with both $S1$ and $S2$ bound to the enzyme, they react to form product P, which immediately unbinds from the enzyme.

$$\begin{array}{cc} S1+E \rightleftarrows ES1 & S2+E \rightleftarrows ES2 \\ \multicolumn{2}{c}{S2+ES1 \rightleftarrows ES1S2 \leftrightarrows S1+ES2} \\ \multicolumn{2}{c}{\downarrow} \\ \multicolumn{2}{c}{P+E} \end{array} \tag{12.4}$$

Remark 12.6.1 (Rate functions for overall reactions). Although elementary reactions, such as those shown in (12.4), are usually deemed to be governed by mass action kinetics, an approximate kinetic rate function is often associated with an enzyme-catalyzed overall reaction such as $S1 + S2 \rightarrow P$. Such a function has as its independent variables only the concentrations of its reactants (e.g., c_{S1}, c_{S2}) and sometimes its products (e.g., c_P), with the total enzyme concentration appearing as a parameter. These overall kinetic rate functions are derived from the presumed multi-reaction underlying mass action mechanism (e.g., (12.4)) by artfully employing various approximations, such as the *pseudo-steady-state approximation* or the *rapid-equilibrium approximation*, which are not discussed here.

The book by I. H. Segel [150] is a classic tome on these approximations. It contains many results that give approximate rate functions for single overall reactions, sorted according to the presumed underlying catalytic mechanism. The rate function for the overall reaction $S1 + S2 \rightarrow P$, when presumed driven by the mechanism (12.4)), will usually take the form

$$\mathscr{K}_{S1+S2\rightarrow P}\left(c_{S1}, c_{S2}\right) = \frac{k\,[E]\,c_{S1}\,c_{S2}}{\alpha + \beta\, c_{S1} + \gamma\, c_{S2} + c_{S1}\, c_{S2}}, \tag{12.5}$$

where k, α, β and γ are positive parameters and where $[E]$ represents the total enzyme concentration, $c_E + c_{ES1} + c_{ES2} + c_{ES1S2}$. That total remains unchanged when the chemical reactions in (12.4) are the only sources of concentration change of the individual enzyme-containing species.

12.6.2 Sources of Rich Behavior Can Lurk in the Most Fundamental Mechanisms of Enzyme Catalysis

It will be useful to consider a simple thought experiment along lines already discussed in Section 4.2.2. Imagine a semi-open stirred vessel having entrapped within it an enzyme E that acts as a catalyst for the reaction $S1 + S2 \rightarrow P$. Substrates $S1$ and $S2$ are added continuously and at fixed rate through an inlet port, while $S1$, $S2$, and the product P are withdrawn from an exit port at rates proportional to their concentrations within the vessel. (We can imagine that the enzyme entrapment is accomplished by means of membranes at the inlet and outlet ports that pass the small molecules $S1$, $S2$, and P but not the large molecule E.)

12.6.2.1 The Coarse-Grained Description

Were we to model the reactor coarsely, at the level of description commonly invoked in pathway diagrams, we would view $S1 + S2 \rightarrow P$ as the sole chemical reaction, and we might associate with that reaction a kinetic rate function $\mathscr{K}_{S1+S2\to P}(\cdot)$ such as (12.5). The differential equations governing the reactor would take the simple form (12.6), where F_{S1} and F_{S2} are the molar supply rates per unit reactor volume of $S1$ and $S2$, and ξ_{S1}, ξ_{S2}, and ξ_P are the proportionality constants describing the respective species molar effluent rates per unit reactor volume.

$$\begin{aligned}\dot{c}_{S1} &= F_{S1} - \xi_{S1}c_{S1} - \mathscr{K}_{S1+S2\to P}(c_{S1}, c_{S2})\\ \dot{c}_{S2} &= F_{S2} - \xi_{S2}c_{S2} - \mathscr{K}_{S1+S2\to P}(c_{S1}, c_{S2})\\ \dot{c}_P &= -\xi_P c_P + \mathscr{K}_{S1+S2\to P}(c_{S1}, c_{S2})\end{aligned} \tag{12.6}$$

We can ask about the capacity of the differential equations (12.6) to admit multiple positive stoichiometrically compatible equilibria. In the sense of Chapter 4, the reaction network appropriate to these differential equations is (12.7).

$$\begin{array}{c} S1 + S2 \to P \\ S1 \rightleftarrows 0 \leftrightarrows S2 \\ \uparrow \\ P \end{array} \tag{12.7}$$

The very simple Species-Reaction Graph for this fully open (and therefore nondegenerate) network is shown in Figure 12.2. There are no cycles, so the conditions of Theorem 11.6.1 are satisfied trivially. So long as the kinetics for the network is weakly monotonic, multiple stoichiometrically compatible positive equilibria are impossible. Weak monotonicity in the kinetics obtains, for example, when the rate function for $S1 + S2 \to P$ is as in (12.5); this is true regardless of the particular parameter values appearing in $\mathscr{K}_{S1+S2\to P}(\cdot)$ and regardless of the parameter values associated with the feed and effluent streams.

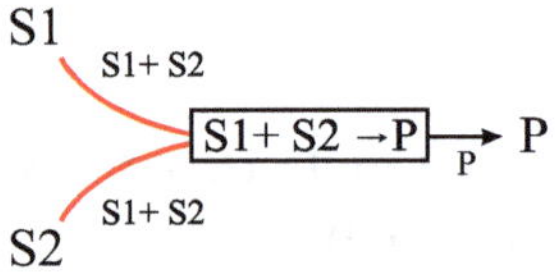

Fig. 12.2. Species-Reaction Graph for network (12.7)

12.6.2.2 Description at a More Refined Mechanistic Level

We indicated earlier that enzyme-catalyzed "reactions" such as $S1 + S2 \to P$, and the approximate rate functions associated with them, are often summaries that mask a great deal of mechanistic detail. Suppose that we chose instead to examine the seemingly simple reactor just considered at a finer, more detailed level of description that takes into account the elementary steps by which enzyme catalysis operates.

In particular, suppose that the details of catalysis are captured by the classical nine-reaction mechanism (12.4), taken with mass action kinetics. In this case, the considerably more complex differential equations governing the reactor are those we exhibited in Chapter 4 as (4.15). As we indicated there, the network appropriate for that system of differential equations consists of the enzymatic mechanism shown in (12.4) taken together with the feed and effluent reactions $S1 \rightleftarrows 0$, $S2 \rightleftarrows 0$, and $P \to 0$. We argued in Section 10.6.5 that the resulting network is nondegenerate.

If our interest is again in the possibility of multiple stoichiometrically compatible positive equilibria, we can ask if the conditions of Theorem 11.6.1 remain satisfied. At this level of description, however, the Species-Reaction Graph—the one appropriate to the more detailed chemistry (12.4)—is far more complicated than it was before. That graph is displayed in Figure 12.3.

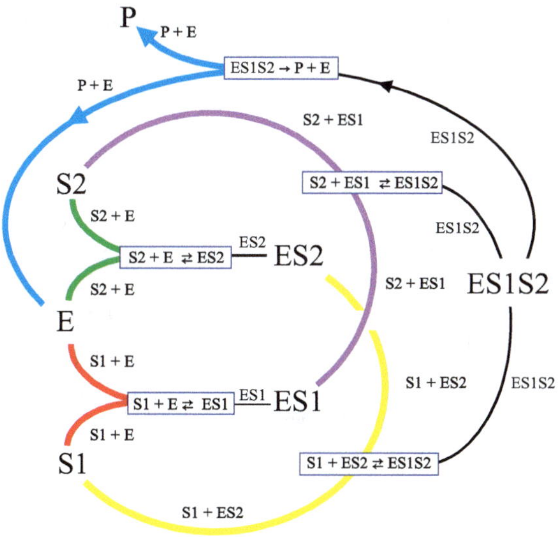

Fig. 12.3. Species-Reaction Graph for the unordered-binding enzyme-catalysis mechanism (12.4)

With respect to the two Species-Reaction Graph conditions in Theorem 11.6.1, the first is satisfied easily: every stoichiometric coefficient is 1, so there is no possibility of a stoichiometrically expansive cycle. Because there are now several cycles, the second condition requires careful scrutiny: Can two *even* cycles be consistently oriented in such a way that their intersection consists of a *single* path, *directed from a species vertex to a reaction vertex*?

The answer is *yes*. In Figure 12.4 we show two *even* cycles (indicated by a thickening of their arcs), each cycle containing the same two c-pairs (purple and green). Given the consistent orientations shown in the figure, the two cycles have as their intersection the directed path, shown in Figure 12.5, originating at the species node

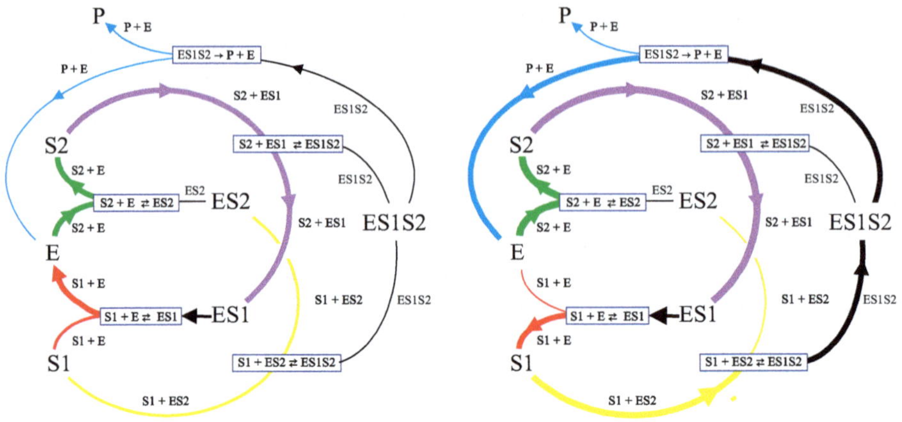

Fig. 12.4. Two even cycles in Figure 12.3, consistently oriented. Cycle edges are thickened.

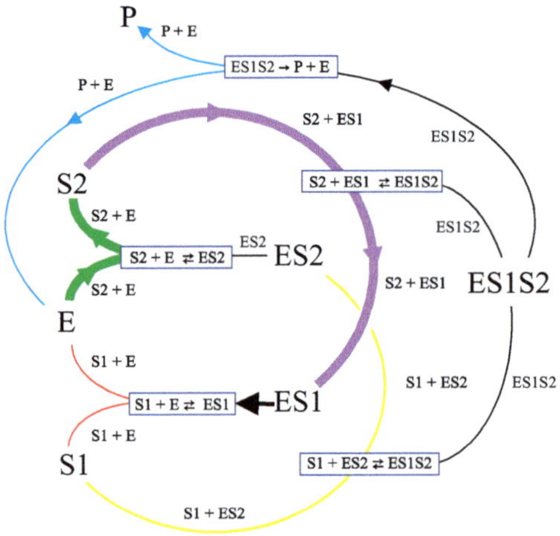

Fig. 12.5. The directed species-to-reaction intersection of the cycles in Figure 12.4.

E and terminating at reaction node $S1 + E \rightleftarrows ES1$. Condition *(ii)* in Theorems 11.5.1 and 11.6.1 is *not* satisfied.

These theorems, then, stand silent about concordance and about the possibility of multiple positive stoichiometrically compatible equilibria. In fact, the network (including the feed and effluent reactions) is *discordant*, as can be confirmed by means of [62]. Theorem 10.5.10 tells us that there is *some* weakly monotonic kinetics for which there are multiple positive stoichiometrically compatible equilibria, but it is silent about whether such equilibria *are consistent with mass action kinetics*, the kinetic class almost always deemed to be operative in consideration of catalytic mechanisms comprised of elementary reactions.

In fact, *for this simple reactor model—a model that invokes one of the most classical mechanisms of enzyme catalysis—there are rate constants such that the resulting mass action system admits multiple stoichiometrically compatible positive equilibria.* One such rate constant set is exhibited in (12.8); some composition trajectories that result are shown in Figure 12.6. Note that in the figure there are two stable equilibria (green)—hence, *bistability*—and one unstable equilibrium (red), with approach to the various equilibria dependent on initial conditions. Other sets of rate constants consistent with bistability are given in [52] and in [105], which also includes some rudimentary bifurcation analysis.

$$\begin{array}{c}
S1 + E \underset{0.15}{\overset{0.25}{\rightleftarrows}} ES1 \qquad\qquad S2 + E \underset{830}{\overset{40}{\rightleftarrows}} ES2 \\
S2 + ES1 \underset{.001}{\overset{.0016}{\rightleftarrows}} ES1S2 \underset{17.5}{\overset{975}{\leftrightarrows}} S1 + ES2 \\
265\downarrow \\
P + E \\
S1 \underset{85}{\overset{1}{\rightleftarrows}} 0 \underset{115}{\overset{1}{\leftrightarrows}} S2 \\
1\uparrow \\
P
\end{array} \tag{12.8}$$

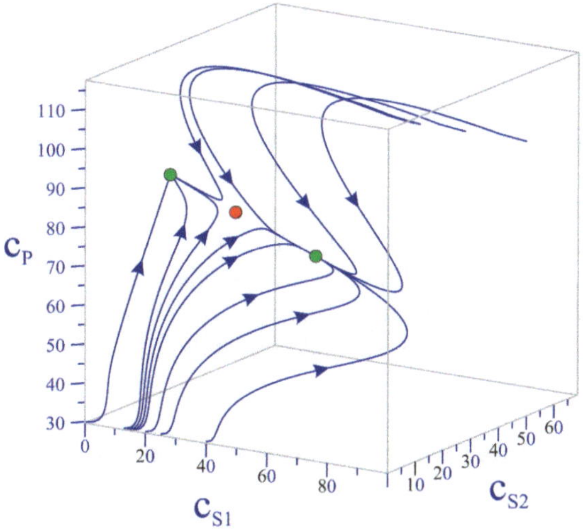

Fig. 12.6. Three equilibria and some c_{S1}–c_{S2}–c_P composition trajectories for the mass action system (12.8). All trajectories shown begin with $c_E = 1$ and $c_{ES1} = c_{ES2} = c_{ES1S2} = 0$.

Remark 12.6.2 (Mechanism matters). There are mechanisms other than (12.4) for catalysis of the overall reaction $S1 + S2 \rightarrow P$ by the enzyme E. One of these requires that a certain substrate, say $S1$, bind to the enzyme in advance of the binding of $S2$, which is presumed to bind only to $ES1$; in particular, a molecule of $S2$ is presumed to attach directly to the $S1$ molecule residing on the enzyme E. In this case the catalytic mechanism takes the form (12.9).

$$\begin{gathered} S1 + E \rightleftarrows ES1 \\ S2 + ES1 \rightleftarrows ES1S2 \rightarrow P + E \end{gathered} \tag{12.9}$$

If, in the thought experiment considered in this section, the ordered-binding mechanism (12.9) is operative rather than the unordered-binding mechanism (12.4), then the network appropriate for the description of the reactor is the one shown in (12.10).

$$\begin{gathered} S1 + E \rightleftarrows ES1 \\ S2 + ES1 \rightleftarrows ES1S2 \rightarrow P + E \\ S1 \rightleftarrows 0 \leftrightarrows S2 \\ \uparrow \\ P \end{gathered} \tag{12.10}$$

We can ask whether, for the entrapped-species network (12.10), there are rate constants such that the resulting mass action differential equations admit multiple positive stoichiometrically compatible equilibria. The network's Species-Reaction Graph is a simplification of the far more elaborate one displayed in Figure 12.3. In particular, there is no vertex corresponding to species $ES2$, nor are there vertices corresponding to the reactions $S2 + E \rightleftarrows ES2$ and $ES2 + S1 \rightleftarrows ES1S2$. Edges adjacent to these vertices are absent as well. Only a single cycle remains, and in it all stoichiometric coefficients are 1.

From Theorems 11.5.1 and 11.8.1, network (12.10) is not only concordant, it is strongly concordant.[3] Theorems 11.6.1 and 11.9.1 tell us that multiple stoichiometrically compatible positive equilibria are impossible not only for mass action kinetics, regardless of rate constant values, but also for every choice of a weakly monotonic kinetics (or even of a two-way weakly monotonic kinetics).

12.6.3 Another Example: Single-Substrate Enzyme Catalysis with an Inhibitor

Here we consider yet another example of a classical mechanism for enzyme catalysis, shown below as (12.11). In this case the single-substrate overall reaction $S \rightarrow P$ is presumed to be catalyzed by the enzyme E. In the first line of (12.11), the substrate S binds reversibly to E and then forms the product P. However, an inhibitor I also binds reversibly to E (and to ES) at a different binding site, the effect of which is to render the enzyme incapable of catalyzing the formation of product P.

[3] By virtue of Theorem 10.6.14, the network is nondegenerate.

$$\begin{aligned} E+S &\rightleftarrows ES \to E+P \\ E+I &\rightleftarrows EI \\ EI+S &\rightleftarrows EIS \leftrightarrows ES+I \end{aligned} \qquad (12.11)$$

We can contemplate a thought experiment very much like the one studied in the preceding section: Again, we imagine a semi-open stirred vessel having entrapped within it the enzyme E. Substrate S and inhibitor I are added continuously at a fixed rate through an inlet port, while S, I, and P are withdrawn from an exit port at rates proportional to their concentrations within the vessel. If we presume mass action kinetics for the mechanism (12.11), the differential equations governing the reactor are similar to, but significantly different from, those shown in (4.15) for the two-substrate enzymatic reactor studied in Chapter 4.

If our interest is in whether there can be parameter values such that the differential equations admit multiple stoichiometrically compatible positive equilibria, it is natural to see what concordance theory might tell us. The network appropriate to the reactor under consideration is the mechanism (12.11) augmented by the reactions $S \rightleftarrows 0$, $I \rightleftarrows 0$, and $P \to 0$. That this entrapped-species network is nondegenerate is an easy consequence of Theorem 10.6.14. Its Species-Reaction Graph is displayed in Figure 12.7.

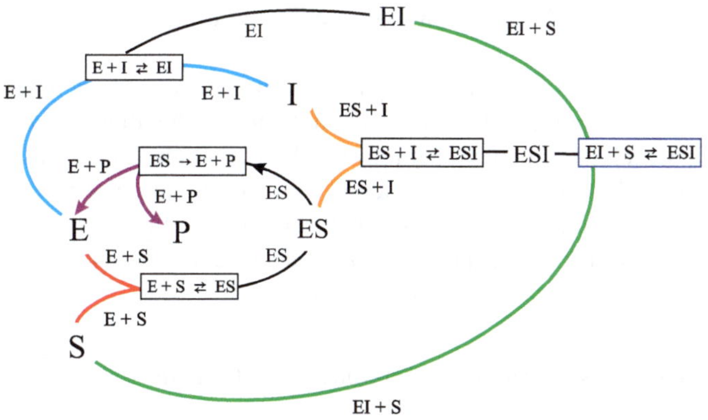

Fig. 12.7. Species-Reaction Graph for the single-substrate enzyme-catalysis-with-inhibition mechanism (12.11). Figure adapted from [52].

Note that the large outermost cycle and the small inner cycle (containing the species nodes E and ES) are both even. Moreover, if both cycles are given a counterclockwise orientation, their intersection is a (single-edge) directed species-to-reaction path beginning at species E and terminating at reaction $E+S \rightleftarrows ES$. In this case, then, condition *(ii)* of Theorems 11.5.1 and 11.6.1 is *not* satisfied, so those theorems are silent about concordance and about the possibility of multiple stoichiometrically compatible positive equilibria. In fact, the network (including the feed and effluent reactions) is *discordant*, as can be confirmed by means of *The Chemical*

Reaction Network Toolbox [62], which also tells us that *there are rate constant values such that the resulting mass action differential equations do indeed admit multiple positive equilibria that are stoichiometrically compatible.*

Here again we have an example of a classical enzyme-catalysis mechanism, written at a very fundamental elementary-step level, that circumvents the powerful strictures *against* rich behavior imposed by Theorem 11.6.1.

Remark 12.6.3. Examples of multiple equilibria emerging from still other and very different elementary mass-action enzyme-catalysis mechanisms can be found in studies of multi-site protein phosphorylation by Markevich, Hoek, and Kholodenko [127] and by Gunawardena [100].

12.6.4 The Example of Dihydrofolate Reductase, a Chemotherapy Target, with Rate Constants Taken from the Literature

We have argued that some of the most fundamental mechanisms of enzyme catalysis circumvent the powerful Species-Reaction Graph theorems that impose concordance and the resulting dullness of behavior. This circumvention gives rise to the possibility that rich behavior can result, at least in the context of the thought experiment we imagined. Moreover, for the examples considered, we asserted that there are indeed mass action rate constants for the elementary reactions such that a choice of experimental parameters (feed concentrations, supply and effluent rates) will give rise to bistability.

But if these classical enzyme-catalysis mechanisms are to serve as a source of rich behavior in real biological settings, it is not enough that there be *some* set of rate constants for which such behavior might be elicited. Rather, the question becomes one of whether rich behavior can be elicited for a combination of rate constants that biochemists might deem to be realistic. In particular, can bistability emerge, at least in the context of our thought experiment, for a set of experimentally determined rate constants *reported in the biochemical literature*?

12.6.4.1 Seeking Experimentally Determined Rate Constants

This question is more difficult to answer than might be supposed. The fact is that complete sets of rate constants determined for fundamental enzyme-catalysis mechanisms *at the mass action level* are not so easy to find. Perhaps the reason for this is that, for practical purposes, it suffices to infer from experiments parameters in coarse-grained overall-reaction kinetic rate functions such as (12.5). These functions are useful approximations that derive their shape from the presumed form of an underlying elementary-step mass action mechanism, but in the process much of the original mass action detail is lost.

It is in the literature of *enzymes targeted by anticancer agents* that fine mechanistic detail, including experimentally determined numerical detail, seems to be retained. It is not an accident that in Section 12.5 we used the thymine-synthesis cycle depicted in Figure 12.1 to motivate our discussion of biochemistry, for that cycle bears an intimate relationship with the workings of the classical chemotherapy drug, methotrexate.

Recall that thymine is an important constituent of DNA. Production of DNA is essential for replication of cells, in particular for cancer cells which replicate at a rapid rate. Thus, inhibition of thymine production would serve to thwart the reproduction of all cells; with luck, proliferation of cancer cells would suffer the most. Methotrexate is an inhibitor of the particular thymine-synthesis overall reaction

$$\mathrm{H_2Folate} + \mathrm{NADPH}\ \ [+\,\mathrm{H}^+] \rightarrow \mathrm{H_4Folate} + \mathrm{NADP}^+, \tag{12.12}$$

one of three overall reactions shown in Figure 12.1. That reaction is catalyzed by the enzyme dihydrofolate reductase (DHFR), to which both reactants (dihydrofolate and NADPH) bind. Methotrexate binds tightly to the site on the DHFR molecule at which dihydrofolate normally binds, thereby inhibiting the reaction (12.12) and thwarting more generally the entire thymine-synthesis cycle depicted in Figure 12.1. Because details of binding and unbinding of reactants and products to and from the DHFR molecule are of special interest, there are compelling reasons to understand DHFR-induced catalysis of overall reaction (12.12) at a fine mechanistic level, even in the absence of methotrexate.

12.6.4.2 The Red Dot in Figure 12.1 Conceals a Lot of Chemistry

Figure 12.8 is the result of heroic efforts in this direction by Appleman et al. [8]. It depicts a very highly detailed elementary-step mechanism thought to underlie catalysis of reaction (12.12) by human DHFR, along with mass action rate constants either measured in experiments or inferred from them.

Note that the overall reaction (12.12) (abbreviated $H2F + NH \rightarrow H4F + N$) resembles our earlier, more symbolic, two-substrate single-product prototype $S1 + S2 \rightarrow P$, but now there are two distinct product molecules. Moreover, the mechanism shown in Figure 12.8 is *by far* more ambitious in its account of fine detail than was our earlier mechanism (12.4). Section (A) of Figure 12.8 contains elementary reactions that mostly describe the reversible binding and unbinding of *reactants* to the enzyme, in some cases to enzyme that already has bound to it the complementary reactant. Only one of the reversible reactions in section (A)—the vertical one—represents a more substantive chemical transformation, whereby enzyme-bound reactants (NH and H2F) are converted to enzyme-bound products (H4F and N). Section (B) represents reactions whereby *products* unbind (and bind) to naked enzyme or to enzyme containing the complementary product. Finally, section (C) depicts nonproductive binding-unbinding steps whereby the enzyme comes to harbor one reactant molecule and one product molecule.

12.6.4.3 The Species-Reaction Graph for the Putative Elementary-Step DHFR Mechanism

We shall want to probe the capacity of the DHFR mass action system of Figure 12.8 to underlie rich behavior—in particular bistability—in the context of a thought experiment of the kind considered earlier in Section 12.6.2.2. We shall have in mind

$$E + H2F \underset{14}{\overset{264}{\rightleftarrows}} EH2F \qquad E + NH \underset{1.7}{\overset{38}{\rightleftarrows}} ENH$$

$$EH2F + NH \underset{19}{\overset{24}{\rightleftarrows}} ENHH2F \underset{98}{\overset{94}{\rightleftarrows}} ENH + H2F$$

$$37 \ \uparrow\!\downarrow \ 1360$$

$$ENH4F$$

(A)

$$EH4F + N \underset{84}{\overset{0.7}{\rightleftarrows}} ENH4F \underset{24}{\overset{46}{\rightleftarrows}} EN + H4F$$

$$EN \underset{17}{\overset{32}{\rightleftarrows}} E + N \qquad EH4F \underset{117}{\overset{5.1}{\rightleftarrows}} E + H4F$$

(B)

$$ENH + H4F \underset{225}{\overset{14}{\rightleftarrows}} ENHH4F \underset{4.4}{\overset{100}{\rightleftarrows}} EH4F + NH$$

$$EH2F + N \underset{4.6}{\overset{20}{\rightleftarrows}} ENH2F \underset{1.3}{\overset{110}{\leftrightarrows}} EN + H2F$$

(C)

Fig. 12.8. A mechanism with rate constants [8] for the action of human dihydrofolate reductase. E = DHFR, H2F = dihydrofolate, H4F = tetrahydrofolate, NH = NADPH, N = $NADP^+$. Rate constants measured at 20^0 C, pH 7.65. First- and second-order rate constants have units of sec^{-1} and $\mu M^{-1} sec^{-1}$. Figure adapted from [52].

a similar semi-open reactor, one in which DHFR is entrapped in a well-stirred vessel to which reactants H2F and NH are added continuously at fixed rates and from which those same reactants and the products, H4F and N, are withdrawn at rates proportional to their concentrations within the vessel.

The network description of the reactor consists of all reactions shown in Figure 12.8 taken together with the feed and effluent reactions $H2F \rightleftarrows 0$, $NH \rightleftarrows 0$, $H4F \to 0$, and $N \to 0$. That this semi-open network is nondegenerate follows quickly from Theorem 10.6.14. A Species-Reaction Graph fragment for the network is shown in Figure 12.9. (Reactions in section (C) are omitted.)

Here again, one of the most fundamental mechanisms of enzyme catalysis for a *single* overall reaction escapes the powerful bistability-precluding sweep of Theorems 11.5.1 and 11.9.1: The small cycle traversing species vertices NH, E, and $EH2F$ (with a return to NH) contains two c-pairs, colored green and purple; thus, it is even. The much larger cycle that traverses species vertices E, $H2F$, $ENHH2F$, $ENH4F$, and $H4F$ (with a return to E) also contains two c-pairs, colored red and blue; it too is even. When both cycles are oriented counterclockwise, their intersection consists of a single edge originating at species vertex E and terminating at reaction vertex $H2F + E \rightleftarrows EH2F$. Condition (ii) of Theorems 11.5.1 and 11.9.1 is not satisfied, so both theorems are silent. Bistability is not precluded.

12.6.4.4 A Hypothetical Experiment That Elicits Bistability for the DHFR Mass-Action Enzyme-Catalysis System of Figure 12.8

In fact, the network that describes our hypothetical experiment is discordant. This determination can be made by means of [62], which also gives much more information: *there is an assignment of rate constants to the various reactions such that*

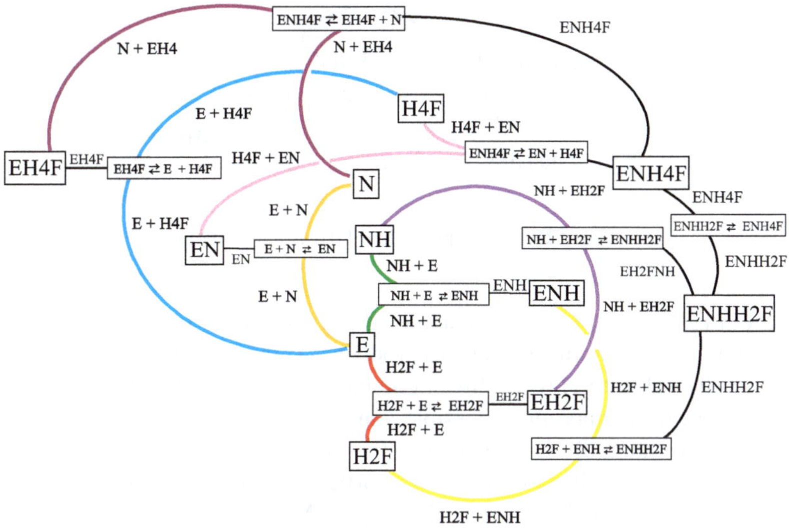

Fig. 12.9. A Species-Reaction Graph fragment for the DHFR mechanism shown in Figure 12.8. This figure is adapted from [52].

the resulting mass action system admits two distinct stoichiometrically compatible positive equilibria.

Recall, however, that our interest is in much more refined question: Do there exist design parameters for the hypothetical experiment under consideration (e.g., flow rates of the feed and effluent, concentrations of reactants in the feed stream) such that bistability is elicited *when the rate constants*[4] *are those measured or inferred by Appleman et al. [8] for the catalytic steps—that is, when the rate constants for the catalytic steps are those exhibited in Figure 12.8*?

The answer is *yes*. With F denoting the (identical) volumetric flow rates of feed and effluent stream solutions, with V denoting the reactor chamber volume, and with c_E^{TOT} denoting the total concentration of enzyme (with and without attached reactant and product molecules), we show in Figure 12.11 a hypothetical design for which simulations indicate the presence of at least two stable stoichiometrically compatible positive equilibria.[5] The equilibrium realized will, as in Figure 12.6, depend on the initial mixture composition within the chamber.

In Figure 12.11 we show the computed[6] dependence of the output equilibrium concentrations of H4F on the total concentration of DHFR within the reactor chamber (with all remaining parameters in Figure 12.10 held at their indicated values).

---

[4] In the sense of Section 4.2.2, the design parameters manifest themselves in rate constants for the feed and effluent reactions $H2F \rightleftarrows 0$, $NH \rightleftarrows 0$, $H4F \to 0$, and $N \to 0$.

[5] The chamber mixture is presumed to be perfectly stirred, and the concentrations of species in the effluent stream are presumed to be identical to their concentrations in the chamber.

[6] Computations were made with the help of *XPP*, software provided by Bard Ermentrout at http://www.math.pitt.edu/ bard/xpp/xpp.html.

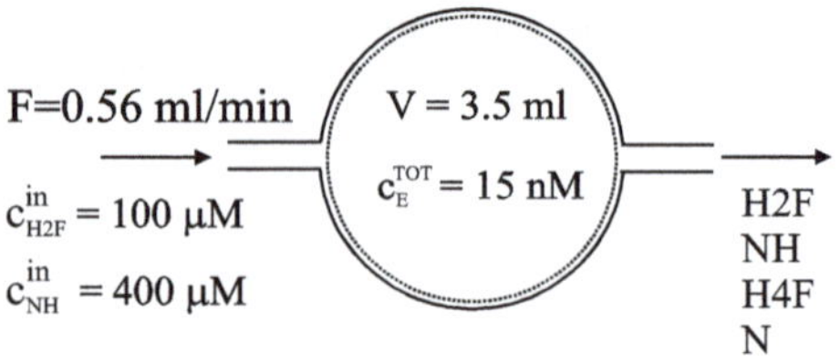

Fig. 12.10. A hypothetical DHFR experiment

Note that at $c_E^{TOT} = 15\ nM$ the two stable (green) equilibria indicate a considerable difference in conversion of reactants to products. At the higher equilibrium, conversion of supplied H2F to H4F is almost complete, with the effluent concentration of H4F very close to its maximum possible value, 100 μM. At the lower equilibrium, conversion is substantially less. Note also that the range of total enzyme concentrations over which bistability exists (for the particular values of the other parameters) is very narrow, with a width of approximately 2 nM.

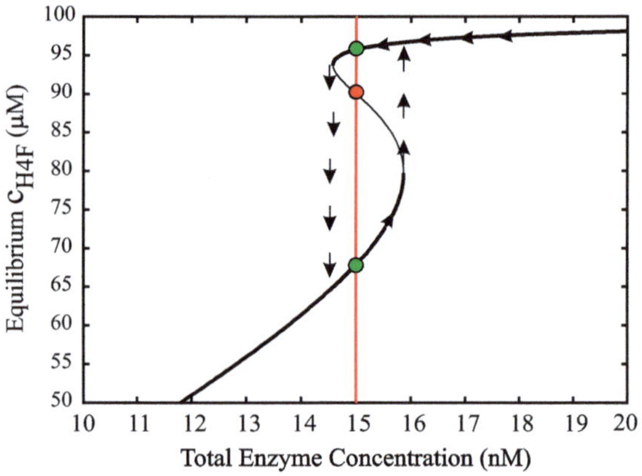

Fig. 12.11. Dependence of output equilibrium H4F concentrations on the total DHFR concentration

This narrow width, however, makes for a very sharp switch: Think of the reactor chamber as a metaphor for a biological cell, one that contains within it separate biochemical machinery for the synthesis and destruction of DHFR at rates that are considerably slower than the occurrence rates of the reactions appearing in Figure 12.8. (In the metaphorical spirit, we will refer to the reactor chamber as "the cell.") We shall suppose that, in response to slow changes in the DHFR concentration, there is a relatively rapid re-equilibration of the cell's constituents in response to each new concentration of enzyme.

Now imagine that the instantaneous concentration of DHFR in the cell is 17 nM, in which case Figure 12.11 indicates that the equilibrium value of the H4F concentration is very high, slightly above 95 μM. If the concentration of DHFR in the cell is caused to decline slowly, the equilibrium concentration of H4F will at first suffer a gentle almost imperceptible decline. When, however, the concentration of DHFR reaches 14 nM, re-equilibration will require that the H4F concentration drop precipitously to approximately 65 μM. (See Figure 12.11.) Thereafter, a slight elevation of the DHFR concentration to a little more than 16 nM will cause a jump in the H4F equilibrium concentration back to a level exceeding 95 μM.

Very small changes in the DHFR concentration can cause a dramatic switch between the high and low conversion states.

Remark 12.6.4. The purpose of this subsection has been a limited one: to show, by means of simulations, that for a classical mechanism underlying a single overall reaction, bistability can be elicited not just for *some* set of mass action rate constants but, more specifically, *for rate constants taken from the biochemical literature*. We attribute no physiological significance to the simulated bistability itself nor to the dramatic switch-like behavior that the simulations indicate in Figure 12.11.

Nevertheless, it is natural to wonder about these phenomena in light of DHFR's role as chemotherapy target. Recall that the purpose of methotrexate, a venerable anticancer drug, is to bind to DHFR in place of H2F so that conversion of H2F to H4F might be thwarted, thereby impeding the thymine-synthesis cycle depicted in Figure 12.1. However, *even in the absence of methotrexate*, the simulations indicate that a tiny change of the concentration of DHFR can result in a dramatic loss of conversion of H2F to H4F. It would seem, then, that experiments aimed at understanding the thwarting of conversion *in the presence of methotrexate* could become confounded.

Remark 12.6.5. Again, the highly idealized reactor considered in this section was meant only to make a mathematical point: that bistability can result from simple physicochemical models of single-enzyme systems invoking rate constant values taken from the literature. In fact, the reactor, as described, is very difficult to implement in the laboratory. Attempts were made by students (Y. Tang and Y. Fan), with guidance from highly gifted colleagues (J. Chalmers , S.-T. Yang, and T. Mitchison). Membranes that entrap the enzyme but pass smaller-molecule reactants and products are, however, fraught with difficulties (e.g., accumulation of enzyme near the exit membrane, concentration polarization).

Bistability can also be elicited in a model DHFR reactor having a fully open membrane-free CFSTR structure—that is, one in which the enzyme is in both the feed and effluent streams. In such models, however, separation of the two stable equilibria is not very sharp. Beautifully designed laboratory attempts by Daniel Knight and David Wood to elicit bistability in a fully open setting were, as in the semi-open case, inconclusive.

Remark 12.6.6. One sometimes gets the impression from reading discussions of rich dynamical behavior in biochemistry that such behavior typically (if not invariably) results from feedback in cycle-containing pathways. The feedback considered is usually of the kind whereby the product of one enzyme-catalyzed overall reaction acts as an inhibitor or inducer of a *second* enzyme, one that catalyzes a *different* overall reaction elsewhere along the pathway.

If there is a lesson to be learned from this section, it is that sources of rich behavior are *already* present in classical mechanisms of enzyme catalysis of a *single* overall reaction. The interplay of two or more overall reactions along a pathway might be sufficient to produce rich behavior, but it is not necessary.

Remark 12.6.7. Readers should be aware that the biological cell contains relatively large molecules in small spaces. At least certain of the very largest molecules are present in fairly small numbers. For some purposes, then, probabilistic models will be more apt than the deterministic mass action models studied here. There is a developing literature along these lines. See, for example, [4, 5, 63].

In the next section, we turn to another part of chemistry, one important to commerce, in which the stabilizing theorems of the preceding chapters are, again, disconcertingly circumvented.

12.7 Rich Behavior Rooted in Classical Mechanisms of Catalysis on Metal Surfaces

Substantial parts of the economies of some industrialized nations depend upon the workings of metal catalysts. In particular, metal catalysts play a central role in major industries that produce prized substances derived from petroleum. Even at the level of the individual gasoline-powered automobile, a metal-based catalytic converter is used to combine toxic carbon monoxide with oxygen to form the more benign carbon dioxide.

Metals used as catalysts for chemical reactions are often uncommon ones such as platinum, palladium, and rhodium. Unlike catalysis by enzymes, which almost always takes place in liquid solutions, catalysis on metals takes place on the metal's solid surface, typically in contact with a gas phase containing reactant molecules. (For this reason the term *heterogeneous catalysis* is often used as a descriptor.) The crystalline metallic surface is deemed to contain *active sites* on which reactants adsorb. Following adsorption, site-bound reactants are then in a position to form products, which then desorb from the catalyst surface.

For a given overall reaction, such as the oxidation of carbon monoxide to carbon dioxide ($2CO + O_2 \to 2CO_2$), the underlying catalytic mechanism—that is, the sequence of underlying elementary reactions—is often a subject of controversy. Here, for example, are two candidates [139] for the mechanism underlying carbon monoxide oxidation on platinum (or palladium), with S denoting an active site on the metallic surface:

$$
\begin{aligned}
CO + S &\rightleftarrows COS \\
O_2 + 2S &\rightleftarrows 2OS \\
COS + OS &\rightarrow CO_2 + 2S
\end{aligned}
\tag{12.13}
$$

$$
\begin{aligned}
CO + S &\rightleftarrows COS \\
O_2 + S &\rightleftarrows O_2S \\
2COS + O_2S &\rightarrow 2CO_2 + 3S
\end{aligned}
\tag{12.14}
$$

An S-containing species name, such as COS, indicates a molecule, such as CO, bound to an active site. In both mechanisms CO binds to an active site, but the two mechanisms differ thereafter with regard to the adsorption of oxygen. In (12.13) O_2 binds reversibly to two sites, yielding an oxygen atom bound to each site. Site-bound CO and O then form carbon dioxide, which desorbs, leaving behind two unoccupied active sites. In (12.14) O_2 binds to a single site, with O_2 otherwise retaining its molecular identity. Bound O_2 then reacts with two copies of bound CO to form two molecules of carbon dioxide, which depart the surface, leaving behind three unoccupied sites.

For each overall reaction catalyzed by a particular metal, the catalysis literature is replete with competing proposals for different mechanistic scenarios (having varying degrees of detail and plausibility). These candidate mechanisms, almost always taken with mass action kinetics, in turn guide mathematical models, which can then be tested against experiments for affirmation of one mechanistic proposal against another. (In [139], Razon and Schmitz provide a heroic summary of the many mechanistic proposals for carbon monoxide oxidation on platinum, as of 1986.) For a different overall reaction and a different metal, such as the hydrogenation of ethylene to ethane on rhodium, there will be yet another slew of competing mechanistic proposals.

Our interest here is in the extent to which mathematical models that derive from these various mechanisms of heterogeneous catalysis might give rise to rich behavior when considered in a simple isothermal semi-open setting similar to the enzyme-driven reactor considered earlier in Section 4.2.2 and then again in Section 12.6. In particular, we shall have in mind a semi-open continuous-flow stirred-tank reactor in which a gaseous solution consisting of reactants (such as carbon monoxide and carbon monoxide) and inerts are supplied to the reactor at a fixed volumetric flow rate and from which mixture within the chamber (containing reactants, products, and inerts) is withdrawn at the same fixed volumetric flow rate. In the chamber is a metallic surface bearing active catalytic sites.

Remark 12.7.1. Our aim is to explore dynamical behavior that can result from chemistry itself, isolated from other confounding physical phenomena. For the purpose of the thought experiments under consideration, we shall assume that there are no complications resulting from transport phenomena. In particular, we shall suppose that in the toy reactor under consideration, the gas phase and catalytic surface are each spatially uniform at every instant.

Remark 12.7.2. As is usually the case with putative metal-catalytic mechanisms, we shall assume mass action kinetics. In the study of reactions occurring on metallic surfaces, concentrations of active sites and of site-bound species are often considered on a per-unit-surface-area basis rather than on the per-unit-volume basis we have invoked so far. The same is true of surface reaction occurrence rates. Nevertheless, these can be converted to a per-unit-volume basis for the particular reactor at hand. As a result, reaction rates, concentrations of gas-phase species, and concentrations of surface-bound species can be treated in a common way.

We will begin with one of the simplest catalytic mechanisms imaginable, a mechanism for the overall reaction $A+B \to C$, whereby A and B both bind to catalyst sites and there combine to form C. The product C desorbs from the metal surface immediately, leaving behind two vacant sites formerly occupied by A and B. For the semi-open reactor we have in mind, the appropriate network description is given in (12.15).

$$\begin{array}{ccc} A+S \rightleftarrows AS & & B+S \rightleftarrows BS \\ AS+BS & \rightarrow & C+2S \\ A \rightleftarrows & 0 & \leftrightarrows B \\ & \uparrow & \\ & C & \end{array} \tag{12.15}$$

Does a network as simple as (12.15) have the capacity to support rich behavior, in particular bistability, when taken with mass action kinetics? Or is it restricted structurally to the behavioral dullness enforced widely across the reaction network landscape by, for example, Theorem 11.6.1?

We exhibit in Figure 12.12 the Species-Reaction Graph for the nondegenerate semi-open network (12.15). There are three cycles, the large outer one and the two smaller cycles labeled I and II. There is one c-pair in the outer cycle, so it is odd. There are no c-pairs in cycles I and II, so they are even. When cycles I and II are oriented in the only way possible, their intersection relative to that orientation is a directed *reaction-to-species* path. Condition (ii) of Theorem 11.6.1 is satisfied. However, condition (i) is not satisfied: each of the even cycles I and II admits an orientation that is stoichiometrically expansive. Thus, Theorem 11.6.1 is silent. In particular, the possibility of multiple stoichiometrically compatible positive equilibria is not precluded.

In fact, the Deficiency One Algorithm [74, 77] indicates quickly [62] that there exists for the very simple network (12.15) an assignment of rate constants such that the resulting mass action differential equations do indeed admit multiple stoichiometrically compatible positive equilibria.[7] The same is true of semi-open networks derived from the putative carbon monoxide oxidation mechanisms (12.13) and (12.14), considered in a similar flow reactor setting. In fact, in a semi-open isothermal setting of the kind considered here, *putative mechanisms commonly invoked to explain*

[7] We will say more about the Deficiency One Algorithm in Part III.

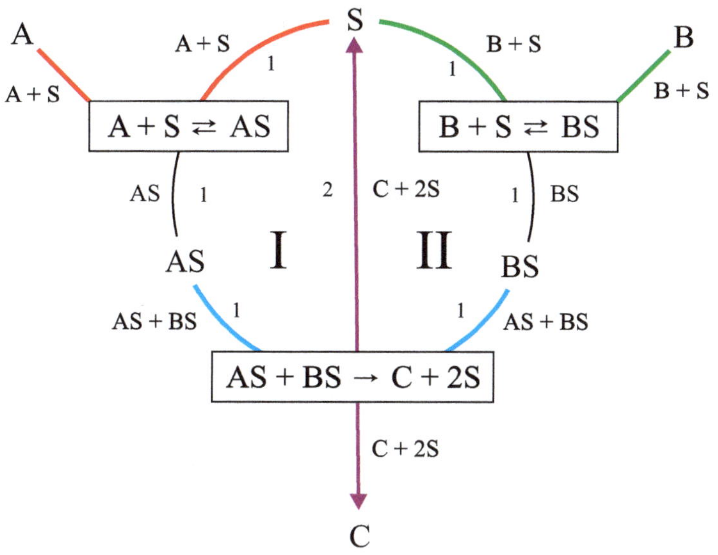

Fig. 12.12. Species-Reaction Graph for the classical metal-catalyst mechanism (12.15)

catalysis on solid surfaces seem to have the structural capacity for multiple stoichiometrically compatible positive equilibria far more often than not [60].

Remark 12.7.3. Semi-open versions of classical mechanisms for catalysis on metal surfaces are often networks *of deficiency one*. Reasons for this are discussed in [123].

The behavior of semi-open networks derived from putative metal-catalytic mechanisms can, when taken with mass action kinetics, be surprisingly rich. For a variant of network (12.15) in which the surface reactions are $A+S \rightleftarrows AS$, $B+2S \rightleftarrows BS_2$, and $AS+BS_2 \to C+3S$, there are rate constants such that there are *five* positive stoichiometrically compatible equilibria, three stable and two unstable. For an even more interesting example, consider the mass action system, in which the surface mechanism is structurally identical to the carbon monoxide oxidation mechanism (12.14) considered earlier ($A = CO$, $B = O$).

$$
\begin{array}{c}
A+S \underset{32.5}{\overset{328.6}{\rightleftarrows}} AS \qquad B_2+S \underset{0.0823}{\overset{79.64}{\rightleftarrows}} B_2S \\
2AS+B_2S \overset{277.8}{\to} 2AB+3S \\
\begin{array}{ccccc}
A & \underset{1}{\overset{1}{\rightleftarrows}} & 0 & \underset{0.6}{\overset{1}{\leftrightarrows}} & B_2 \\
 & & 1\uparrow & & \\
 & & AB & &
\end{array}
\end{array}
\tag{12.16}
$$

In Figure 12.13 we show some composition trajectories for the resulting differential equations, projected onto a $c_{AB}-c_{AS}-c_{B_2S}$ view. Initial concentrations were chosen to satisfy the condition $c_{AS}+c_{B_2S}+c_S=1$. There are two stable equilibria (green), an unstable equilibrium (red), and an unstable limit cycle (red).

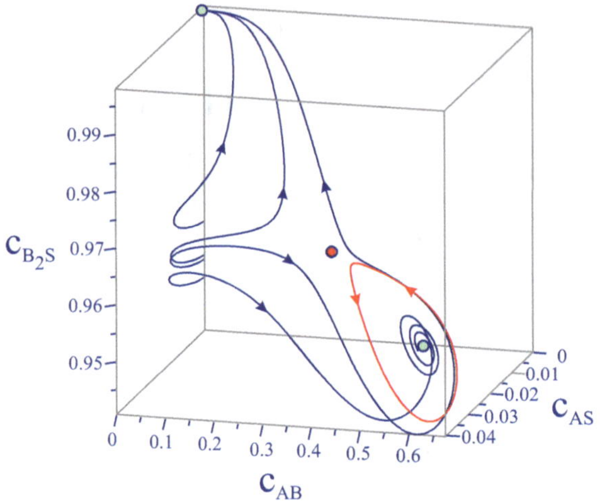

Fig. 12.13. Some trajectories for the mass action system (12.16) in a $c_{AS}-c_{AB}-c_{B_2S}$ projection

Remark 12.7.4 (False catalyst poisoning). The topmost stable equilibrium in Figure 12.13 appears to be on the boundary, but it is not: c_{AB} and c_{AS} are both positive but very small; c_{B_2S} is very close to but less than 1. It is an equilibrium in which almost all active sites are occupied by the reactant B_2, with very few active sites occupied by the co-reactant A. There is very little generation of the desired product AB, and almost no AB appears in the effluent stream. An engineer who saw the bottommost (highly productive) equilibrium after start-up on Monday and the topmost (unproductive) equilibrium after start-up on Tuesday might conclude erroneously that by Tuesday the catalyst itself had been poisoned. ("Poisoning" is a serious concern of catalysis scientists.)

Remark 12.7.5 (Multiple equilibria as a basis for mechanism discrimination). We indicated earlier that conjectured catalytic mechanisms for a particular overall reaction are often controversial. Mechanism discrimination sometimes revolves around the capacity of one candidate or another to explain, through derived mathematical models, dynamical behavior observed in experiments.

Because the capacity for multiple equilibria in semi-open settings is so common across a broad variety of mechanistic models, the mere qualitative *existence* of multiple equilibria in experiments cannot provide a strong basis for mechanism discrimination. In some instances, however, ever meager *quantitative* information about measured concentrations of some (but not necessarily all) species in two distinct stoichiometrically compatible equilibria can serve to discriminate very incisively between rival mechanisms on the basis of their capacity to explain the numerical data [60, 61, 74, 131, 144].

Remark 12.7.6 (Traveling composition waves on isothermal catalyst surfaces). Experiments have shown the presence of traveling composition waves along catalytic surfaces [43]. These waves seem to be accompanied by, and have come to be associated with, wave-like changes in the surface's crystallographic structure. It is probably not the case, however, that such surface restructuring is necessary for the existence of concentration waves. In [83] it is shown by means of Conley index methods [41] and Deficiency one theory that reaction-diffusion partial differential equations corresponding to a great variety of catalytic mechanisms already have the capacity to admit stable traveling concentration waves, even in the absence of such crystallographic restructuring.

12.8 Remarks About the Big Picture

The road we began in Chapter 1 has been a long one. Along the way, though, we acquired a considerable amount of theory that makes far less mysterious some of the observations we made then. On the whole, the theorems we have accumulated enforce, very powerfully, a certain dullness of behavior very widely across the reaction network landscape, sometimes when the kinetics is mass action but also for other very broad classes of kinetics of which the mass action class is only a special case. In the immediately preceding sections, about both biology and metal catalysis, we digressed to zoom in on parts of the reaction network landscape where the enforcement of dull behavior is circumvented. Before closing this chapter, we should zoom out once again to assess the big picture in its entirety.

12.8.1 Some Things We Don't Know

Although the big picture has been revealed to a great extent, it is not entirely unveiled. Apart from networks embraced by the Deficiency Zero and Star-Like Theorems, we have said little—either for mass action or more general kinetics—about where in the reaction network landscape cyclic composition trajectories could *not* be located.[8] While the principal Species-Reaction Graph theorems (and the underlying concordance theorems) ensure, for very general kinetics, uniqueness of positive equilibria and the negativity of real eigenvalues over vast parts of the reaction network landscape, they do not preclude complex eigenvalues with nonnegative real parts, nor do they preclude periodic orbits.

Recall from Sections 10.7.3–10.7.4 that even for certain *concordant* networks—those with discordant fully open extensions[9]—the possibility of rich dynamics is not

[8] Throughout this subsection, we have in mind only networks with positively dependent reaction vectors, for those are the only ones that might admit a positive equilibrium or a periodic orbit containing a positive composition.

[9] Note, though, that a network with a discordant fully open extension must have a Species-Reaction Graph that violates at least one of the two conditions given in Theorem 11.6.1.

only present, it is actually *ensured*, so long as the kinetics is not severely constrained: There is invariably a differentiably monotonic kinetics for which the resulting kinetic system admits an unstable positive equilibrium, in fact one having a positive real eigenvalue. And, at least if the network is weakly reversible, there are very good mathematical reasons to expect the existence of a differentiably monotonic kinetics that gives rise to a cyclic composition trajectory.

Might we hope that, *for every concordant network having a concordant fully open extension*, unstable positive equilibria and periodic orbits are denied for all choices of differentiably monotonic kinetics? Alas, that hope is dashed by the example discussed in Remark 10.7.5. Still, it is reasonable to suppose that, within this same large network class, instabilities are very rare and can actually be precluded for a still large and easily described subclass, yet to be determined.

12.8.2 Why the Big Picture Is Already Satisfying, Even Beautiful

A precision machine on a factory floor will not function well if the table upon which it rests wobbles badly with every approach of a worker, if the floor itself undulates with every passing automobile, if the earth beneath the factory is in daily upheaval, and if the operator, experiencing all of this, turns to despair and drink. No, the complex functioning of the machine requires of its environment stability and dull reliability.

In this respect, and with particular reference to biology, the picture of the entire reaction network landscape that has emerged—however incompletely—is already a satisfying one. Within that landscape there are small isolated regions in which certain exotic phenomena such as bistability might appear and far larger regions in which those same phenomena are impossible, even for very broad categories of kinetics. It is in those small isolated regions that some of the most classical mechanisms of enzyme catalysis are to be found. These are special reaction network structures that play their roles within the cell over and over again in different ways and in varying contexts.

In biology, then, there are myriad opportunities for rich behavior to emerge from those isolated parts of the reaction network landscape where exotic dynamics are located. At the same time, that rich behavior can be played out against the duller dynamical background typical of chemistries located almost everywhere else. Like our factory machine, a particular enzyme-driven cellular network, evolved perhaps for precise metabolic switching, can play its role in a complex chemical environment that is dynamic but not constantly subject to turmoil.

If classical mechanisms for enzyme catalysis serve to make life itself interesting, what are we to make of the similarly behaved mechanisms for catalysis on metals? Perhaps all that can be said of these is that they make life just a little more interesting for chemical engineers.

Part III

Going Deeper

In the study of literature, much usually depends on direct confrontation with a work. Who would dare to approach A Farewell to Arms *by a synopsis? It is only natural to distrust a literary experience if we have been guided too carefully through it, for the act of reading should provide as its product a feeling of truth upon which our senses will later work.*

But the study of science is different. Much like the study of history, it begins with legends and over simplifications. Then the same ground is revisited, details are added, complexities are engaged, unanswerable questions begin to be posed. A scientific account is a story which can always be retold, for the line of the narrative in scientific writing is to be found in the deepening of the concept.

Norman Mailer
Of a Fire on the Moon
1970

In Part III we will revisit several theorems stated in Part II, with an eye toward deepening the narrative. It is not our intention to prove all theorems stated in Part II, for in some cases the journal literature is reasonably accessible. Instead, the emphasis is on theory for which the existing literature is not so self-contained or could benefit from a more consolidated, better-motivated presentation. In fact, the very first chapter of Part III is devoted entirely to thermodynamic motivation for a conceptual path that will lead, ultimately, to both the Deficiency Zero and Deficiency One Theorems.

13

Quasi-Thermodynamic Kinetic Systems

In Chapter 7 we discussed, briefly, origins of the Deficiency Zero Theorem. There we introduced the idea of *complex balancing*, a major generalization by Horn and Jackson [109] of an earlier related idea, *detailed balancing*. We also hinted at connections of both ideas to classical thermodynamics. In this chapter we will elaborate on thermodynamic roots underlying arguments to come, largely to provide motivation for purely mathematical proof techniques that might otherwise seem improvisatory. In the next two chapters, we will consider detailed balancing and complex balancing more thoroughly.

13.1 A Little Classical Thermodynamics

Readers need not know anything about thermodynamics to understand this book's theorems or their proofs. The mathematics is self-contained.

Nevertheless, classical thermodynamics lurks in the background, out of view. That is because in [109] Horn and Jackson posed questions which, while motivated by thermodynamics, eventually led to theorems that make no explicit reference to thermodynamic ideas: the Deficiency Zero Theorem, the Star-like Network Theorem, and Deficiency One Theorem. In this book, a few thermodynamic artifacts do appear in the foreground, but only in the form of terminology. Words such as *quasi-thermostatic* and *quasi-thermodynamic* will be defined explicitly in mathematical terms, requiring no thermodynamic education. Nevertheless, it will be useful—and perhaps, for some readers, inspiring—to know something of the thermodynamic background. That is the aim of this section.

For our limited purposes, we will, throughout this section, confine our attention to a mixture that fills a closed rigid vessel (of fixed volume). We will suppose that the mixture remains spatially homogeneous at all times and that its (spatially uniform) temperature is maintained constant by means of the addition or removal of heat when that becomes necessary. Operative in the mixture is a kinetic system $\{\mathscr{S},\mathscr{C},\mathscr{R},\mathscr{K}\}$.

The original version of this chapter was revised: Equation 13.11 has been corrected. The correction to this chapter is available at https://doi.org/10.1007/978-3-030-03858-8_20

M. Feinberg, *Foundations of Chemical Reaction Network Theory*,
Applied Mathematical Sciences 202,
https://doi.org/10.1007/978-3-030-03858-8_13

Thus, the mixture composition is subject to temporal variations by virtue of the occurrence of chemical reactions.

13.1.1 The First Law

The First Law of Thermodynamics is about energy, heat, and work. From the First Law comes the idea of the *internal energy* of a body, which depends solely on the body's instantaneous state. With $U(t)$ denoting the internal energy of our mixture at time t and $Q(t)$ denoting its instantaneous rate of heat receipt, the First Law takes the simple form

$$\dot{U}(t) = Q(t), \tag{13.1}$$

where, as usual, the overdot indicates time differentiation. (For the situation under study, no work is done on the mixture.)

13.1.2 The Second Law

The Second Law of Thermodynamics has various formulations, but it is essentially a prohibition against certain ways in which a body might receive heat from its exterior. From the Second Law comes the idea of the *entropy* of a body, which, like the internal energy, depends only on the body's instantaneous state. From the Second Law also comes the idea of an *absolute temperature scale* that codifies hotness. With $S(t)$ denoting the instantaneous entropy of our mixture at time t, $Q(t)$ the instantaneous rate at which it receives heat, and Θ its fixed and spatially uniform temperature, the Second Law gives, for our mixture, a *Clausius-Duhem inequality*, holding at each instant:

$$\dot{S}(t) \geq \frac{Q(t)}{\Theta}. \tag{13.2}$$

Remark 13.1.1. The mathematical foundations of thermodynamics remain somewhat murky, at least to me. In [81, 82] Richard Lavine and I offered a modern attempt, based on the Hahn-Banach theorem, to deduce the existence and properties of entropy and absolute temperature—in particular their interplay in the Clausius-Duhem inequality—from a Kelvin-Planck statement of the Second Law, a version that makes no mention of entropy or temperature but, instead, refers only to prohibitions against certain types of heat receipt.

13.1.3 The Non-increasing Helmholtz Free Energy

For our mixture, the instantaneous *Helmholtz free energy*, $H(t)$, is defined by

$$H(t) := U(t) - \Theta S(t). \tag{13.3}$$

From the First and Second Laws, represented in (13.1) and (13.2), it follows that for our mixture we should always have

$$\dot{H}(t) \leq 0. \tag{13.4}$$

This is to say that, as reactions progress and the mixture composition changes, values of the Helmholtz free energy cannot increase.

13.1.4 Laws vs. Constitutive Equations

Most thermodynamics books unfortunately place at center stage, alongside the First and Second Laws, certain other "laws," such as the "Ideal Gas Law," that are not laws at all. Instead, they are what are now called *constitutive equations*: mathematical models for the behavior of a particular class of materials in particular circumstances. The so-called "Law of Mass Action," in the context of chemical kinetics, provides another example. It is a rule for relating, up to a parameter value, the rate of a chemical reaction to the concentrations of its reactant species. The result is a constitutive equation, not a law; it is a mathematical model deemed approximately valid for a particular reaction in circumscribed settings.

Laws of physics tell us what all bodies have in common, regardless of their composition. Constitutive equations tell us how, because of their differing compositions and differing circumstances, one body is different from another.

13.1.5 A Constitutive Description of the Helmholtz Free Energy

It is generally supposed that, for a particular mixture, there is a constitutive function that relates the Helmholtz free energy density (i.e., the Helmholtz free energy per unit mixture volume) to the mixture composition. That is, there is a function $h : \mathbb{R}_+^{\mathscr{S}} \to \mathbb{R}$ such that $h(c)$ is the Helmholtz free energy density when the mixture composition is c.[1] We assume hereafter that $h(\cdot)$ is smooth, say twice continuously differentiable.

Example 13.1.2 (The Helmholtz free energy density for ideal gas mixtures). For an ideal gas mixture with species set $\mathscr{S}$, the Helmholtz free energy density takes the form (13.5), where $c^* \in \mathbb{R}_+^{\mathscr{S}}$ depends on the particular substances of which the mixture is composed. (For a given substance, there is a degree of arbitrariness in how c^* might be chosen.)

$$h(c) := \sum_{s \in \mathscr{S}} c_s (\ln c_s - \ln c_s^* - 1) \tag{13.5}$$

A brief article [69] gives an indication of how constitutive equations for ideal gas mixtures—in particular the Ideal Gas Law and the Gibbs free energy function (a variant of the Helmholtz free energy function)—derive in full detail from surprisingly simple postulates that seem to say nothing at all.

13.1.6 Some Consequences of a Non-increasing Helmholtz Free Energy

From (13.4) (and the constancy of our mixture volume), we must have, at every instant along any solution of the differential equation for the kinetic system $\{\mathscr{S}, \mathscr{C}, \mathscr{R}, \mathscr{K}\}$,

$$\frac{d}{dt} h(c(t)) = \nabla h(c(t)) \cdot \dot{c}(t) = \nabla h(c(t)) \cdot f(c(t)) \leq 0, \tag{13.6}$$

[1] We are suppressing the dependence of the Helmholtz free energy density on temperature, which is fixed in the reactor under consideration. To avoid unnecessary technical considerations, we are restricting the domain of $h(\cdot)$ to compositions at which all species concentrations are strictly positive.

where $f(\cdot)$ is the kinetic system's species-formation-rate function. Because we can choose any initial composition, we must in fact have

$$\nabla h(c) \cdot f(c) \leq 0, \quad \forall c \in \mathbb{R}_+^{\mathscr{S}}. \tag{13.7}$$

We will denote by $\mu(c)$ the gradient at composition c of the Helmholtz free energy function. That is,

$$\mu(c) := \nabla h(c), \quad \forall c \in \mathbb{R}_+^{\mathscr{S}}. \tag{13.8}$$

The component of $\mu(c)$ corresponding to species s—that is,

$$\mu_s(c) := \frac{\partial h}{\partial c_s}(c), \tag{13.9}$$

is called the *chemical potential* of species s at composition c. Thus, we have

$$\mu(c) \cdot f(c) \leq 0, \quad \forall c \in \mathbb{R}_+^{\mathscr{S}}. \tag{13.10}$$

By the *chemical potential function* for the mixture under study, we mean the function $\mu(\cdot) : \mathbb{R}_+^{\mathscr{S}} \to \mathbb{R}^{\mathscr{S}}$ constructed from (13.8) in the obvious way. By $d\mu(c) : \mathbb{R}^{\mathscr{S}} \to \mathbb{R}^{\mathscr{S}}$ we mean the derivative of $\mu(\cdot)$ at $c \in \mathbb{R}_+^{\mathscr{S}}$; that is,

$$d\mu(c)\gamma = \left.\frac{d\mu(c+\theta\gamma)}{d\theta}\right|_{\theta=0}, \quad \forall \gamma \in \mathbb{R}^{\mathscr{S}}. \tag{13.11}$$

To avoid a more technical discussion, *we shall assume for the remainder of this section that $d\mu(c)$ is nonsingular at every $c \in \mathbb{R}_+^{\mathscr{S}}$.*

Example 13.1.3 (The ideal-gas-mixture chemical potential). For an ideal gas mixture with Helmholtz free energy density function (13.5), the chemical potential function is given by

$$\mu(c) = \ln c - \ln c^*, \quad \forall c \in \mathbb{R}_+^{\mathscr{S}}. \tag{13.12}$$

In this case

$$d\mu(c)\gamma = \frac{\gamma}{c}, \quad \forall \gamma \in \mathbb{R}^{\mathscr{S}}. \tag{13.13}$$

Clearly, $d\mu(c)$ is nonsingular at every $c \in \mathbb{R}_+^{\mathscr{S}}$.

Note that there are two very clear, easily delineated, classes of compositions at which equality holds in (13.10): Equality will hold for any composition $c^\dagger$ at which $f(c^\dagger) = 0$—that is, at any kinetic equilibrium. The second class is just a little less apparent. Because $f(\cdot)$ takes values in S, the stoichiometric subspace for the operative reaction network, equality will hold in (13.10) at any composition $c^\dagger$ for which $\mu(c^\dagger)$ is a member of $S^\perp$. Thus, the two classes are

$$E_{\mathscr{K}} := \{c \in \mathbb{R}_+^{\mathscr{S}} \ : \ f(c) = 0\} \quad \text{and} \quad E_\mu := \{c \in \mathbb{R}_+^{\mathscr{S}} \ : \ \mu(c) \in S^\perp\}. \tag{13.14}$$

Apart from the network $\{\mathscr{S},\mathscr{C},\mathscr{R}\}$, the first of these classes derives solely from the kinetics $\mathscr{K}$ that gives rise to the species-formation-rate function $f(\cdot)$. The second class appears to be entirely divorced from the kinetics $\mathscr{K}$, as it derives directly from the Helmholtz free energy density function. In the case of the ideal gas Helmholtz function (13.5), for example, there is no explicit reference to a kinetics.

Despite the seemingly complete disconnection of their origins, the following theorem tells us that the two classes are, for the most part, *identical.* Ideas underlying the theorem appeared in an article by Bowen [29] and in a far more general setting. See also an article by Krambeck [119] for a formulation closer in spirit to ours. His was also inspired in part by Bowen, but the arguments in [119], in particular for *(i)* in the theorem below, are different from the ones given here.

Theorem 13.1.4. *Let $\{\mathscr{S},\mathscr{C},\mathscr{R},\mathscr{K}\}$ be a kinetic system with stoichiometric subspace S and species-formation-rate function $f(\cdot):\overline{\mathbb{R}}_+^{\mathscr{S}} \to S$. Moreover, let $\mu(\cdot):\mathbb{R}_+^{\mathscr{S}} \to \mathbb{R}^{\mathscr{S}}$ be a smooth function having a nonsingular derivative at each $c \in \mathbb{R}_+^{\mathscr{S}}$. If $\mu(\cdot)$ and $f(\cdot)$ together satisfy the condition*

$$\mu(c)\cdot f(c) \leq 0 \quad \forall c \in \mathbb{R}_+^{\mathscr{S}} \tag{13.15}$$

then the following statements hold true:

(i) If $\mu(c^\dagger)$ lies in $S^\perp$, then $c^\dagger$ is an equilibrium (i.e., $f(c^\dagger)=0$).

(ii) If $c^\dagger \in \mathbb{R}_+^{\mathscr{S}}$ is a nondegenerate equilibrium, then $\mu(c^\dagger)$ lies in $S^\perp$.

Proof. Before beginning the proof, we will need to address a small matter of mathematical hygiene. In Section 3.6 we regarded the derivative at c of the species-formation-rate function as a map $df(c):S \to S$ *with domain restricted to the stoichiometric subspace*. Moreover, we regarded an equilibrium $c^\dagger$ to be nondegenerate if the linear map $df(c^\dagger):S \to S$ is nonsingular. That is the sense in which the term *nondegenerate* is used in the theorem statement. In such a case, $df(c^\dagger)$ is surjective onto S.

Here we will find it convenient to temporarily regard the derivative of $f(\cdot)$ as a map $\bar{d}f(c):\mathbb{R}^{\mathscr{S}} \to S$, having domain $\mathbb{R}^{\mathscr{S}}$. (Note the overbar.) That is,

$$\bar{d}f(c)\gamma = \left.\frac{df(c+\theta\gamma)}{d\theta}\right|_{\theta=0}, \quad \forall \gamma \in \mathbb{R}^{\mathscr{S}}. \tag{13.16}$$

Clearly, if $c^\dagger$ is a nondegenerate equilibrium, then $\bar{d}f(c^\dagger)$, like $df(c^\dagger)$, is surjective onto S.

To begin the proof, we define $\phi:\mathbb{R}_+^{\mathscr{S}} \to \mathbb{R}$ by

$$\phi(c) := \mu(c)\cdot f(c), \quad \forall c \in \mathbb{R}_+^{\mathscr{S}}. \tag{13.17}$$

Now suppose that $c^\dagger \in \mathbb{R}_+^{\mathscr{S}}$ is such that $\phi(c^\dagger)=0$. By virtue of (13.15), $\phi(\cdot)$ has a local maximum at $c^\dagger$, in which case we must have $d\phi(c^\dagger)\gamma = 0$ for all $\gamma \in \mathbb{R}^{\mathscr{S}}$. This implies that

$$[d\mu(c^{\dagger})\gamma]\cdot f(c^{\dagger}) + \mu(c^{\dagger})\cdot \bar{d}f(c^{\dagger})\gamma = 0, \quad \forall \gamma \in \mathbb{R}^{\mathscr{S}}. \tag{13.18}$$

To prove *(i)* we suppose that $\mu(c^{\dagger})$ lies in $S^{\perp}$, in which case $\phi(c^{\dagger}) = 0$. Because $\bar{d}f(c^{\dagger})$ takes values in S, the local maximum condition (13.18) reduces to

$$[d\mu(c^{\dagger})\gamma]\cdot f(c^{\dagger}) = 0, \quad \forall \gamma \in \mathbb{R}^{\mathscr{S}}. \tag{13.19}$$

Recall the supposition that $d\mu(c): \mathbb{R}^{\mathscr{S}} \to \mathbb{R}^{\mathscr{S}}$ is nonsingular at all $c \in \mathbb{R}^{\mathscr{S}}_{+}$. In particular, $d\mu(c^{\dagger})$ is nonsingular and therefore surjective onto $\mathbb{R}^{\mathscr{S}}$. For this reason (13.19) can hold only if $f(c^{\dagger}) = 0$.

Remark 13.1.5. In a more fussy treatment, we would have required only that $d\mu(\cdot)$ be nonsingular at the particular $c^{\dagger}$ under consideration or, weaker still, that the image of $d\mu(c^{\dagger})$ contain S fully.

To prove *(ii)* we suppose that $c^{\dagger}$ is a nondegenerate equilibrium. Because $f(c^{\dagger}) = 0$, $\phi(c^{\dagger}) = 0$. The local maximum condition (13.18) reduces to

$$\mu(c^{\dagger})\cdot \bar{d}f(c^{\dagger})\gamma = 0, \quad \forall \gamma \in \mathbb{R}^{\mathscr{S}}. \tag{13.20}$$

Because $\bar{d}f(c^{\dagger})$ takes values in S and (from nondegeneracy) is actually surjective onto S, (13.20) can hold only if $\mu(c^{\dagger})$ lies in $S^{\perp}$. □

Corollary 13.1.6. *If, in the statement of Theorem 13.1.4, $\mu(\cdot)$ is the chemical potential function of an ideal gas mixture—that is, if for some $c^{*} \in \mathbb{R}^{\mathscr{S}}_{+}$*

$$\mu(c) = \ln c - \ln c^{*}, \quad \forall c \in \mathbb{R}^{\mathscr{S}}_{+}, \tag{13.21}$$

then c^{*} must be a kinetic equilibrium; that is, $f(c^{*}) = 0$.

Proof. Note that $\mu(c^{*}) = 0$, so $\mu(c^{*})$ is trivially a member of $S^{\perp}$. In this case, Theorem 13.1.4 requires that $f(c^{*}) = 0$. □

13.1.7 Monday-Wednesday-Friday Equilibria vs. Tuesday-Thursday Equilibria

For a reaction network $\{\mathscr{S}, \mathscr{C}, \mathscr{R}\}$ with stoichiometric subspace S, to say that $\mu(c)$ is a member of $S^{\perp}$ is to say that, for each $y \to y' \in \mathscr{R}$, $(y' - y)\cdot \mu(c) = 0$ or, equivalently, that $y' \cdot \mu(c) = y \cdot \mu(c)$.

In a thermodynamics course taught to chemical engineering students—say on Mondays, Wednesdays, and Fridays—such equations[2] are typically referred to as the "equations of chemical equilibrium." For example, if the reaction network under study is

$$A + B \rightleftarrows C \qquad 2A \rightleftarrows B \tag{13.22}$$

then the so-called equations of chemical equilibrium are

[2] Recall that, for the narrow isothermal reactor context we have in mind, dependence of the chemical potential on temperature is being suppressed.

$$\mu_A(c) + \mu_B(c) = \mu_C(c) \qquad 2\mu_A(c) = \mu_B(c). \tag{13.23}$$

With the chemical potential functions in hand, such a set of equations would then be used to locate "thermodynamic equilibria"—that is, $c \in \mathbb{R}^{\mathscr{S}}_{+}$ that satisfy (13.23).

In a reactor design course taught to the very same chemical engineering students—say on Tuesdays and Thursdays—an equilibrium is regarded to be a rest point of the differential equations of a kinetic system $\{\mathscr{S}, \mathscr{C}, \mathscr{R}, \mathscr{K}\}$. Thus, if the network (13.22) is endowed with a kinetics $\mathscr{K}$, the Tuesday-Thursday equilibria are solutions of the equations

$$\begin{aligned}
0 &= -\mathscr{K}_{A+B\to C}(c) + \mathscr{K}_{C\to A+B}(c) - 2\mathscr{K}_{2A\to B}(c) + 2\mathscr{K}_{B\to 2A}(c)\\
0 &= -\mathscr{K}_{A+B\to C}(c) + \mathscr{K}_{C\to A+B}(c) + \mathscr{K}_{2A\to B}(c) - \mathscr{K}_{B\to 2A}(c)\\
0 &= \mathscr{K}_{A+B\to C}(c) - \mathscr{K}_{C\to A+B}(c).
\end{aligned} \tag{13.24}$$

Except in small colleges, the MWF thermodynamics course and the TTh reactor design course are likely to be taught by different professors with different research interests. In particular, it can be—and often is—the case that the MWF thermodynamics professor has little or no interest in nonlinear differential equations (and perhaps a scanty education in the subject). In the MWF thermodynamics course, which might in its own way be excellent, the differential equations for a reactor rarely make an appearance.

The MWF equations of equilibria rely only on the chemical potentials and make no mention of the kinetics $\mathscr{K}$, while the TTh equations of equilibria derive explicitly from $\mathscr{K}$ and not at all from the chemical potentials. It is a wonder, then, that the two sets of equilibria have any relation to each other. Yet, Theorem 13.1.4 tells us that, so long as the chemical potential functions and the kinetics are orchestrated by the First and Second Laws to satisfy (13.15), the MWF and the TTh equilibrium sets, calculated in very different ways, are virtually identical: All of the MWF thermodynamic equilibria are TTh kinetic equilibria, and the only TTh kinetic equilibria that might be missed in the MWF thermodynamic calculations are degenerate ones.

13.1.8 The Second Law Imposes an Orchestration of Constitutive Equations

Theorem 13.1.4 provides a dramatic demonstration, in a very narrow setting, of a much wider truth about the Second Law: *The various constitutive functions for a particular material, though they might seem to bear on different aspects of material behavior, cannot be independent of each other. Rather, the Second Law requires that they be finely orchestrated.* In our case, the Helmholtz free energy function (and the chemical potential function that derives from it) cannot be independent of the kinetic rate functions for the various reactions.

Example 13.1.7 (A Toy Example). Suppose that an ideal gas mixture, comprised of species A and B, is contained in the closed vessel we have been studying in this section. In particular, suppose that the mixture's Helmholtz free energy function is that given by (13.5) with $c_A^* = 1$ and $c_B^* = 1$. In this case, the chemical potentials take the simple form $\mu_A(c) \equiv \ln c_A$ and $\mu_B(c) \equiv \ln c_B$.

Imagine also that the operative reaction network consists of two reversible reactions $A \rightleftarrows B$ and $2A \rightleftarrows 2B$. We will suppose that the kinetics is mass action but, for now, with unspecified rate constants. If the kinetics is to be consistent with the specified Helmholtz free energy function (and the corresponding chemical potentials), Corollary 13.1.6 requires that the rate constants be such that the compositions $c_A^* = 1$ and $c_B^* = 1$ be an equilibrium—that is, a rest point of the differential equations

$$\begin{aligned} \dot{c}_A &= -k_{A\to B}c_A + k_{B\to A}c_B - 2k_{2A\to 2B}c_A^2 + 2k_{2B\to 2A}c_B^2 \\ \dot{c}_B &= \ \ k_{A\to B}c_A - k_{B\to A}c_B + 2k_{2A\to 2B}c_A^2 - 2k_{2B\to 2A}c_B^2. \end{aligned} \tag{13.25}$$

This already restricts values that the rate constants might take, but Theorem 13.1.4 requires far more. Note that each of the mass action systems

$$A \underset{1}{\overset{1}{\rightleftarrows}} B \qquad 2A \underset{1}{\overset{1}{\rightleftarrows}} 2B \tag{13.26}$$

and

$$A \underset{3}{\overset{1}{\rightleftarrows}} B \qquad 2A \underset{1}{\overset{2}{\rightleftarrows}} 2B \tag{13.27}$$

admits $c_A^* = 1$ and $c_B^* = 1$ as an equilibrium. Nevertheless, only the first of these is consistent with the specified Helmholtz free energy function:

In each case there are many equilibria of (13.25) distributed across the various positive stoichiometric compatibility classes (in fact, one in each), and they are all nondegenerate.[3] If c^{**} is any such equilibrium, it follows from Theorem 13.1.4 that $\ln c^{**}$ must be a member of $S^\perp$, which in our case is the set of all vectors $M \in \mathbb{R}^{\mathscr{S}}$ with $M_A = M_B$. Thus, if c^{**} is an equilibrium of (13.25), Theorem 13.1.4 requires that $c_A^{**} = c_B^{**}$.

Although all equilibria for the mass action system (13.26) comply with this requirement, that is not true of the mass action system (13.27). For example, when the rate constants are those shown in (13.27), the composition given by $c_A^{**} = 7/4$ and $c_B^{**} = 2$ is an equilibrium. Thus, the rate constants shown in (13.27) are inconsistent with the Helmholtz free energy function indicated.

The example demonstrates how the Second Law imposes an orchestration on constitutive functions. In this case, specification of a Helmholtz free energy function served to severely circumscribe the kinetics, *in effect denying the possibility of certain combinations of rate constant values.*

Remark 13.1.8. That consistency of a mass action system with thermodynamics might require constraints on the rate constants was noticed by Wegscheider [171] in 1901. (Thermodynamics was still a relatively young subject.) To make the point, he studied toy examples such as the one just considered and also the network

[3] The nondegeneracy and uniqueness (within a stoichiometric compatibility class) of positive equilibria for the network under consideration, regardless of rate constant values, follow quickly from the Deficiency One Algorithm, implemented in [62] and discussed more fully in Chapter 17.

$$\begin{gathered} A \rightleftarrows B \qquad C \rightleftarrows D \\ A+C \rightleftarrows B+D. \end{gathered} \tag{13.28}$$

It will have some significance later on that network (13.28) and the network studied in Example 13.1.7 both have deficiencies of one.

Remark 13.1.9. It will have occurred to some readers that the Helmholtz free energy function, because it is non-increasing along solutions of the kinetic differential equations, can serve as a Lyapunov function, giving rise to important qualitative information about dynamics. In fact, the ideal gas mixture Helmholtz free energy function will serve as a Lyapunov function later in this chapter and then in proofs of the Deficiency Zero Theorem and the Star-like Network Theorem.

Remark 13.1.10. When I was an undergraduate student, classical thermodynamics appeared to be a beautiful (and somewhat kabbalistic) subject, but its purpose was not clear. It began with much talk about the limited efficiency of engines and refrigerators—important considerations, no doubt, for mechanical engineers—and then it veered toward chemical reactions, solubility, the number of phases that could coexist at an equilibrium, and so on. I didn't really understand what was happening.

It was only much later that I came to understand the power of the Second Law (in the form of the Clausius-Duhem inequality) to impose orchestration on the seemingly unrelated constitutive equations for a particular material. For this delayed comprehension, I could not be faulted because, in neither the textbooks nor in the classroom, was there even a terminological distinction made between true laws of physics and constitutive equations that were called laws. For me, and for many others, real clarity derived, ultimately, from a paper [40] by Bernard Coleman and Walter Noll.

13.2 Quasi-Thermodynamics

The preceding section was about the thermodynamics of a *closed* reactor whose contents are homogeneous, have fixed volume, and are held at a constant temperature. For the remainder of this chapter, thermodynamics will provide the motivation for, but not the substance of, some crucial ideas. No laws of thermodynamics are invoked.

We will maintain the viewpoint taken throughout most of this book, that kinetic systems—in particular mass action systems—are *mathematical objects* that can serve to model a variety of physicochemical circumstances, including *open* reactors such as those considered in Chapter 4. In *some* cases, those mathematical objects *might* happen to enjoy the very same mathematical properties that thermodynamics imposes in more severely circumscribed closed reactor settings. When that occurs, behavior is constrained in ways reminiscent of behavior enforced by a non-increasing Helmholtz free energy.[4]

[4] Keep in mind that the open reactors considered in Chapter 4 are not confined to CFSTRs. They included also reactors such as the one considered in Section 4.17, for which certain species concentrations are deemed constant in time.

With this in mind, Horn and Jackson [109] chose their vocabulary suggestively but carefully. For them, *quasi-thermostatic* and *quasi-thermodynamic* kinetic systems are *mathematical objects* that have *mathematical* properties very much like those of thermodynamically constrained kinetic systems discussed in the preceding section. More precisely, quasi-thermostatic and quasi-thermodynamic kinetic systems are kinetic systems that have *mathematical* properties very much like those of thermodynamically constrained closed-reactor *ideal gas mixtures*. As we shall see, such mathematical properties have powerful behavioral consequences.

First we will consider attributes of what Horn and Jackson called quasi-*thermostatic* kinetic systems, in particular the way in which their equilibria are distributed across positive stoichiometric compatibility classes. With that behind us, we will then study the temporal behavior of the more sharply constrained kinetic systems Horn and Jackson called quasi-*thermodynamic*.

13.3 Quasi-Thermostatic Kinetic Systems

Definition 13.3.1 ([109]). *A kinetic system* $\{\mathscr{S},\mathscr{C},\mathscr{R},\mathscr{K}\}$ *with stoichiometric subspace* S *and species-formation-rate function* $f : \overline{\mathbb{R}}^{\mathscr{S}}_{+} \to S$ *is* **quasi-thermostatic** *if there is a* $c^* \in \mathbb{R}^{\mathscr{S}}_{+}$ *such that the set of positive equilibria—that is, the set* $\{c \in \mathbb{R}^{\mathscr{S}}_{+} : f(c) = 0\}$*—is identical to the set*

$$E := \{c \in \mathbb{R}^{\mathscr{S}}_{+} : \ln c - \ln c^* \in S^{\perp}\}. \tag{13.29}$$

In this case we say that the kinetic system is quasi-thermostatic with respect to c^*.

Remark 13.3.2. Suppose that, for a quasi-thermostatic kinetic system, c^* satisfies the requirements of Definition 13.3.1. Then it is clearly a member of E and is therefore an equilibrium. Moreover, it is easy to see that if $c^{**} \in \mathbb{R}^{\mathscr{S}}_{+}$ is a member of E or, equivalently, is an equilibrium, then

$$E = \{c \in \mathbb{R}^{\mathscr{S}}_{+} : \ln c - \ln c^{**} \in S^{\perp}\}. \tag{13.30}$$

This follows from the fact that

$$\ln c - \ln c^{**} = (\ln c - \ln c^*) + (\ln c^* - \ln c^{**}) \tag{13.31}$$

and that, by supposition, $\ln c^{**} - \ln c^*$ is a member of $S^{\perp}$. Thus, if a kinetic system is quasi-thermostatic with respect to c^*, it is also quasi-thermostatic with respect to every other member of E.

For us, the central fact about quasi-thermostatic kinetic systems is the way in which their equilibria are distributed across positive stoichiometric compatibility classes. That distribution is described in the following theorem.

Theorem 13.3.3 ([109]). *In a quasi-thermostatic kinetic system, each positive stoichiometric compatibility class contains precisely one equilibrium.*

The proof given here is different from the one given by Horn and Jackson [109]. The uniqueness argument is presented below, while the existence argument, which has some separate mathematical interest, is deferred until Appendix 13.A.

Proof (Uniqueness). For a quasi-thermostatic kinetic system $\{\mathscr{S},\mathscr{C},\mathscr{R},\mathscr{K}\}$, with stoichiometric subspace S and positive equilibrium set E (as given by (13.29)), suppose that two members of E, say c and c', reside in the same positive stoichiometric compatibility class. From stoichiometric compatibility $c' - c$ resides in S, and from properties of E we have that $\ln c' - \ln c$ lies in $S^{\perp}$. Thus,

$$0 = (c' - c)\cdot(\ln c' - \ln c) = \sum_{s \in \mathscr{S}} (c'_s - c_s)(\ln c'_s - \ln c_s). \tag{13.32}$$

Because the function $\ln : \mathbb{R}_+ \to \mathbb{R}$ is strictly increasing, (13.32) can hold only if $c'_s = c_s$ for all $s \in \mathscr{S}$, which is to say only if $c' = c$. □

13.4 Quasi-Thermodynamic Kinetic Systems

Definition 13.4.1 ([109]). *A kinetic system $\{\mathscr{S},\mathscr{C},\mathscr{R},\mathscr{K}\}$ with stoichiometric subspace S and species-formation-rate function $f : \overline{\mathbb{R}}_+^{\mathscr{S}} \to S$ is* **quasi-thermodynamic** *if there is a $c^* \in \mathbb{R}_+^{\mathscr{S}}$ such that the system in quasi-thermostatic with respect to c^* and, in addition,*

$$(\ln c - \ln c^*)\cdot f(c) \leq 0, \quad \forall c \in \mathbb{R}_+^{\mathscr{S}}, \tag{13.33}$$

with equality holding only if $f(c) = 0$ or, equivalently, only if $\ln c - \ln c^*$ lies in $S^{\perp}$. In this case we say that the kinetic system is quasi-thermodynamic with respect to c^*.

Remark 13.4.2. If, for a kinetic system, we knew only that the bound (13.33) holds for some $c^* \in \mathbb{R}_+^{\mathscr{S}}$, we could *almost* be certain that the system is quasi-thermostatic. Indeed, Theorem 13.1.4 tells us that the system is quasi-thermostatic relative to c^* *provided that all its positive equilibria are nondegenerate*. To see that nondegeneracy of equilibria is important, consider the highly pathological mass action system

$$2A \overset{1}{\leftarrow} A + B \overset{1}{\rightarrow} 2B. \tag{13.34}$$

In this case, *all* positive compositions are equilibria, including $c_A^* = 1, c_B^* = 1$. For this choice of c^*, (13.33) holds trivially, but it is certainly not true that (13.34) is quasi-thermostatic.

Remark 13.4.3. In Remark 13.3.2 we observed that if a kinetic system $\{\mathscr{S},\mathscr{C},\mathscr{R},\mathscr{K}\}$ with stoichiometric subspace S and species-formation-rate function $f : \overline{\mathbb{R}}_+^{\mathscr{S}} \to S$ is quasi-thermostatic with respect to c^* and if c^{**} is a member of

$$E = \{c \in \mathbb{R}_+^{\mathscr{S}} : \ln c - \ln c^* \in S^{\perp}\}, \tag{13.35}$$

then the kinetic system is also quasi-thermostatic with respect to c^{**}. The same observation extends to quasi-thermodynamicity: If the kinetic system is quasi-thermodynamic with respect to c^*, then it is also quasi-thermodynamic with respect to each $c^{**} \in E$. In fact, for each such c^{**}, we have

$$(\ln c - \ln c^*) \cdot f(c) = (\ln c - \ln c^{**}) \cdot f(c), \quad \forall c \in \mathbb{R}_+^{\mathscr{S}}. \tag{13.36}$$

This follows from (13.31), the supposition that $\ln c^{**} - \ln c^*$ is a member of $S^{\perp}$, and the fact that $f(\cdot)$ takes values in S.

Theorem 13.4.4 ([109]). *For a quasi-thermodynamic kinetic system, there is within each positive stoichiometric compatibility class precisely one equilibrium; that equilibrium is asymptotically stable; and there is no nontrivial cyclic composition trajectory along which all species concentrations are positive.*

Proof. Let $\{\mathscr{S}, \mathscr{C}, \mathscr{R}, \mathscr{K}\}$ be the quasi-thermodynamic kinetic system of the theorem statement, let S be its stoichiometric subspace, and let $f : \overline{\mathbb{R}}_+^{\mathscr{S}} \to S$ be its species-formation-rate function. That there is precisely one equilibrium in each positive stoichiometric compatibility class follows from Theorem 13.3.3 and the fact that every quasi-thermodynamic system is, by definition, also quasi-thermostatic.

We consider an arbitrary positive stoichiometric compatibility class, and we suppose that c^* is the equilibrium within it. From Remark 13.4.3 it follows that the kinetic system under study is quasi-thermodynamic with respect to c^*. Thus, we have

$$(\ln c - \ln c^*) \cdot f(c) \le 0, \quad \forall c \in \mathbb{R}_+^{\mathscr{S}}, \tag{13.37}$$

with equality holding only if c is an equilibrium. Because there is no other equilibrium in the positive stoichiometric compatibility class containing c^*, strict inequality obtains there for all $c \ne c^*$. That is, we actually have

$$(\ln c - \ln c^*) \cdot f(c) < 0, \quad \forall c \in (c^* + S) \cap \mathbb{R}_+^{\mathscr{S}}, \ c \ne c^*. \tag{13.38}$$

Now let $h : \mathbb{R}_+^{\mathscr{S}} \to \mathbb{R}$ be defined by

$$h(c) := \sum_{s \in \mathscr{S}} [c_s (\ln c_s - \ln c_s^* - 1) + c_s^*]. \tag{13.39}$$

Clearly,

$$h(c^*) = 0. \tag{13.40}$$

Moreover, from the strict concavity of the logarithm function, we have, for each $s \in \mathscr{S}$ and each $c_s > 0$,

$$\ln c_s - \ln c_s^* \ge \frac{1}{c_s}(c_s - c_s^*) \tag{13.41}$$

with equality holding if and only if $c_s = c_s^*$. From this we obtain

$$h(c) > 0, \quad \forall c \neq c^*. \tag{13.42}$$

Moreover, straightforward computation gives

$$\nabla h(c) = \ln c - \ln c^*, \quad \forall c \in \mathbb{R}_+^{\mathscr{S}}. \tag{13.43}$$

From this and (13.38), we have

$$\nabla h(c) \cdot f(c) < 0, \quad \forall c \in (c^* + S) \cap \mathbb{R}_+^{\mathscr{S}},\ c \neq c^*. \tag{13.44}$$

This tells us that, along a solution of the differential equation $\dot{c} = f(c)$ residing in the positive stoichiometric compatibility class containing c^*, the function $h(\cdot)$ takes strictly decreasing values (except for the constant solution that is forever at the equilibrium c^*). In fact, along such a nonconstant solution

$$\frac{d}{dt}h(c(t)) = \nabla h(c(t)) \cdot \dot{c}(t) = \nabla h(c(t)) \cdot f(c(t)) < 0. \tag{13.45}$$

To prove asymptotic stability of c^*, we first note that the positive stoichiometric compatibility class containing c^* is open in the relative topology on the full stoichiometric compatibility class containing c^*. Now let $\bar{h}(\cdot)$ be the restriction of $h(\cdot)$ to the positive stoichiometric compatibility class containing c^*. From (13.40), (13.42), (13.44), and (13.45), it follows that $\bar{h}(\cdot)$ is a strict Lyapunov function [104] for c^* on the positive stoichiometric compatibility class containing it. Thus, c^* is asymptotically stable relative to initial conditions in its positive stoichiometric compatibility class.

To show that the positive stoichiometric compatibility class containing c^* cannot contain a nontrivial (i.e., nonconstant) periodic solution, we suppose that $c : [0,T] \to \mathbb{R}_+^{\mathscr{S}}$ is such a solution with $c(T) = c(0)$. Then, with $h(\cdot)$ as before,

$$\begin{aligned} h(c(T)) - h(c(0)) &= \int_0^T \frac{d}{dt}h(c(t))dt \\ &= \int_0^T \nabla h(c(t)) \cdot f(c(t))dt. \end{aligned} \tag{13.46}$$

Because the solution is nonconstant, (13.44) and (13.45) tell us that the integrand is negative, in which case we have

$$h(c(T)) < h(c(0)). \tag{13.47}$$

But this contradicts the supposition that $c(T) = c(0)$. □

13.5 Application: Proof of the Star-Like Network Theorem

In this section we will prove Theorem 7.6.1 of Chapter 7—the Star-Like Network Theorem—by showing that every mass action system in which the underlying reaction network is reversible and star-like is quasi-thermodynamic, regardless of rate constant values. The dull behavior described in Theorem 7.6.1 then becomes a direct consequence of Theorem 13.4.4.

Recall that a reversible star-like network is a network of the form shown in Figure 13.1, in which each reversible reaction pair has as one of its complexes a common "central" complex, which we will label y_0. When $\{\mathscr{S},\mathscr{C},\mathscr{R}\}$ is a star-like network, we will denote by $\mathscr{C}^*$ the complex set $\mathscr{C} \setminus \{y_0\}$—that is, the set of all complexes other than y_0, those that lie at the points of the star.

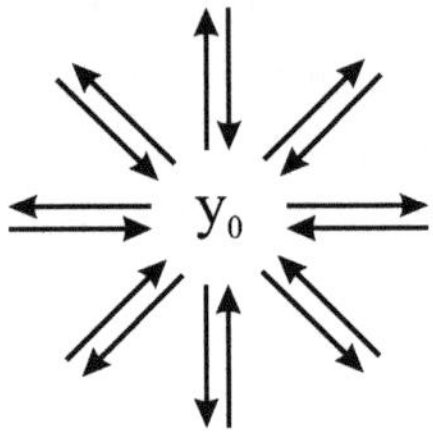

Fig. 13.1. The shape of a reversible star-like network

Theorem 13.5.1. *If $\{\mathscr{S},\mathscr{C},\mathscr{R}\}$ is a reversible star-like network, then, for any choice of $k \in \mathbb{R}^{\mathscr{R}}_+$, the mass action system $\{\mathscr{S},\mathscr{C},\mathscr{R},k\}$ is quasi-thermodynamic.*

Proof. For the mass action system $\{\mathscr{S},\mathscr{C},\mathscr{R},k\}$ we will have to show, among other things, that there is a positive equilibrium. Because the network is reversible, and therefore weakly reversible, the existence of a positive equilibrium is already ensured by a theorem in [55]. For this special case, however, we provide a dedicated proof in Appendix 13.B.

Suppose, then, that $c^* \in \mathbb{R}^{\mathscr{S}}_+$ is an equilibrium, in which case

$$\sum_{y \in \mathscr{C}^*} k_{y \to y_0}(c^*)^y (y_0 - y) + \sum_{y \in \mathscr{C}^*} k_{y_0 \to y}(c^*)^{y_0}(y - y_0) = 0 \tag{13.48}$$

or, equivalently,

$$\sum_{y \in \mathscr{C}^*} k_{y_0 \to y}(c^*)^{y_0}(y_0 - y) = \sum_{y \in \mathscr{C}^*} k_{y \to y_0}(c^*)^{y}(y_0 - y). \tag{13.49}$$

Now we let $\mu : \mathbb{R}^{\mathscr{S}}_+ \to \mathbb{R}^{\mathscr{S}}$ be defined by

$$\mu(c) := \ln c - \ln c^*. \tag{13.50}$$

For every $c \in \mathbb{R}^{\mathscr{S}}_+$, the species formation rate can be written in the form

$$\begin{aligned} f(c) &= \sum_{y \to y' \in \mathscr{R}} k_{y \to y'} c^y (y' - y) \\ &= \sum_{y \to y' \in \mathscr{R}} k_{y \to y'} (c^*)^y e^{y \cdot \mu(c)} (y' - y) \\ &= \sum_{y \in \mathscr{C}^*} k_{y \to y_0}(c^*)^y e^{y \cdot \mu(c)} (y_0 - y) - e^{y_0 \cdot \mu(c)} \sum_{y \in \mathscr{C}^*} k_{y_0 \to y}(c^*)^{y_0}(y_0 - y). \end{aligned} \tag{13.51}$$

From this and (13.49), we can write, for every $c \in \mathbb{R}_+^{\mathscr{S}}$,

$$f(c) = \sum_{y \in \mathscr{C}^*} k_{y \to y_0} (c^*)^y (e^{y \cdot \mu(c)} - e^{y_0 \cdot \mu(c)})(y_0 - y) \tag{13.52}$$

and, therefore,

$$(\ln c - \ln c^*) \cdot f(c) = \sum_{y \in \mathscr{C}^*} k_{y \to y_0} (c^*)^y (e^{y \cdot \mu(c)} - e^{y_0 \cdot \mu(c)})(y_0 \cdot \mu(c) - y \cdot \mu(c)). \tag{13.53}$$

Because the exponential function is strictly increasing, we have, for all real x and x_0,

$$(e^x - e^{x_0})(x_0 - x) \leq 0 \tag{13.54}$$

with equality holding if and only if $x = x_0$. Thus, from (13.53) it follows that

$$(\ln c - \ln c^*) \cdot f(c) \leq 0, \quad \forall c \in \mathbb{R}_+^{\mathscr{S}} \tag{13.55}$$

with equality holding if and only if $(y - y_0) \cdot \mu(c) = 0$ for all $y \in \mathscr{C}^*$. This last condition is equivalent to the requirement that $\mu(c) = \ln c - \ln c^*$ be a member of $S^\perp$, where S is the stoichiometric subspace for the network $\{\mathscr{S}, \mathscr{C}, \mathscr{R}\}$.

We still need to argue that the mass action system $\{\mathscr{S}, \mathscr{C}, \mathscr{R}, k\}$ is quasi-thermostatic, which is to say that set of positive equilibria is identical to the set

$$E := \{c \in \mathbb{R}_+^{\mathscr{S}} : \ln c - \ln c^* \in S^\perp\}. \tag{13.56}$$

Clearly, if $c \in \mathbb{R}_+^{\mathscr{S}}$ is such that $f(c) = 0$, then equality holds in (13.55), in which case we know that $\mu(c) = \ln c - \ln c^*$ lies in $S^\perp$. On the other hand, if $\mu(c)$ lies in $S^\perp$, then $(y - y_0) \cdot \mu(c) = 0$ for all $y \in \mathscr{C}^*$. From (13.52) it follows that $f(c) = 0$. □

13.6 What's Coming?

As was the case in the proof of the Star-like Network Theorem, Theorems 13.3.3 and 13.4.4 will provide stepping stones for proofs of some of the principal theorems introduced earlier. In particular, proofs of the weakly reversible part of the Deficiency Zero Theorem will amount to showing that any mass action system that derives from a weakly reversible deficiency zero network is quasi-thermodynamic, no matter what values the rate constants take. Proof of the Deficiency One Theorem will amount to showing that any mass action system in which the underlying network satisfies the theorem's structural requirements is quasi-thermostatic, again regardless of rate constant values.

Proof of the Deficiency Zero Theorem will benefit from prior discussion of *detailed balancing* and *complex balancing*, the subjects of the next two chapters.

Appendix 13.A Proof of Theorem 13.3.3

Our aim in this appendix is to complete the proof of Theorem 13.3.3. We will begin with Proposition 13.A.1, which appeared in 1979 [71] and which has some interest in its own right, divorced from its use in chemical reaction network theory. Although by $\mathscr{S}$ and S in the proposition statement we obviously have in mind the species set and the stoichiometric subspace, the proposition is stated more abstractly. Theorem 13.3.3 will amount to a corollary of Proposition 13.A.1. The same proposition will find a different use in Appendix 13.B.

Proposition 13.A.1 ([71, 76]). *Let $\mathscr{S}$ be any finite set, let $\mathbb{R}^{\mathscr{S}}$ be the vector space generated by $\mathscr{S}$, let S be a linear subspace of $\mathbb{R}^{\mathscr{S}}$, and let a and b be members of $\mathbb{R}^{\mathscr{S}}_{+}$. There is a unique vector $\mu \in S^{\perp}$ such that*

$$ae^{\mu} - b$$

is an element of S.

Proof. Let $\phi : \mathbb{R}^{\mathscr{S}} \to \mathbb{R}$ be defined by

$$\phi(x) := \sum_{s \in \mathscr{S}} (a_s e^{x_s} - b_s x_s)\,. \tag{13.A.1}$$

Straightforward computation shows that the gradient of ϕ at x is given by

$$\nabla\phi(x) \equiv ae^{x} - b \tag{13.A.2}$$

and that the Hessian of ϕ at x, $H(x) : \mathbb{R}^{\mathscr{S}} \to \mathbb{R}^{\mathscr{S}}$, is given by

$$H(x)\gamma \equiv (ae^{x})\gamma. \tag{13.A.3}$$

Moreover, for each $x \in \mathbb{R}^{\mathscr{S}}$, $H(x)$ is positive-definite: For all nonzero $\gamma \in \mathbb{R}^{\mathscr{S}}$

$$\gamma \cdot H(x)\gamma = \gamma \cdot ae^{x}\gamma = \sum_{s \in \mathscr{S}} a_s e^{x_s} (\gamma_s)^2 > 0. \tag{13.A.4}$$

Thus, the function ϕ is strictly convex [143].

Next we want to show that, for any nonzero $x \in \mathbb{R}^{\mathscr{S}}$,

$$\lim_{\lambda \to \infty} \phi(\lambda x) = \infty. \tag{13.A.5}$$

Note that

$$\phi(\lambda x) = \sum_{s \in \mathscr{S}} (a_s e^{\lambda x_s} - \lambda\, b_s x_s). \tag{13.A.6}$$

Note also that, for $x_s \neq 0$, the positivity of a_s and b_s gives

$$\lim_{\lambda \to \infty} (a_s e^{\lambda x_s} - \lambda\, b_s x_s) = \infty, \tag{13.A.7}$$

while for $x_s = 0$ we have

$$(a_s\, e^{\lambda\, x_s} - \lambda\, b_s x_s) = a_s, \quad \forall \lambda \in \mathbb{R}. \tag{13.A.8}$$

Thus, for $x \neq 0$, (13.A.6)–(13.A.8) imply (13.A.5).

Now let $\bar{\phi} : S^{\perp} \to \mathbb{R}$ be the restriction of ϕ to $S^{\perp}$. Since ϕ is continuous and convex, so is $\bar{\phi}$. The continuity of $\bar{\phi}$ and a standard result for convex functions ensure that the set

$$C := \{x \in S^{\perp} : \bar{\phi}(x) \leq \bar{\phi}(0)\} \tag{13.A.9}$$

is closed, convex (and obviously contains the zero vector). Moreover, it follows from (13.A.5) that C contains no half line with endpoint 0. Since, in a finite-dimensional vector space, every unbounded closed convex set containing 0 must contain a half line with endpoint 0 (see [163], p. 105), it follows that C is bounded and therefore compact.

Thus, there exists $\mu \in C$ such that

$$\bar{\phi}(\mu) \leq \bar{\phi}(x), \quad \forall x \in C. \tag{13.A.10}$$

In fact, from the definition of C, we have

$$\bar{\phi}(\mu) \leq \bar{\phi}(x), \quad \forall x \in S^{\perp}. \tag{13.A.11}$$

Therefore, for all $\gamma \in S^{\perp}$,

$$0 = \frac{d}{d\theta}\bar{\phi}(\mu + \theta\gamma)\,|_{\theta=0} \tag{13.A.12}$$

$$= \frac{d}{d\theta}\phi(\mu + \theta\gamma)\,|_{\theta=0} \tag{13.A.13}$$

$$= \nabla\phi(\mu)\cdot\gamma. \tag{13.A.14}$$

It follows that $\nabla\phi(\mu)$ must lie in S so that, from (13.A.2), we have the inclusion

$$a\,e^{\mu} - b \in S. \tag{13.A.15}$$

Thus, $\mu \in S^{\perp}$ satisfies the requirements of the proposition.

To prove uniqueness we presume that $\mu' \in S^{\perp}$ also satisfies the inclusion

$$a\,e^{\mu'} - b \in S. \tag{13.A.16}$$

From (13.A.15) and (13.A.16), we have

$$a\,(e^{\mu'} - e^{\mu}) \in S, \tag{13.A.17}$$

and, since $\mu' - \mu$ lies in $S^{\perp}$, we also have

$$0 = (\mu' - \mu)\cdot[a\,(e^{\mu'} - e^{\mu})] = \sum_{s \in \mathscr{S}} a_s(\mu'_s - \mu_s)(e^{\mu'_s} - e^{\mu_s})\,. \tag{13.A.18}$$

Since each a_s is positive and the exponential function is strictly monotonically increasing, (13.A.18) can hold only if $\mu'_s = \mu_s$ for all $s \in \mathscr{S}$—that is, only if $\mu' = \mu$. □

Remark 13.A.2. There is an interesting observation that can be made here. With $\mathbb{R}^{\mathscr{S}}$ and S as in Proposition 13.A.1, it is a standard result in linear algebra that any $b \in \mathbb{R}^{\mathscr{S}}$ has a unique representation

$$b = x_1 + x_2, \quad x_1 \in S^{\perp},\ x_2 \in S.$$

Taking $a_{\mathscr{s}} = 1$ for all $\mathscr{s} \in \mathscr{S}$ in Proposition 13.A.1, we obtain a similar (but deeper) result: Any *positive* $b \in \mathbb{R}^{\mathscr{S}}$—that is, any $b \in \mathbb{R}^{\mathscr{S}}_+$—admits a unique representation

$$b = e^{\mu_1} + \mu_2, \quad \mu_1 \in S^{\perp},\ \mu_2 \in S.$$

The following corollary of Proposition 13.A.1 is essentially a restatement of Theorem 13.3.3, proved in 1972 by Horn and Jackson [109] but in a different way.

Corollary 13.A.3. *Let $\{\mathscr{S},\mathscr{C},\mathscr{R}\}$ be a reaction network with stoichiometric subspace S. For any $c^* \in \mathbb{R}^{\mathscr{S}}_+$ the set*

$$E := \{c \in \mathbb{R}^{\mathscr{S}}_+ : \ln c - \ln c^* \in S^{\perp}\} \tag{13.A.19}$$

meets each positive stoichiometric compatibility class in precisely one point.

Proof. Let p be an arbitrary element of $\mathbb{R}^{\mathscr{S}}_+$. We will show that E meets the positive stoichiometric compatibility class containing p in precisely one point. That there can be at most one such point was proved just after the statement of Theorem 13.3.3. To prove the existence of such a point, we note that Proposition 13.A.1 ensures the existence of $\mu \in S^{\perp}$ such that

$$c^* e^{\mu} - p \in S. \tag{13.A.20}$$

Now let $c^{\dagger}$ be defined by

$$c^{\dagger} := c^* e^{\mu}. \tag{13.A.21}$$

From (13.A.20) and (13.A.21), it follows that $c^{\dagger}$ is in the positive stoichiometric compatibility class containing p. Taking logarithms in (13.A.21), we obtain

$$\ln c^{\dagger} - \ln c^* = \mu \in S^{\perp} \tag{13.A.22}$$

Thus, $c^{\dagger}$ is a member of E as well. □

Appendix 13.B Existence of a Positive Equilibrium for a Reversible Star-Like Mass Action System

Our objective here is to prove that, when $\{\mathscr{S},\mathscr{C},\mathscr{R}\}$ is a reversible star-like network and $k \in \mathbb{R}^{\mathscr{R}}_+$ is an arbitrary rate constant assignment, the mass action system $\{\mathscr{S},\mathscr{C},\mathscr{R},k\}$ admits a positive equilibrium. As in Section 13.5, we denote by y_0 the complex central to the star network and by $\mathscr{C}^*$ the remaining complexes.

What we want to show is that there exists $c \in \mathbb{R}^{\mathscr{S}}_+$ such that

$$\sum_{y \in \mathscr{C}^*} k_{y \to y_0} c^y (y_0 - y) + \sum_{y \in \mathscr{C}^*} k_{y_0 \to y} c^{y_0} (y - y_0) = 0. \tag{13.B.1}$$

This can be written in the form

$$e^{y_0 \cdot \ln c} \sum_{y \in \mathscr{C}^*} (k_{y \to y_0} e^{(y - y_0) \cdot \ln c} - k_{y_0 \to y})(y - y_0) = 0. \tag{13.B.2}$$

Clearly, it will be enough to show the existence of $x \in \mathbb{R}^{\mathscr{S}}$ such that

$$\sum_{y \in \mathscr{C}^*} (k_{y \to y_0} e^{(y - y_0) \cdot x} - k_{y_0 \to y})(y - y_0) = 0, \tag{13.B.3}$$

for then we can take $c = e^x$.

With these things in mind, we let $T : \mathbb{R}^{\mathscr{C}^*} \to \mathbb{R}^{\mathscr{S}}$ be the linear transformation defined by

$$T\alpha = \sum_{y \in \mathscr{C}^*} \alpha_y (y - y_0), \quad \forall \alpha \in \mathbb{R}^{\mathscr{C}^*}. \tag{13.B.4}$$

Before proceeding further we record the following remark:

Remark 13.B.1. Suppose that V and W are finite-dimensional vector spaces, each with a scalar product, and that $L : V \to W$ is a linear transformation. By the *transpose* (sometimes called the *adjoint*) of L, we mean the linear transformation $L^T : W \to V$ such that

$$(Lv) \cdot w = v \cdot L^T w \quad \forall v \in V, w \in W. \tag{13.B.5}$$

It is a standard fact of linear algebra that $\operatorname{im} L^T = (\ker L)^\perp$ [98].

Note that the transpose of T, $T^T : \mathbb{R}^{\mathscr{S}} \to \mathbb{R}^{\mathscr{C}^*}$, is given by

$$T^T z = \sum_{y \in \mathscr{C}^*} [(y - y_0) \cdot z]\, \omega_y, \quad \forall z \in \mathbb{R}^{\mathscr{S}}, \tag{13.B.6}$$

where $\{\omega_y\}_{y \in \mathscr{C}^*}$ is the standard basis for $\mathbb{R}^{\mathscr{C}^*}$. (Recall Section 2.2.2.) Let a and b in $\mathbb{R}^{\mathscr{C}^*}$ be defined by

$$a_y = k_{y \to y_0} \quad \text{and} \quad b_y = k_{y_0 \to y}, \quad \forall y \in \mathscr{C}^*. \tag{13.B.7}$$

In view of (13.B.3), (13.B.6), and (13.B.7), our proof will be complete if we can show that there is a $\mu \in \operatorname{im} T = (\ker T)^\perp$ such that

$$a e^{\mu} - b$$

is a member of $\ker T$. In fact, the existence of such a μ is ensured by Proposition 13.A.1. □

14

Detailed Balancing

Early in the 20th century, there came into being a strongly held belief, called sometimes *the principle of microscopic reversibility* and sometimes *the principle of detailed balance* [31, 57, 85, 124, 165]. In rough terms, the principle asserts that when, in a naturally occurring physical system, a collection of distinct and varied reversible molecular processes gives rise to the system's dynamics, a state of equilibrium can result only when, *for each such process*, the occurrence rate of the process and the occurrence rate of its reverse are identical.

It was generally recognized, certainly by Dirac [57], that the principle does not derive from classical thermodynamics. Rather, the supporting arguments are usually molecular and statistical in nature, involving the reversal of particle velocities. In any case, the principle of detailed balance is not deemed applicable to all stationary states of physical processes but only to "states of thermodynamic equilibrium," presumably for closed systems such as the one considered at the beginning of the preceding chapter.

We will continue to view a kinetic system as a *mathematical* object that, in the spirit of Chapter 4, might serve to model a reactor *open* to transport of species to and from its exterior. For kinetic systems, then, detailed balancing at equilibria is not a constraint we will regard to be broadly compelling. For a mass action system, however, we will see that there are remarkable consequences of the existence of even a single positive equilibrium at which detailed balancing obtains. We shall also see that for a mass action system to admit such an equilibrium, there are sometimes, depending on the structure of the underlying network, severe restrictions that require a tight orchestration of the rate constants. In fact, we present in Section 14.4 necessary and sufficient conditions that rate constants must respect if they are to be consistent with detailed balancing.

For us, the idea of detailed balancing will largely serve as a gateway to the broader, more far-reaching idea of *complex balancing*, introduced in 1972 by Horn and Jackson [109]. Parts of this chapter are meant only to highlight more or less well-known consequences of detailed balancing in mass action systems that, as we shall see, extend to the larger class of complex balanced mass action systems. In particular, at least parts of Section 14.2 seem to have been well-established before

M. Feinberg, *Foundations of Chemical Reaction Network Theory*,
Applied Mathematical Sciences 202,
https://doi.org/10.1007/978-3-030-03858-8_14

1972, although perhaps not in the form stated here. See, for example, papers by Wei [173–175] and by Krambeck [119].

14.1 Detailed Balancing in Kinetic Systems

We begin with a definition.

Definition 14.1.1. *For a kinetic system* $\{\mathscr{S},\mathscr{C},\mathscr{R},\mathscr{K}\}$ *in which the reaction network* $\{\mathscr{S},\mathscr{C},\mathscr{R}\}$ *is reversible,* **detailed balancing** *obtains at composition* $c^* \in \overline{\mathbb{R}}_+^{\mathscr{S}}$ *if*

$$\mathscr{K}_{y\to y'}(c^*) = \mathscr{K}_{y'\to y}(c^*), \quad \forall y \to y' \in \mathscr{R}. \tag{14.1}$$

To see the connection between detailed balancing and equilibria, it helps to recall that, in vector form, the differential equation for a kinetic system $\{\mathscr{S},\mathscr{C},\mathscr{R},\mathscr{K}\}$ is

$$\dot{c} = \sum_{y\to y'\in\mathscr{R}} \mathscr{K}_{y\to y'}(c)(y'-y). \tag{14.2}$$

Throughout this section it will be understood that the reaction network $\{\mathscr{S},\mathscr{C},\mathscr{R}\}$ under consideration is reversible. We can give the network an orientation by designating for each reversible reaction pair a choice of a "forward" reaction in that pair. This amounts to choosing a subset $\mathscr{R}^{\rightarrow} \subset \mathscr{R}$, containing half as many reactions as there are in $\mathscr{R}$, such that if $y \to y'$ is contained in $\mathscr{R}^{\rightarrow}$ then $y' \to y$ is not. Hereafter we will suppose that such a choice of an orientation has been made.

Equation (14.2) can be rewritten in the form

$$\dot{c} = \sum_{y\to y'\in\mathscr{R}^{\rightarrow}} [\mathscr{K}_{y\to y'}(c) - \mathscr{K}_{y'\to y}(c)]\,(y'-y). \tag{14.3}$$

Clearly, if detailed balancing obtains at $c^* \in \overline{\mathbb{R}}_+^{\mathscr{S}}$, then c^* is an equilibrium, for then each term in (14.2) vanishes separately. On the other hand, if c^* is an equilibrium, it is not necessarily true that detailed balancing obtains there, for the reaction vector set

$$\{y'-y \in \mathbb{R}^{\mathscr{S}} : y \to y' \in \mathscr{R}^{\rightarrow}\} \tag{14.4}$$

need not be linearly independent. If the set is dependent, then an equilibrium does not require that the multipliers in (14.3) all be zero. However, if the set (14.4) is independent, then detailed balancing *must* obtain at *every* equilibrium. We will return to this last observation later on.

A standard example indicates that, for mass action systems, it can happen that detailed balancing obtains at a positive equilibrium only under very special circumstances: Consider a mass action system in which (14.5) is the underlying reaction network.

$$\begin{array}{ccc} A & \rightleftarrows & B \\ & \nwarrow\!\searrow \quad \nearrow\!\swarrow & \\ & C & \end{array} \tag{14.5}$$

For $c \in \overline{\mathbb{R}}_+^{\mathscr{S}}$ to be an equilibrium, it is enough that c satisfy the equations

$$\begin{aligned} 0 &= -(k_{A\to B} + k_{A\to C})c_A + k_{B\to A}c_B + k_{C\to A}c_C \\ 0 &= k_{A\to B}c_A - (k_{B\to A} + k_{B\to C})c_B + k_{C\to B}c_C \\ 0 &= k_{A\to C}c_A + k_{B\to C}c_B - (k_{C\to A} + k_{C\to B})c_C. \end{aligned} \tag{14.6}$$

On the other hand, detailed balancing at composition c requires that c satisfy the stronger conditions

$$k_{A\to B}c_A = k_{B\to A}c_B, \qquad k_{B\to C}c_B = k_{C\to B}c_C, \qquad k_{C\to A}c_C = k_{A\to C}c_A. \tag{14.7}$$

It is easily verified that for all choices of (positive) rate constants, there is a positive solution of (14.6), in fact one in each positive stoichiometric compatibility class. On the other hand, for there to exist $c \neq 0$ at which detailed balancing obtains, it is necessary and sufficient that the rate constants satisfy the highly restrictive condition

$$k_{A\to B}\, k_{B\to C}\, k_{C\to A} = k_{A\to C}\, k_{C\to B}\, k_{B\to A}. \tag{14.8}$$

Another example is afforded by a network studied by Wegscheider [172]:

$$\begin{gathered} A \rightleftarrows B \qquad C \rightleftarrows D \\ A + C \rightleftarrows B + D. \end{gathered} \tag{14.9}$$

When the kinetics is mass action, detailed balancing obtains at positive equilibria if and only if the rate constants satisfy the condition

$$\frac{k_{A\to B}k_{C\to D}}{k_{A+C\to B+D}} = \frac{k_{B\to A}k_{D\to C}}{k_{B+D\to A+C}}. \tag{14.10}$$

These examples demonstrate that, for mass action systems in which the set (14.4) is dependent, we should not expect detailed balancing to obtain at equilibria unless the rate constants are suitably orchestrated. The precise nature of that orchestration will be the subject of Section 14.4. First, however, we shall consider the remarkable consequences for mass action systems of the existence of even a single positive equilibrium at which detailed balancing obtains.

14.2 Consequences of Detailed Balancing in Mass Action Systems

Throughout this section the reaction network $\{\mathscr{S}, \mathscr{C}, \mathscr{R}\}$ is understood to be reversible and to have stoichiometric subspace S. In the theorem below, $f : \overline{\mathbb{R}}_+^{\mathscr{S}} \to S$ denotes the species-formation-rate function of the mass action system $\{\mathscr{S}, \mathscr{C}, \mathscr{R}, k\}$, and $df(c) : S \to S$ is the derivative of $f(\cdot)$ at $c \in \mathbb{R}_+^{\mathscr{S}}$. Recall that if V is a vector space with scalar product "$\cdot$," a linear transformation $T : V \to V$ is *symmetric* if

$$u \cdot Tw = w \cdot Tu, \quad \forall u, w \in V$$

and *negative-definite* if

$$v \cdot Tv < 0, \quad \forall v \in V, v \neq 0.$$

Theorem 14.2.1. *If, for $k \in \mathbb{R}_+^{\mathscr{R}}$, the mass action system $\{\mathscr{S}, \mathscr{C}, \mathscr{R}, k\}$ admits a positive equilibrium at which detailed balancing obtains, then the following hold true:*

(i) The mass action system $\{\mathscr{S}, \mathscr{C}, \mathscr{R}, k\}$ is quasi-thermodynamic.
(ii) Detailed balancing obtains at every positive equilibrium, regardless of stoichiometric compatiblity class.
(iii) If $c \in \mathbb{R}_+^{\mathscr{S}}$ is an equilibrium, then $df(c) : S \to S$ is symmetric and negative-definite relative to the scalar product "$*$" in $\mathbb{R}^{\mathscr{S}}$ given by

$$v * w := \sum_{s \in \mathscr{S}} \frac{v_s w_s}{c_s}, \quad \forall v, w \in \mathbb{R}^{\mathscr{S}}. \tag{14.11}$$

As a result, all roots of the characteristic polynomial of $df(c)$ are real and negative, and there exists for S a $*$-orthonormal basis consisting of entirely eigenvectors of $df(c)$.

Proof. Suppose that, for the mass action system $\{\mathscr{S}, \mathscr{C}, \mathscr{R}\}$, detailed balancing obtains at the equilibrium $c^* \in \mathbb{R}_+^{\mathscr{S}}$. We give the reaction network an orientation $\mathscr{R}^{\rightarrow} \subset \mathscr{R}$ and write the species-formation-rate function in the following form:

$$f(c) \equiv \sum_{y \to y' \in \mathscr{R}^{\rightarrow}} [k_{y \to y'} c^y - k_{y' \to y} c^{y'}](y' - y) \tag{14.12}$$

$$= \sum_{y \to y' \in \mathscr{R}^{\rightarrow}} \kappa_{y \to y'} \left[\left(\frac{c}{c^*}\right)^y - \left(\frac{c}{c^*}\right)^{y'} \right] (y' - y),$$

where

$$\kappa_{y \to y'} := k_{y \to y'} (c^*)^y = k_{y' \to y} (c^*)^{y'}. \tag{14.13}$$

Hereafter we shall understand c to be a composition in $\mathbb{R}_+^{\mathscr{S}}$, and we let

$$\mu(c) := \ln c - \ln c^*. \tag{14.14}$$

In this case we can write

$$f(c) \equiv \sum_{y \to y' \in \mathscr{R}^{\rightarrow}} \kappa_{y \to y'} [\, e^{y \cdot \mu(c)} - e^{y' \cdot \mu(c)}](y' - y) \tag{14.15}$$

and

$$(\ln c - \ln c^*) \cdot f(c) = \sum_{y \to y' \in \mathscr{R}^{\rightarrow}} \kappa_{y \to y'} [\, e^{y \cdot \mu(c)} - e^{y' \cdot \mu(c)}][y' \cdot \mu(c) - y \cdot \mu(c)]. \tag{14.16}$$

Because the exponential function $x \to e^x$ is strictly monotonically increasing, the term in (14.16) corresponding to $y \to y' \in \mathscr{R}^{\to}$ is nonpositive and is zero precisely when $(y'-y)\cdot\mu(c) = 0$. Thus, we have

$$(\ln c \; - \; \ln c^*)\cdot f(c) \le 0, \quad \forall c \in \mathbb{R}_+^{\mathscr{S}}, \tag{14.17}$$

with equality holding if and only $\mu(c) = \ln c \; - \; \ln c^*$ lies in $S^{\perp}$.

In particular if c is an equilibrium—that is, if $f(c) = 0$—then equality must hold in (14.17), in which case $\ln c \; - \; \ln c^*$ lies in $S^{\perp}$. On the other hand, if c is such that $\mu(c) = \ln c \; - \; \ln c^*$ lies in $S^{\perp}$, then we have $y \cdot \mu(c) = y' \cdot \mu(c)$ for each $y \to y' \in \mathscr{R}^{\to}$. From (14.15) it follows immediately that c is an equilibrium. These considerations tell us that, for the mass action system $\{\mathscr{S}, \mathscr{C}, \mathscr{R}, k\}$, the positive equilibrium set is

$$E := \{c \in \mathbb{R}_+^{\mathscr{S}} : \ln c - \ln c^* \in S^{\perp}\}, \tag{14.18}$$

which is to say that the system is quasi-thermostatic. From this and (14.17), it follows that the system is quasi-thermodynamic. This completes the proof of (i).

To prove (ii) we suppose that $c \in \mathbb{R}_+^{\mathscr{S}}$ is an equilibrium, in which case it is a member of E. Thus, $\mu(c) = \ln c \; - \; \ln c^*$ lies in $S^{\perp}$, whereupon $y \cdot \mu(c) = y' \cdot \mu(c)$ for each $y \to y' \in \mathscr{R}^{\to}$. From this it follows that

$$\kappa_{y \to y'}[\, e^{y\cdot\mu(c)} \; - \; e^{y'\cdot\mu(c)} \,] = 0, \quad \forall y \to y' \in \mathscr{R}^{\to}. \tag{14.19}$$

This, however, is equivalent to

$$k_{y \to y'}\, c^y = k_{y' \to y}\, c^{y'}, \quad \forall y \to y' \in \mathscr{R}^{\to}, \tag{14.20}$$

which is to say that detailed balancing obtains at c.

To prove (iii) we suppose that $c \in \mathbb{R}_+^{\mathscr{S}}$ is an equilibrium. Some calculation will indicate that $df(c) : S \to S$ is given by

$$df(c)\sigma = \sum_{y \to y' \in \mathscr{R}^{\to}} \eta_{y \to y'}[y * \sigma \; - \; y' * \sigma]\,(y'-y), \quad \forall \sigma \in S, \tag{14.21}$$

where

$$\eta_{y \to y'} := k_{y \to y'} c^y = k_{y' \to y} c^{y'} \tag{14.22}$$

and "$*$" is the scalar product in $\mathbb{R}^{\mathscr{S}}$ given by (14.11). The symmetry of $df(c)$ relative to this scalar product follows from the fact that for all $\sigma, \sigma' \in S$

$$\sigma' * df(c)\sigma = \sigma * df(c)\sigma' = \sum_{y \to y' \in \mathscr{R}^{\to}} \eta_{y \to y'}[(y-y') * \sigma]\,[(y'-y) * \sigma']. \tag{14.23}$$

That $df(c)$ is negative-definite relative to the "$*$" scalar product follows from the fact that for all $\sigma \in S$

$$\sigma * df(c)\sigma = -\sum_{y \to y' \in \mathscr{R}^{\to}} \eta_{y \to y'}[(y'-y) * \sigma]^2. \tag{14.24}$$

Note that the right side is nonpositive and can take the value zero only if $\sigma \in S$ is $*$-orthogonal to each member of the set $\{y'-y : y \to y' \in \mathscr{R}^{\to}\}$. Since this set generates S, such a σ could only be the zero vector. The remainder of (iii) is a consequence of the well-known Theorem 11.A.4 [98, 120]. □

Because of item (ii), it makes sense to speak of a *detailed balanced mass action system*. This is a mass action system in which the underlying reaction network is reversible and for which there exists a detailed balanced positive equilibrium. Item (ii) ensures that in such a case, detailed balancing will obtain at *all* positive equilibria, regardless of stoichiometric compatibility class. With this in mind, we record a definition:

Definition 14.2.2. *A mass action system* $\{\mathscr{S},\mathscr{C},\mathscr{R},k\}$ *is* **detailed balanced** *if the network* $\{\mathscr{S},\mathscr{C},\mathscr{R}\}$ *is reversible and there exists for the system a positive equilibrium at which detailed balancing obtains.*

By virtue of Theorem 14.2.1, a detailed balanced mass action system inherits all the dynamical properties of quasi-thermodynamic systems given by Theorem 13.4.4. Theorem 14.2.1 confers additional qualities as well. The theorem below provides a partial summary.

Theorem 14.2.3. *For a detailed balanced mass action system, there is within each positive stoichiometric compatibility class precisely one equilibrium; that equilibrium is asymptotically stable; and there is no nontrivial cyclic composition trajectory along which all species concentrations are positive. Moreover, at each positive equilibrium, the derivative of the species-formation-rate function has a characteristic polynomial that admits only real negative roots. In particular, the derivative is nonsingular.*

Remark 14.2.4 (Every reversible network is nondegenerate.). Theorem 14.2.3 provides a basis for the assertion that every reversible reaction network is nondegenerate in the sense of Definition 10.6.1: If $\{\mathscr{S},\mathscr{C},\mathscr{R}\}$ is a reversible network, we can certainly choose rate constants $k \in \mathbb{R}_+^{\mathscr{R}}$ such that the mass action system $\{\mathscr{S},\mathscr{C},\mathscr{R},k\}$ is detailed balanced. In fact, let $c^* \in \mathbb{R}_+^{\mathscr{S}}$ be any positive composition, and then choose $k \in \mathbb{R}_+^{\mathscr{R}}$ to satisfy, for each $y \to y' \in \mathscr{R}$, the equation

$$k_{y\to y'}(c^*)^y = k_{y'\to y}(c^*)^{y'}. \tag{14.25}$$

The resulting mass action system is detailed balanced at the equilibrium c^*, so the derivative of the species-formation-rate function at c^* is nonsingular.

14.3 Reversible Deficiency Zero Forest-Like Reaction Networks

Recall from Definition 6.7.4 that a reaction network is *forest-like* if, roughly speaking, for each pair of directly linked complexes y and y', removal of the reaction arrow(s) that directly link y and y' serves to disconnect the linkage class containing those complexes. Thus, network (13.28) is forest-like, but network (14.5) is not. Other examples were provided in Section 6.7. There we also asserted that a reversible network is forest-like precisely when, in the standard reaction diagram, there are no directed cycles traversing more than two complexes.

Now let $\{\mathscr{S},\mathscr{C},\mathscr{R},\mathscr{K}\}$ be a kinetic system in which the network $\{\mathscr{S},\mathscr{C},\mathscr{R}\}$ is reversible. We give $\{\mathscr{S},\mathscr{C},\mathscr{R}\}$ an orientation $\mathscr{R}^{\rightarrow} \subset \mathscr{R}$. Earlier in this chapter, we made the easy observation that if the reaction vector set

$$\{y' - y \in \mathbb{R}^{\mathscr{S}} : y \rightarrow y' \in \mathscr{R}^{\rightarrow}\} \tag{14.26}$$

is linearly independent, then detailed balancing *must* obtain at every equilibrium. Since (14.26) is a set of generators for the stoichiometric subspace, it will be independent precisely when the number of vectors in (14.26) is identical to the dimension of the stoichiometric subspace—that is, to the rank s of the network.

If $\{\mathscr{S},\mathscr{C},\mathscr{R}\}$ is forest-like, the number of vectors in (14.26) is exactly $n-\ell$, where n is the number of complexes and ℓ is the number of linkage classes.[1] Thus, if $\{\mathscr{S},\mathscr{C},\mathscr{R}\}$ is forest-like, then the set (14.26) is independent if and only if $n-\ell=s$ or, equivalently, *if and only if the network's deficiency* ($\delta := n-\ell-s$) *is zero.*

Remark 14.3.1. It is easy to see that if the network is *not* forest-like, then there is no hope that the set (14.26) can be independent. The presence of a not-necessarily-directed cycle in $\mathscr{R}^{\rightarrow}$ would obviate that.

Our conclusions are summarized in the following proposition.

Proposition 14.3.2. *Let* $\{\mathscr{S},\mathscr{C},\mathscr{R}\}$ *be a reversible forest-like deficiency zero network. Then for any kinetics* $\mathscr{K}$, *detailed balancing obtains at every equilibrium of the kinetic system* $\{\mathscr{S},\mathscr{C},\mathscr{R},\mathscr{K}\}$.

It is tempting to assert that, for a reversible forest-like deficiency zero network, any assignment of rate constants to the various reactions will give rise to a detailed balanced mass action system. However, we need be concerned about the *existence* of a positive equilibrium. The existence follows as a corollary of far more general (and far more difficult) theorems [55, 76, 107], but for the very narrow circumstances considered here, the proof is sufficiently simple that we should give it.

Proposition 14.3.3. *Let* $\{\mathscr{S},\mathscr{C},\mathscr{R}\}$ *be a reversible forest-like deficiency zero network. Then for any choice of* $k \in \mathbb{R}_+^{\mathscr{R}}$, *the mass action system* $\{\mathscr{S},\mathscr{C},\mathscr{R},k\}$ *is detailed balanced.*

Proof. Let $\mathscr{R}^{\rightarrow} \subset \mathscr{R}$ be an orientation for the network. For the stipulated rate constants, we want to show the existence of $c \in \mathbb{R}_+^{\mathscr{S}}$ at which detailed balancing obtains. That is, we want the existence of c that satisfies the system of equations

$$k_{y\rightarrow y'}\, c^{y} = k_{y'\rightarrow y}\, c^{y'}, \quad \forall y \rightarrow y' \in \mathscr{R}^{\rightarrow} \tag{14.27}$$

or, equivalently, a vector $\ln c \in \mathbb{R}^{\mathscr{S}}$ that satisfies the system

$$(y' - y)\cdot \ln c = \ln \frac{k_{y\rightarrow y'}}{k_{y'\rightarrow y}}, \quad \forall y \rightarrow y' \in \mathscr{R}^{\rightarrow}. \tag{14.28}$$

[1] This follows from a standard result in graph theory, relating the number of edges in a forest to the number of vertices and the number of connected components.

Because the network is forest-like and has a deficiency of zero, we know that the set

$$\{y' - y \in \mathbb{R}^{\mathscr{S}} : y \to y' \in \mathscr{R}^{\rightarrow}\} \tag{14.29}$$

is independent. It is a standard fact of linear algebra that if, in a vector space with scalar product, $\{v_1, v_2, \ldots, v_k\}$ is a linearly independent set and $\{\alpha_1, \alpha_2, \ldots, \alpha_k\}$ is a set of scalars, then the set of equations $v_i \cdot \xi = \alpha_i$, $i = 1, 2, \ldots, k$ admits a solution. Thus, there is a vector $\ln c \in \mathbb{R}^{\mathscr{S}}$ that is a solution of (14.28). As a result, $c := \exp(\ln c)$ is the desired positive composition at which detailed balancing obtains. □

Taken together, the various facts we have accumulated in this chapter provide the basis for a version of the Deficiency Zero Theorem restricted to reversible forest-like reaction networks.

Theorem 14.3.4 (Deficiency Zero Theorem Restricted to Reversible Forest-like Networks). *Let $\{\mathscr{S}, \mathscr{C}, \mathscr{R}\}$ be a reversible forest-like deficiency zero reaction network with stoichiometric subspace S. For any assignment of rate constants $k \in \mathbb{R}_{+}^{\mathscr{R}}$, the resulting mass action system $\{\mathscr{S}, \mathscr{C}, \mathscr{R}, k\}$ has the following properties:*

(i) There is within each positive stoichiometric compatibility class precisely one equilibrium; that equilibrium is asymptotically stable; and there is no nontrivial cyclic composition trajectory along which all species concentrations are positive.

(ii) If c^* is a positive equilibrium, then the derivative $df(c^*) : S \to S$ of the species-formation-rate function at c^* is nonsingular. In fact, relative to the scalar product "$*$" in $\mathbb{R}^{\mathscr{S}}$ given by

$$u * w := \sum_{s \in \mathscr{S}} \frac{u_s w_s}{c_s^*}, \tag{14.30}$$

$df(c^*)$ is symmetric and negative-definite. As a result, all roots of the characteristic polynomial of $df(c^*)$ are real and negative, and there exists in S a $*$-orthonormal basis consisting of eigenvectors of $df(c^*)$.

Remark 14.3.5. This version of the Deficiency Zero Theorem is already fairly broad in its scope, but it fails to embrace deficiency zero weakly reversible networks (in particular reversible networks) that are not forest-like. The very simple triangle (14.5) provides an example.

Nevertheless, Theorem 14.3.4 gives properties of forest-like deficiency zero networks that are not generally shared by deficiency zero networks that are not forest-like. For example, there are mass action rate constants for network (14.5) for which, at a positive equilibrium c^*, the characteristic polynomial of $df(c^*)$ has non-real roots (but with negative real parts).

14.4 Necessary and Sufficient Conditions for Detailed Balancing in Mass Action Systems: Rate Constant Orchestration

We know that for any reversible deficiency zero forest-like network, *every* assignment of rate constants results in a mass action system that is detailed balanced. On the other hand, we also know that there are examples of reaction networks, such as (14.5) and (14.9), for which mass action detailed balancing will result only when the rate constants are tightly orchestrated. It is the purpose of this section to relate that orchestration precisely to reaction network structure. That is, for a reversible reaction network $\{\mathscr{S},\mathscr{C},\mathscr{R}\}$, we provide necessary and sufficient conditions that a rate constant assignment $k \in \mathbb{R}_+^{\mathscr{R}}$ must satisfy in order that the mass action system $\{\mathscr{S},\mathscr{C},\mathscr{R},k\}$ be detailed balanced. The proof is given in [75].

Hereafter in this section $\{\mathscr{S},\mathscr{C},\mathscr{R}\}$ is a reversible network. As usual, we denote by n the number of complexes, by ℓ the number of linkage classes, by s the rank, and by $\delta := n - \ell - s$ the deficiency. Because the network is reversible, its reactions occur in pairs, and we shall often refer to this or that reaction pair. We denote by p the number of reaction pairs, which is just half the number of reactions in $\mathscr{R}$.

The orchestration of rate constants equivalent to detailed balancing will proceed by way of two distinct sets of constraining equations, called here the *independent cycle conditions* and the *spanning forest conditions*. Both derive from a single choice of a "spanning forest" for the reaction network under consideration.

14.4.1 Spanning Forests in the Standard Reaction Diagram

In the network $\{\mathscr{S},\mathscr{C},\mathscr{R}\}$ there will be *at least* $n-\ell$ reaction pairs. For the purposes of this section,[2] a *spanning forest* for $\{\mathscr{S},\mathscr{C},\mathscr{R}\}$ is a reversible subnetwork $\{\mathscr{S},\mathscr{C},\mathscr{F}\}$, perhaps identical to $\{\mathscr{S},\mathscr{C},\mathscr{R}\}$, having the same n complexes, the same ℓ linkage classes, and precisely $n-\ell$ reaction pairs. Each spanning forest will have the same rank and deficiency as the original network.

Some toy examples will be helpful. We begin with network (14.31), for which $n = 9$, $\ell = 3$, $s = 6$, and $\delta = 9 - 3 - 6 = 0$. There are nine reaction pairs, so $p = 9$, and $p > n - \ell$.

$$\begin{array}{ccc ccc ccc}
A & \rightleftarrows & B+C & & A+D & \rightleftarrows & E & & A+E \rightleftarrows H \\
& \searrow\!\!\nwarrow \quad \swarrow\!\!\nearrow & & & \downarrow\uparrow & \swarrow\!\!\nearrow & \uparrow\downarrow & & \\
& D & & & G & \leftrightarrows & 2F & &
\end{array} \qquad (14.31)$$

A choice of spanning forest, indicated by red arrows, is shown in (14.32). The linkage classes remain what they were before: $\{A,\ B+C,\ D\}$, $\{A+D,\ E,\ 2F,\ G\}$, and $\{A+E,\ H\}$, but now $p = n - \ell = 6$.

$$\begin{array}{ccc ccc ccc}
A & \rightleftarrows & B+C & & A+D & \rightleftarrows & E & & A+E \rightleftarrows H \\
& \searrow\!\!\nwarrow \quad \swarrow\!\!\nearrow & & & \downarrow\uparrow & \swarrow\!\!\nearrow & \uparrow\downarrow & & \\
& D & & & G & \leftrightarrows & 2F & &
\end{array} \qquad (14.32)$$

For reasons that will become apparent later, a far richer example is afforded by network (14.33), a variant of the Horn and Jackson network considered in

[2] In this section we are temporarily viewing a reversible reaction network's standard reaction diagram as a graph having the complexes as its vertices and reaction *pairs* $\rightleftarrows$ as its undirected edges.

Section 5.1.1. Here $n = 4$, $\ell = 2$, $s = 1$, and $\delta = 4 - 1 - 1 = 2$. In this case $p = 5$. In (14.34) we show a choice of a spanning forest.

$$\begin{array}{ccc} 3A & \rightleftarrows & A+2B \\ \downarrow\uparrow & \nwarrow\searrow & \uparrow\downarrow \\ 2A+B & \leftrightarrows & 3B \end{array} \tag{14.33}$$

$$\begin{array}{ccc} 3A & \rightleftarrows & A+2B \\ \downarrow\uparrow & \nwarrow\searrow & \uparrow\downarrow \\ 2A+B & \leftrightarrows & 3B \end{array} \tag{14.34}$$

Note that for a network with $p > n - \ell$, a particular choice of spanning forest will omit $p - (n - \ell)$ reaction pairs of the original network. These are the *out-of-forest reaction pairs* relative to the particular choice of spanning forest. Relative to the spanning forest (14.32) for network (14.31), the out-of-forest reaction pairs in (14.31) are $D \rightleftarrows B + C$, $G \rightleftarrows E$, and $E \rightleftarrows 2F$. Relative to the spanning forest (14.34) for network (14.33), the out-of-forest reaction pairs are $3A \rightleftarrows 2A + B$ and $A + 2B \rightleftarrows 3B$.

Remark 14.4.1. For the reversible reaction network $\{\mathscr{S}, \mathscr{C}, \mathscr{R}\}$ under consideration in this section, it will be understood that a particular choice $\{\mathscr{S}, \mathscr{C}, \mathscr{F}\}$ of spanning forest has been made. When we consider network (14.31), it will be understood that the spanning forest choice is (14.32), and for network (14.33) the spanning forest choice is (14.34).

14.4.2 Independent Cycle Conditions

If, for the reversible reaction network $\{\mathscr{S}, \mathscr{C}, \mathscr{R}\}$, p exceeds $n - \ell$, the standard reaction diagram will contain at least one cycle—that is, a reversible subnetwork having a circular structure. In more precise terms, by a *cycle* in $\{\mathscr{S}, \mathscr{C}, \mathscr{R}\}$ we shall mean (in this section[3]) a connected reversible subnetwork in which each complex is directly linked by reversible reaction pairs to precisely two other complexes. The cycles in network (14.31) are displayed in (14.35).

$$\begin{array}{ccc} A & \rightleftarrows & B+C \\ & \nwarrow\searrow \quad \nearrow\swarrow & \\ & D & \end{array} \qquad \begin{array}{ccc} A+D & \rightleftarrows & E \\ \downarrow\uparrow & & \uparrow\downarrow \\ G & \leftrightarrows & 2F \end{array} \qquad \begin{array}{cc} A+D & \rightleftarrows E \\ \downarrow\uparrow & \nearrow\swarrow \\ G & \end{array} \qquad \begin{array}{ccc} & & E \\ & \nearrow\swarrow & \uparrow\downarrow \\ G & \leftrightarrows & 2F \end{array} \tag{14.35}$$

A particular choice of a spanning forest will give rise to a set of $p - (n - \ell)$ *independent cycles* obtained in the following way: Corresponding to each out-of-forest reaction pair, choose in the network the (unique) cycle containing that reaction pair

[3] Recall that here we are regarding the standard reaction diagram of $\{\mathscr{S}, \mathscr{C}, \mathscr{R}\}$ to be a graph having the complexes as its vertices and the reaction pairs $\rightleftarrows$ as its edges. Thus, a reaction pair $y \rightleftarrows y'$ is not a cycle.

but no other out-of-forest reaction pair. Thus, for network (14.31) with spanning forest chosen as in (14.32), the $p-(n-\ell)=9-(9-3)=3$ independent cycles are shown in (14.36). In each cycle the out-of-forest reaction pair is green, and the in-forest reaction pairs are red.

$$
\begin{array}{ccc}
A & \rightleftarrows & B+C \\
 \searrow\!\!\nwarrow & & \swarrow\!\!\nearrow \\
 & D &
\end{array}
\qquad
\begin{array}{ccc}
A+D & \rightleftarrows & E \\
\downarrow\!\uparrow & & \uparrow\!\downarrow \\
G & \leftrightarrows & 2F
\end{array}
\qquad
\begin{array}{ccc}
A+D & \rightleftarrows & E \\
\downarrow\!\uparrow & \swarrow\!\!\nearrow & \\
G & &
\end{array}
\tag{14.36}
$$

In the case of network (14.33) with spanning forest chosen as in (14.37), the $p-(n-\ell)=5-(4-1)=2$ independent cycles are shown in (14.37). Again, the out-of-forest reaction pair is green, and the in-forest reaction pairs are red.

$$
\begin{array}{ccc}
3A & & \\
\downarrow\!\uparrow & \nwarrow\!\!\searrow & \\
2A+B & \leftrightarrows & 3B
\end{array}
\qquad
\begin{array}{ccc}
3A & \rightleftarrows & A+2B \\
 & \nwarrow\!\!\searrow & \uparrow\!\downarrow \\
 & & 3B
\end{array}
\tag{14.37}
$$

We are ready to indicate how, for a reversible reaction network containing cycles, the detailed balancing cycle conditions on rate constants are constructed. We suppose that a spanning forest has been chosen and that, relative to that spanning forest, the $p-(n-\ell)$ independent cycles have been isolated. Then, for each such cycle, we write an equation which asserts that the product of rate constants associated with the clockwise direction is equal to the product of the rate constants associated with the counterclockwise direction.[4] *The resulting $p-(n-\ell)$ equations are the detailed balance cycle conditions for the network (relative to the choice of spanning forest).*

Example 14.4.2. For network (14.31) with spanning forest chosen as in (14.32), giving rise to the independent cycles (14.36), the detailed balance cycle conditions are displayed in (14.38).

$$
\begin{aligned}
k_{A\to B+C}\,k_{B+C\to D}\,k_{D\to A} &= k_{A\to D}\,k_{D\to B+C}\,k_{B+C\to A} \\
k_{A+D\to E}\,k_{E\to 2F}\,k_{2F\to G}\,k_{G\to A+D} &= k_{A+D\to G}\,k_{G\to 2F}\,k_{2F\to E}\,k_{E\to A+D} \\
k_{A+D\to E}\,k_{E\to G}\,k_{G\to A+D} &= k_{A+D\to G}\,k_{G\to E}\,k_{E\to A+D}
\end{aligned}
\tag{14.38}
$$

Example 14.4.3. For network (14.33) with spanning forest chosen as in (14.34), giving rise to the independent cycles (14.37), the detailed balance cycle conditions are displayed in (14.39).

$$
\begin{aligned}
k_{3A\to 3B}\,k_{3B\to 2A+B}\,k_{2A+B\to 3A} &= k_{3A\to 2A+B}\,k_{2A+B\to 3B}\,k_{3B\to 3A} \\
k_{3A\to A+2B}\,k_{A+2B\to 3B}\,k_{3B\to 3A} &= k_{3A\to 3B}\,k_{3B\to A+2B}\,k_{A+2B\to 3A}
\end{aligned}
\tag{14.39}
$$

Remark 14.4.4. If, for the set of independent cycles, the clockwise and counterclockwise rate constant products are equal, then those products will be equal on all other cycles, not just the independent cycles generated by the choice of a spannning forest.

[4] The assignment to a cycle of clockwise and oppositely oriented counterclockwise directions is arbitrary and inconsequential.

14.4.3 Spanning Forest Conditions

For reaction networks having deficiency greater than zero, the $p-(n-\ell)$ cycle conditions constitute only part of the constraints on rate constants that are necessary and sufficient for a mass action system to be detailed balanced. In fact, for a reversible network of deficiency δ, there will be δ more equations, called the *detailed balance spanning forest conditions*, that complete the rate-constant-orchestration requirements.

We suppose that $\{\mathscr{S},\mathscr{C},\mathscr{R}\}$ is a reversible reaction network of deficiency $\delta > 0$ and that, for the network, $\{\mathscr{S},\mathscr{C},\mathscr{F}\}$ is a fixed but arbitrary choice of spanning forest. Thus, $\mathscr{F}$ contains $n-\ell$ reversible reaction pairs taken from $\mathscr{R}$. We now give the spanning forest an *orientation* by selecting from each reversible reaction pair in $\mathscr{F}$ a "forward" reaction. This amounts to a choice of a subset $\mathscr{F}^{\rightarrow} \subset \mathscr{F}$ to be regarded as the spanning forest's $n-\ell$ forward reactions. The subnetwork $\{\mathscr{S},\mathscr{C},\mathscr{F}^{\rightarrow}\}$ is an *oriented spanning forest* for the original network $\{\mathscr{S},\mathscr{C},\mathscr{R}\}$. For example, in consideration of network (14.33), we can give the spanning forest (14.34) the orientation shown in (14.40).

$$\begin{array}{ccc} 3A & \rightarrow & A+2B \\ & & \searrow \\ 2A+B & \leftarrow & 3B \end{array} \tag{14.40}$$

The detailed balance spanning forest conditions are then formulated in the following way: For the equation

$$\sum_{y\rightarrow y' \in \mathscr{F}^{\rightarrow}} \alpha_{y\rightarrow y'}(y'-y) = 0 \tag{14.41}$$

we find δ independent solutions.[5] For each such solution $\{\alpha_{y\rightarrow y'}\}_{y\rightarrow y' \in \mathscr{F}^{\rightarrow}}$, we write a rate constant constraint of the form

$$\prod_{y\rightarrow y' \in \mathscr{F}^{\rightarrow}} (k_{y\rightarrow y'})^{\alpha_{y\rightarrow y'}} = \prod_{y\rightarrow y' \in \mathscr{F}^{\rightarrow}} (k_{y'\rightarrow y})^{\alpha_{y\rightarrow y'}}. \tag{14.42}$$

The resulting δ equations are the *detailed balance spanning forest conditions* for the network.

Example 14.4.5. Consider the deficiency *two* network (14.33), taken with the oriented spanning forest (14.40). In this case, (14.41) takes the form

$$\begin{aligned} &\alpha_{3A\rightarrow A+2B}\,(A+2B-3A) + \alpha_{3A\rightarrow 3B}(3B-3A) + \alpha_{3B\rightarrow 2A+B}(2A+B-3B) \\ &= (2\alpha_{3A\rightarrow A+2B} + 3\alpha_{3A\rightarrow 3B} - 2\alpha_{3B\rightarrow 2A+B})(B-A) \\ &= 0. \end{aligned} \tag{14.43}$$

[5] By a solution we mean of course a set of numbers $\{\alpha_{y\rightarrow y'}\}_{y\rightarrow y' \in \mathscr{F}^{\rightarrow}}$ (or, equivalently, a vector $\alpha \in \mathbb{R}^{\mathscr{F}^{\rightarrow}}$) that satisfies (14.41). We say that a set of solutions is independent if the set is independent, viewed as vectors in $\mathbb{R}^{\mathscr{F}^{\rightarrow}}$. That there will be an independent set of δ solutions (but not more) is a consequence of elementary considerations in linear algebra.

An independent set of two solutions is given by

$$\alpha_{3A\to A+2B} = 1, \quad \alpha_{3A\to 3B} = 0, \quad \alpha_{3B\to 2A+B} = 1 \tag{14.44}$$

and

$$\alpha_{3A\to A+2B} = 0, \quad \alpha_{3A\to 3B} = 2, \quad \alpha_{3B\to 2A+B} = 3. \tag{14.45}$$

These give rise to the spanning forest conditions

$$\begin{aligned} k_{3A\to A+2B}k_{3B\to 2A+B} &= k_{A+2B\to 3A}k_{2A+B\to 3B} \\ (k_{3A\to 3B})^2(k_{3B\to 2A+B})^3 &= (k_{3B\to 3A})^2(k_{2A+B\to 3B})^3. \end{aligned} \tag{14.46}$$

14.4.4 The Detailed Balance Rate Constant Theorem

We are now prepared to state conditions on rate constants that are both necessary and sufficient to ensure detailed balancing in mass action systems.

Theorem 14.4.6 (The Detailed Balance Rate Constant Theorem [75]**).** *Consider a reversible reaction network of deficiency δ in which there are p reaction pairs, n complexes, and ℓ linkage classes. An assignment of rate constants gives rise to a detailed balanced mass action system if and only if, for a fixed but arbitrary choice of spanning forest, those rate constants satisfy the $p-(n-\ell)$ independent cycle conditions and the δ spanning forest conditions that the spanning forest induces. (The spanning forest conditions can be formulated for any choice of forest orientation and for any choice of independent solutions of (14.41).)*

Example 14.4.7. Network (14.31) has a deficiency of zero ($n=9, \ell=3, s=6$), so there are no spanning forest conditions. There are, however, independent cycle conditions that the rate constants must satisfy. The number of these is $p-(n-\ell)=9-(9-3)=3$. Three independent cycle conditions were given in (14.38).

Example 14.4.8. For the deficiency two network (14.33), there were two spanning forest conditions that the rate constants must satisfy, given by (14.46). There were two independent cycle conditions given by (14.39). It is instructive to display a full set of rate constant constraints that are necessary and sufficient for detailed balancing:

$$\begin{aligned} k_{3A\to 3B}\,k_{3B\to 2A+B}\,k_{2A+B\to 3A} &= k_{3A\to 2A+B}\,k_{2A+B\to 3B}\,k_{3B\to 3A} \\ k_{3A\to A+2B}\,k_{A+2B\to 3B}\,k_{3B\to 3A} &= k_{3A\to 3B}\,k_{3B\to A+2B}\,k_{A+2B\to 3A} \\ k_{3A\to A+2B}k_{3B\to 2A+B} &= k_{A+2B\to 3A}k_{2A+B\to 3B} \\ (k_{3A\to 3B})^2(k_{3B\to 2A+B})^3 &= (k_{3B\to 3A})^2(k_{2A+B\to 3B})^3. \end{aligned} \tag{14.47}$$

For the seemingly simple two-species network (14.33), the rate constant orchestration required for detailed balancing is somewhat elaborate.

In the next two examples, we return to elementary but instructive networks considered in Section 14.1. In both cases we already articulated necessary and sufficient conditions that rate constants must satisfy in order that there be detailed balancing. Now, however, we can see how those conditions emerge systematically from Theorem 14.4.6.

Example 14.4.9. The simple triangular network (14.48) has a deficiency of zero ($n = 3, \ell = 1, s = 2$), so there are no spanning forest conditions that the rate constants must satisfy. There is however a single independent cycle ($p - (n - \ell) = 3 - (3 - 1) = 1$) that gives rise to the rate constant constraint (14.49). This is precisely the sole constraint indicated in Section 14.1 as necessary and sufficient for mass action detailed balancing.

$$\begin{array}{ccccc} A & & \rightleftarrows & & B \\ & \nwarrow\!\!\searrow & & \nearrow\!\!\swarrow & \\ & & C & & \end{array} \tag{14.48}$$

$$k_{A\to B}\, k_{B\to C}\, k_{C\to A} = k_{A\to C}\, k_{C\to B}\, k_{B\to A} \tag{14.49}$$

Example 14.4.10. Recall the network (14.50) studied by Wegscheider. Here there are no cycles—in fact, $p - (n - \ell) = 3 - (6 - 3) = 0$—so there are no independent cycle conditions that rate constants must satisfy in order for detailed balancing to obtain.

$$\begin{array}{c} A \rightleftarrows B \qquad C \rightleftarrows D \\ A + C \rightleftarrows B + D \end{array} \tag{14.50}$$

However, the network has a deficiency of *one* ($n = 6, \ell = 3, s = 2$), so there will be a single spanning forest constraint. Network (14.50) is already a spanning forest for itself, one we can choose to orient as in (14.51).

$$\begin{array}{c} A \to B \qquad C \to D \\ A + C \to B + D. \end{array} \tag{14.51}$$

In this case, (14.41) takes the form

$$\alpha_{A\to B}\,(B - A) + \alpha_{C\to D}\,(D - C) + \alpha_{A+C\to B+D}\,(B + D - A - C) = 0, \tag{14.52}$$

a solution of which is $\alpha_{A\to B} = 1$, $\alpha_{C\to D} = 1$, $\alpha_{A+C\to B+D} = -1$. For this choice of a solution, the lone spanning forest condition (14.42) becomes

$$\frac{k_{A\to B} k_{C\to D}}{k_{A+C\to B+D}} = \frac{k_{B\to A} k_{D\to C}}{k_{B+D\to A+C}}. \tag{14.53}$$

This is the orchestration of rate constants we asserted in Section 14.1 to be necessary and sufficient for mass action detailed balancing.

Remark 14.4.11. There is a less nuanced (and far less structure-related) way to formulate conditions on rate constants that are necessary and sufficient for mass action detailed balancing [75].[6] Suppose that $\{\mathscr{S},\mathscr{C},\mathscr{R}\}$ is a reversible reaction network having p reaction pairs and rank s. We give the full network an orientation by choosing for each reversible reaction pair a forward direction. In this way we are selecting a subset $\mathscr{R}^{\rightarrow} \subset \mathscr{R}$ containing p reactions. Next, we consider the equation

$$\sum_{y \to y' \in \mathscr{R}^{\rightarrow}} \alpha_{y \to y'}(y' - y) = 0, \tag{14.54}$$

and we find for it a linearly independent set of $m := p - s$ solutions, $\{\alpha^1, \alpha^2, \ldots, \alpha^m\}$ in $\mathbb{R}^{\mathscr{R}^{\rightarrow}}$. Then the mass action system $\{\mathscr{S},\mathscr{C},\mathscr{R},k\}$ is detailed balanced if and only if $k \in \mathbb{R}_+^{\mathscr{R}}$ satisfies the m equations

$$\prod_{y \to y' \in \mathscr{R}^{\rightarrow}} (k_{y \to y'})^{\alpha^\theta_{y \to y'}} = \prod_{y \to y' \in \mathscr{R}^{\rightarrow}} (k_{y' \to y})^{\alpha^\theta_{y \to y'}}, \quad \theta = 1, 2, \ldots, m. \tag{14.55}$$

Theorem 14.2.1 is, in effect, an elaboration of this procedure, one that connects rate constant constraints to two quite separate aspects of a network's standard reaction diagram: (i) its cycle structure, which is entirely separate from the algebraic nature of the complexes, and (ii) its deficiency, which is largely independent of the diagram's fine graphical structure, in particular its cycle structure. (Recall Remark 6.3.3.)

[6] As I indicated in [75], I think I learned the less structure-related conditions formulated in this remark from F. Horn.

15

Complex Balancing

We come now to the idea of complex balancing, which serves as a precursor to the Deficiency Zero Theorem. It is an idea that was hardly obvious at the time that Horn and Jackson [109] proposed it and examined its deep consequences. At the very least, it required that they take seriously a reaction network's standard diagram as a directed graph. This might seem natural now, especially in the context of this book. But hardly obvious in 1972 was the notion that the *vertices* (complexes) of that graph are important entities with a mathematical life of their own, having flows into and out of them—flows that, at an equilibrium, might or might not be in balance. It is a wonder that Horn and Jackson could first imagine and then prove that the presence of such balance has profound ramifications.

15.1 Complex Balancing in Kinetic Systems

In Section 7.7 we described complex balancing briefly. Here we make the definition formal. For a reaction network $\{\mathscr{S},\mathscr{C},\mathscr{R}\}$, not necessarily reversible, we denote by $\mathscr{R}_{\to y}$ (respectively, $\mathscr{R}_{y\to}$) the set of all reactions having y as the product (reactant) complex.

Definition 15.1.1. *For a kinetic system* $\{\mathscr{S},\mathscr{C},\mathscr{R},\mathscr{K}\}$ *we say that* **complex balancing** *obtains at composition* $c \in \overline{\mathbb{R}}_+^{\mathscr{S}}$ *if*

$$\sum_{y'\to y\,\in\,\mathscr{R}_{\to y}} \mathscr{K}_{y'\to y}(c) = \sum_{y\to y'\,\in\,\mathscr{R}_{y\to}} \mathscr{K}_{y\to y'}(c), \quad \forall y \in \mathscr{C}. \tag{15.1}$$

This is to say that there is complex balancing at composition c if, for every complex y, the sum of the rates at c of all reactions having y as the product complex is exactly equal to the sum of the rates at c of all reactions having y as the reactant complex.

Remark 15.1.2 (Preview: Two striking fact about complex balancing). We can preview here two striking facts about complex balancing in general kinetic systems that

M. Feinberg, *Foundations of Chemical Reaction Network Theory*,
Applied Mathematical Sciences 202,
https://doi.org/10.1007/978-3-030-03858-8_15

will be proved later on. Both play a strong role in the proof of the Deficiency Zero Theorem. Consider a kinetic system $\{\mathscr{S},\mathscr{C},\mathscr{R},\mathscr{K}\}$. First, *complex balancing can obtain at a positive composition only if the network* $\{\mathscr{S},\mathscr{C},\mathscr{R}\}$ *is weakly reversible* [107]. Second, *if the network* $\{\mathscr{S},\mathscr{C},\mathscr{R}\}$ *has a deficiency of zero, then complex balancing obtains at all equilibria* [67].

Remark 15.1.3 (A piece of the Deficiency Zero Theorem). Taken together, the two assertions made in Remark 15.1.2 give rise to a corollary: *For a kinetic system in which the underlying reaction network has a deficiency of zero and is not weakly reversible, there can be no positive equilibrium. That is, in any equilibrium one or more species must be absent.* Note that this corollary makes no mention of complex balancing. Rather, complex balancing acts as a conceptual tool in the underlying argument.

Remark 15.1.4. The weak reversibility requirement for complex balancing at positive equilibria is more subtle than might first appear. It is easy to see that complex balancing at a positive equilibrium requires that each complex act in the network as both a reactant complex in some reaction *and* as a product complex in a different reaction. This, however, does not imply weak reversibility; network (15.2) is a counterexample.

$$\begin{array}{ccccc} A & \rightleftarrows & B & \rightarrow & C \\ \upharpoonleft\downharpoonright & & \downharpoonleft\upharpoonright & & \upharpoonleft\downharpoonright \\ D & \rightleftarrows & E & \rightarrow & F \end{array} \tag{15.2}$$

Remark 15.1.5 (Detailed balancing and complex balancing). For a reversible network, it is not difficult to see that detailed balancing at a composition implies complex balancing and, if the network is forest-like, that complex balancing implies detailed balancing. If a network is not forest-like, then there can be complex balancing at a composition even when detailed balancing does not obtain there. Indeed, there can be complex balancing at a positive composition even when the network is not reversible, so long as it is weakly reversible. For a discussion of other relationships between detailed and complex balancing with special reference to mass action kinetics, see an article by Dickenstein and Pérez Millán [56].

It is perhaps intuitively apparent that complex balancing can obtain only at equilibria. To see this in a formal way, it suffices to rearrange[1] the species-formation-rate function for the kinetic system $\{\mathscr{S},\mathscr{C},\mathscr{R},\mathscr{K}\}$ as in (15.3). Clearly, $f(c) = 0$ when complex balancing obtains at c.

$$\begin{aligned} f(c) &:= \sum_{\mathscr{R}} \mathscr{K}_{y\to y'}(c)\,(y'-y) \\ &= \sum_{y\in\mathscr{C}} \Big[\, \sum_{\mathscr{R}_{\to y}} \mathscr{K}_{y'\to y}(c) - \sum_{\mathscr{R}_{y\to}} \mathscr{K}_{y\to y'}(c)\,\Big]\, y \end{aligned} \tag{15.3}$$

[1] The symbols $\mathscr{R}_{\to y}$ and $\mathscr{R}_{y\to}$ below summations are understood to be abbreviations for $y'\to y \in \mathscr{R}_{\to y}$ and $y \to y' \in \mathscr{R}_{y\to}$.

Remark 15.1.6 (The complex-formation-rate function). For use later on, we shall find it convenient to have available, for a kinetic system $\{\mathscr{S},\mathscr{C},\mathscr{R},\mathscr{K}\}$, the functions $g_y(\cdot) : \overline{\mathbb{R}}_+^{\mathscr{S}} \to \mathbb{R}$ defined by

$$g_y(c) := \sum_{\mathscr{R}_{\to y}} \mathscr{K}_{y' \to y}(c) - \sum_{\mathscr{R}_{y \to}} \mathscr{K}_{y \to y'}(c), \quad \forall y \in \mathscr{C}. \tag{15.4}$$

If, for each $y \to y' \in \mathscr{R}$ and each composition $c \in \overline{\mathbb{R}}_+^{\mathscr{S}}$, we view $\mathscr{K}_{y \to y'}(c)$ as a "flow" in the standard reaction diagram from complex y to complex y', then $g_{\bar{y}}(c)$ is the net flow to complex $\bar{y}$. We have complex balancing at composition c when $g_y(c) = 0$ for every $y \in \mathscr{C}$.

We will also find useful a vector formulation: By $g(\cdot) : \overline{\mathbb{R}}_+^{\mathscr{S}} \to \mathbb{R}^{\mathscr{C}}$ we mean the function defined by

$$g(c) := \sum_{y \in \mathscr{C}} g_y(c)\omega_y. \tag{15.5}$$

Recall from Section 2.2.2 that $\{\omega_y\}_{y \in \mathscr{C}}$ is the standard basis for $\mathbb{R}^{\mathscr{C}}$. Note that we have complex balancing at composition c precisely when $g(c) = 0$.

Although we shall refer to $g_y(\cdot)$ as the "complex-formation-rate function" for complex y, this is somewhat misleading. The complex y is an abstract notion, not a physical entity (such as a molecule of a particular species) that is produced or consumed with the occurrence of chemical reactions.

15.2 Consequences of Complex Balancing in Mass Action Systems

To illustrate the greatly extended range that the complex balancing idea provides, it suffices to revisit the simple triangular network (15.6). In Theorem 14.2.1 we established that any mass action system that gives rise to a *detailed balanced* positive equilibrium is, among other things, quasi-thermodynamic. The quasi-thermodynamic property, in turn, ensures largely dull behavior (Theorem 13.4.4).

$$\begin{array}{ccc} A & \rightleftarrows & B \\ & \nwarrow\!\!\searrow \quad \swarrow\!\!\nearrow & \\ & C & \end{array} \tag{15.6}$$

Recall, however, that mass action detailed balancing obtains for network (15.6) only when the rate constants satisfy the highly constraining condition

$$k_{A\to B}\, k_{B\to C}\, k_{C\to A} = k_{A\to C}\, k_{C\to B}\, k_{B\to A}. \tag{15.7}$$

Theorem 14.2.1 tells us nothing about the presence of the quasi-thermodynamic property for any rate constant assignments other than ones narrowly circumscribed by (15.7).

Horn and Jackson not only conceived the idea of complex balancing—a substantially weaker condition than detailed balancing—they also showed in [109] that *the quasi-thermodynamic property is present in all members of the far larger class of mass action systems in which there is a positive equilibrium at which complex balancing obtains*. See Theorem 15.2.2 below.

In the case of network (15.6), it is not difficult to see that, *no matter what values the rate constants take*, the resulting mass action differential equations admit positive equilibria and that at all such equilibria *complex balancing* obtains.[2] Thus, for *all* assignments of rate constants to (15.6)—*not just those satisfying* (15.7)—the resulting mass action system is quasi-thermodynamic and has the dull dynamical qualities that the quasi-thermodynamic property imposes.

Remark 15.2.1. It is important to remember that for us (and for Horn and Jackson in [109]), kinetic systems—and mass action systems in particular—are viewed as models for physicochemical systems of great variety, not just for closed reactors in which detailed balancing at equilibria might be deemed mandatory by virtue of statistical-mechanical arguments. It is for this reason that we attach considerable importance to quasi-thermodynamicity that occurs for rate constants beyond those consistent with detailed balancing.

Parts (i) and (ii) of Theorem 15.2.2 are contained in [109], but the proof we will give [70, 71] is considerably simpler than the earlier one provided by Horn and Jackson. Their proof, which has its own attractions, is more graph-theoretical in spirit, involving decomposition of a network into cycles. In the theorem statement, S is as usual the stoichiometric subspace of the network $\{\mathscr{S},\mathscr{C},\mathscr{R}\}$, $f : \overline{\mathbb{R}}_+^{\mathscr{S}} \to S$ is the species-formation-rate function of the mass action system $\{\mathscr{S},\mathscr{C},\mathscr{R},k\}$, and $df(c) : S \to S$ is the derivative of $f(\cdot)$ at $c \in \mathbb{R}_+^{\mathscr{S}}$.

Theorem 15.2.2. *If, for $k \in \mathbb{R}_+^{\mathscr{R}}$, the mass action system $\{\mathscr{S},\mathscr{C},\mathscr{R},k\}$ admits a positive equilibrium at which complex balancing obtains, then the following hold true:*

(i) The mass action system $\{\mathscr{S},\mathscr{C},\mathscr{R},k\}$ is quasi-thermodynamic.
(ii) Complex balancing obtains at every positive equilibrium, regardless of stoichiometric compatibility class.
(iii) If $c \in \mathbb{R}_+^{\mathscr{S}}$ is an equilibrium, then $df(c) : S \to S$ is negative-definite relative to the scalar product "$*$" in $\mathbb{R}^{\mathscr{S}}$ given by

$$v * w := \sum_{s \in \mathscr{S}} \frac{v_s w_s}{c_s}, \quad \forall v, w \in \mathbb{R}^{\mathscr{S}}. \tag{15.8}$$

As a result, all roots of the characteristic polynomial of $df(c)$ have negative real part. In particular $df(c)$ is nonsingular, and the equilibrium c is hyperbolic.

[2] As we shall see, this is true of all networks which, like network (15.6), are weakly reversible and have a deficiency of zero [67, 107].

Proof. Throughout the proof we will suppose that, for the mass action system, $\{\mathscr{S},\mathscr{C},\mathscr{R},k\}$, $c^* \in \mathbb{R}_+^{\mathscr{S}}$ is an equilibrium at which complex balancing obtains. Thus, we have

$$\sum_{\mathscr{R}_{\to y}} k_{y'\to y}(c^*)^{y'} - \sum_{\mathscr{R}_{y\to}} k_{y\to y'}(c^*)^{y} = 0, \quad \forall y \in \mathscr{C}. \tag{15.9}$$

Hereafter we let $\mu : \mathbb{R}_+^{\mathscr{S}} \to \mathbb{R}^{\mathscr{S}}$ be defined by

$$\mu(c) := \ln c - \ln c^*. \tag{15.10}$$

To prove (i) we note that, for every $c \in \mathbb{R}_+^{\mathscr{S}}$,

$$\begin{aligned}(\ln c - \ln c^*) \cdot f(c) &= \sum_{y\to y' \in \mathscr{R}} k_{y\to y'} c^y (y' - y) \cdot \mu(c) \qquad (15.11)\\ &= \sum_{y\to y' \in \mathscr{R}} k_{y\to y'}(c^*)^y e^{y\cdot\mu(c)} (y'\cdot\mu(c) - y\cdot\mu(c)).\end{aligned}$$

We recall an elementary property of the exponential function:

$$e^x(x' - x) \le e^{x'} - e^x, \tag{15.12}$$

with equality holding if and only if $x' = x$. Thus, from (15.11) we can write

$$\begin{aligned}(\ln c - \ln c^*) \cdot f(c) &\le \sum_{y\to y' \in \mathscr{R}} k_{y\to y'}(c^*)^y (e^{y'\cdot\mu(c)} - e^{y\cdot\mu(c)}) &(15.13)\\ &= \sum_{y\in\mathscr{C}} [\sum_{\mathscr{R}_{\to y}} k_{y'\to y}(c^*)^{y'} - \sum_{\mathscr{R}_{y\to}} k_{y\to y'}(c^*)^y]\, e^{y\cdot\mu(c)} &(15.14)\\ &= 0. &(15.15)\end{aligned}$$

Note that equality holds in the estimate (15.13) if and only if $y' \cdot \mu(c) = y \cdot \mu(c)$ for every $y \to y' \in \mathscr{R}$. Stated differently, equality holds in the estimate (15.13) precisely when c is such that $\mu(c)$ is orthogonal to every reaction vector—that is, precisely when $\mu(c)$ lies in $S^\perp$.

In summary, then, we have so far

$$(\ln c - \ln c^*) \cdot f(c) \le 0, \quad \forall c \in \mathbb{R}_+^{\mathscr{S}}, \tag{15.16}$$

with equality holding if and only if $\ln c - \ln c^*$ lies in $S^\perp$. Note in particular that if $f(c) = 0$, in which case equality holds in (13.37), then $\ln c - \ln c^*$ lies in $S^\perp$. It remains to be shown that if $\ln c - \ln c^*$ lies in $S^\perp$, then $f(c) = 0$. (Recall that a quasi-thermodynamic kinetic system is required to be quasi-thermostatic.)

Adapting the form for $f(\cdot)$ given by (15.3) to the mass action system under study, we can write for each $c \in \mathbb{R}_+^{\mathscr{S}}$

$$\begin{aligned}f(c) &= \sum_{y\in\mathscr{C}} [\sum_{\mathscr{R}_{\to y}} k_{y'\to y} c^{y'} - \sum_{\mathscr{R}_{y\to}} k_{y\to y'} c^y]\, y \qquad (15.17)\\ &= \sum_{y\in\mathscr{C}} [\sum_{\mathscr{R}_{\to y}} k_{y'\to y}(c^*)^{y'} e^{y'\cdot\mu(c)} - \sum_{\mathscr{R}_{y\to}} k_{y\to y'}(c^*)^y e^{y\cdot\mu(c)}]\, y.\end{aligned}$$

Now if $c \in \mathbb{R}_+^{\mathscr{S}}$ is such that $\mu(c) := \ln c - \ln c^*$ is in $S^\perp$, then, for each $y \in \mathscr{C}$, we have $(y' - y) \cdot \mu(c) = 0$ whenever y' is such that $y \to y'$ or $y' \to y$ is a reaction. This, (15.9), and (15.17) tell us that if $\ln c - \ln c^*$ lies in $S^\perp$, we have

$$f(c) = \sum_{y \in \mathscr{C}} e^{y \cdot \mu(c)} \left[\sum_{\mathscr{R}_{\to y}} k_{y' \to y} (c^*)^{y'} - \sum_{\mathscr{R}_{y \to}} k_{y \to y'} (c^*)^{y} \right] y = 0. \tag{15.18}$$

That is, c is an equilibrium. Thus, the set of positive equilibria coincides with the set

$$E := \{ c \in \mathbb{R}_+^{\mathscr{S}} : \ln c - \ln c^* \in S^\perp \}. \tag{15.19}$$

This completes the proof of (i).

To prove (ii) we suppose that $c^{**} \in \mathbb{R}_+^{\mathscr{S}}$ is an equilibrium other than c^*, in which case we know that c^{**} is a member of E. Stated differently, $\mu(c^{**})$ is a member of $S^\perp$. Thus, for each $y \in \mathscr{C}$ we know that $(y' - y) \cdot \mu(c^{**}) = 0$ for any $y' \in \mathscr{C}$ for which $y \to y'$ or $y' \to y$ is a reaction. This prepares us for a calculation: For each $y \in \mathscr{C}$

$$\begin{aligned} g_y(c^{**}) &= \sum_{\mathscr{R}_{\to y}} k_{y' \to y} (c^{**})^{y'} - \sum_{\mathscr{R}_{y \to}} k_{y \to y'} (c^{**})^{y} \\ &= \sum_{\mathscr{R}_{\to y}} k_{y' \to y} (c^*)^{y'} e^{y' \cdot \mu(c^{**})} - \sum_{\mathscr{R}_{y \to}} k_{y \to y'} (c^*)^{y} e^{y \cdot \mu(c^{**})} \\ &= e^{y \cdot \mu(c^{**})} \left[\sum_{\mathscr{R}_{\to y}} k_{y' \to y} (c^*)^{y'} - \sum_{\mathscr{R}_{y \to}} k_{y \to y'} (c^*)^{y} \right] \qquad (15.20) \\ &= 0. \end{aligned}$$

Thus, we have complex balancing at c^{**}.

To prove (iii) we suppose that $c \in \mathbb{R}_+^{\mathscr{S}}$ is an equilibrium of the mass action system $\{\mathscr{S}, \mathscr{C}, \mathscr{R}, k\}$. By virtue of (ii), complex balancing obtains at c. Recall from Remark 3.6.2 that the derivative of $f(\cdot)$ at c takes the form

$$df(c)\sigma = \sum_{y \to y' \in \mathscr{R}} k_{y \to y'} c^y (y * \sigma)(y' - y), \quad \forall \sigma \in S, \tag{15.21}$$

where "*" is the scalar product in $\mathbb{R}^{\mathscr{S}}$ given by (15.8). Thus, we have for all $\sigma \in S$

$$\sigma * df(c)\sigma = \sum_{y \to y' \in \mathscr{R}} k_{y \to y'} c^y (y * \sigma)(y' * \sigma - y * \sigma). \tag{15.22}$$

Now we invoke an inequality:

$$x(x' - x) \le \frac{1}{2}[(x')^2 - (x)^2], \tag{15.23}$$

with equality holding if and only if $x' = x$. Thus, we can write for all $\sigma \in S$

$$\sigma * df(c)\sigma \le \frac{1}{2} \sum_{y \to y' \in \mathscr{R}} k_{y \to y'} (c)^y \, [(y' * \sigma)^2 - (y * \sigma)^2)] \tag{15.24}$$

$$= \frac{1}{2} \sum_{y \in \mathscr{C}} \left[\sum_{\mathscr{R}_{\to y}} k_{y' \to y} (c)^{y'} - \sum_{\mathscr{R}_{y \to}} k_{y \to y'} (c)^{y} \right] (y * \sigma)^2 \tag{15.25}$$

$$= 0. \tag{15.26}$$

The equality (15.26) follows from complex balancing at c. Note that equality holds in the bound (15.24) only if σ is such that in each term $y' * \sigma = y * \sigma$ or, equivalently, only if $(y' - y) * \sigma = 0$ for each $y \to y' \in \mathscr{R}$. This requires that $\sigma \in S$ be $*$-orthogonal to S, in which case $\sigma = 0$. Thus, $df(c) : S \to S$ is negative-definite relative to the scalar product "$*$". The remainder of (iii) results from standard properties of linear transformations that are negative-definite relative to a scalar product. □

Because of item (ii) in Theorem 15.2.2, it makes sense to speak of a *complex balanced mass action system*. This is a mass action system for which there exists a positive equilibrium at which complex balancing obtains. Item (ii) ensures that in such a case, complex balancing will obtain at *all* positive equilibria, regardless of stoichiometric compatibility class. With this in mind, we put on record the following definition:

Definition 15.2.3. *A mass action system is* **complex balanced** *if there exists for it a positive equilibrium at which complex balancing obtains.*

By virtue of Theorem 15.2.2, a complex balanced mass action system inherits all the dynamical properties of quasi-thermodynamic systems given by Theorem 13.4.4. Theorem 15.2.2 gives still more information. The following theorem summarizes some of this.

Theorem 15.2.4. *For a complex balanced mass action system, there is within each positive stoichiometric compatibility class precisely one equilibrium; that equilibrium is asymptotically stable; and there is no nontrivial cyclic composition trajectory along which all species concentrations are positive. Moreover, at each positive equilibrium, the derivative of the species-formation-rate function has a characteristic polynomial that admits only roots with negative real part. In particular, the derivative is nonsingular.*

Remark 15.2.5. Theorem 15.2.4 does not assert that all composition trajectories beginning in a positive stoichiometric compatibility class will invariably approach its unique positive equilibrium in the limit with increasing time. Indeed, the analysis we provided does not preclude the possibility that such a trajectory might approach a boundary equilibrium at which one or more species are absent. However, in very recent (and difficult) work [45], it appears that Gheorghe Craciun proved that, for a complex balanced mass action system, the unique equilibrium in each positive stoichiometric compatibility class is in fact approached in the limit by all trajectories that begin within it. See also Section 7.8.

15.3 Multicell Mass Action Systems in Which the Intracellular Chemistry Is Complex Balanced

We are in a position to describe, informally, how the idea of complex balancing provides a path to the Multicell Deficiency Zero Theorem stated in Chapter 7. Again, the argument is due to Shapiro and Horn [152, 153]. Our short-term goal is to provide support for the following proposition, in which it is understood that a *multicell system* adheres to the Shapiro-Horn constraints imposed in Section 7.5. These will not be repeated here.

Proposition 15.3.1. *A multicell system in which the intracellular chemistry is mass action and complex balanced is itself complex balanced and therefore quasi-thermodynamic.*

There is an unavoidable complexity in a full proof of Proposition 15.3.1, for one must contend with the notational awkwardness of dealing simultaneously with what might be a multitude of species and a multitude of interconnected cells. For this reason we will forgo great generality and focus instead on a system consisting of just two cells, with transport between them. In this limited context, the main idea of the Shapiro-Horn argument will be far more transparent. Then we will consider briefly a three-cell system that provides a richer range of mathematical possibilities more typical of systems with still larger numbers of cells.

15.3.1 A Two-Cell Example

Consider a two-cell system subject to the Shapiro-Horn constraints described in Section 7.5. So that we might have a concrete example of an intracellular chemistry to keep in mind, it will be useful to imagine that the chemical reactions occurring within the cells are those shown in the toy network (15.27). The kinetics is presumed to be mass action, with rate constants identical in both cells. Moreover, *we will suppose that the rate constants are such that the intracellular mass action system is complex balanced.*[3] This we will call the *base mass action system.*

$$\begin{gathered} B \rightleftarrows A \leftrightarrows C \\ B + C \rightleftarrows 2A \end{gathered} \tag{15.27}$$

In the spirit of Section 4.2.4, we can model the two-cell configuration under consideration as an augmented mass action system in which the underlying reaction network is that shown in (15.28). Recall that the added reactions such as $A1 \rightleftarrows A2$ are incorporated to model intercellular transport. In light of the constraints imposed on multicell systems in Section 7.5, we suppose that $k_{A1\to A2} = k_{A2\to A1}$, $k_{B1\to B2} = k_{B2\to B1}$, and $k_{C1\to C2} = k_{C2\to C1}$.

[3] Because network (15.27) is forest-like, complex balancing is synonymous with detailed balancing (Remark 15.1.5), whereupon the required rate constant orchestration is given by Theorem 14.2.3.

$$
\begin{array}{ccc}
B1 & \rightleftarrows & B2 \\
\uparrow\!\downarrow & & \downarrow\!\uparrow \\
A1 & \leftrightarrows & A2 \\
\downarrow\!\uparrow & & \uparrow\!\downarrow \\
C1 & \rightleftarrows & C2
\end{array}
\tag{15.28}
$$

$$
B1 + C1 \rightleftarrows 2A1 \qquad B2 + C2 \rightleftarrows 2A2
$$

To argue that the mass action system so constructed is complex balanced, we need to show that there is a positive equilibrium at which complex balancing obtains. For this purpose let c^* be a positive equilibrium of the complex balanced base mass action system associated with the original three-species network (15.27). Now consider a composition $c^\dagger$ for the two-cell mass action system associated with six-species network (15.28), chosen so that

$$
c^\dagger_{A1} = c^\dagger_{A2} = c^*_A, \quad c^\dagger_{B1} = c^\dagger_{B2} = c^*_B, \quad \text{and} \quad c^\dagger_{C1} = c^\dagger_{C2} = c^*_C. \tag{15.29}
$$

It is easy to see that $c^\dagger$ is a positive equilibrium for the two-cell mass action system *and that at that equilibrium complex balancing obtains*. Although complex balancing is obvious in this case, it will be useful to consider just why complex balancing obtains. The reasons will extend to multicell systems for which complex balancing is less obvious, in particular to the three-cell system we will discuss later.

There are two kinds of complexes to consider, those such as $A1$ and $B2$ that are monomolecular and those such as $B1 + C1$ and $2A2$ that are not monomolecular.

15.3.1.1 Complexes That Are Not Monomolecular

The reactions leading to and from a non-monomolecular complex are, apart from nomenclature, no different from those present in the base complex balanced mass action system, and at $c^\dagger$ their rates in the multicell case are precisely what they were at c^* in the base system. Thus, for a non-monomolecular complex, the net complex-formation rate in the multicell case remains *zero*, just as it was in the base *complex balanced* mass action system.

15.3.1.2 Monomolecular Complexes

The situation for a monomolecular complex, for example, $A2$, is a little more complicated. In the multicell case there are intercellular reactions, in addition to the intracellular reactions, that must be taken into account. At composition $c^\dagger$ the total of all rates of *intracellular* reactions of the form $A2 \to y$ must balance those of the form $y' \to A2$, for these rates are identical to their counterparts at composition c^* in the base *complex balanced* mass action system.

As for the intercellular reactions that originate and terminate at $A2$, their rates must also balance by virtue of the Shapiro-Horn constraints in Section 7.5 on intercellular flows. In particular, the Shapiro-Horn constraints imply that, *in the absence of intracellular reactions*, any multicell state such as $c^{\dagger}$ that is uniform across cells must be an equilibrium. This in turn requires that, in any uniform multicell state such as $c^{\dagger}$, the net rate of all *intercellular* reactions adjacent to a monomolecular complex be zero.

Taking account of both intracellular and intercellular reactions, then, we can assert that at $c^{\dagger}$ the rates of *all* reactions—inter- and intracellular—that have a particular monomolecular complex as their product are balanced by the rates of *all* reactions that have that complex as their reactant.

15.3.1.3 Two-Cell Summary

In summary, at $c^{\dagger}$ we have balancing at all complexes, those that are monomolecular and those that are not. This is to say that the multicell mass action system associated with network (15.28) is complex balanced.

15.3.2 A Three-Cell Example

Although it is not difficult to see how the two-cell argument in support of Proposition 15.3.1 extends to a general multicell setting, we need to be aware that, when there are more than two cells, the possibilities for intercellular transport become richer. Consider, for example, a three-cell system in which the base intracellular complex balanced mass action system is what it was before. Suppose, however, that the patterns of intercellular transport of species A, B, and C are now those captured in the intercellular reactions (and rate constants) shown in (15.30).

$$
\begin{array}{ccc}
A1 \underset{5}{\overset{6}{\rightleftarrows}} A2 & B1 \overset{3}{\leftarrow} B2 & C1 \underset{1}{\overset{1}{\rightleftarrows}} C2 \\
{}_1\nwarrow \quad \swarrow_1 & {}_3\searrow \quad \nearrow_3 & {}_5 \nwarrow\!\!\!\searrow_5 \quad {}_7 \nearrow\!\!\!\swarrow_7 \\
A3 & B3 & C3
\end{array}
\tag{15.30}
$$

The cell-to-cell flows are more complicated than they were in the two-cell case, but the rate constants shown are such that the Shapiro-Horn constraints on intercellular transport are satisfied. In particular, any *uniform* state of the multicell system is an equilibrium of the intercellular-reactions-only mass action system (15.30).

Thus, for any positive equilibrium of the full multicell mass action system in which there are no cell-to-cell differences in composition, there is no net contribution from intercellular reactions to the net formation rate of any unimolecular complex. This is what the argument in the two-cell case required. With this in mind, we can see how the two-cell argument in support of Proposition 15.3.1 extends generally, so long as the Shapiro-Horn constraints on intercellular transport are respected.

15.3.3 A Multicell-Dynamics Theorem Based on Complex Balancing

Taken with Theorem 13.4.4, Proposition 15.3.1 gives the following Shapiro-Horn result.

Theorem 15.3.2. *For a multicell system in which the intracellular kinetics is mass action and complex balanced, the governing differential equations have the following properties: In each positive stoichiometric compatibility class for the multicell network, there is precisely one multicell equilibrium; that equilibrium is asymptotically stable; and at that equilibrium the concentration of each species is the same in every cell. Moreover, the differential equations cannot admit a cyclic composition trajectory along which all species concentrations are positive.*

There is one special feature of multicell systems asserted in Theorem 15.3.2 that requires some elaboration: *We need to argue that, when the hypothesis is satisfied, there can be no positive equilibrium that exhibits cell-to-cell variations in composition.* Recall the symmetry-breaking example in Section 5.2.2.

When the hypothesis of the theorem is satisfied, the corresponding multicell mass action system is, by virtue of Proposition 15.3.1, complex balanced and therefore quasi-thermodynamic. In particular, it is quasi-thermostatic. Suppose that c^* is a positive equilibrium of the multicell mass action system in which there are no cell-to-cell composition variations. (Such an equilibrium clearly exists.) The quasi-thermostatic property tells us that the positive equilibrium set takes the form

$$E := \{c \in \mathbb{R}_+^{\mathscr{S}} : \ln c - \ln c^* \in S^\perp\}, \tag{15.31}$$

where it is understood that $\mathscr{S}$ is the species set for the full multicell mass action system—for example, $\{A1, A2, B1, B2, C1, C2\}$ in the case of Section 15.3.1—and S is the stoichiometric subspace for the underlying network—for example, (15.28).

We want to argue that in any positive equilibrium, the concentration of a species, say A, is the same in cell I and cell K. From the Shapiro-Horn constraints on intercellular transport, it follows that, in the multicell mass action system, there must be an intercellular reaction sequence of the form

$$A_I \to A_{J(1)} \to A_{J(2)} \to \cdots \to A_K. \tag{15.32}$$

For this reason $A_K - A_I$ is a member of the stoichiometric subspace for the multicell mass action system. Thus, if c is some other positive equilibrium (and therefore a member of E), we must have

$$(\ln \frac{c}{c^*}) \cdot (A_K - A_I) = \ln \frac{c_{A_K}}{c^*_{A_K}} - \ln \frac{c_{A_I}}{c^*_{A_I}} = 0. \tag{15.33}$$

Because, by supposition, c^* is a uniform multicell equilibrium, we have $c^*_{A_K} = c^*_{A_I}$. Therefore, $c_{A_K} = c_{A_I}$. □

Remark 15.3.3. As we noted at the beginning of Chapter 14, at least some chemists hold to the doctrine that all naturally occurring reaction networks (as distinct from networks constructed to model open systems) are reversible and that their kinetics should conform to detailed balancing at (closed system) equilibria. If in a multicell system the base chemistry is mass action and complies with that doctrine, Theorem 15.3.2 tells us that we should not expect exotic behavior.

It should be kept in mind, however, that the base mass action system might in fact be a model derived from consideration of a physical setting resembling those that gave rise to the Edelstein system (Sections 4.2.3 and 5.1.2) or to the Brusselator (Remark 4.2.3)—settings in which certain species are deemed to have approximately constant concentrations by virtue of their availability in large excess. Moreover, the "cells" in a multicell system might resemble continuous-flow stirred-tank reactors admitting *extracellular* transport of material to and from an environment exterior to the multicell system itself. In such cases the detailed balancing doctrine is no longer apt.

15.3.4 The Path to the Deficiency Zero Multicell Theorem

Theorem 15.3.2 will provide the basis for a proof of the Deficiency Zero Multicell Theorem of Chapter 7. In the next chapter, we will start on the road to showing that complex balancing obtains in every mass action system in which the underlying reaction network is weakly reversible and has a deficiency of zero. If, in a multicell system, the base intracellular network has these attributes, the intracellular complex balancing required in Theorem 15.3.2 is ensured. The Deficiency Zero Multicell Theorem will then follow as a result.

16

Deficiency Zero Theory Foundations and Some Key Propositions

At the end of this chapter, we will complete the proofs of the Deficiency Zero Theorem and some of its variants. A considerable amount of the work has already been done: In Chapter 13 we described the dynamical consequences that accrue to a kinetic system having the quasi-thermodynamic property. In Chapter 15, we indicated how, for a mass action system, complex balancing at a positive equilibrium ensures quasi-thermodynamicity. In this chapter we will, among other things, connect complex balancing in a kinetic system to architectural properties of the underlying reaction network—in particular to the network's deficiency and to its graphical structure. The Deficiency Zero Theorem derives from these connections.

Along the way we will accumulate some powerful technical tools that will find additional use when we delve more deeply into higher-deficiency theory and when we prove the robustness theorems stated earlier in Chapter 9. This chapter, then, serves also as a repository of tools from which we will draw later on.

Perhaps the most important of these tools, Theorem 16.4.2, I call the *First Salt Theorem*[1] because of a simple mental picture used to motivate it, consisting of interconnected barrels containing salt solutions of different concentrations. This theorem will be applied in a variety of ways. One such application, for example, appears in Appendix 16.B, where the salt theorem will permit us to prove the previously stated Theorem 8.A.2, about the coincidence of stoichiometric and kinetic subspaces for mass action systems.

Another wide-ranging tool, Proposition 16.2.1, indicates how the structure of a reaction network, as reflected in its deficiency, can place severe constraints on what complex formation rates might be at an equilibrium. As we shall see, not only in this chapter but also subsequent ones, this proposition and the salt theorem work hand in hand to give powerful information about reaction network behavior.

[1] There will be two more salt theorems in the next chapter, motivated by the same picture.

M. Feinberg, *Foundations of Chemical Reaction Network Theory*,
Applied Mathematical Sciences 202,
https://doi.org/10.1007/978-3-030-03858-8_16

16.1 A Decomposition of the Species-Formation-Rate Function

A reaction network, viewed in terms of its standard reaction diagram, is in part a directed graph, with the complexes playing the role of vertices and the reaction arrows playing the role of directed edges. Certain aspects of a reaction network—for example, its strong-linkage classes—depend *only* on this graphical structure and are uninfluenced by the precise nature of the stoichiometric coefficients appearing in the complexes, except insofar as they serve to label the graph's vertices. Beyond its graphical structure, however, a reaction network has an additional *algebraic* structure residing in the stoichiometric coefficients that the complexes carry.

These are quite separate aspects of reaction network structure, with differing degrees of influence in different parts of reaction network theory. For example, in the hypothesis of the Star-Like Reaction Network Theorem, stoichiometry plays no role at all. On the other hand, in deciding the rank or deficiency of a network, stoichiometry is dominant: Graphical structure plays a role only to the extent that it determines how the complexes are partitioned into linkage classes. The precise way in which reaction arrows connect complexes within linkage classes is inconsequential. (Recall Remarks 6.2.2 and 6.3.3.)

There are advantages, then, in formulating the species-formation-rate function in such a way that the different roles of stoichiometry and graphical structure might be seen more clearly. That is our objective in the next few sections. We begin by introducing the *stoichiometric map*.

16.1.1 The Stoichiometric Map

The stoichiometric map for a reaction network $\{\mathscr{S},\mathscr{C},\mathscr{R}\}$ is the linear transformation $Y : \mathbb{R}^{\mathscr{C}} \to \mathbb{R}^{\mathscr{S}}$ defined by its action on $\{\omega_y\}_{y\in\mathscr{C}}$, the standard basis for $\mathbb{R}^{\mathscr{C}}$, as follows:

$$Y\omega_y = y, \quad \forall y \in \mathscr{C}. \tag{16.1}$$

For a kinetic system $\{\mathscr{S},\mathscr{C},\mathscr{R},\mathscr{K}\}$ such a map permits us to "factor out" the most overt influences of stoichiometry on the species-formation-rate function $f(\cdot)$:

$$\begin{aligned} f(c) &:= \sum_{y\to y'\in\mathscr{R}} \mathscr{K}_{y\to y'}(c)(y'-y) \\ &= Y \sum_{y\to y'\in\mathscr{R}} \mathscr{K}_{y\to y'}(c)(\omega_{y'}-\omega_y). \end{aligned} \tag{16.2}$$

16.1.2 The Complex-Formation-Rate Function

The decomposition (16.2) provides motivation for a definition, to which we turn next.

Definition 16.1.1. *For a kinetic system* $\{\mathscr{S},\mathscr{C},\mathscr{R},\mathscr{K}\}$ *the* **complex-formation-rate function** $g(\cdot) : \overline{\mathbb{R}}_+^{\mathscr{S}} \to \mathbb{R}^{\mathscr{C}}$ *is defined by*

$$g(c) := \sum_{y\to y' \in \mathscr{R}} \mathscr{K}_{y\to y'}(c)(\omega_{y'} - \omega_y). \tag{16.3}$$

We say that **complex balancing** *obtains at* $c^* \in \overline{\mathbb{R}}_+^{\mathscr{S}}$ *if* $g(c^*) = 0$.

To see that this function deserves its name, we can rearrange the sum in (16.3):

$$g(c) = \sum_{y\in\mathscr{C}} \Big[\sum_{\mathscr{R}_{\to y}} \mathscr{K}_{y'\to y}(c) - \sum_{\mathscr{R}_{y\to}} \mathscr{K}_{y\to y'}(c) \Big] \omega_y, \tag{16.4}$$

where $\mathscr{R}_{\to y}$ and $\mathscr{R}_{y\to}$ are the sets of reactions having y as, respectively, the product and reactant complex. Thus, the component of $g(c)$ corresponding to complex $y \in \mathscr{C}$ is

$$g_y(c) = \sum_{\mathscr{R}_{\to y}} \mathscr{K}_{y'\to y}(c) - \sum_{\mathscr{R}_{y\to}} \mathscr{K}_{y\to y'}(c). \tag{16.5}$$

This is precisely what in Remark 15.1.6 we called the complex formation rate for complex y at composition c. From (16.2) and (16.3), it follows that the species-formation-rate function is related to the complex-formation-rate function through the equation[2]

$$f(c) = Yg(c), \quad \forall c \in \overline{\mathbb{R}}_+^{\mathscr{S}}. \tag{16.6}$$

16.1.3 The Linear Subspace Span (Δ) and Its Properties

Note that for a kinetic system derived from a reaction network $\{\mathscr{S},\mathscr{C},\mathscr{R}\}$ the complex-formation-rate function takes values in the span of the set

$$\Delta_\to := \{\omega_{y'} - \omega_y \in \mathbb{R}^{\mathscr{C}} : y \to y' \in \mathscr{R}\}. \tag{16.7}$$

Since $y \to y'$ implies that y and y' are linked[3] (denoted $y \sim y'$), it follows that $\Delta_\to$ is contained in the set

$$\Delta := \{\omega_{y'} - \omega_y \in \mathbb{R}^{\mathscr{C}} : y \sim y'\} \tag{16.8}$$

and that

$$\mathrm{span}\,(\Delta_\to) \subset \mathrm{span}\,(\Delta). \tag{16.9}$$

In fact, though, we have the following:

Lemma 16.1.2. *For a reaction network* $\{\mathscr{S},\mathscr{C},\mathscr{R}\}$ *let* $\Delta_\to$ *and* Δ *be defined as in* (16.7) *and* (16.8). *Then*

$$\mathrm{span}\,(\Delta_\to) = \mathrm{span}\,(\Delta). \tag{16.10}$$

Moreover, for a reaction network having n *complexes and* ℓ *linkage classes*

$$\dim\,\mathrm{span}\,(\Delta_\to) = \dim\,\mathrm{span}\,(\Delta) = n - \ell. \tag{16.11}$$

---

[2] This factorization was introduced in [109] and was used heavily in [67, 71, 107].

[3] Recall Definition 6.1.1.

Proof. In light of (16.9), it is enough to show that span (Δ) is contained in span $(\Delta_{\rightarrow})$ or, equivalently, that Δ is contained in span $(\Delta_{\rightarrow})$. Suppose that $\omega_{y'} - \omega_y$ is a member of Δ. Then y' and y are linked ($y \sim y'$), and one of the three possibilities listed in Definition 6.1.1 must obtain. If $y = y'$, then $\omega_{y'} - \omega_y = 0$, which lies in span $(\Delta_{\rightarrow})$. If y and y' are directly linked ($y \leftrightarrow y'$), then either $y \rightarrow y'$ or $y' \rightarrow y$ is a member of $\mathscr{R}$, so that either $\omega_{y'} - \omega_y$ or $\omega_y - \omega_{y'}$ is a member of $\Delta_{\rightarrow}$; in either case $\omega_{y'} - \omega_y$ is a member of span $(\Delta_{\rightarrow})$. Finally, if there is a sequence $y(l), y(2), \ldots, y(k) \subset \mathscr{C}$ such that

$$y \leftrightarrow y(1) \leftrightarrow y(2) \leftrightarrow \cdots \leftrightarrow y(k) \leftrightarrow y'$$

then, by the argument immediately preceding, the vectors

$$\{\omega_{y'} - \omega_{y(k)},\ \omega_{y(k)} - \omega_{y(k-1)},\ \ldots,\ \omega_{y(2)} - \omega_{y(1)},\ \omega_{y(1)} - \omega_y\} \tag{16.12}$$

are members of span $(\Delta_{\rightarrow})$, and so then is their sum, $\omega_{y'} - \omega_y$.

To complete the proof, it suffices to ascertain the dimension of span (Δ). Let $\{\mathscr{L}^1, \mathscr{L}^2, \ldots \mathscr{L}^\ell\}$ be the linkage classes of the network, and for $\theta = 1, 2, \ldots, \ell$, let

$$\Delta^\theta := \{\omega_{y'} - \omega_y \in \mathbb{R}^{\mathscr{C}} : y \in \mathscr{L}^\theta,\ y' \in \mathscr{L}^\theta\}. \tag{16.13}$$

Then

$$\Delta = \bigcup_{\theta=1}^{\ell} \Delta^\theta \tag{16.14}$$

and

$$\operatorname{span}(\Delta) = \operatorname{span}(\Delta^1) \oplus \operatorname{span}(\Delta^2) \oplus \cdots \oplus \operatorname{span}(\Delta^\ell). \tag{16.15}$$

Let n_θ be the number of complexes in linkage class $\mathscr{L}^\theta$, and let those complexes be denoted $y_1, y_2, \ldots, y_{n_\theta}$. Then any member of Δ^θ (and therefore any member of span (Δ^θ)) can be written as a linear combination of the linearly independent set

$$\{\omega_{y_2} - \omega_{y_1}, \omega_{y_3} - \omega_{y_1}, \ldots, \omega_{y_{n_\theta}} - \omega_{y_1}\}. \tag{16.16}$$

Thus, dim span $(\Delta^\theta) = n_\theta - 1$. From (16.15) it then follows that

$$\dim \operatorname{span}(\Delta) = \sum_{\theta=1}^{\ell} (n_\theta - 1) = n - \ell. \quad \square \tag{16.17}$$

In preparation for our next lemma, it will be useful to recall some notation given in Chapter 2. Consider a reaction network $\{\mathscr{S}, \mathscr{C}, \mathscr{R}\}$ with linkage classes $\{\mathscr{L}^1, \mathscr{L}^2, \ldots, \mathscr{L}^\ell\}$. Then $\omega_{\mathscr{L}^\theta}$ is the vector of $\mathbb{R}^{\mathscr{C}}$ whose y^{th} component is 1 for $y \in \mathscr{L}^\theta$ and is 0 otherwise. That is,

$$\omega_{\mathscr{L}^\theta} := \sum_{y \in \mathscr{L}^\theta} \omega_y. \tag{16.18}$$

Recall also that $\mathbb{R}^{\mathscr{C}}$ carries the standard scalar product (Section 2.2.2).

Lemma 16.1.3. *Let* $\{\mathscr{S},\mathscr{C},\mathscr{R}\}$ *be a reaction network having linkage classes* $\{\mathscr{L}^\theta\}_{\theta=1,2,\ldots,\ell}$, *and let* Δ *be as in* (16.8). *Then the set*

$$\{\omega_{\mathscr{L}^\theta}\}_{\theta=1,2,\ldots,\ell} \subset \mathbb{R}^{\mathscr{C}} \tag{16.19}$$

is a basis for $[\text{span}(\Delta)]^\perp$.

Proof. Clearly, every vector in the set (16.19) is orthogonal to each vector of Δ and therefore to each vector of span (Δ). Because the vectors in (16.19) have disjoint support, the set is independent. Since $\dim \mathbb{R}^{\mathscr{C}} = n$, where n is the number of complexes, it follows from Lemma 16.1.2 that the dimension of the orthogonal complement of span (Δ) is $n-(n-\ell)=\ell$. Thus, (16.19) is a basis for $[\text{span}(\Delta)]^\perp$. □

Remark 16.1.4. From Lemma 16.1.3 we know that, for a chemical reaction network $\{\mathscr{S},\mathscr{C},\mathscr{R}\}$, a vector $g \in \mathbb{R}^{\mathscr{C}}$ lies in span (Δ) if and only if

$$g \cdot \omega_{\mathscr{L}^\theta} = 0, \quad \theta = 1,2,\ldots,\ell. \tag{16.20}$$

That is, a vector $g \in \mathbb{R}^{\mathscr{C}}$ lies in span (Δ) or, equivalently, span $(\Delta_\rightarrow)$ if and only if its components corresponding to complexes within each linkage class sum to zero:

$$\sum_{y \in \mathscr{L}^\theta} g_y = 0, \quad \theta = 1,2,\ldots,\ell. \tag{16.21}$$

16.2 The Deficiency as a Measure of How Tightly an Equilibrium Complex-Formation-Rate Vector Is Constrained

When we first introduced the deficiency in Chapter 6, we provided an interpretation of what it means for a reaction network to have a deficiency of zero: *Deficiency zero networks are ones for which the reaction vectors are as independent as the partition of complexes into linkage classes will allow*. In a sense, then, the deficiency of a reaction network is a measure of how dependent the reaction vectors are, given the network's linkage class structure. Here we will provide another interpretation of the deficiency, albeit a more technical one, in terms of the extent to which reaction network structure, quite apart from kinetics, restricts values that an equilibrium complex-formation-rate vector might have. As we shall see, for a deficiency *zero* network, an equilibrium complex-formation-rate vector is restricted very tightly: *It can only be the zero vector*, so at every equilibrium complex balancing *must* obtain.

16.2.1 Another Interpretation of the Deficiency

To begin we recall that, for a kinetic system $\{\mathscr{S},\mathscr{C},\mathscr{R},\mathscr{K}\}$ with stoichiometric subspace S, the species-formation-rate function $f(\cdot) : \overline{\mathbb{R}}_+^{\mathscr{S}} \to S$ has a decomposition $f(c) = Yg(c)$, where $Y : \mathbb{R}^{\mathscr{C}} \to \mathbb{R}^{\mathscr{S}}$ is the stoichiometric map for the network and $g(\cdot) : \overline{\mathbb{R}}_+^{\mathscr{S}} \to \mathbb{R}^{\mathscr{C}}$ is the complex-formation-rate function. Thus if $c^* \in \overline{\mathbb{R}}_+^{\mathscr{S}}$ is an equilibrium—that is, if $f(c^*) = 0$—we must have

$$Yg(c^*) = 0. \tag{16.22}$$

Now it might happen that $g(c^*) = 0$, in which case we have complex balancing at c^*, but (16.22) requires only that $g(c^*)$ be a member of $\ker Y$. However, we can say more. Recall from the preceding section that $g(c^*)$ must be a member of $\mathrm{span}(\Delta)$, where $\Delta \subset \mathbb{R}^{\mathscr{C}}$ is given by (16.8). Thus, when c^* is an equilibrium, we actually have the inclusion

$$g(c^*) \in \ker Y \cap \mathrm{span}(\Delta). \tag{16.23}$$

Note that the right side of (16.23) is a linear subspace of $\mathbb{R}^{\mathscr{C}}$ determined by the reaction network alone and not at all by the kinetics. We will want to know that subspace's dimension.

Proposition 16.2.1 **([68, 71]).** *If, for reaction network* $\{\mathscr{S},\mathscr{C},\mathscr{R}\}$, $Y : \mathbb{R}^{\mathscr{C}} \to \mathbb{R}^{\mathscr{S}}$ *is the stoichiometric map, if*

$$\Delta := \{\omega_{y'} - \omega_y \in \mathbb{R}^{\mathscr{C}} : y \sim y'\}, \tag{16.24}$$

and if δ *is the network's deficiency, then*

$$\delta = \dim\,[\ker Y \cap \mathrm{span}(\Delta)]. \tag{16.25}$$

In particular, if the network has a deficiency of zero, then

$$\ker Y \cap \mathrm{span}(\Delta) = \{0\}. \tag{16.26}$$

Proof. Let $\bar{Y} : \mathrm{span}(\Delta) \to \mathbb{R}^{\mathscr{S}}$ be the restriction of $Y : \mathbb{R}^{\mathscr{C}} \to \mathbb{R}^{\mathscr{S}}$ to $\mathrm{span}(\Delta)$. Then, from the standard theorem relating the dimensions of the domain, kernel, and image of a linear map, we have

$$\dim \mathrm{span}(\Delta) = \dim \ker \bar{Y} + \dim \mathrm{im}\,\bar{Y}. \tag{16.27}$$

From Lemma 16.1.2 we know that $\dim \mathrm{span}(\Delta) = n - \ell$, where n is the number of complexes and ℓ is the number of linkage classes. To determine $\dim \mathrm{im}\,\bar{Y}$ we first note that, for each reaction $y \to y'$, the corresponding reaction vector is given by

$$y' - y = \bar{Y}(\omega_{y'} - \omega_y). \tag{16.28}$$

Thus, the set of reaction vectors for the network is identical to the image under $\bar{Y}$ of the set $\Delta_{\to}$ defined in (16.7). From this it follows easily that the span of the set of reaction vectors—that is, the stoichiometric subspace S—is the image under $\bar{Y}$ of $\mathrm{span}(\Delta_{\to})$ or, from Lemma 16.1.2, $\mathrm{span}(\Delta)$. This is to say that $\mathrm{im}\,\bar{Y} = S$, and, as a result,

$$\dim \mathrm{im}\,\bar{Y} = \dim S = s, \tag{16.29}$$

where s is the rank of the network. Finally, we note that

$$\ker \bar{Y} = \ker Y \cap \mathrm{span}\,\Delta. \tag{16.30}$$

Rearrangement of (16.27) then gives

$$\dim\,[\ker Y \cap \mathrm{span}(\Delta)] = n - \ell - s = \delta. \quad \square \tag{16.31}$$

16.2.2 Deficiency Zero Implies Complex Balancing at Every Equilibrium

Proposition 16.2.1 has an immediate and powerful corollary, which we state as a theorem:

Theorem 16.2.2 ([67]). *Complex balancing obtains at every equilibrium of a kinetic system in which the underlying reaction network has a deficiency of zero.*

Proof. From Proposition 16.2.1 and the inclusion (16.23), it follows that, when the deficiency is zero, the complex formation rate at an equilibrium can only be the zero vector, which is to say that complex balancing obtains at that equilibrium. □

Theorem 16.2.2 has an important corollary of its own, which derives also from Theorem 15.2.2.

Corollary 16.2.3. *Consider a mass action system in which the underlying reaction network has a deficiency of zero. If that mass action system admits a positive equilibrium, it is quasi-thermodynamic.*

Remark 16.2.4 (The path forward). Theorem 16.2.2 and Corollary 16.2.3 lie at the heart of the proof of the Deficiency Zero Theorem. They provide starting points for a logical cascade that connects reaction network structure first to compulsory complex balancing at equilibria, then to quasi-thermodynamicity, and finally to the highly circumscribed dynamics that the quasi-thermodynamic property imposes.

The theorem and its corollary suggest an important question: *For a network of deficiency zero class, what general connections are there between the existence of* positive *equilibria and graphical features of the network, especially when the kinetics is mass action.* In the deficiency zero case we will prove that, when the kinetics is mass action, weak reversibility is sufficient for a positive equilibrium and that, regardless of kinetics, weak reversibility is necessary. (Recall that weak reversibility is no longer necessary when the deficiency is one.)

Although we will have interest in arbitrary kinetics, it will be mathematically advantageous to first develop tools motivated by mass action considerations. As we shall see, those tools will have powerful consequences that extend in later chapters to networks of higher deficiency, and they will even find use in the study of kinetic systems that are not mass action.[4] Tool development will begin in the next section after a brief remark.

Remark 16.2.5 (Some useful observations). So that we can have them available in later chapters, it will be helpful to record here some consequences of material just accumulated. For a reaction network $\{\mathscr{S},\mathscr{C},\mathscr{R}\}$ let

$$g = \sum_{y \in \mathscr{C}} g_y \, \omega_y \tag{16.32}$$

be an element of $\mathbb{R}^{\mathscr{C}}$. In Remark 16.1.4 we observed that g lies in $\operatorname{span}(\Delta)$ if and only if

[4] See, for example, Section 16.5.

$$\sum_{y \in \mathscr{L}^\theta} g_y = 0, \quad \theta = 1,2,\ldots,\ell, \tag{16.33}$$

where $\mathscr{L}^1, \mathscr{L}^2, \ldots, \mathscr{L}^\ell$ are the linkage classes of the network. If Y is the stoichiometric map for the network, we can act with Y on both sides of (16.32) to see that g lies in ker Y if and only if

$$\sum_{y \in \mathscr{C}} g_y\, y = 0. \tag{16.34}$$

Thus, g lies in ker $Y \cap \text{span}(\Delta)$ if and only if it satisfies both (16.33) and (16.34). If the deficiency of the network is zero, we have from Proposition 16.2.1 that the only solution to (16.33) and (16.34) is the trivial $g = 0$. If the deficiency of the network is one, we have the existence of a nonzero g that solves (16.33) and (16.34), and the only other solutions are scalar multiples of that g.

16.3 Decomposing the Mass Action Species-Formation-Rate Function

Recall that in Section 16.1 we decomposed the species-formation-rate function of a general kinetic system in order to indicate more clearly the interplay of stoichiometry (carried by the stoichiometric map) and the complex-formation-rate function. Here we carry that decomposition somewhat further in the special case of mass action kinetics. Along the way, we will introduce a new linear map, denoted A_k, that has qualitative properties very closely tied to a reaction network's structure as a directed graph, qualitative properties that make several important results of chemical reaction network theory possible.

Recall that a kinetic system $\{\mathscr{S}, \mathscr{C}, \mathscr{R}, \mathscr{K}\}$ is mass action if the individual reaction-rate functions take the form

$$\mathscr{K}_{y\to y'}(c) = k_{y\to y'} c^y, \quad \forall c \in \overline{\mathbb{R}}^{\mathscr{S}}_+, \ \forall y \to y' \in \mathscr{R}, \tag{16.35}$$

where $k_{y\to y'} > 0$ is a rate constant for reaction $y \to y'$and

$$c^y := \prod_{s \in \mathscr{S}} c_s^{y_s}. \tag{16.36}$$

In this case the complex-formation-rate function takes the form

$$g(c) := \sum_{y\to y' \in \mathscr{R}} k_{y\to y'} c^y (\omega_{y'} - \omega_y), \tag{16.37}$$

and complex balancing obtains at composition c^* precisely when

$$\sum_{y\to y' \in \mathscr{R}} k_{y\to y'} (c^*)^y (\omega_{y'} - \omega_y) = 0. \tag{16.38}$$

We will find it helpful to write (16.37) in a more structured form. For a mass action system $\{\mathscr{S}, \mathscr{C}, \mathscr{R}, k\}$, let $A_k : \mathbb{R}^{\mathscr{C}} \to \mathbb{R}^{\mathscr{C}}$ be the linear transformation defined by

$$A_k\, x := \sum_{y \to y' \in \mathscr{R}} k_{y \to y'}\, x_y\, (\omega_{y'} - \omega_y). \tag{16.39}$$

Moreover, let $\psi : \overline{\mathbb{R}}_+^{\mathscr{S}} \to \overline{\mathbb{R}}_+^{\mathscr{C}}$ be defined by

$$\psi(c) = \sum_{y \in \mathscr{C}} c^y\, \omega_y\, ; \tag{16.40}$$

that is, the yth component of $\psi(c)$ is given by

$$\psi_y(c) = c^y. \tag{16.41}$$

With this as background, *the complex-formation-rate function for a mass action system* $\{\mathscr{S}, \mathscr{C}, \mathscr{R}, k\}$ *takes the form*

$$g(c) = A_k\, \psi(c), \quad \forall c \in \overline{\mathbb{R}}_+^{\mathscr{S}}, \tag{16.42}$$

and, in view of (16.6), the species-formation-rate function can be written[109]

$$f(c) = YA_k\, \psi(c), \quad \forall c \in \overline{\mathbb{R}}_+^{\mathscr{S}}. \tag{16.43}$$

Thus, the differential equation for the mass action system can be written as

$$\boxed{\dot{c} = YA_k\, \psi(c).} \tag{16.44}$$

This form has the advantage of isolating a network's stoichiometry, captured in the map Y and in the function $\psi(\cdot)$, from the network's structure as a directed graph, captured in the map A_k.

If, for a mass action system $\{\mathscr{S}, \mathscr{C}, \mathscr{R}, k\}$, we want to study the equilibria of the differential equation (16.44), we must ask for which $c \in \overline{\mathbb{R}}_+^{\mathscr{S}}$ it is the case that $\psi(c)$ lies in $\ker YA_k$. Because $\psi(\cdot)$ takes values in $\overline{\mathbb{R}}_+^{\mathscr{C}}$, we shall be interested in the nature of

$$\ker YA_k \cap \overline{\mathbb{R}}_+^{\mathscr{C}}. \tag{16.45}$$

Note that when c is a positive equilibrium—that is, when c is an equilibrium in $\mathbb{R}_+^{\mathscr{S}}$—$\psi(c)$ lies in $\mathbb{R}_+^{\mathscr{C}}$. For this reason we will be especially interested in the nature of

$$\ker YA_k \cap \mathbb{R}_+^{\mathscr{C}}. \tag{16.46}$$

There is a case of special interest: If $c \in \overline{\mathbb{R}}_+^{\mathscr{S}}$ is such that $\psi(c)$ lies in $\ker A_k$, then, from (16.42), complex balancing obtains at c. For this reason and because we always have the inclusion

$$\ker A_k \subset \ker YA_k, \tag{16.47}$$

we will want to understand the nature of $\ker A_k$, in particular its intersection with $\overline{\mathbb{R}}_+^{\mathscr{C}}$ and especially with $\mathbb{R}_+^{\mathscr{C}}$. In fact, our interest in $\ker A_k$ is made even more compelling by the following lemma:

Lemma 16.3.1. *Suppose that reaction network $\{\mathscr{S},\mathscr{C},\mathscr{R}\}$ has a deficiency of zero. Then for any $k \in \mathbb{R}^{\mathscr{R}}$*

$$\ker YA_k = \ker A_k. \tag{16.48}$$

Proof. We need to show that $\ker YA_k$ is contained in $\ker A_k$. Suppose that x is a member of $\ker YA_k$. Since $YA_kx = 0$, A_kx lies in $\ker Y$. From (16.39) it is clear that A_kx also lies in $\operatorname{span}(\Delta)$, where Δ is given by (16.8). Thus, A_kx lies in $\ker Y \cap \operatorname{span}(\Delta)$, which Proposition 16.2.1 tells us consists only of the zero vector. Thus x is a member of $\ker A_k$. □

Before proceeding further, it is perhaps worth re-emphasizing the separate roles played by a network's stoichiometric and graphical attributes. The mappings Y and $\psi(\cdot)$ for a mass action system $\{\mathscr{S},\mathscr{C},\mathscr{R},k\}$ are heavily influenced by stoichiometry—that is, by the nature of the complexes contained in $\mathscr{C}$—but not at all by the "reacts to" relation $\mathscr{R}$. Just the reverse is true of the mapping A_k, for while $\mathscr{R}$ enters its construction in a very direct way, $\mathscr{C}$ merely plays the role of an index set. In this sense, A_k relates to the structure of a network essentially through its graphical character, and we can study A_k without paying much attention to the stoichiometric information carried by the set $\mathscr{C}$. As we shall see in the next section, certain qualities of A_k are even independent of k. To a great extent, it is this insensitivity to parameter values that gives parts of chemical reaction network their range and power.

16.4 The First Salt Theorem

One of the most useful ideas in chemical reaction network theory is that, for a mass action system $\{\mathscr{S},\mathscr{C},\mathscr{R},k\}$, ker A_k has a definite *qualitative* structure determined solely by the nature of the underlying network $\{\mathscr{S},\mathscr{C},\mathscr{R}\}$ as a directed graph *and not at all on k—that is, not at all on values of the rate constants*. We are about to state a theorem that will describe the structure of $\ker A_k$. Although the theorem is not so easy to prove, what it says has simple and obvious intuitive appeal when, in a contrived physical setting, the result is given a physical interpretation.

To set the stage, we will begin by writing out a system of scalar equations that characterize the kernel of A_k. Then we will associate those equations with the dynamics of a cluster of interconnected barrels, each containing a salt solution. What we call the *Salt Theorem*—actually a theorem about ker A_k—will tell us how, at steady state, the salt is distributed among the barrels. Viewed in terms of the salt picture, the result will be intuitively obvious. Even so, this not-so-easy-to-prove theorem will have far-reaching reaction network theory implications that are far from transparent.

16.4.1 Scalar Equations Describing ker A_k

Suppose that, for some $k \in \mathbb{R}^{\mathscr{R}}_+$, the vector $b \in \mathbb{R}^{\mathscr{C}}$ lies in the kernel of the map $A_k : \mathbb{R}^{\mathscr{C}} \to \mathbb{R}^{\mathscr{C}}$ defined as in (16.39). Then,

$$A_k\, b = \sum_{y \to y' \in \mathscr{R}} k_{y \to y'}\, b_y\, (\omega_{y'} - \omega_y) = 0 \tag{16.49}$$

or, equivalently,

$$\sum_{y \in \mathscr{C}} \Big[\sum_{\mathscr{R}_{\to y}} k_{y' \to y} b_{y'} - \sum_{\mathscr{R}_{y \to}} k_{y \to y'}\, b_y \Big]\, \omega_y = 0. \tag{16.50}$$

This amounts to a set of scalar equations, one for each complex:

$$\sum_{\mathscr{R}_{\to y}} k_{y' \to y} b_{y'} - \sum_{\mathscr{R}_{y \to}} k_{y \to y'}\, b_y = 0, \quad \forall y \in \mathscr{C}. \tag{16.51}$$

16.4.2 The Salt-Barrel Picture

Now imagine a set of n barrels of fixed identical volume, where n is the number of complexes, with each barrel labeled uniquely with the name of one of the complexes. The barrels are designed to contain a saline solution. The instantaneous concentrations of salt, say in grams per liter, are not necessarily the same in the various barrels. Solution is constantly pumped from some of the barrels to certain other of the barrels: There is a directed pipeline from barrel y to barrel y' for each $y \to y' \in \mathscr{R}$. The pumping rate of solution (in liters per minute) from barrel y to barrel y' is $k_{y \to y'}$. We will suppose that within each barrel the salt mixture spontaneously homogenizes, so that the salt concentration within each barrel is always spatially uniform. The concentrations of salt in the various barrels will be our focus.

Remark 16.4.1. For barrel y the net rate of solution received, in liters per minute, is

$$\sum_{\mathscr{R}_{\to y}} k_{y' \to y} - \sum_{\mathscr{R}_{y \to}} k_{y \to y'}. \tag{16.52}$$

This might be negative, in which case the barrel would eventually empty, or positive, in which case the barrel would eventually explode. We will suppose, therefore, that each barrel has built into it a pump-containing safety duct, connected to an external water reservoir, that provides for the addition or removal of water *but not salt* to compensate for the net addition of solution through the inter-barrel connections. (Each such safety duct is equipped with a filter that prevents flow of salt through it.) Thus, whatever salt is contained within the entire barrel system at any given time is contained there for all time.

Suppose that during some initiating period before the pumps are switched on, the various barrels are each filled with saline solution. The initial salt concentrations within the barrels might differ, perhaps greatly; however, it will help to imagine that, initially, there is at least *some* salt in each barrel. When the pumps are switched on, salt will be carried from one barrel to another. In fact, if the instantaneous salt concentration in barrel y is x_y grams per liter, then the instantaneous rate at which salt is carried from barrel y to barrel y' is $k_{y \to y'} x_y$ grams per minute.

16.4.3 Salt Concentrations at Steady State

Let us consider the distribution of salt concentrations across the barrel system at steady state, when the concentrations no longer change. Such a state requires that, for each barrel, the rate at which salt is pumped to the barrel is exactly balanced by the rate at which salt is pumped from that barrel. If, for each $y \in \mathscr{C}$, b_y denotes the steady-state salt concentration in barrel y, this balance requirement is expressed in the system of equations already written as (16.51). In vector terms, the steady-state salt distribution requires that $b \in \overline{\mathbb{R}}_+^{\mathscr{C}}$ be a member of $\ker A_k$.

What might that distribution be, at least qualitatively? In particular, where will salt be found at steady state? It will be helpful to think about a definite barrel-system example derived from consideration of a mass action system in which the underlying reaction network is that shown in (16.53). The complexes are displayed only symbolically, for the details of their stoichiometry are irrelevant to the example.

$$\begin{array}{ccccc} y_1 & \rightleftarrows & y_2 & \rightarrow & y_3 \\ \upharpoonleft\downharpoonright & & \downharpoonleft\upharpoonright & & \upharpoonleft\downharpoonright \\ y_4 & \rightleftarrows & y_5 & \rightarrow & y_6 \end{array} \qquad y_9 \leftarrow y_7 \rightarrow y_8 \qquad y_{10} \rightleftarrows y_{11} \rightleftarrows y_{12} \tag{16.53}$$

It is not difficult to see that no salt initially in barrels y_3 and y_6 can find its way to any of the other barrels. Thus, salt initially contained in those barrels can only be shared between them at steady state, and, moreover, at steady state each of those two barrels will contain *some* salt. It is only slightly more difficult to see that any salt initially in barrels y_1, y_2, y_4 or y_5 will eventually find its way to barrels y_3 and y_6 and be trapped within those two barrels forever after. Thus, over a long time period, all salt will be depleted from barrels y_1, y_2, y_4 and y_5. At steady state the salt concentrations within them can only be zero.

Similarly, at steady state there will be salt in barrels y_8 and y_9 but none in barrel y_7. As for barrels y_{10}, y_{11}, and y_{12}, we expect that at steady state the three barrels will share, to at least *some* extent, the salt initially contained within them, though perhaps not in equal amounts.

In summary, then, we expect that $b_{y_1}, b_{y_2}, b_{y_4}, b_{y_5}$, and b_{y_7} are all zero. We also expect that $b_{y_3}, b_{y_6}, b_{y_8}, b_{y_9}, b_{y_{10}}, b_{y_{11}}$, and $b_{y_{12}}$ are all strictly positive. *This is to say that we expect the support of b to be the union of the terminal strong-linkage classes of network* (16.53).

There is something else we might intuit, perhaps with less confidence. Imagine that, after executing the salt experiment just described, we add more salt to each of the barrels, again not necessarily in equal amounts. What might the new steady state be? The qualitative situation would certainly be the same as before, with respect to the presence or absence of salt in the various barrels.

But what about quantities? It is natural to expect that, where there is salt, the new steady-state salt concentrations will be higher than they were previously. However, there is also a sharper semiquantitative guess we might make about them: Although the steady-state concentrations are higher, the relative *proportions* of salt in the terminal barrels should be the same as they were in the original steady state, at least in those terminal barrels that are strongly linked.

16.4.4 The First Salt Theorem and Some Corollaries

All of these intuitive guesses are substantiated by the following theorem. Although in the physical context just considered the intuition is clear and devoid of real sophistication, the theorem itself is not so easy to prove without resort to prior machinery.[5] We provide a "from scratch" proof in Appendix 16.A. The theorem will serve as a workhorse for much that we will do.[6]

When we develop deficiency one theory, there will be more salt theorems, motivated by the same salt-barrel picture.

Theorem 16.4.2 (The First Salt Theorem [80]**).** *Let* $\{\mathscr{S},\mathscr{C},\mathscr{R}\}$ *be a reaction network with terminal strong-linkage classes* $\{\Lambda^1, \Lambda^2, \ldots, \Lambda^t\}$*, let* α *be any element of* $\mathbb{R}_+^{\mathscr{R}}$*, and let* $A_\alpha : \mathbb{R}^{\mathscr{C}} \to \mathbb{R}^{\mathscr{C}}$ *be the linear transformation defined by*

$$A_\alpha x := \sum_{y \to y' \in \mathscr{R}} \alpha_{y\to y'} x_y (\omega_{y'} - \omega_y). \tag{16.54}$$

Then the kernel of A_α *has a basis* $\{b^1, b^2, \ldots, b^t\}$ *residing in* $\overline{\mathbb{R}}_+^{\mathscr{C}}$ *and having the property that*

$$\text{supp } b^\theta = \Lambda^\theta, \quad \theta = 1, 2, \ldots, t. \tag{16.55}$$

In particular, the dimension of $\ker A_\alpha$ *is independent of* α *and is equal to the number of terminal strong-linkage classes in the network* $\{\mathscr{S},\mathscr{C},\mathscr{R}\}$*.*

Remark 16.4.3. Theorem 16.4.2, an indispensable tool, tells us that *no matter what values the rate constants take* in our original mass-action invocation of the map A_k, its kernel maintains an inviolable relationship to the graphical structure of the network $\{\mathscr{S},\mathscr{C},\mathscr{R}\}$ from which A_k derives. In particular, $\ker A_k$ will always have dimension t, where t is the number of terminal strong-linkage classes in the network. Moreover, $\ker A_k$ will invariably have a basis with certain attractive features: In that basis there will be a vector b^θ corresponding to terminal strong-linkage class Λ^θ wherein $b_y^\theta > 0$ for all $y \in \Lambda^\theta$ and $b_y^\theta = 0$ for all $y \notin \Lambda^\theta$. These *qualitative* certainties are important to the *qualitative* understanding of chemistry and biology, in which values of rate constants are often known poorly when they are known at all.

In the following corollary of Theorem 16.4.2, we address the question of when $\ker A_\alpha$ contains a positive vector—that is, a vector in $\mathbb{R}_+^{\mathscr{C}}$.

Corollary 16.4.4. *Let* $\{\mathscr{S},\mathscr{C},\mathscr{R}\}$ *be a reaction network, let* α *be any vector of* $\mathbb{R}_+^{\mathscr{R}}$*, and let* A_α *be as in Theorem 16.4.2. Then* $\ker A_\alpha \cap \mathbb{R}_+^{\mathscr{R}}$ *is nonempty if and only if the network is weakly reversible.*

[5] For the special case of weakly reversible networks, Horn [107] gave a proof of an analogous result based on the Perron-Frobenius Theorem.

[6] In fact, in the theorem statement, we have replaced the rate constant symbol k by the more neutral symbol α so that we can later adapt the map A_α to a variety of situations.

Proof. Suppose that there is a positive vector p in $\ker A_\alpha$. If the network is not weakly reversible, then there is a complex y that is not terminal. Theorem 16.4.2 tells us that any vector in $\ker A_\alpha$ has support in the terminal strong-linkage classes, in which case we must have $p_y = 0$, a contradiction.

Suppose on the other hand that the network is weakly reversible, in which case the set of terminal complexes is identical to the set of complexes. The sum of all the basis vectors b^θ, $\theta = 1,2,\ldots,\mathfrak{t}$, in Theorem 16.4.2 lies in $\ker A_\alpha$ and has as its support the set of all terminal complexes, which is to say the set of all complexes. That sum, therefore, is a member of $\ker A_\alpha \cap \mathbb{R}^{\mathscr{R}}_+$. □

For a mass action system $\{\mathscr{S},\mathscr{C},\mathscr{R},k\}$ it is clear that the linear transformation A_k takes values in $\text{span}(\Delta)$, which is to say that the image of A_k lies in $\text{span}(\Delta)$. The following corollary tells us that in robust models it will be the case that $\text{im}\, A_k$ actually *coincides* with $\text{span}(\Delta)$. This is most fortunate, for reasons explained just after the proof, in Remark 16.4.6.

Corollary 16.4.5 ([71, 80]). *Let $\{\mathscr{S},\mathscr{C},\mathscr{R}\}$ be a reaction network, let k be any element of $\mathbb{R}^{\mathscr{R}}_+$, let $A_k : \mathbb{R}^{\mathscr{C}} \to \mathbb{R}^{\mathscr{C}}$ be as in* (16.39), *and let Δ be as in* (16.8). *Then*

$$\text{im}\, A_k = \text{span}(\Delta) \tag{16.56}$$

if and only if each linkage class of the network contains precisely one terminal strong-linkage class. In particular (16.56) *holds if the network is weakly reversible.*

Proof. Since $\text{im}\, A_k$ is contained in $\text{span}(\Delta)$, we need only examine circumstances under which the dimensions of these linear subspaces of $\mathbb{R}^{\mathscr{C}}$ are identical. From Lemma 16.1.2 we have that $\dim \text{span}(\Delta) = n - \ell$, where n is the number of complexes and ℓ is the number of linkage classes in the network. Since $\dim \mathbb{R}^{\mathscr{C}} = n$, it follows from Theorem 16.4.2 and the standard theorem relating the dimensions of the domain, kernel, and image of a linear transformation that

$$\dim \text{im}\, A_k = n - \mathfrak{t} \tag{16.57}$$

where $\mathfrak{t}$ is the number of terminal strong-linkage classes in the network. Thus,

$$\dim \text{span}(\Delta) - \dim \text{im}\, A_k = \mathfrak{t} - \ell, \tag{16.58}$$

so (16.56) holds if and only if $\mathfrak{t} = \ell$—that is, if and only if each linkage class contains precisely one terminal strong-linkage class. This condition holds trivially for weakly reversible networks. □

Remark 16.4.6 (Im A_k vs. Span (Δ)). Note that when the graphical condition in Corollary 16.4.5 is satisfied, the image of A_k is in fact independent of any particular choice of $k \in \mathbb{R}^{\mathscr{R}}_+$. In that case $\text{im}\, A_k$ depends only on the way complexes are partitioned into linkage classes. To some extent it is this fact that makes networks for which $\mathfrak{t} = \ell$ (and weakly reversible networks in particular) relatively easy to study. When $\mathfrak{t}$ exceeds ℓ, the location of $\text{im}\, A_k$ in $\mathbb{R}^{\mathscr{C}}$ will generally be influenced by k.

A primary aim of chemical reaction network theory is to make assertions about connections between the behavior of a kinetic system $\{\mathscr{S},\mathscr{C},\mathscr{R},\mathscr{K}\}$ and properties of its underlying reaction network $\{\mathscr{S},\mathscr{C},\mathscr{R}\}$, assertions that require no knowledge of often-unknown parameter values lurking in the kinetics $\mathscr{K}$. In particular, for a mass action system $\{\mathscr{S},\mathscr{C},\mathscr{R},k\}$, one would like to make behavioral assertions (such as those in the Deficiency Zero Theorem) that are independent of the rate constants carried by k.

There are certain instances in which, for a mass action system, the location of im A_k in $\mathbb{R}^{\mathscr{C}}$ has some importance; for example, in [80] the nature of its intersection with $\ker Y$ plays an important role. Were it usually the case that im A_k is delicately dependent on k, the goal of rate-constant-free dynamical assertions would be seriously obstructed.[7] Corollary 16.4.5, however, carries this good news: *Except perhaps when the underlying network is "forked"—that is, except perhaps when $\mathfrak{t} > \ell$—*im A_k *does not depend on k at all! It is identical to* span (Δ), *which is determined by the reaction network alone.*

And even when a forked mass action system exhibits (usually peculiar) behavior derived solely from the fact that im A_k is smaller than span (Δ), that behavior is *never* robust: A minuscule perturbation of the model—obtained by adding enough reverse reactions (with tiny rate constants) to make the number of terminal strong-linkage classes equal to the number of linkage classes—will cause the new im$A_{k'}$ to coincide with span (Δ). Moreover, addition of such reverse reactions would be greeted with high praise from chemists, who generally advise earnestly that all reactions should be deemed reversible. Corollary 16.4.5 tells us that when im A_k is smaller than span (Δ), it is because those admonishments have gone unheeded to a perhaps reckless extent.

16.5 Salt Theorem Consequences for General Kinetic Systems

The First Salt Theorem (Theorem 16.4.2) derived from consideration of mass action systems. As we shall see in this section, the theorem's range is wider than might be supposed, for it gives important information even for general kinetic systems.

16.5.1 Some Key Propositions, Useful for General Kinetic Systems [71]

We begin with a proposition that is actually a corollary of Theorem 16.4.2, although A_α appears nowhere in it.

Proposition 16.5.1. *For reaction network* $\{\mathscr{S},\mathscr{C},\mathscr{R}\}$ *there exists* $\alpha \in \mathbb{R}_+^{\mathscr{R}}$ *such that*

$$\sum_{y \to y' \in \mathscr{R}} \alpha_{y \to y'} (\omega_{y'} - \omega_y) = 0 \qquad (16.59)$$

if and only if the network is weakly reversible.

[7] In fact, the rate-constant-dependent kinetic subspace would often be smaller than the stoichiometric subspace, and all the associated difficulties described in Appendices 3.A and 8.A would ensue. We will say more about this in Appendix 16.B.

Proof. Suppose first that the network is weakly reversible, and let k be any vector of $\mathbb{R}_+^{\mathscr{R}}$. From Corollary 16.4.4 we know that $\ker A_k$ contains a positive vector $p \in \mathbb{R}_+^{\mathscr{C}}$. Thus,

$$A_k p = \sum_{y \to y' \in \mathscr{R}} k_{y \to y'} p_y (\omega_{y'} - \omega_y) = 0. \tag{16.60}$$

Taking $\alpha_{y \to y'} = k_{y \to y'} p_y$ for all $y \to y' \in \mathscr{R}$, we obtain $\alpha \in \mathbb{R}_+^{\mathscr{R}}$ that satisfies (16.59).

Now suppose, on the other hand, that there is $\alpha \in \mathbb{R}_+^{\mathscr{R}}$ that satisfies (16.59). In this case $\ker A_\alpha$ contains the vector $\omega_{\mathscr{C}} \in \mathbb{R}_+^{\mathscr{C}}$ defined by $(\omega_{\mathscr{C}})_y = 1,\ \forall y \in \mathscr{C}$. From Corollary 16.4.4, the existence of such a positive vector in $\ker A_\alpha$ implies that the network is weakly reversible. □

Recall our discussion in Chapter 3, in reference to a general kinetic system $\{\mathscr{S}, \mathscr{C}, \mathscr{R}, \mathscr{K}\}$, of necessary conditions for the existence of a positive equilibrium or the existence of a cyclic composition trajectory containing a positive composition. In particular, Lemma 3.5.2 said this: For the differential equation of a kinetic system to admit either a positive equilibrium or a cyclic composition trajectory that passes through a positive composition, it is necessary that the reaction vectors for the underlying reaction network be positively dependent. That is, it is necessary that there be an $\alpha \in \mathbb{R}_+^{\mathscr{R}}$ that satisfies

$$\sum_{y \to y' \in \mathscr{R}} \alpha_{y \to y'} (y' - y) = 0. \tag{16.61}$$

The following proposition begins to address this possibility. (In the proposition we do not insist that α be positive.)

Proposition 16.5.2. *For a reaction network $\{\mathscr{S}, \mathscr{C}, \mathscr{R}\}$, $\alpha \in \mathbb{R}^{\mathscr{R}}$ satisfies the equation*

$$\sum_{y \to y' \in \mathscr{R}} \alpha_{y \to y'} (y' - y) = 0 \tag{16.62}$$

if it satisfies the equation

$$\sum_{y \to y' \in \mathscr{R}} \alpha_{y \to y'} (\omega_{y'} - \omega_y) = 0. \tag{16.63}$$

Moreover, if the deficiency of the network is zero, then α satisfies (16.62) ***only if*** *it also satisfies* (16.63).

Proof. It is a trivial matter to see that when (16.63) holds so does (16.62); just act on both sides of (16.63) with Y, the stoichiometric map for the network. On the other hand, when (16.62) holds we have

$$Y\Big[\sum_{y \to y' \in \mathscr{R}} \alpha_{y \to y'} (\omega_{y'} - \omega_y)\Big] = 0, \tag{16.64}$$

in which case

$$\sum_{y \to y' \in \mathscr{R}} \alpha_{y \to y'} (\omega_{y'} - \omega_y) \tag{16.65}$$

lies in ker Y and, clearly, span(Δ). When the network has a deficiency of zero, Proposition 16.2.1 tells us that the zero vector is the only member of $\mathbb{R}^{\mathscr{C}}$ that can lie in both. □

The preceding proposition is stated in terms of α in $\mathbb{R}^{\mathscr{R}}$, not necessarily in $\mathbb{R}^{\mathscr{R}}_{+}$ or even $\overline{\mathbb{R}}^{\mathscr{R}}_{+}$. In the following proposition, we focus on *strictly* positive solutions of (16.61).

Proposition 16.5.3. *Let $\{\mathscr{S}, \mathscr{C}, \mathscr{R}\}$ be a reaction network of deficiency zero. Then there exists $\alpha \in \mathbb{R}^{\mathscr{R}}_{+}$ that satisfies*

$$\sum_{y \to y' \in \mathscr{R}} \alpha_{y \to y'} (y' - y) = 0 \tag{16.66}$$

if and only if the network is weakly reversible.

Proof. This is an immediate consequence of Propositions 16.5.2 and 16.5.1. □

Remark 16.5.4. A network that has a nonzero deficiency might admit a strictly positive solution to (16.66) even if the network is not weakly reversible. For example, the network

$$2A \to A + B \leftarrow 2B \tag{16.67}$$

is not weakly reversible, but, for the network, equation (16.66) admits the strictly positive solution $\alpha_{2A \to A+B} = 1$, $\alpha_{2B \to A+B} = 1$. In this case, however, the network's deficiency is *one* ($n = 3$, $\ell = 1$, $s = 1$).

Taken together, Propositions 16.5.2 and 16.5.3 tell us that for all weakly reversible networks of deficiency zero, there will exist strictly positive solutions to (16.62) and that all such solutions will also satisfy (16.63). In our next proposition, we address the situation for weakly reversible networks of nonzero deficiency.

The proposition will eventually tell us that, for a weakly reversible network of nonzero deficiency, there is invariably a rate constant assignment such that the resulting mass action system is complex balanced and also a different rate constant assignment such that the resulting mass action system is not complex balanced. (See Proposition 16.5.8.) From this it will follow that the weakly reversible deficiency zero networks are the only ones for which complex balancing is ensured regardless of kinetics.

Proposition 16.5.5. *Let $\{\mathscr{S}, \mathscr{C}, \mathscr{R}\}$ be a weakly reversible network. Then there exists $\alpha \in \mathbb{R}^{\mathscr{R}}_{+}$ that satisfies* (16.63) *and, therefore,* (16.62). *If the deficiency of the network is not zero, then there exists $\bar{\alpha} \in \mathbb{R}^{\mathscr{R}}_{+}$ that satisfies* (16.62) *but not* (16.63).

Proof. The existence of $\alpha \in \mathbb{R}^{\mathscr{R}}_{+}$ satisfying (16.63) is already given by Proposition 16.5.1. If the deficiency of the network is not zero, it follows from Proposition 16.2.1 that there exists a nonzero vector g in ker $Y \cap$ span(Δ). From

Lemma 16.1.2 we have $\text{span}(\Delta_{\rightarrow}) = \text{span}(\Delta)$ where, it will be recalled, $\Delta_{\rightarrow}$ is given by (16.7). Because g lies in $\text{span}(\Delta_{\rightarrow})$, there exists $\beta \in \mathbb{R}^{\mathscr{R}}$ satisfying

$$g = \sum_{y \rightarrow y' \in \mathscr{R}} \beta_{y \rightarrow y'} (\omega_{y'} - \omega_y). \tag{16.68}$$

Now let $\alpha \in \mathbb{R}^{\mathscr{R}}_{+}$ satisfy (16.63) and choose $\lambda > 0$ to be sufficiently large that

$$\bar{\alpha} := \beta + \lambda\, \alpha \tag{16.69}$$

is a member of $\mathbb{R}^{\mathscr{R}}_{+}$. From (16.63) and (16.68), it follows that

$$g = \sum_{y \rightarrow y' \in \mathscr{R}} \bar{\alpha}_{y \rightarrow y'} (\omega_{y'} - \omega_y). \tag{16.70}$$

Since g is not the zero vector, it follows from (16.70) that $\bar{\alpha}$ does not satisfy (16.63). To see that $\bar{\alpha}$ does satisfy (16.62), it is only necessary to recall that g is a member of $\ker Y$. Acting with Y on both sides of (16.70), we obtain

$$0 = \sum_{y \rightarrow y' \in \mathscr{R}} \bar{\alpha}_{y \rightarrow y'} (y' - y). \qquad \square \tag{16.71}$$

16.5.2 Proof of the Deficiency Zero Theorem, Part (i)

The material we have accumulated so far in this section provides the foundation for the first part of the Deficiency Zero Theorem.

Proposition 16.5.6 (Part (i) of the Deficiency Zero Theorem). *Let $\{\mathscr{S}, \mathscr{C}, \mathscr{R}\}$ be a reaction network of deficiency zero. If the network is not weakly reversible, then, for an arbitrary kinetics $\mathscr{K}$, the differential equations for the kinetic system $\{\mathscr{S}, \mathscr{C}, \mathscr{R}, \mathscr{K}\}$ cannot admit a positive equilibrium, nor can they admit a cyclic composition trajectory containing a composition at which all species concentrations are positive.*

Proof. In Chapter 3, Lemma 3.5.2 told us that, in order for the differential equations of a general kinetic system to admit either of the two phenomena indicated, the reaction vectors must be positively dependent. This is to say that for the underlying network, there must be a strictly positive α that satisfies (16.66). For a deficiency zero network, however, Proposition 16.5.3 tells us that this can happen only if the network is weakly reversible. $\square$

16.5.3 Complex Balancing and Weak Reversibility

The following proposition asserts that, for a general kinetic system, weak reversibility of the underlying network, regardless of its deficiency, is a necessary condition for the existence of a positive equilibrium at which there is complex balancing.[8]

Proposition 16.5.7 ([107]). *For a kinetic system to admit a positive equilibrium at which complex balancing obtains, it is necessary that the underlying reaction network be weakly reversible.*

Proof. Let $\{\mathscr{S},\mathscr{C},\mathscr{R},\mathscr{K}\}$ be a kinetic system, and suppose that $c^* \in \mathbb{R}_+^{\mathscr{S}}$ is an equilibrium at which complex balancing obtains. This requires that at c^* the complex-formation-rate function take the value zero; that is,

$$g(c^*) := \sum_{y \to y' \in \mathscr{R}} \mathscr{K}_{y \to y'}(c^*)(\omega_{y'} - \omega_y) = 0. \tag{16.72}$$

From the general requirements we imposed on kinetic rate functions—recall Remark 3.2.2—it follows that $\mathscr{K}_{y \to y'}(c^*)$ is positive for all $y \to y' \in \mathscr{R}$. Proposition 16.5.1 then requires that the network $\{\mathscr{S},\mathscr{C},\mathscr{R}\}$ be weakly reversible. □

16.5.4 Complex Balancing in Networks of Nonzero Deficiency

Although the following proposition is nominally about mass action kinetics, it actually tells us something more: We know that for any kinetic system in which the underlying network has a deficiency of zero, there must be complex balancing at every equilibrium. Part *(ii)* of the proposition below tells us that there is no hope that, among the weakly reversible networks, there is a wider class of networks having that same property. In fact, so long as a weakly reversible network has a nonzero deficiency, it can be assigned a kinetics—in particular a mass action kinetics—for which there is a positive equilibrium at which complex balancing does *not* obtain.

Proposition 16.5.8 ([107]). *Let $\{\mathscr{S},\mathscr{C},\mathscr{R}\}$ be a weakly reversible reaction network of arbitrary deficiency.*

(i) There exists for the network a mass action kinetics k such that the resulting mass action system $\{\mathscr{S},\mathscr{C},\mathscr{R},k\}$ admits a positive equilibrium at which complex balancing obtains, in which case complex balancing obtains at all positive equilibria.

[8] The proof provided here emerges as a simple consequence of what we called the First Salt Theorem, a theorem that will find many other uses. Horn provided a different and dedicated "from scratch" proof of Proposition 16.5.7 in [107].

(ii) If the deficiency of the network is greater than zero, then there exists for the network a mass action kinetics $\bar{k}$ such that the resulting mass action system $\{\mathscr{S},\mathscr{C},\mathscr{R},\bar{k}\}$ admits a positive equilibrium at which complex balancing does not obtain, in which case complex balancing does not obtain at any other positive equilibrium.

Proof. This proposition is an almost direct corollary of Proposition 16.5.5. Because the network is weakly reversible, we have the existence of $\alpha \in \mathbb{R}_+^{\mathscr{R}}$ that satisfies

$$\sum_{y \to y' \in \mathscr{R}} \alpha_{y \to y'}(\omega_{y'} - \omega_y) = 0. \tag{16.73}$$

Now let c^* be any composition in $\mathbb{R}_+^{\mathscr{S}}$. For each $y \to y' \in \mathscr{R}$ choose $k_{y \to y'}$ to satisfy

$$\alpha_{y \to y'} = k_{y \to y'}\,(c^*)^y. \tag{16.74}$$

With $k \in \mathbb{R}_+^{\mathscr{R}}$ chosen this way, the complex-formation-rate function for the mass action system $\{\mathscr{S},\mathscr{C},\mathscr{R},k\}$ takes the value zero at c^*, which is to say that complex balancing obtains at c^*, in which case c^* is an equilibrium. That complex balancing obtains at all other positive equilibria follows from Theorem 15.2.2. Thus, *(i)* is proved.

To prove *(ii)* we invoke Proposition 16.5.5 to choose $\bar{\alpha} \in \mathbb{R}_+^{\mathscr{R}}$ that satisfies both

$$\sum_{y \to y' \in \mathscr{R}} \bar{\alpha}_{y \to y'}(y' - y) = 0 \tag{16.75}$$

and

$$\sum_{y \to y' \in \mathscr{R}} \bar{\alpha}_{y \to y'}(\omega_{y'} - \omega_y) \neq 0. \tag{16.76}$$

Now let $c^* \in \mathbb{R}_+^{\mathscr{S}}$ be an arbitrary choice of a positive composition. For each $y \to y' \in \mathscr{R}$ choose $\bar{k}_{y \to y'}$ to satisfy

$$\bar{\alpha}_{y \to y'} = \bar{k}_{y \to y'}\,(c^*)^y. \tag{16.77}$$

With $\bar{k} \in \mathbb{R}_+^{\mathscr{R}}$ constructed in this way, the mass action system $\{\mathscr{S},\mathscr{C},\mathscr{R},\bar{k}\}$ will, by virtue of (16.75), admit c^* as an equilibrium. However, by virtue of (16.76), complex balancing does not obtain at c^*. That complex balancing does not obtain at any other positive equilibrium follows from Theorem 15.2.2. □

Remark 16.5.9 (For every weakly reversible network there is dull behavior for some choice of rate constants). In Chapter 5 our "toy zoo" contained two examples of networks that, taken with mass action kinetics, admit exotic behavior *regardless of rate constant values*. In particular, any mass action system derived from network (5.3) exhibits multiple equilibria within a positive stoichiometric compatibility class, and any mass action system derived from the Lotka network (5.8) exhibits a cyclic composition trajectory along which all species concentrations are positive. Note that neither network (5.3) nor the Lotka network is weakly reversible.

Proposition 16.5.8 tells us that there can be no example of a *weakly reversible* network having either of the two properties described. That is, there can be no weakly reversible network such that, *regardless of rate constant values*, the resulting mass action system will invariably admit multiple positive equilibria or invariably admit a periodic orbit. For a weakly reversible network, there is always an assignment of rate constants for which the resulting mass action system is complex balanced, in which case the dull behavior ensured by Theorem 15.2.4 must result.

Thus, for a *weakly reversible* reaction network, there is a bifurcation problem that arises whenever there are rate constants that give rise to multiple positive equilibria or periodic orbits, for there are invariably different rate constants for which those same phenomena are absent.

16.5.5 Weakly Reversible Networks Are Invariably Nondegenerate: Proof

In Chapters 10 and 11, some of the most important assertions were made for networks that are *nondegenerate*. Degenerate networks are those that are so pathological that, for *every* differentiably monotonic kinetics, the derivative of the species-formation-rate function is singular at *every* positive composition. In Theorem 10.6.6 we asserted without proof the important fact that every weakly reversible network is nondegenerate. We are now in a position to give a proof.

Theorem 16.5.10. *Every weakly reversible network is nondegenerate.*

Proof. Weak reversibility and Proposition 16.5.8 ensure that, for the network under consideration, there is a rate constant assignment such that the resulting mass action system admits a positive equilibrium at which complex balancing obtains. For that system, Theorem 15.2.4 tells us that at any positive equilibrium the derivative of the species formation function is nonsingular. Thus, the network is nondegenerate. □

16.5.6 Which Species Can Coexist at an Equilibrium of a Deficiency Zero Kinetic System?

Earlier in the book, in Section 3.7, we addressed questions about where, for a general kinetic system $\{\mathscr{S},\mathscr{C},\mathscr{R},\mathscr{K}\}$, an equilibrium might be found on the boundary of $\overline{\mathbb{R}}_+^{\mathscr{S}}$. Our interest was in the nature of the set of species that might coexist at an equilibrium. We also addressed the question of where on the boundary a cyclic composition trajectory might reside. In both cases we argued that the support of an equilibrium or of a cyclic composition trajectory—that is, the set of species actually present—is sharply restricted: it must be *reaction-transitive* (Definition 3.7.3).

At the close of Section 3.7, we asserted that significantly sharper statements could be made when the deficiency of the underlying reaction network is *zero*. In Section 7.4, when there was more language available, we made just such a statement about the nature of deficiency zero equilibria in the form of Theorem 7.4.1.

In informal terms the theorem said this, among other things: *If c^* is an equilibrium, then each nonterminal complex must contain at least one species that is absent at c^*, and each reaction having a nonterminal reactant complex must be "switched* *off." Moreover, at c^* all reactions having their reactant complexes in a particular terminal strong-linkage class are "switched on" or else all are "switched off."* The theorem was stated without proof. It is our aim in this section to provide its theoretical foundations.

We begin by recalling that our definition of a kinetics in Chapter 3 required that the rate of a reaction be positive at a particular composition if and only if, at that composition, all species in the reaction's *reactant* complex are present. Thus, at a given composition, either *all* reactions sharing the same reactant complex are switched on or *none* are. This idea provides motivation for the following definition, where as usual $\mathscr{R}_{y\to}$ denotes the set of all reactions having y as reactant complex.

Definition 16.5.11. *For a chemical reaction network $\{\mathscr{S},\mathscr{C},\mathscr{R}\}$ an element $\alpha \in \overline{\mathbb{R}}_+^{\mathscr{R}}$ is* **reactant-complex-toggled** *if, for each $y \in \mathscr{C}$,* supp α *either contains all members of $\mathscr{R}_{y\to}$ or else it contains none of them. In this case, members of the set*

$$\mathscr{C}_\alpha^\circ := \{y \in \mathscr{C} : \exists\, y \to y' \in \text{supp } \alpha\} \tag{16.78}$$

are the **toggled-on reactant complexes**, *and members of the set $\mathscr{R}_\alpha^\circ :=$ supp α are the* **toggled-on reactions**.

Proposition 16.5.12. *Suppose that, for reaction network $\{\mathscr{S},\mathscr{C},\mathscr{R}\}$, $\alpha \in \overline{\mathbb{R}}_+^{\mathscr{R}}$ is a nonzero reactant-complex-toggled solution of the equation*

$$\sum_{y\to y' \in \mathscr{R}} \alpha_{y\to y'}(\omega_{y'} - \omega_y) = 0. \tag{16.79}$$

Then the set of toggled-on reactant complexes is the union of terminal strong-linkage classes. That is, there are terminal strong-linkage classes Λ^1, ,Λ^2, ... Λ^k such that

$$\mathscr{C}_\alpha^\circ = \bigcup_{\theta=1}^{k} \Lambda^\theta. \tag{16.80}$$

Moreover, the subnetwork $\{\mathscr{S},\mathscr{C}_\alpha^\circ,\mathscr{R}_\alpha^\circ\}$ corresponding to the toggled-on reactions is weakly reversible.

Proof. Suppose that $\alpha \in \overline{\mathbb{R}}_+^{\mathscr{R}}$ is a nonzero reactant-complex-toggled solution of (16.79). Choose α^* to be any vector in $\mathbb{R}_+^{\mathscr{R}}$ that coincides with α on supp α and is positive for all $y \to y' \notin$ supp α. Let $A_{\alpha^*} : \mathbb{R}^{\mathscr{C}} \to \mathbb{R}^{\mathscr{C}}$ be defined in the usual way:

$$A_{\alpha^*}x := \sum_{y\to y' \in \mathscr{R}} \alpha^*_{y\to y'}x_y(\omega_{y'} - \omega_y). \tag{16.81}$$

Moreover let

$$\omega_{\mathscr{C}_\alpha^\circ} := \sum_{y\in\mathscr{C}_\alpha^\circ} \omega_y. \tag{16.82}$$

Thus, $(\omega_{\mathscr{C}_\alpha^\circ})_y = 1$ if y is a toggled-on complex and is zero otherwise.

It is not difficult to see from (16.79) that $\omega_{\mathscr{C}_\alpha^\circ}$ lies in ker A_{α^*}. As a result the salt theorem (Theorem 16.4.2) tells us that the support of $\omega_{\mathscr{C}_\alpha^\circ}$—that is, $\mathscr{C}_\alpha^\circ$—is the union of terminal strong-linkage classes. For the subnetwork $\{\mathscr{S}, \mathscr{C}_\alpha^\circ, \mathscr{R}_\alpha^\circ\}$ the linkage classes are then the terminal strong-linkage classes $\Lambda^1, ,\Lambda^2, \dots \Lambda^k$ of $\{\mathscr{S},\mathscr{C},\mathscr{R}\}$ described in the proposition statement. Because the complexes within each Λ^θ are strongly linked relative to the reactions in $\mathscr{R}_\alpha^\circ$, the subnetwork is weakly reversible. □

For networks of deficiency zero, Proposition 16.5.12 has an important and almost immediate corollary. The corollary is similar to the proposition itself, the crucial difference residing in the distinction between equations (16.79) and (16.83).

Proposition 16.5.13. *Suppose that, for a reaction network $\{\mathscr{S},\mathscr{C},\mathscr{R}\}$ of deficiency zero, $\alpha \in \overline{\mathbb{R}}_+^{\mathscr{R}}$ is a nonzero reactant-complex-toggled solution of the equation*

$$\sum_{y\to y' \in \mathscr{R}} \alpha_{y\to y'}(y'-y) = 0. \tag{16.83}$$

Then the set of toggled-on reactant complexes is the union of terminal strong-linkage classes. That is, there are terminal strong-linkage classes $\Lambda^1, ,\Lambda^2, \dots \Lambda^k$ such that

$$\mathscr{C}_\alpha^\circ = \bigcup_{\theta=1}^{k} \Lambda^\theta. \tag{16.84}$$

Moreover, the subnetwork $\{\mathscr{S}, \mathscr{C}_\alpha^\circ, \mathscr{R}_\alpha^\circ\}$ corresponding to the toggled-on reactions is weakly reversible.

Proof. With Y denoting the stoichiometric map for network $\{\mathscr{S},\mathscr{C},\mathscr{R}\}$, equation (16.83) can be written in the form

$$Y\Big[\sum_{y\to y' \in \mathscr{R}} \alpha_{y\to y'}(\omega_{y'}-\omega_y)\Big] = 0. \tag{16.85}$$

Because the sum in the bracket clearly lies in span(Δ), it must be a member of $\ker Y \cap \operatorname{span}(\Delta)$. Because the deficiency of network $\{\mathscr{S},\mathscr{C},\mathscr{R}\}$ is zero, Proposition 16.2.1 tells us that the sum in the bracket can only be the zero vector. Proposition 16.5.12 then gives the desired result . □

We are now in a position to prove Theorem 7.4.1, which we repeat below as Theorem 16.5.14.

Theorem 16.5.14 (Deficiency Zero Boundary Equilibria Theorem). *Suppose that $\{\mathscr{S},\mathscr{C},\mathscr{R}\}$ is a deficiency zero reaction network, that $\mathscr{K}$ is a kinetics for the network (not necessarily mass action), and that $c^* \in \overline{\mathbb{R}}_+^{\mathscr{S}}$ is an equilibrium of the kinetic system $\{\mathscr{S},\mathscr{C},\mathscr{R},\mathscr{K}\}$.*

(i) If $y \in \mathscr{C}$ is not a terminal complex, then supp $y \not\subset$ supp c^*. *In particular, there is at least one species in* supp y *that is absent at composition c^*, and, for any reaction $y \to y'$ having y as its reactant complex, it must be the case that $\mathscr{K}_{y\to y'}(c^*) = 0$.*

(ii) If $y^* \in \mathscr{C}$ is a member of terminal strong-linkage class Λ and $\operatorname{supp} y^* \subset \operatorname{supp} c^*$, *then* $\operatorname{supp} y \subset \operatorname{supp} c^*$ *for every complex $y \in \Lambda$. Moreover, $\mathscr{K}_{y\to y'}(c^*) > 0$ for every reaction $y \to y'$ such that y is in Λ.*

Proof. Suppose that a kinetic system $\{\mathscr{S},\mathscr{C},\mathscr{R},\mathscr{K}\}$ admits an equilibrium $c^* \in \overline{\mathbb{R}}_+^{\mathscr{S}}$, not necessarily positive. Then

$$\sum_{\mathscr{R}} \mathscr{K}_{y\to y'}(c^*)(y'-y) = 0. \tag{16.86}$$

Consider $\alpha \in \overline{\mathbb{R}}_+^{\mathscr{R}}$ defined by

$$\alpha_{y\to y'} := \mathscr{K}_{y\to y'}(c^*), \quad \forall y \to y' \in \mathscr{R}. \tag{16.87}$$

From properties of a kinetics given in Definition 3.2.1, it follows that $y \to y'$ is a member of supp α if and only if supp $y \subset \operatorname{supp} c^*$. Thus α is reactant-complex-toggled solution of

$$\sum_{y\to y' \in \mathscr{R}} \alpha_{y\to y'}(y'-y) = 0 \tag{16.88}$$

with

$$\mathscr{C}_\alpha^\circ = \{y \in \mathscr{C} : \operatorname{supp} y \subset \operatorname{supp} c^*\} \tag{16.89}$$

and

$$\mathscr{R}_\alpha^\circ = \{y \to y' \in \mathscr{R} : \operatorname{supp} y \subset \operatorname{supp} c^*\}. \tag{16.90}$$

Hereafter we will presume that $\mathscr{C}_\alpha^\circ$ is not empty, for otherwise the theorem is true vacuously.

Now suppose that the deficiency of network $\{\mathscr{S},\mathscr{C},\mathscr{R}\}$ is zero. In this case, Proposition 16.5.13 tells us that $\mathscr{C}_\alpha^\circ$, given by (16.89), is the union of terminal strong-linkage classes. Therefore, if y is a nonterminal complex, it must be the case that supp $y \not\subset$ supp c^*. On the other hand, if y^* is such that $\operatorname{supp} y^* \subset$ supp c^* and y^* is a member of terminal strong-linkage class Λ, then every member of Λ lies in $\mathscr{C}_\alpha^\circ$. This is to say that if y is a member of Λ, then supp $y \subset$ supp c^*. From this it follows that $\mathscr{K}_{y\to y'}(c^*) > 0$ for every reaction $y \to y'$ such that y is in Λ. □

Remark 16.5.15. In Remark 7.4.2 of Chapter 7, we asserted that results of the kind given in Theorem 16.5.14 carry over from equilibria to cyclic composition trajectories. To see that this is so, we need only recall that if $c(\cdot) : [t_1,t_2] \to \overline{\mathbb{R}}_+^{\mathscr{S}}$ is a periodic solution of the differential equations for a kinetic system $\{\mathscr{S},\mathscr{C},\mathscr{R},\mathscr{K}\}$, then from (3.21) we have

$$\sum_{y\to y' \in \mathscr{R}} \Big(\int_{t_1}^{t_2} \mathscr{K}_{y\to y'}(c(t))\, dt\Big)(y'-y) = 0. \tag{16.91}$$

Let $\alpha \in \overline{\mathbb{R}}_+^{\mathscr{R}}$ be defined by

$$\alpha_{y\to y'} := \int_{t_1}^{t_2} \mathscr{K}_{y\to y'}(c(t))\, dt, \quad \forall y \to y' \in \mathscr{R}. \tag{16.92}$$

If the cyclic composition trajectory has support $\mathscr{S}^*$—see Remark 7.4.2—then $\alpha_{y \to y'}$ is positive if supp$y$ is contained in $\mathscr{S}^*$ and is zero otherwise. Thus, $\alpha \in \overline{\mathbb{R}}_+^{\mathscr{R}}$ is a reaction-complex-toggled solution of

$$\sum_{y \to y' \in \mathscr{R}} \alpha_{y \to y'}(y' - y) = 0 \tag{16.93}$$

with

$$\mathscr{C}_\alpha^\circ = \{y \in \mathscr{C} : \text{supp}\, y \subset \mathscr{S}^*\} \tag{16.94}$$

and

$$\mathscr{R}_\alpha^\circ = \{y \to y' \in \mathscr{R} : \text{supp}\, y \subset \mathscr{S}^*\}. \tag{16.95}$$

If the reaction network $\{\mathscr{S}, \mathscr{C}, \mathscr{R}\}$ has a deficiency of zero, then the reasoning in the proof of Theorem 16.5.14 applies, with supp c^* replaced by $\mathscr{S}^*$. As a result, we can make the following statement, repeated from Remark 7.4.2: *If $y \in \mathscr{C}$ is not terminal, then there is at least one species in* supp y *that is not present in* $\mathscr{S}^*$. *On the other hand, if y^* is a terminal complex with* supp y^* *contained in* $\mathscr{S}^*$, *then, for every other complex y in the terminal strong-linkage class containing y^*,* $\mathscr{S}^*$ *contains* supp y.

16.6 Positive Equilibria for Weakly Reversible Deficiency Zero Mass Action Systems

To complete the proof of the Deficiency Zero Theorem, we must establish the existence of a positive equilibrium for any mass action system in which the underlying reaction network is weakly reversible and has a deficiency of zero. In fact, this is already a consequence of work by Horn in 1972 [107]. Moreover, what we need is subsumed by the later and far broader results of Deng, Nachman, et al. [55] and Boros [28], to the effect that *any* weakly reversible mass action system, *regardless of deficiency*, has a positive equilibrium. (Recall Remark 8.7.2.)

Here we present a relatively short and transparent proof in the deficiency zero case, different from Horn's. It is based on the First Salt Theorem, an elementary property of deficiency zero networks, and this simple consequence of standard facts in linear algebra: If, in a real vector space with scalar product, $\{v_1, v_2, \ldots, v_k\}$ is a linearly independent set and $\{\xi_1, \xi_2, \ldots, \xi_k\}$ is a set of real numbers, then there is a vector x such that $v_i \cdot x = \xi_i,\ i = 1, 2, \ldots, k$. For the record, here is the proposition we want to prove:

Proposition 16.6.1. *There is a positive equilibrium for any mass action system in which the underlying reaction network is weakly reversible and has a deficiency of zero.*

Proof. For a mass action system $\{\mathscr{S},\mathscr{C},\mathscr{R},k\}$ in which reaction network $\{\mathscr{S},\mathscr{C},\mathscr{R}\}$ is weakly reversible and has a deficiency of zero, our aim is to show that there is a positive equilibrium, which is to say that there is a $c \in \mathbb{R}_+^{\mathscr{S}}$ such that

$$YA_k\,\Big(\sum_{y\,\in\mathscr{C}} c^y\omega_y\Big) = 0. \tag{16.96}$$

Here $Y : \mathbb{R}^{\mathscr{C}} \to \mathbb{R}^{\mathscr{S}}$ is the stoichiometric map for the network and $A_k : \mathbb{R}^{\mathscr{C}} \to \mathbb{R}^{\mathscr{C}}$ is as before. For a deficiency zero network, Lemma 16.3.1 tells us that c will satisfy (16.96) if and only if

$$A_k\,\Big(\sum_{y\,\in\mathscr{C}} c^y\omega_y\Big) = 0. \tag{16.97}$$

Our objective will be achieved if we can show that there is an $x \in \mathbb{R}^{\mathscr{S}}$ such that

$$\sum_{y\,\in\mathscr{C}} e^{y\cdot x}\omega_y \in \ker A_k\,, \tag{16.98}$$

for then we can take $c \in \mathbb{R}_+^{\mathscr{S}}$ to satisfy $x = \ln c$, in which case (16.97) *will be satisfied.*

As usual we denote by n the number of complexes in the network $\{\mathscr{S},\mathscr{C},\mathscr{R}\}$, by ℓ the number of linkage classes, and by s its rank or, equivalently, the dimension of the stoichiometric subspace. Suppose that the network has linkage classes $\{\mathscr{L}^1,\mathscr{L}^2,\ldots,\mathscr{L}^\ell\}$, which for a weakly reversible network are identical to the terminal strong-linkage classes. Let $\{b^1,\ b^2,\ \ldots,\ b^\ell\}$ be the basis for $\ker A_k$ described in Theorem 16.4.2. We want to show, then, that there is an $x \in \mathbb{R}^{\mathscr{S}}$ and (necessarily positive) numbers $\lambda_1,\ \lambda_2,\ \ldots,\ \lambda_\ell$ such that

$$\sum_{y\,\in\mathscr{C}} e^{y\cdot x}\omega_y = \lambda_1 b^1 + \lambda_2 b^2 + \cdots + \lambda_\ell\, b^\ell. \tag{16.99}$$

Because each b^θ has support $\mathscr{L}^\theta$, this is equivalent to the requirement that there exist x and positive $\lambda_1,\ \lambda_2,\ \ldots,\ \lambda_\ell$ that satisfy the following system of equations: For $\theta = 1,2,\ldots,\ell$

$$y\cdot x = \ln\,\lambda_\theta + \ln\,b_y^\theta, \quad \forall y \in \mathscr{L}^\theta. \tag{16.100}$$

To show that this requirement can be met, we choose within each linkage class $\mathscr{L}^\theta$ a particular complex that we will denote by y_*^θ. It is not difficult to see that the $n-\ell$ vectors

$$\bigcup_{\theta=1}^{\ell}\{y - y_*^\theta :\ y \in\ \mathscr{L}^\theta \setminus y_*^\theta\} \tag{16.101}$$

constitute a set of generators for the stoichiometric subspace, S. From this and the fact that, for a deficiency zero network, $\dim\, S = s = n - \ell$, it follows that the set (16.101) is linearly independent. Thus, there exists $x \in \mathbb{R}^{\mathscr{S}}$ such that for $\theta = 1,2,\ldots,\ell$

$$(y - y_*^\theta)\cdot x = \ln\,b_y^\theta - \ln\,b_{y_*^\theta}^\theta, \quad \forall y \in \mathscr{L}^\theta \setminus y_*^\theta. \tag{16.102}$$

With x chosen this way, we can then choose positive $\lambda_1,\ \lambda_2,\ \ldots,\ \lambda_\ell$ to satisfy

$$y_*^\theta\cdot x = \ln\,\lambda_\theta + \ln\,b_{y_*^\theta}^\theta, \quad \theta = 1,2,\ldots,\ell. \tag{16.103}$$

These choices suffice to satisfy (16.100) for $\theta = 1,2,\ldots,\ell$. □

16.7 Every Weakly Reversible Deficiency Zero Mass Action System Is Complex Balanced and Quasi-Thermodynamic

Having established the existence of a positive equilibrium for every weakly reversible deficiency zero mass action system, we are in a position to posit the following theorem, which is a direct consequence of Theorem 16.2.2, Corollary 16.2.3, and Proposition 16.6.1:

Theorem 16.7.1. *Every mass action system derived from a weakly reversible deficiency zero network is complex balanced and therefore quasi-thermodynamic.*

16.8 Proofs of the Deficiency Zero Theorem and Its Offshoots: Summing Up

We are now quite close to the completion of proofs for several theorems presented in Chapter 7, all related to networks of deficiency zero. To a great extent, the proofs to follow amount to little more than making simple connections between theorems and propositions already accumulated.

16.8.1 Proof of the Deficiency Zero Theorem, Part (ii)

In Section 16.5.2 we proved part (i) of Theorem 7.1.1, the Deficiency Zero Theorem. We have now accumulated the essential material for a proof of part (ii), which we state below as Proposition 16.8.1.

Proposition 16.8.1 (The Deficiency Zero Theorem, Part (ii)). *Consider a mass action system derived from a weakly reversible deficiency zero reaction network. Regardless of rate constant values, the resulting differential equations have the following properties: There exists within each positive stoichiometric compatibility class precisely one equilibrium; that equilibrium is asymptotically stable; and there cannot exist a nontrivial cyclic composition trajectory along which all species concentrations are positive.*

Proof. From Theorem 16.7.1 it follows that the mass action system under consideration is quasi-thermodynamic. The conclusion of Proposition 16.8.1 derives from properties of quasi-thermodynamic systems given in Theorem 13.4.4. This completes the proof of the Deficiency Zero Theorem. □

16.8.2 Proof of the Derivative Version of the Deficiency Zero Theorem

We are also in a position to prove the derivative version of the Deficiency Zero Theorem, stated in Chapter 7 and which we repeat below as Theorem 16.8.2.

Theorem 16.8.2 (Derivative Version of the Deficiency Zero Theorem). *Let $\{\mathscr{S},\mathscr{C},\mathscr{R}\}$ be a weakly reversible deficiency zero reaction network with stoichiometric subspace S. For any assignment of rate constants $k \in \mathbb{R}_+^{\mathscr{R}}$, the resulting mass action system $\{\mathscr{S},\mathscr{C},\mathscr{R},k\}$ has the following properties: If c^* is a positive equilibrium, then the derivative $df(c^*) : S \to S$ of the species-formation-rate function at c^* is nonsingular. In fact, relative to the scalar product "$*$" in $\mathbb{R}^{\mathscr{S}}$ given by*

$$u * w := \sum_{s \in \mathscr{S}} \frac{u_s w_s}{c_s^*}, \tag{16.104}$$

$df(c^*)$ is negative-definite. If the network is not only weakly reversible but also forest-like (and therefore reversible), then $df(c^*)$ is symmetric relative to the "$*$" scalar product. In this case, all roots of the characteristic polynomial of $df(c^*)$ are real and negative; moreover, there exists in S a basis consisting of eigenvectors of $df(c^*)$.

Proof. From Theorem 16.2.2, complex balancing obtains at the positive equilibrium c^*. The result for the general weakly reversible case then follows from Theorem 15.2.2. For the narrower forest-like (reversible) case, the stronger result is already included in Theorem 14.3.4, which was proved in Chapter 14, on detailed balancing. □

16.8.3 Proof of the Deficiency Zero Multicell Theorem

Finally, we can complete the argument for the Shapiro-Horn Deficiency Zero Multicell Theorem of Chapter 7. We repeat the theorem below.

Theorem 16.8.3 (Deficiency Zero Multicell Theorem). *Consider a multicell system in which the intracellular kinetics is mass action and the underlying intracellular reaction network is weakly reversible and has a deficiency of zero. Then, regardless of the deficiency of the full multicell network and regardless of parameter values, the corresponding differential equations for the multicell system have the following properties: In each positive stoichiometric compatibility class, there is precisely one multicell equilibrium; that equilibrium is asymptotically stable; and at that equilibrium the concentration of each species is the same in every cell. Moreover, the differential equations cannot admit a cyclic composition trajectory along which all species concentrations are positive.*

Proof. If the intracellular kinetic system is mass action and its underlying reaction network is weakly reversible and has a deficiency of zero, then the intracellular kinetic system is complex balanced, by virtue of Theorem 16.7.1. The conclusion of Theorem 16.8.3 then follows from Theorem 15.3.2. □

Appendix 16.A Proof of the First Salt Theorem

In this appendix we prove Theorem 16.4.2, which is repeated below as Theorem 16.A.1.

Theorem 16.A.1 (The First Salt Theorem). *Let $\{\mathscr{S},\mathscr{C},\mathscr{R}\}$ be a reaction network with terminal strong-linkage classes $\{\Lambda^1,\Lambda^2,\dots,\Lambda^t\}$, let α be any element of $\mathbb{R}_+^{\mathscr{R}}$, and let $A_\alpha:\mathbb{R}^{\mathscr{C}}\to\mathbb{R}^{\mathscr{C}}$ be the linear transformation defined by*

$$A_\alpha x := \sum_{y\to y'\in\mathscr{R}} \alpha_{y\to y'} x_y(\omega_{y'}-\omega_y). \tag{16.A.1}$$

Then the kernel of A_α has a basis $\{b^1,b^2,\dots,b^t\}$ residing in $\overline{\mathbb{R}}_+^{\mathscr{C}}$ and having the property that

$$\text{supp } b^\theta = \Lambda^\theta, \quad \theta = 1,2,\dots,t. \tag{16.A.2}$$

In particular, the dimension of $\ker A_\alpha$ is independent of α and is equal to the number of terminal strong-linkage classes in the network $\{\mathscr{S},\mathscr{C},\mathscr{R}\}$.

The proof [80] of Theorem 16.A.1 will proceed by means of a series of lemmas. As before, we denote by $\mathscr{R}_{\to y}$ and $\mathscr{R}_{y\to}$, respectively, the set of all reactions having y as the product complex and the set of all reactions having y as the reactant complex. That is, for each $y\in\mathscr{C}$,

$$\mathscr{R}_{\to y} = \{y^*\to y^{**}\in\mathscr{R} : y^{**}=y\} \quad\text{and}\quad \mathscr{R}_{y\to} = \{y^*\to y^{**}\in\mathscr{R} : y^*=y\}. \tag{16.A.3}$$

Note that

$$\mathscr{R} = \bigcup_{y\in\mathscr{C}} \mathscr{R}_{\to y} = \bigcup_{y\in\mathscr{C}} \mathscr{R}_{y\to}. \tag{16.A.4}$$

Some additional notation will set the stage for the first lemma. For each $x\in\mathbb{R}^{\mathscr{C}}$, we denote by $|\,x\,|$ the vector in $\mathbb{R}^{\mathscr{C}}$ obtained by taking componentwise absolute values:

$$|\,x\,|_y := |\,x_y\,|, \quad \forall y\in\mathscr{C}. \tag{16.A.5}$$

Lemma 16.A.2. *If x is a member of $\ker A_\alpha$, then so is $|\,x\,|$.*

Proof. For each $x\in\mathbb{R}^{\mathscr{C}}$ we can rewrite (16.A.1) in the form

$$A_\alpha x = \sum_{y\in\mathscr{C}}\Big\{\sum_{\mathscr{R}_{\to y}} \alpha_{y'\to y}x_{y'} - \sum_{\mathscr{R}_{y\to}} \alpha_{y\to y'}x_y\Big\}\omega_y. \tag{16.A.6}$$

If either $\mathscr{R}_{\to y}$ or $\mathscr{R}_{y\to}$ is empty, the corresponding sum is understood to be zero.

Now suppose that x is a member of $\ker A_\alpha$. Because of the independence of the set $\{\omega_y\}_{y\in\mathscr{C}}$, each braced quantity in (16.A.6) is zero, whereupon we can write

$$\sum_{\mathscr{R}_{\to y}} \alpha_{y'\to y}\,|\,x_{y'}\,| \ge \sum_{\mathscr{R}_{y\to}} \alpha_{y\to y'}\,|\,x_y\,|, \quad \forall y\in\mathscr{C}. \tag{16.A.7}$$

From this it follows that, for $x \in \ker A_\alpha$,

$$\sum_{y\in\mathscr{C}} \sum_{\mathscr{R}_{\to y}} \alpha_{y'\to y} \mid x_{y'} \mid \geq \sum_{y\in\mathscr{C}} \sum_{\mathscr{R}_{y\to}} \alpha_{y\to y'} \mid x_y \mid \tag{16.A.8}$$

with equality holding if and only if equality holds in (16.A.7) for every $y \in \mathscr{C}$. However, equality *must* hold in (16.A.8), for both sides amount to a sum of $\alpha_{y\to y'} \mid x_y \mid$ over all $y \to y' \in \mathscr{R}$. Thus, for $x \in \ker A_\alpha$ we have

$$\sum_{\mathscr{R}_{\to y}} \alpha_{y'\to y} \mid x_{y'} \mid = \sum_{\mathscr{R}_{y\to}} \alpha_{y\to y'} \mid x_y \mid, \quad \forall y \in \mathscr{C}. \tag{16.A.9}$$

As a consequence,

$$A_\alpha \mid x \mid = \sum_{y\in\mathscr{C}} \{ \sum_{\mathscr{R}_{\to y}} \alpha_{y'\to y} \mid x_{y'} \mid - \sum_{\mathscr{R}_{y\to}} \alpha_{y\to y'} \mid x_y \mid \} \omega_y = 0. \tag{16.A.10}$$

Thus, $\mid x \mid$ lies in $\ker A_\alpha$. □

In preparation for our next lemma, we recall from Definition 6.5.2 that in the reaction network $\{\mathscr{S}, \mathscr{C}, \mathscr{R}\}$ a complex y *ultimately reacts to* a complex y', denoted $y \Rightarrow y'$, if either $y = y'$ or there is a directed reaction path leading from y to y'.

Lemma 16.A.3. *If x is a member of* $\ker A_\alpha$, *if $x_y = 0$, and if $y' \Rightarrow y$, then $x_{y'} = 0$.*

Proof. If $y = y'$, the result follows trivially. If there is a direct reaction $y' \to y$, the result follows from (16.A.9). Suppose, on the other hand, that there is a reaction sequence $y' \to y_1 \to y_2 \cdots \to y_k \to y$. We know that $x_y = 0$ implies that $x_{y_k} = 0$, which then implies that $x_{y_{k-1}} = 0$, and so on up the chain to finally give $x_{y'} = 0$. □

Lemma 16.A.4. *Let Λ be a strong-linkage class (not necessarily terminal). If x is in* $\ker A_\alpha$ *and if $x_y = 0$ for some $y \in \Lambda$, then $x_{y'} = 0$ for all $y' \in \Lambda$. Equivalently, $\Lambda \cap \operatorname{supp} x \neq \emptyset$ implies that $\Lambda \subset \operatorname{supp} x$.*

Proof. If y' is in the same strong-linkage class as y, then $y' \Rightarrow y$. The result follows from the preceding lemma. □

Lemma 16.A.5. *Let $\mathscr{J} \subset \mathscr{C}$ be the set of all complexes that lie in terminal strong-linkage classes. If x lies in* $\ker A_\alpha$, *then* $\operatorname{supp} x$ *lies in $\mathscr{J}$. Stated differently, $x_y = 0$ if $y \notin \mathscr{J}$.*

Proof. For $\mathscr{J} = \mathscr{C}$ there is nothing to prove. Suppose, then, that $\mathscr{J} \neq \mathscr{C}$. For each $y \in \mathscr{J}$ we partition the set $\mathscr{R}_{\to y}$ of all reactions leading directly to y into two parts, those having reactant complex internal to $\mathscr{J}$ and those having reactant complex external to $\mathscr{J}$:

$$\mathscr{R}_{\to y}^{\text{int}} := \{y' \to y \in \mathscr{R}_{\to y} : y' \in \mathscr{J}\} \quad \text{and} \quad \mathscr{R}_{\to y}^{\text{ext}} := \{y' \to y \in \mathscr{R}_{\to y} : y' \notin \mathscr{J}\}. \tag{16.A.11}$$

Note that $\mathscr{R}_{\to y}$ is the disjoint union of these sets. Now we sum both sides of (16.A.9) over $\mathscr{J}$ to obtain

$$\sum_{y \in \mathscr{J}} \sum_{\mathscr{R}^{\text{int}}_{\to y}} \alpha_{y' \to y} \mid x_{y'} \mid + \sum_{y \in \mathscr{J}} \sum_{\mathscr{R}^{\text{ext}}_{\to y}} \alpha_{y' \to y} \mid x_{y'} \mid = \sum_{y \in \mathscr{J}} \sum_{\mathscr{R}_{y \to}} \alpha_{y \to y'} \mid x_y \mid . \quad (16.A.12)$$

The first term on the left of (16.A.12) is equal to the quantity on the right, for both amount to the sum of all reactions of the form $y \to y'$ with both y and y' residing in $\mathscr{J}$. Thus, we have

$$\sum_{y \in \mathscr{J}} \sum_{\mathscr{R}^{\text{ext}}_{\to y}} \alpha_{y' \to y} \mid x_{y'} \mid = 0. \quad (16.A.13)$$

Now let

$$\mathscr{G} := \{y' \in \mathscr{C} \setminus \mathscr{J} : \exists\, y' \to y \in \mathscr{R} \text{ with } y \in \mathscr{J}\}. \quad (16.A.14)$$

This is the set of all nonterminal complexes that react directly to a terminal complex. From (16.A.13) it follows that $x_{y'} = 0$ for all $y' \in \mathscr{G}$. It remains to be shown that $x_{y^*} = 0$ for every nonterminal complex y^* that is not in $\mathscr{G}$—that is, for every nonterminal complex that does not react *directly* to a terminal complex. However, because $\mathscr{C}$ is finite, it can be argued without great difficulty that for such a y^*, there is a complex $y' \in \mathscr{G}$ to which y^* ultimately reacts. As a result, we have the existence of y' such that $x_{y'} = 0$ and $y^* \Rightarrow y'$. That $x_{y^*} = 0$ then follows from Lemma 16.A.3. □

Recall that in the statement of Theorem 16.A.1, $\{\Lambda^1, \Lambda^2, \ldots, \Lambda^t\}$ are the terminal strong-linkage classes of the network under consideration. We denote by Γ^θ the linear subspace of $\mathbb{R}^{\mathscr{C}}$ consisting of all vectors with support in $\Lambda^\theta, \theta = 1, 2, \ldots, t$. From Lemma 16.A.5 we have

$$\ker A_\alpha \subset \Gamma^1 \oplus \Gamma^2 \oplus \cdots \oplus \Gamma^t. \quad (16.A.15)$$

Lemma 16.A.6. *For $\theta = 1, 2, \ldots, t$, Γ^θ is invariant under A_α. That is, if x lies in Γ^θ, then $A_\alpha x$ lies in Γ^θ.*

Proof. We denote by $\mathscr{R}^\theta$ the set of those reactions having reactant complex in Λ^θ. That is,

$$\mathscr{R}^\theta := \{y \to y' \in \mathscr{R} : y \in \Lambda^\theta\}. \quad (16.A.16)$$

Let x lie in Γ^θ, in which case $x_y = 0$ for all $y \notin \Lambda^\theta$. From (16.A.1) it follows that

$$A_\alpha x := \sum_{y \to y' \in \mathscr{R}^\theta} \alpha_{y \to y'} x_y (\omega_{y'} - \omega_y). \quad (16.A.17)$$

If $\mathscr{R}^\theta$ is empty, as when Λ^θ consists of a single (product) complex, $A_\alpha x = 0$, and the conclusion of the lemma obtains. Suppose that $\mathscr{R}^\theta$ is not empty. From the fact that Λ^θ is a *terminal* strong-linkage class (Definition 6.6.3), it follows that, for every $y \to y' \in \mathscr{R}^\theta$, *both* y and y' are members of Λ^θ. In each term of (16.A.1), then, both ω_y and $\omega_{y'}$ are members of Γ^θ. From this it follows that $A_\alpha x$ lies in Γ^θ. □

Lemma 16.A.7. *For each $\theta = 1, 2, \ldots, t$ there is a $b^\theta \in \Gamma^\theta$ such that*

(i) $A_\alpha b^\theta = 0$

(ii) $b_y^\theta > 0, \quad \forall y \in \Lambda^\theta$

(iii) If x^θ *lies in* Γ^θ *and* $A_\alpha x^\theta = 0$, *then* $x^\theta = \xi b^\theta$ *for some number* ξ.

Proof. Let $A_\alpha^\theta : \Gamma^\theta \to \Gamma^\theta$ be the restriction of A_α to Γ^θ. Then

$$A_\alpha^\theta x \equiv \sum_{y \to y' \in \mathscr{R}^\theta} \alpha_{y \to y'} x_y (\omega_{y'} - \omega_y). \tag{16.A.18}$$

Note that $\operatorname{im} A_\alpha^\theta$ is smaller than Γ^θ; in fact, it is easy to see that the vector

$$\omega_{\Lambda^\theta} := \sum_{y \in \Lambda^\theta} \omega_y \tag{16.A.19}$$

in Γ^θ is orthogonal to im A_α^θ, relative to the standard scalar product in $\mathbb{R}^{\mathscr{C}}$. Because dim $\operatorname{im} A_\alpha^\theta < \dim \Gamma^\theta$ and because, from a standard theorem in linear algebra,

$$\dim \Gamma^\theta = \dim \operatorname{im} A_\alpha^\theta + \dim \ker A_\alpha^\theta,$$

we know that dim $\ker A_\alpha^\theta > 0$. Thus, there is a nonzero vector $b^\theta \in \ker A_\alpha \cap \Gamma^\theta$. Clearly, supp b^θ is contained in Λ^θ. Moreover, because supp $b^\theta \cap \Lambda^\theta \neq \varnothing$, Lemma 16.A.4 tells us that Λ^θ is contained in supp b^θ. Therefore, supp $b^\theta = \Lambda^\theta$, which is to say that $b_y^\theta \neq 0$ for every $y \in \Lambda^\theta$. Lemma 16.A.2 tells us that we can take each b_y^θ to be positive. This completes the proof of parts (i) and (ii).

To prove (iii) we consider a vector $x^\theta \in \Gamma^\theta$ such that $A_\alpha x^\theta = 0$. For some particular $y^* \in \Lambda^\theta$, let ξ be such that $x_{y^*}^\theta - \xi b_{y^*}^\theta = 0$. Because both x^θ and b^θ lie in ker A_α, so does the vector $x^\theta - \xi b^\theta$, in which case Lemma 16.A.4 tells us that $x^\theta - \xi b^\theta = 0$ or, equivalently, that $x^\theta = \xi b^\theta$. □

To prove Theorem 16.A.1, we will show that the set of vectors $\{b^1, b^2, \ldots, b^t\}$ described in Lemma 16.A.7 is a basis for $\ker A_\alpha$. Because these vectors have disjoint support, they are linearly independent. It remains to be shown that they constitute a set of generators for $\ker A_\alpha$. Suppose, then, that x is a member of $\ker A_\alpha$. From (16.A.15) we have the decomposition

$$x = x^1 + x^2 + \cdots + x^t, \quad x^\theta \in \Gamma^\theta, \quad \theta = 1, 2, \ldots, t. \tag{16.A.20}$$

Because x is contained in ker A_α, we also have

$$A_\alpha x = A_\alpha x^1 + A_\alpha x^2 + \cdots + A_\alpha x^t = 0. \tag{16.A.21}$$

From Lemma 16.A.6 we know that $A_\alpha x^\theta$ is a member of Γ^θ, $\theta = 1, 2, \ldots, t$. Because the Γ^θ are independent, it must be the case that

$$A_\alpha x^\theta = 0, \quad \theta = 1, 2, \ldots, t. \tag{16.A.22}$$

From this and part (iii) of Lemma 16.A.7, it follows that there are numbers $\xi_1, \xi_2, \ldots, \xi_t$ such that $x^\theta = \xi_\theta b^\theta$, $\theta = 1, 2, \ldots, t$. Therefore,

$$x = \xi_1 b^1 + \xi_2 b^2 + \cdots + \xi_t b^t. \tag{16.A.23}$$

Because $x \in \ker A_\alpha$ was arbitrary, (16.A.23) tells us that $\{b^1, b^2, \ldots, b^t\}$ is a (linearly independent) set of generators for $\ker A_\alpha$, so it is a basis. This completes the proof of Theorem 16.A.1. □

Appendix 16.B The Kinetic and Stoichiometric Subspaces: Salt-Theorem Insights

In two appendices earlier in the book, one in Chapter 3 and another in Chapter 8, we discussed the kinetic subspace for a kinetic system $\{\mathscr{S},\mathscr{C},\mathscr{R},\mathscr{K}\}$, which resides within (and might coincide with) the stoichiometric subspace for the underlying network $\{\mathscr{S},\mathscr{C},\mathscr{R}\}$. With mass action systems in mind, we described in Appendix 8.A problems that ensue when the kinetic subspace is actually smaller than the stoichiometric subspace. To some extent, these problems result from the fact that, while the stoichiometric subspace is an attribute of the network alone, the kinetic subspace is not. It depends on the particular kinetics at hand, which might be poorly known.

We also indicated in Appendix 8.A that, fortunately, for mass action systems, it will typically be the case that, no matter what values the rate constants take, the kinetic and stoichiometric subspaces coincide. In particular this coincidence will occur whenever, in the underlying reaction network, each linkage class contains no more than one terminal strong-linkage class. This is always the case for weakly reversible networks and, in particular, for reversible networks.

We also argued in Appendix 8.A that a mass-action-model phenomenon that derives solely from lack of coincidence of the kinetic and stoichiometric subspaces will not be robust. Perturbations of the model that simply add reverse reactions (with tiny rate constants) will cause the resulting kinetic subspace to coincide with the stoichiometric subspace.

Much of the discussion in Appendix 8.A derived from a theorem stated there without proof. We now have the equipment, in particular the First Salt Theorem, to give the proof. That is the purpose of this appendix. The theorem we want to prove is repeated below:

Theorem 16.B.1 ([73, 80]). *Consider a reaction network $\{\mathscr{S},\mathscr{C},\mathscr{R}\}$ having ℓ linkage classes, $\mathfrak{t}$ terminal strong-linkage classes, deficiency δ, and stoichiometric subspace $S \subset \mathbb{R}^{\mathscr{S}}$. Suppose that for a particular $k \in \mathbb{R}^{\mathscr{R}}_{+}$, the mass action system $\{\mathscr{S},\mathscr{C},\mathscr{R},k\}$ gives rise to the kinetic subspace $K \subset \mathbb{R}^{\mathscr{S}}$. Then the following statements hold true:*

(i) If $\mathfrak{t} = \ell$, then, regardless of rate constant values, $K = S$. That is, if each linkage class in the underlying network contains no more than one terminal strong-linkage class, there is no distinction between the kinetic and stoichiometric subspaces.

(ii) If $\mathfrak{t} - \ell > \delta$, then, regardless of rate constant values, K is smaller than S. In fact,

$$\dim S - \dim K \geq \mathfrak{t} - \ell - \delta. \tag{16.B.1}$$

In particular, if $\delta = 0$, then

$$\dim S - \dim K = \mathfrak{t} - \ell. \tag{16.B.2}$$

(iii) If $t > \ell$, if $t - \ell \geq \delta$, and if the rate constants are such that the mass action system $\{\mathscr{S},\mathscr{C},\mathscr{R},k\}$ admits a positive equilibrium, then K is smaller than S. In fact,

$$\dim S - \dim K > t - \ell - \delta. \tag{16.B.3}$$

In particular, if $\delta = 1$, then

$$\dim S - \dim K = t - \ell. \tag{16.B.4}$$

Remark 16.B.2. In item *(iii)* the supposition that there be a positive equilibrium is stronger than it need be. As we shall see, it is enough that there be an equilibrium c^* such that $\operatorname{supp} y \subset \operatorname{supp} c^*$ for some nonterminal complex y.

Proof (of Theorem 16.B.1). We will begin by recalling the definition of the kinetic subspace, focused however on the mass action system $\{\mathscr{S},\mathscr{C},\mathscr{R},k\}$ of the theorem statement. In this case the species-formation-rate function $f(\cdot) : \overline{\mathbb{R}}_+^{\mathscr{S}} \to S$ takes the form given earlier in (16.43):

$$f(c) = YA_k\,\psi(c), \quad \forall c \in \overline{\mathbb{R}}_+^{\mathscr{S}}. \tag{16.B.5}$$

Here $Y : \mathbb{R}^{\mathscr{C}} \to \mathbb{R}^{\mathscr{S}}$ is the stoichiometric map for the network, $A_k : \mathbb{R}^{\mathscr{C}} \to \mathbb{R}^{\mathscr{C}}$ is, as usual, defined by

$$A_k\,x := \sum_{y \to y' \in \mathscr{R}} k_{y\to y'}\, x_y\,(\omega_{y'} - \omega_y), \tag{16.B.6}$$

and $\psi : \overline{\mathbb{R}}_+^{\mathscr{S}} \to \overline{\mathbb{R}}_+^{\mathscr{C}}$ is defined by

$$\psi(c) := \sum_{y \in \mathscr{C}} c^y\,\omega_y. \tag{16.B.7}$$

It will be useful to remember that

$$\operatorname{im} A_k \subset \operatorname{span} \Delta \tag{16.B.8}$$

where

$$\Delta := \{\omega_{y'} - \omega_y \in \mathbb{R}^{\mathscr{C}} : y \sim y'\}, \tag{16.B.9}$$

with $y \sim y'$ indicating that y and y' are in the same linkage class.

The *kinetic subspace* $K \subset \mathbb{R}^{\mathscr{S}}$ for the mass action system $\{\mathscr{S},\mathscr{C},\mathscr{R},k\}$ is defined by

$$K := \operatorname{span}(\operatorname{im} f). \tag{16.B.10}$$

That is, K is the smallest linear subspace in $\mathbb{R}^{\mathscr{S}}$ containing the image of $f(\cdot)$. From this and (16.B.5), it follows that

$$K := YA_k\,[\operatorname{span}(\operatorname{im}\psi)], \tag{16.B.11}$$

which is to say that K is the result of acting with YA_k on $\operatorname{span}(\operatorname{im}\psi)$, the smallest linear subspace of $\mathbb{R}^{\mathscr{C}}$ containing the image of $\psi(\cdot)$.

Lemma 16.B.3. $\operatorname{span}(\operatorname{im}\psi) = \mathbb{R}^{\mathscr{C}}$.

Proof. Suppose that $\operatorname{span}(\operatorname{im}\psi)$ is smaller than $\mathbb{R}^{\mathscr{C}}$. Then there is a nonzero vector $a \in \mathbb{R}^{\mathscr{C}}$ orthogonal to $\operatorname{span}(\operatorname{im}\psi)$, relative to the standard scalar product in $\mathbb{R}^{\mathscr{C}}$. In that case, we have

$$a \cdot \psi(c) = \sum_{y \in \mathscr{C}} a_y \prod_{s \in \mathscr{S}} c_s^{y_s} = 0, \quad \forall c \in \overline{\mathbb{R}}_+^{\mathscr{S}}. \tag{16.B.12}$$

This amounts to the existence of a nontrivial polynomial that is identically zero on $\overline{\mathbb{R}}_+^{\mathscr{S}}$, an impossibility. □

The lemma and (16.B.11) allow us to write[9]

$$K = \operatorname{im} YA_k \tag{16.B.13}$$

or, equivalently,

$$K = Y[\operatorname{im} A_k]. \tag{16.B.14}$$

On the other hand, from the proof of Proposition 16.2.1, it follows that

$$S = Y[\operatorname{span}(\Delta)]. \tag{16.B.15}$$

We are now in a position to prove part *(i)* of Theorem 16.B.1. Corollary 16.4.5 of the salt theorem tells us that $\operatorname{im} A_k = \operatorname{span}(\Delta)$ when $t = \ell$. In this case, (16.B.14) and (16.B.15) tell us that K and S are identical. Part *(i)* is proved.

When t exceeds ℓ, Corollary 16.4.5 tells us that $\operatorname{im} A_k$ is smaller than $\operatorname{span}(\Delta)$. It might still happen, though, that $K = S$, depending on the location of $\operatorname{im} A_k$ in $\mathbb{R}^{\mathscr{C}}$ relative to $\ker Y$. The following lemma will help to explore this possibility and to provide a foothold for proof of part *(ii)* of Theorem 16.B.1.

Lemma 16.B.4. *Let n be the number of complexes in the reaction network $\{\mathscr{S},\mathscr{C},\mathscr{R}\}$ of the theorem statement. Then*

$$\dim S = n - \ell - \dim\left[\ker Y \cap \operatorname{span}(\Delta)\right] \tag{16.B.16}$$

and, for the mass action system $\{\mathscr{S},\mathscr{C},\mathscr{R},k\}$,

$$\dim K = n - t - \dim\left[\ker Y \cap \operatorname{im} A_k\right]. \tag{16.B.17}$$

Proof. Equation (16.B.16) is a simple consequence of equation (16.31) and the fact that s, the rank of the network, is identical to $\dim S$. To prove (16.B.17) we let $\hat{Y} : \operatorname{im} A_k \to \mathbb{R}^{\mathscr{S}}$ be the restriction of $Y : \mathbb{R}^{\mathscr{C}} \to \mathbb{R}^{\mathscr{S}}$ to $\operatorname{im} A_k$. From the standard theorem relating the dimensions of the domain, kernel, and image of a linear transformation, we obtain

---

[9] Because of this equation, the following analysis is largely about the image of YA_k, a linear transformation that has a life of its own, apart from its role in determining the kinetic subspace. (See, for example, its separate role in the proof of Proposition 16.B.9.) The symbol K can, in what follows, be replaced everywhere with $\operatorname{im} YA_k$.

$$\dim \operatorname{im} A_k = \dim \ker \hat{Y} + \dim \operatorname{im} \hat{Y}. \tag{16.B.18}$$

From the salt theorem—see in particular (16.57) in the proof of Corollary 16.4.5—we know that $\dim \operatorname{im} A_k = n - t$. Moreover, $\operatorname{im} \hat{Y} = Y[\operatorname{im} A_k] = K$. Finally, $\ker \hat{Y} = \ker Y \cap \operatorname{im} A_k$. From these, the formula (16.B.17) follows. □

From the preceding lemma, we obtain the equation

$$\dim S - \dim K = t - \ell - \dim [\ker Y \cap \operatorname{span}(\Delta)] + \dim [\ker Y \cap \operatorname{im} A_k]. \tag{16.B.19}$$

From Proposition 16.2.1 we have

$$\delta = \dim [\ker Y \cap \operatorname{span}(\Delta)], \tag{16.B.20}$$

so we can also write

$$\dim S - \dim K = t - \ell - \delta + \dim [\ker Y \cap \operatorname{im} A_k], \tag{16.B.21}$$

from which (16.B.1) follows.

To complete the proof of part *(ii)* of Theorem 16.B.1 we need to show that if the reaction network $\{\mathscr{S},\mathscr{C},\mathscr{R}\}$ has a deficiency of zero, then, regardless of rate constant values, we have

$$\dim S - \dim K = t - \ell. \tag{16.B.22}$$

To see that this is so, we note that $\delta = 0$ implies that $\dim [\ker Y \cap \operatorname{span}(\Delta)] = 0$. Because $\operatorname{im} A_k$ resides in $\operatorname{span}(\Delta)$, we must also have $\dim [\ker Y \cap \operatorname{im} A_k] = 0$. The result then follows from (16.B.21).

We turn now to proof of part *(iii)* of Theorem 16.B.1. We begin with a lemma.

Lemma 16.B.5. *Suppose that for a mass action system $\{\mathscr{S},\mathscr{C},\mathscr{R},k\}$ there exists $x \in \mathbb{R}^{\mathscr{C}}$ for which $YA_kx = 0$ and for which $x_y \neq 0$ for some $y \in \mathscr{C}$ not in a terminal strong-linkage class. Then*

$$\dim [\ker Y \cap \operatorname{im} A_k] \geq 1. \tag{16.B.23}$$

In fact, if the network is not weakly reversible, such an x exists when $\ker YA_k$ *contains a positive vector (i.e., a vector in $\mathbb{R}^{\mathscr{C}}_+$), in particular when the mass action system $\{\mathscr{S},\mathscr{C},\mathscr{R},k\}$ admits a positive equilibrium.*

Proof. From the salt theorem (Theorem 16.4.2), it follows that all nonzero vectors in $\ker A_k$ have support entirely on complexes residing in terminal strong-linkage classes. Thus, for the x in the lemma statement, we have $A_kx \neq 0$. Because $YA_kx = 0$, the nonzero vector A_kx lies in $\ker Y \cap \operatorname{im} A_k$, the dimension of which cannot then be zero. The last sentence of the lemma is a consequence of the following: If the network is not weakly reversible, there is at least one complex y that is not terminal. If $x \in \ker YA_k$ is positive, then $x_y \neq 0$. If c^* is a positive equilibrium, then $\psi(c^*)$ is a positive vector in $\ker YA_k$. □

Remark 16.B.6. If, for a mass action system $\{\mathscr{S},\mathscr{C},\mathscr{R},k\}$, the hypothesis of Lemma 16.B.5 is satisfied and the deficiency of network $\{\mathscr{S},\mathscr{C},\mathscr{R}\}$ is *one*, then we actually have the equality

$$\dim[\ker Y \cap \operatorname{im} A_k] = 1. \tag{16.B.24}$$

This is because $\operatorname{im} A_k$ is contained in $\operatorname{span}(\Delta)$, so, from Proposition 16.2.1 and Lemma 16.B.5,

$$\delta = 1 = \dim[\ker Y \cap \operatorname{span}(\Delta)] \geq \dim[\ker Y \cap \operatorname{im} A_k] \geq 1, \tag{16.B.25}$$

from which (16.B.24) follows. In this case,

$$\ker Y \cap \operatorname{span}(\Delta) = \ker Y \cap \operatorname{im} A_k, \tag{16.B.26}$$

and, from (16.B.21) and Proposition 16.2.1, we have the *equality*

$$\dim S - \dim K = t - \ell. \tag{16.B.27}$$

To prove part *(iii)* of Theorem 16.B.1, we observe that if $t > \ell$, the network is not weakly reversible. If the mass action system $\{\mathscr{S},\mathscr{C},\mathscr{R},k\}$ admits a positive equilibrium, then the hypothesis of Lemma 16.B.5 is satisfied, in which case the inequality (16.B.23) results. This and (16.B.21) give

$$\dim S - \dim K > t - \ell - \delta. \tag{16.B.28}$$

If $\delta = 1$, then (16.B.27) follows from Remark 16.B.6. This completes the proof of Theorem 16.B.1 □

Remark 16.B.7. The assumption in part *(iii)* of Theorem 16.B.1 that there be a positive equilibrium is stronger than it need be. We can instead suppose only that there exists for the mass action system $\{\mathscr{S},\mathscr{C},\mathscr{R},k\}$ an equilibrium $c^* \in \overline{\mathbb{R}}_+^{\mathscr{S}}$ such that, for some *nonterminal* complex y, supp y is contained in supp c^*. Because c^* is an equilibrium, we have from (16.B.5) the inclusion $\psi(c^*) \in \ker YA_k$. Because supp y is contained in supp c^*, we also have $\psi_y(c^*) \neq 0$. In this case, the hypothesis of Lemma 16.B.5 is satisfied. The proof of part *(iii)* can then proceed in the same way as before.

Remark 16.B.8 (Network-structural bounds on dim ker YA_k*).* If, for a mass action system $\{\mathscr{S},\mathscr{C},\mathscr{R},k\}$, $c^* \in \overline{\mathbb{R}}_+^{\mathscr{S}}$ is an equilibrium, we must have $\psi(c^*) \in \ker YA_k$. It is for this reason that we have stressed the importance of understanding the nature of ker YA_k. Not only its location in $\mathbb{R}^{\mathscr{C}}$ but even its dimension can depend on k. The following proposition, related to the theorem just proved, gives dimension bounds *that are independent of k*. They depend on reaction network structure alone.

Proposition 16.B.9 ([71]). *Let $\{\mathscr{S},\mathscr{C},\mathscr{R}\}$ be a reaction network having ℓ linkage classes, t terminal strong-linkage classes, and deficiency δ. Moreover, let $Y : \mathbb{R}^{\mathscr{C}} \to \mathbb{R}^{\mathscr{S}}$ be the stoichiometric map for the network, and, for an arbitrary $k \in \mathbb{R}_+^{\mathscr{R}}$, let $A_k : \mathbb{R}^{\mathscr{C}} \to \mathbb{R}^{\mathscr{C}}$ be as in* (16.B.6). *Then*

$$\ell + \delta \le \dim\ker YA_k \le t + \delta. \tag{16.B.29}$$

If $\delta = 0$, *then* $\dim\ker YA_k = t$. *If* $\delta = 1$ *and* $\ker YA_k$ *contains a positive vector (in particular, if the mass action system* $\{\mathscr{S},\mathscr{C},\mathscr{R},k\}$ *admits a positive equilibrium), then* $\dim\ker YA_k = t + 1$.

The following is an obvious corollary of Proposition 16.B.9, which has some importance of its own.

Corollary 16.B.10. *If* $t = \ell$ *for the reaction network* $\{\mathscr{S},\mathscr{C},\mathscr{R}\}$, *then for any* $k \in \mathbb{R}_+^{\mathscr{R}}$

$$\dim\ker YA_k = \ell + \delta = t + \delta. \tag{16.B.30}$$

We turn now to proof of Proposition 16.B.9.

Proof (of Proposition 16.B.9). We begin by noting that, from a standard theorem of linear algebra relating the dimensions of the domain, kernel, and image of a linear transformation,

$$\dim\ker YA_k = n - \dim \operatorname{im} YA_k, \tag{16.B.31}$$

where n is the number of complexes. Note that for a mass action system $\{\mathscr{S},\mathscr{C},\mathscr{R},k\}$ the kinetic subspace K is synonymous with $\operatorname{im} YA_k$. Thus, assertions about K in the proof and discussion of Theorem 16.B.1 are apt here as well, with K replaced by $\operatorname{im} YA_k$.

To get the lower bound in the proposition statement, we recall that $\operatorname{im} YA_k$ is contained in S, the stoichiometric subspace for the underlying network. For this reason,

$$\dim \operatorname{im} YA_k \le \dim S = s,$$

where s is the rank of the network. Thus, we can write

$$\dim\ker YA_k \ge n - s = \ell + \delta. \tag{16.B.32}$$

To get the upper bound in the proposition statement, we invoke (16.B.13) and (16.B.17) in (16.B.31), and then Proposition 16.2.1, to write

$$\dim\ker YA_k = t + \dim\left[\ker Y \cap \operatorname{im} A_k\right] \le t + \dim\left[\ker Y \cap \operatorname{span}(\Delta)\right] = t + \delta. \tag{16.B.33}$$

Finally, if $\delta = 0$, Lemma 16.3.1 and the salt theorem (Theorem 16.4.2) tell us that $\dim\ker YA_k = t$. If $\delta = 1$ and $\ker YA_k$ contains a positive vector, then the first equality in (16.B.33) and Remark 16.B.6 tell us that $\dim\ker YA_k = t + 1$. □

17

Deficiency One Theory Foundations

This chapter has two goals. First, we want to provide not only a proof of the Deficiency One Theorem but also a guide to the thinking behind the proof. Another salt theorem, motivated by the same salt-barrel picture met in the last chapter, will play a crucial role. Second, we want to elaborate more fully on the so-called Deficiency One Algorithm described in Section 8.5. The algorithm itself will be laid out in far more detail, and its theoretical underpinnings will be explained. In this, the salt-barrel picture will find use yet again.

17.1 Proof of the Deficiency One Theorem

It will help if we first state the Deficiency One Theorem again, slightly rephrased to make more explicit the requirement that each linkage class contain precisely one terminal strong-linkage class:

Theorem 17.1.1 (The Deficiency One Theorem). *Consider a reaction network* $\{\mathscr{S},\mathscr{C},\mathscr{R}\}$ *having* ℓ *linkage classes. Suppose that the deficiency of the entire network and the deficiencies of the individual linkage classes satisfy the following conditions:*

(i) $\delta_\theta \leq 1, \quad \theta = 1,2,\ldots,\ell,$

(ii) $\sum_{\theta=1}^{\ell} \delta_\theta = \delta.$

Moreover, suppose that

(iii) each linkage class contains just one terminal strong-linkage class.

If, for a particular $k \in \mathbb{R}_+^{\mathscr{R}}$, *the mass action system* $\{\mathscr{S},\mathscr{C},\mathscr{R},k\}$ *admits a positive equilibrium, then each positive stoichiometric compatibility class contains precisely one equilibrium, and at every positive equilibrium, the derivative of the species-formation-rate function is nonsingular. If the network is weakly reversible, then, for every* $k \in \mathbb{R}_+^{\mathscr{R}}$, *the mass action system* $\{\mathscr{S},\mathscr{C},\mathscr{R},k\}$ *admits a positive equilibrium.*

M. Feinberg, *Foundations of Chemical Reaction Network Theory*,
Applied Mathematical Sciences 202,
https://doi.org/10.1007/978-3-030-03858-8_17

Remark 17.1.2. The argument will follow closely the one given in [76], but with more motivation provided along the way. We will not be concerned here with proof of the theorem's final sentence, about the existence of a positive equilibrium in the weakly reversible case when conditions *(i)* and *(ii)* of the theorem are satisfied; for a weakly reversible network condition, *(iii)* is satisfied trivially. A proof was given in [76], and a related one was given by Boros [25].[1] Both of these are subsumed by the far more general results of Deng, Nachman et al. [55], and Boros [28] that assert the existence of a positive equilibrium for *any* mass action system in which the underlying reaction network is weakly reversible.

17.1.1 A Little About Proof Strategy

There are two things we need to prove for a mass action system that satisfies conditions *(i)–(iii)* and that admits a positive equilibrium: First, that in each positive stoichiometric compatibility class there is precisely one equilibrium and, second, that at all positive equilibria the derivative of the species-formation-rate function is nonsingular. It is tempting to suppose that the latter might provide the basis for a proof of the former, but in fact both will be proved more or less simultaneously by virtually identical methods.

In this discussion of strategy, let us focus, for the sake of concretion, on proof of existence and uniqueness of equilibria within the various positive stoichiometric compatibility classes. In the case of the Deficiency Zero Theorem, the chain of argument for a weakly reversible mass action system $\{\mathscr{S},\mathscr{C},\mathscr{R},k\}$ connected the deficiency zero condition to the inevitability of complex balancing at a positive equilibrium, which in turn ensured that the mass action system under consideration is quasi-thermodynamic and, in particular, *quasi-thermostatic*. Thus, the chain of reasoning passed through complex balancing to the certainty that the positive equilibrium set E takes a certain form: If $c^* \in \mathbb{R}_+^{\mathscr{S}}$ is a particular equilibrium and $S \subset \mathbb{R}^{\mathscr{S}}$ is the stoichiometric subspace, then

$$E := \{c \in \mathbb{R}_+^{\mathscr{S}} : \ln c - \ln c^* \in S^\perp\}. \tag{17.1}$$

As we showed in Theorem 13.3.3, each positive stoichiometric compatibility class contains precisely one member of the set E given by (17.1).

In the case of the Deficiency One Theorem, there is no hope that the argument can similarly pass through complex balancing, except perhaps in special cases. Note that the theorem embraces mass action systems for which the underlying reaction network is not weakly reversible, and, for those, complex balancing at a positive equilibrium is impossible. (Recall Proposition 16.5.7.) And even if the network *is* weakly reversible, Proposition 16.5.8 tells us that, unless the network has a deficiency of zero, there will inevitably be rate constant assignments for which complex balancing does *not* obtain at any positive equilibrium.

[1] In [27], Boros showed that a positive equilibrium exists for any mass action system in which the underlying network is weakly reversible and has a deficiency of one.

This is not to say that a mass action system that satisfies the conditions of the Deficiency One Theorem cannot be quasi-thermostatic, for complex balancing is merely sufficient for the quasi-thermostatic property, not necessary. Indeed, this possibility will provide our path: *We will show that when a reaction network satisfies conditions (i)–(iii) and the rate constants are such that the resulting mass action system admits a positive equilibrium, then that mass action system is quasi-thermostatic*. From Theorem 13.3.3 it will then follow that each positive stoichiometric compatibility class contains precisely one equilibrium.

Remark 17.1.3. Although we will show that mass action systems that admit positive equilibria and that satisfy the conditions of the Deficiency One Theorem are *quasi-thermostatic*, there is no hope that all such systems will also be *quasi-thermodynamic*. That hope is dashed by the mass action system (8.5). It admits an unstable positive equilibrium and a cyclic composition trajectory along which all species concentrations are positive—behavior which, according to Theorem 13.4.4, is impossible for a quasi-thermodynamic mass action system.

17.1.2 Some Preliminaries

Hereafter in Section 17.1, it will be understood that network $\{\mathscr{S},\mathscr{C},\mathscr{R}\}$, with stoichiometric subspace S, satisfies conditions *(i)–(iii)* of Theorem 17.1.1 and that *the mass action system $\{\mathscr{S},\mathscr{C},\mathscr{R},k\}$ admits a positive equilibrium c^**. In this subsection we will consider and reformulate separately the two assertions we must affirm: that the mass action system $\{\mathscr{S},\mathscr{C},\mathscr{R},k\}$ is quasi-thermostatic and that, at the (generic) positive equilibrium c^*, the derivative of the species-formation-rate function is nonsingular. Then we will show that both assertions, reformulated, are in fact just special cases of a more general assertion, one that we will set out to prove.

Remark 17.1.4 (When complex balancing obtains at c^*). For the mass action system $\{\mathscr{S},\mathscr{C},\mathscr{R},k\}$ it might happen that complex balancing obtains at the positive equilibrium c^* (whereupon the network $\{\mathscr{S},\mathscr{C},\mathscr{R}\}$ must be weakly reversible). In that case, there is nothing to prove, for the quasi-thermostatic property and nonsingularity of the derivative at c^* of the species-formation-rate function are already ensured by Theorem 15.2.2. For this reason we will assume that complex balancing does *not* obtain at c^*.

17.1.2.1 Reformulating the Quasi-Thermostatic Problem

Because $c^* \in \mathbb{R}_+^{\mathscr{S}}$ is an equilibrium of the mass action system $\{\mathscr{S},\mathscr{C},\mathscr{R},k\}$ it must satisfy the equation

$$\sum_{y\to y'\in\mathscr{R}} k_{y\to y'}(c^*)^y(y'-y) = 0. \tag{17.2}$$

To show that $\{\mathscr{S},\mathscr{C},\mathscr{R},k\}$ is quasi-thermostatic, we must show that the positive equilibrium set is of the form (17.1). That is, we must show that $c^{**} \in \mathbb{R}_+^{\mathscr{S}}$ satisfies

$$\sum_{y \to y' \in \mathscr{R}} k_{y \to y'} (c^{**})^y (y' - y) = 0, \tag{17.3}$$

if and only if

$$\ln c^{**} - \ln c^* \in S^\perp. \tag{17.4}$$

Now let $\kappa \in \mathbb{R}_+^{\mathscr{R}}$ be defined by

$$\kappa_{y \to y'} := k_{y \to y'} (c^*)^y, \quad \forall y \to y' \in \mathscr{R}. \tag{17.5}$$

Then equations (17.2) and (17.3) are equivalent to

$$\sum_{y \to y' \in \mathscr{R}} \kappa_{y \to y'} (y' - y) = 0. \tag{17.6}$$

and

$$\sum_{y \to y' \in \mathscr{R}} \kappa_{y \to y'} \exp[y \cdot (\ln c^{**} - \ln c^*)]\,(y' - y) = 0. \tag{17.7}$$

For this reason, we will have shown the mass action system $\{\mathscr{S}, \mathscr{C}, \mathscr{R}, k\}$ to be quasi-thermostatic if we can prove the following assertion:

Assertion 17.1.5 *Let $\{\mathscr{S}, \mathscr{C}, \mathscr{R}\}$ be a reaction network that satisfies conditions (i)–(iii) in Theorem 17.1.1. Moreover, let κ be an element of $\mathbb{R}_+^{\mathscr{R}}$ that satisfies the equation*

$$\sum_{y \to y' \in \mathscr{R}} \kappa_{y \to y'} (y' - y) = 0. \tag{17.8}$$

Then an element $\mu \in \mathbb{R}^{\mathscr{S}}$ satisfies the equation

$$\sum_{y \to y' \in \mathscr{R}} \kappa_{y \to y'} \exp(y \cdot \mu)\,(y' - y) = 0 \tag{17.9}$$

if and only if μ is contained in $S^\perp$.

17.1.2.2 Reformulating the Nonsingularity Problem

For the mass action system $\{\mathscr{S}, \mathscr{C}, \mathscr{R}, k\}$, the derivative $df(c^*) : S \to S$ at the equilibrium $c^* \in \mathbb{R}_+^{\mathscr{S}}$ of the species-formation-rate function is given by

$$df(c^*)\,\sigma = \sum_{y \to y' \in \mathscr{R}} \kappa_{y \to y'} (y * \sigma)(y' - y), \quad \forall \sigma \in S, \tag{17.10}$$

where, for each $y \to y' \in \mathscr{R}$,

$$\kappa_{y \to y'} := k_{y \to y'} (c^*)^y \tag{17.11}$$

and "$*$" indicates the scalar product in $\mathbb{R}^{\mathscr{S}}$ defined by

$$v * w := \sum_{s \in \mathscr{S}} \frac{v_s w_s}{c_s^*}, \quad \forall v, w \in \mathbb{R}^{\mathscr{S}}. \tag{17.12}$$

Because c^* is an equilibrium, $\kappa \in \mathbb{R}_+^{\mathscr{R}}$ again satisfies the equation (17.6). To prove the nonsingularity of $df(c^*)$ when conditions *(i)–(iii)* are satisfied, it is enough to prove the following assertion:

Assertion 17.1.6 *Let $\{\mathscr{S},\mathscr{C},\mathscr{R}\}$ be a reaction network that satisfies conditions (i)–(iii) in Theorem 17.1.1. Moroever, let κ be an element of $\mathbb{R}_+^{\mathscr{R}}$ that satisfies the equation*

$$\sum_{y\to y'\in\mathscr{R}} \kappa_{y\to y'}(y'-y) = 0, \tag{17.13}$$

and let $\mathbb{R}^{\mathscr{S}}$ be equipped with the scalar product (17.12)*, where c^* is some fixed vector of $\mathbb{R}_+^{\mathscr{S}}$. Then an element $\mu \in \mathbb{R}^{\mathscr{S}}$ satisfies the equation*

$$\sum_{y\to y'\in\mathscr{R}} \kappa_{y\to y'}\ (y*\mu)\ (y'-y) = 0 \tag{17.14}$$

only if μ is orthogonal to the stoichiometric subspace.

17.1.2.3 A Proposition That Subsumes Both Assertions 17.1.5 and 17.1.6

Our goal will be to prove the following proposition, which subsumes both Assertions 17.1.5 and 17.1.6. To do so will amount to proof of the Deficiency One Theorem (apart from the existence of a positive equilibrium in the weakly reversible case).

Proposition 17.1.7 (The Deficiency One Theorem recast). *Let $\{\mathscr{S},\mathscr{C},\mathscr{R}\}$ be a reaction network that satisfies conditions (i)–(iii) in Theorem 17.1.1. Moroever, let κ be an element of $\mathbb{R}_+^{\mathscr{R}}$ that satisfies the equation*

$$\sum_{y\to y'\in\mathscr{R}} \kappa_{y\to y'}(y'-y) = 0, \tag{17.15}$$

let $\mathbb{R}^{\mathscr{S}}$ be equipped with a scalar product "$\cdot$" (not necessarily the standard one), and let $\phi:\mathbb{R}\to\mathbb{R}$ be continuous and strictly monotonic. Then an element $\mu\in\mathbb{R}^{\mathscr{S}}$ satisfies the equation

$$\sum_{y\to y'\in\mathscr{R}} \kappa_{y\to y'}\ \phi(y\cdot\mu)\ (y'-y) = 0 \tag{17.16}$$

if and only if μ is orthogonal to the stoichiometric subspace.

Remark 17.1.8. Our interest is in the case in which $\phi(\cdot)$ is increasing, so that is what we shall assume. When $\phi(\cdot)$ is decreasing, the proof is almost identical.

Remark 17.1.9 (A special case). Recall from Remark 17.1.4 that, when complex balancing obtains at the putative positive equilibrium c^*, the conclusions of the Deficiency One Theorem are already consequences of Theorem 15.2.2. With $\kappa\in\mathbb{R}_+^{\mathscr{R}}$ interpreted as in (17.5) or (17.11), complex balancing at c^* is equivalent to the condition[2]

$$\sum_{y\to y'\in\mathscr{R}} \kappa_{y\to y'}(\omega_{y'}-\omega_y) = 0. \tag{17.17}$$

Because they have already been studied, in our proof of Proposition 17.1.7, we will exclude from consideration those $\kappa\in\mathbb{R}_+^{\mathscr{R}}$ for which (17.17) obtains. That is, we will assume that (17.17) does *not* obtain. Proposition 17.1.7 nevertheless remains true in all cases. For a proof of Proposition 17.1.7 when equation (17.17) *does* obtain, see Proposition 5.3 in [76].

[2] Recall Definition 16.1.1 and equation (16.38).

17.1.3 Proof of the "If" Part of Proposition 17.1.7

We denote by $\mathscr{L}^1, \mathscr{L}^2, \ldots, \mathscr{L}^\ell$ the linkage classes of the network under consideration, and we denote by $\mathscr{R}^\theta$ the reactions associated with linkage class $\mathscr{L}^\theta$. That is,

$$\mathscr{R}^\theta := \{y \to y' \in \mathscr{R} : y \in \mathscr{L}^\theta,\ y' \in \mathscr{L}^\theta\}. \tag{17.18}$$

Note that $\{\mathscr{R}^1, \mathscr{R}^2, \ldots, \mathscr{R}^\ell\}$ amounts to a partition of $\mathscr{R}$ into disjoint subsets, which in the sense of Appendix 6.A induces a decomposition of the network $\{\mathscr{S}, \mathscr{C}, \mathscr{R}\}$ into subnetworks $\{\mathscr{S}, \mathscr{L}^\theta, \mathscr{R}^\theta\}_{\theta=1,2,\ldots,\ell}$, with stoichiometric subspaces

$$S_\theta = \operatorname{span}\{y' - y : y \to y' \in \mathscr{R}^\theta\}, \quad \theta = 1, 2, \ldots, \ell. \tag{17.19}$$

Because of condition *(ii)* in Theorem 17.1.1 and Remark 6.A.3 of Appendix 6.A, the stoichiometric subspace for the entire network is the direct sum of the stoichiometric subspaces associated with the linkage classes:

$$S = S_1 \oplus S_2 \oplus \cdots \oplus S_\ell. \tag{17.20}$$

Note that (17.15) can be written as

$$\sum_{\theta=1}^{\ell} \sum_{y \to y' \in \mathscr{R}^\theta} \kappa_{y\to y'}(y' - y) = 0. \tag{17.21}$$

By virtue of the direct sum (17.20), we in fact have

$$\sum_{y \to y' \in \mathscr{R}^\theta} \kappa_{y\to y'}(y' - y) = 0, \quad \theta = 1, 2, \ldots, \ell. \tag{17.22}$$

Now suppose that μ is a member of $S^\perp$. Our aim is to show that (17.16) holds. It is not difficult to see that if complexes y and y' are in the same linkage class, then $y' \cdot \mu = y \cdot \mu$. Thus, there is a number z^θ such that $y \cdot \mu = z^\theta$ whenever y is a member of linkage class $\mathscr{L}^\theta$. In this case, (17.16) takes the form

$$\sum_{\theta=1}^{\ell} \phi(z^\theta) \Big(\sum_{y \to y' \in \mathscr{R}^\theta} \kappa_{y\to y'}(y' - y)\Big) = 0. \tag{17.23}$$

That this equation holds follows from (17.22). This completes proof of the "if" part of Proposition 17.1.7. □

17.1.4 Beginning the "Only If" Part of Proposition 17.1.7

In proving the "only if" part of Proposition 17.1.7, we will find great power in tools accumulated in the preceding chapter. It will be useful, then, to formulate Proposition 17.1.7 in terms of the stoichiometric map Y and the map A_κ, for in Chapter 16 we studied both maps considerably, in particular their interactions. Recall that

$Y : \mathbb{R}^{\mathscr{C}} \to \mathbb{R}^{\mathscr{S}}$ is a linear transformation defined by its action on the standard basis for $\mathbb{R}^{\mathscr{C}}$:

$$Y \omega_y = y, \quad \forall y \in \mathscr{C}. \tag{17.24}$$

Recall also that the map $A_\kappa : \mathbb{R}^{\mathscr{C}} \to \mathbb{R}^{\mathscr{C}}$ is given by

$$A_\kappa x := \sum_{y \to y' \in \mathscr{R}} \kappa_{y \to y'} x_y (\omega_{y'} - \omega_y). \tag{17.25}$$

It follows, then, that

$$Y A_\kappa x \equiv \sum_{y \to y' \in \mathscr{R}} \kappa_{y \to y'} x_y (y' - y). \tag{17.26}$$

In Proposition 17.1.7, equation (17.15) can be expressed in the form

$$Y A_\kappa \, \omega_{\mathscr{C}} = 0, \tag{17.27}$$

where, as usual,

$$\omega_{\mathscr{C}} := \sum_{y \in \mathscr{C}} \omega_y. \tag{17.28}$$

Moreover, equation (17.16) can be written as

$$Y A_\kappa \sum_{y \in \mathscr{C}} \phi(y \cdot \mu)\, \omega_y = 0. \tag{17.29}$$

With these ideas in mind, we can rewrite Proposition 17.1.7 in a form more amenable to tools developed in Chapter 16.

Proposition 17.1.10 (The Deficiency One Theorem recast again). *Let $\{\mathscr{S}, \mathscr{C}, \mathscr{R}\}$ be a reaction network that satisfies conditions (i)–(iii) in Theorem 17.1.1. Moroever, let κ be a vector of $\mathbb{R}^{\mathscr{R}}_+$ such that*

$$\omega_{\mathscr{C}} \in \ker Y A_\kappa, \tag{17.30}$$

let $\mathbb{R}^{\mathscr{S}}$ be equipped with a scalar product "$\cdot$" (not necessarily the standard one), and let $\phi : \mathbb{R} \to \mathbb{R}$ be continuous and strictly monotonic. Then an element $\mu \in \mathbb{R}^{\mathscr{S}}$ satisfies the inclusion

$$\sum_{y \in \mathscr{C}} \phi(y \cdot \mu)\, \omega_y \in \ker Y A_\kappa \tag{17.31}$$

if and only if μ is orthogonal to the stoichiometric subspace.

What remains to be proved is the "only if" part.

Remark 17.1.11. In the spirit of Remark 17.1.9, there is a case that we shall not consider, for it corresponds to complex balancing at the putative equilibrium c^*, a situation for which the results we want are already given by Theorem 15.2.2. In Remark 17.1.9 we indicated that we would restrict our attention only to $\kappa \in \mathbb{R}^{\mathscr{R}}_+$ for which (17.17) is *not* satisfied. It is not difficult to see that this is equivalent to restricting attention to $\kappa \in \mathbb{R}^{\mathscr{R}}_+$ for which

$$\omega_{\mathscr{C}} \notin \ker A_\kappa. \tag{17.32}$$

17.1.5 The Case of One Linkage Class and Motivation for the Second Salt Theorem

It will be useful to consider first the case in which *there is only one linkage class*. In view of the importance of $\ker YA_\kappa$ in Proposition 17.1.10, we will begin by examining its structure when $\ell = 1$ and conditions *(i)* and *(iii)* are satisfied. (When $\ell = 1$, condition *(ii)* is satisfied trivially.) From condition *(iii)* we have $t = \ell = 1$, where as usual t is the number of terminal strong-linkage classes. From *(i)* the deficiency of the network is either zero or one.

Suppose that $\delta = 0$. In this case Lemma 16.3.1 indicates that $\ker YA_\kappa = \ker A_\kappa$. Condition (17.30) then requires that $\omega_{\mathscr{C}} \in \ker A_\kappa$, a case excluded from consideration in Remark 17.1.11.

Remark 17.1.12. The deficiency zero single-linkage-class case requires that the corresponding mass action system in the original Deficiency One Theorem formulation be complex balanced, whereupon the desired result follows from Theorem 15.2.2. Even apart from this, it is fairly easy to see that Proposition 17.1.10 must be true when $\delta = 0$ and $t = \ell = 1$. Because $\ker YA_\kappa = \ker A_\kappa$, the First Salt Theorem (Theorem 16.4.2) tells us that $\dim\ \ker YA_\kappa = 1$. Thus, if μ satisfies (17.31), the vector

$$\sum_{y\in\mathscr{C}} \phi(y\cdot\mu)\,\omega_y \tag{17.33}$$

must be colinear with $\omega_{\mathscr{C}}$. This requires that $y'\cdot\mu = y\cdot\mu$ for each pair of complexes y' and y. From this it follows easily that μ is a member of $S^\perp$.

We turn then to the case $\delta = 1$, $t = \ell = 1$. Because the positive vector $\omega_{\mathscr{C}}$ lies in $\ker YA_\kappa$, it follows from Proposition 16.B.9 that $\dim\ker YA_\kappa = 2$. Clearly, $\ker YA_\kappa$ contains $\ker A_\kappa$. It follows from the First Salt Theorem (Theorem 16.4.2) that $\ker A_\kappa$ is one-dimensional and contains a nonzero vector $b \in \overline{\mathbb{R}}_+^{\mathscr{C}}$ with support $\Lambda \in \mathscr{C}$, the lone terminal strong-linkage class of the network under consideration. Thus we have

$$b_y > 0,\ \forall y \in \Lambda \quad \text{and} \quad b_y = 0,\ \forall y \notin \Lambda. \tag{17.34}$$

By supposition (Remark 17.1.11), the vector $\omega_{\mathscr{C}}$, which resides in $\ker YA_\kappa$, does not also reside in $\ker A_\kappa$. For this reason b and $\omega_{\mathscr{C}}$ are not colinear, whereupon $\{\omega_{\mathscr{C}}, b\}$ is a basis for $\ker YA_\kappa$.

Thus, if μ satisfies the inclusion (17.31), there are numbers ξ and η such that

$$\sum_{y\in\mathscr{C}} \phi(y\cdot\mu)\,\omega_y = \xi\,\omega_{\mathscr{C}} + \eta\, b, \tag{17.35}$$

which is equivalent to

$$\phi(y\cdot\mu) \;=\; \xi \;+\; \eta\, b_y, \quad \forall y \in \mathscr{C}. \tag{17.36}$$

Remark 17.1.13. It follows from (17.34) and (17.36) that $y'\cdot\mu = y\cdot\mu$ whenever y' and y are nonterminal complexes.

Note that *if* $\eta = 0$, then, for each pair of complexes y and y', we have $y' \cdot \mu = y \cdot \mu$. In particular, $(y' - y) \cdot \mu = 0$ for every reaction $y \to y' \in \mathscr{R}$, whereupon μ is a member of $S^{\perp}$. Our aim will be to show that, when the conditions of Proposition 17.1.10 obtain, equation (17.35) can be satisfied *only if* $\eta = 0$. We will show that η can be neither positive nor negative.

To begin, we will arrange the complexes in a sequence $y(1), y(2), \ldots, y(n)$ such that

$$b_{y(1)} \geq b_{y(2)} \geq \ldots \geq b_{y(n)}. \tag{17.37}$$

Remark 17.1.14. We denote by m the number of complexes in the sole terminal strong-linkage class Λ. From (17.34) it follows that the set $\{y(1), y(2), \ldots, y(m)\}$ coincides with Λ. If $n > m$, then we have $b_{y(m)} > b_{y(m+1)}$ and

$$b_{y(m+1)} = b_{y(m+2)} = \cdots = b_{y(n)} = 0.$$

Remark 17.1.15. Along the string (17.37), a *strict* inequality must be present somewhere, for otherwise the nonzero vector b, which lies in $\ker A_\kappa$, would be colinear with $\omega_{\mathscr{C}}$. In this case $\omega_{\mathscr{C}}$ would reside in $\ker A_\kappa$, which belies the supposition invoked in Remark 17.1.11. Of course if $\Lambda \neq \mathscr{C}$, then the presence of a strict inequality is already ensured by (17.34).

Recall the presumption that $\phi(\cdot)$ is strictly increasing. When in (17.36) η is positive, we must have, as a consequence of (17.37), the relations

$$y(1) \cdot \mu \geq y(2) \cdot \mu \geq \ldots \geq y(n) \cdot \mu, \tag{17.38}$$

with strict inequality holding in (17.38) *precisely where strict inequality holds in* (17.37). If η is negative, we must have

$$y(1) \cdot \mu \leq y(2) \cdot \mu \leq \ldots \leq y(n) \cdot \mu, \tag{17.39}$$

again *with strict inequality holding in* (17.39) *precisely where strict inequality holds in* (17.37). By virtue of Remark 17.1.15, strict inequality must hold *somewhere* in both cases.

Our aim will be to show that neither (17.38) nor (17.39) can obtain. As we have seen, the ordering in each is tied in a precise way to the ordering of the components of b, a vector in the one-dimensional linear subspace $\ker A_\kappa$. Just how those components are ordered will depend, in turn, on κ, the only parameter that determines the particular instance of the map A_κ under study. An understanding of that dependence is crucial. In this, the salt-barrel picture described in Section 16.4.2 can guide our thinking. We turn next to our Second Salt Theorem.

17.1.6 The Second Salt Theorem

In our preliminary discussion of the Second Salt Theorem, we will continue to suppose that there is just one linkage class containing just one terminal strong-linkage class Λ. In view of Remark 17.1.11, our main interest will be in κ for which $A_\kappa \omega_{\mathscr{C}} \neq 0$. With this in mind, we let

$$g := A_\kappa \omega_{\mathscr{C}}. \tag{17.40}$$

In terms of components, (17.40) takes the form

$$g_y := \sum_{\mathscr{R}_{\to y}} \kappa_{y' \to y} - \sum_{\mathscr{R}_{y \to}} \kappa_{y \to y'}, \quad \forall y \in \mathscr{C}, \tag{17.41}$$

where as usual $\mathscr{R}_{\to y}$ and $\mathscr{R}_{y\to}$ are, respectively, the sets of reactions having y as the product complex and y as the reactant complex. Note that the presumed inclusion $b \in \ker A_\kappa$ is equivalent to the system of equations

$$\sum_{\mathscr{R}_{\to y}} \kappa_{y' \to y} b_{y'} - \sum_{\mathscr{R}_{y \to}} \kappa_{y \to y'} b_y = 0, \quad \forall y \in \mathscr{C}. \tag{17.42}$$

Recall the salt-barrel picture invoked in Section 16.4.2. There is one salt-solution barrel corresponding to each $y \in \mathscr{C}$ (and labeled as such). The number b_y is identified with the steady-state salt concentration in barrel y. For each $y \to y' \in \mathscr{R}$, we presume that there is a flow of salt solution from barrel y to barrel y' at rate $\kappa_{y\to y'}$. Thus, the steady-state rate of transport of salt from barrel y to barrel y' is $\kappa_{y\to y'} b_y$. The system of equations (17.42) expresses a steady-state salt balance: For each $y \in \mathscr{C}$ the flow rate of salt into barrel y is exactly balanced by the flow rate of salt out of barrel y. Finally, (17.41) gives rise to a salt-barrel interpretation of g_y. It is the *net* volumetric flow rate of solution into barrel y from all the other barrels. (Recall that, even at salt-concentration steady state, g_y might be positive, negative, or zero. See Remark 16.4.1.)

By way of preparation for statement of our Second Salt Theorem, there is one more piece of interpretation that will help guide intuition. Suppose that $\mathscr{P} \subset \mathscr{C}$ is a set of complexes and that $\mathscr{P}'$ is the complement of $\mathscr{P}$ in $\mathscr{C}$. Then

$$\sum_{y \in \mathscr{P}} g_y \tag{17.43}$$

is the *net* volumetric rate of flow of *solution* to barrels labeled with complexes in $\mathscr{P}$ from the remaining barrels—that is, from barrels labeled with complexes in $\mathscr{P}'$. In fact, if we denote by $\mathscr{P} \to \mathscr{P}'$ the set of reactions that have reactant complex in $\mathscr{P}$ and product complex in $\mathscr{P}'$, and if we denote the opposite by $\mathscr{P}' \to \mathscr{P}$ then[3]

$$\sum_{y \in \mathscr{P}} g_y = \sum_{\mathscr{P}' \to \mathscr{P}} \kappa_{y' \to y} - \sum_{\mathscr{P} \to \mathscr{P}'} \kappa_{y \to y'}. \tag{17.44}$$

Suppose, as before, the complexes are arranged in a sequence $y(1), y(2), \ldots, y(n)$ such that

$$b_{y(1)} \geq b_{y(2)} \geq \ldots \geq b_{y(n)}, \tag{17.45}$$

and suppose that the sole terminal strong-linkage class Λ contains m complexes, with m perhaps equal to n. Remark 17.1.14 tells us that $\Lambda = \{y(1), y(2), \ldots, y(m)\}$ and that $b_{y(m)} > b_{y(m+1)}$ when $n > m$. The salt-barrel picture provides motivation for the following assertion:

[3] For an elaboration on this claim, see Lemma 17.A.2 in Appendix 17.A.

Assertion 17.1.16 *For $\theta = 1,2,\ldots,m$,*

$$\sum_{i=1}^{\theta} g_{y(i)} \geq 0. \tag{17.46}$$

Moreover, for $1 \leq \theta < m$ strict inequality holds in (17.46) *if $b_{y(\theta)} > b_{y(\theta+1)}$. If $\theta = m$, strict inequality holds in* (17.46) *if and only if $m < n$—that is, if and only if $\Lambda \neq \mathscr{C}$.*

Here is the intuitive reasoning: For $\theta < m$ let $\mathscr{P} = \{y(1), y(2), \ldots, y(\theta)\}$, and let $\mathscr{P}'$ denote the set of remaining complexes. Because of the manner in which complexes were sequenced, no barrel carrying a label in $\mathscr{P}'$ has a higher salt concentration than does any barrel carrying a label in $\mathscr{P}$. In less formal terms, each $\mathscr{P}$ barrel is at least as salty as every $\mathscr{P}'$ barrel. And when $b_{y(\theta)} > b_{y(\theta+1)}$, each $\mathscr{P}$ barrel is actually saltier than every $\mathscr{P}'$ barrel.

Recall that the salt concentrations carried by b characterize a *steady-state* distribution of salt among the barrels, a distribution consistent with the inter-barrel solution flow rates carried by κ. Note that there are flows of solution from the set of salty $\mathscr{P}$ barrels to the set of not-so-salty $\mathscr{P}'$ barrels and also flows from the not-so-salty barrels to the salty ones. How then can the steady state be maintained? In particular, how can the total amount of salt in the set of $\mathscr{P}$ barrels remain constant if only salty solutions are flowing from the set of $\mathscr{P}$ barrels to the $\mathscr{P}'$ barrels while only not-so-salty solutions are flowing from the set of $\mathscr{P}'$ barrels to the $\mathscr{P}$ barrels?

The answer lies in consideration of the *total amount of solution* flowing in the two directions. *There must be at least as much not-so-salty solution flowing toward the set of $\mathscr{P}$ barrels as there is salty solution flowing away from the set of $\mathscr{P}$ barrels. And when every $\mathscr{P}$ barrel is in fact saltier than every $\mathscr{P}'$ barrel, there must in fact be more solution flowing to the set of $\mathscr{P}$ barrels than there is solution flowing away from them.* For the case $\theta < m$, this is what Assertion 17.1.16 says.

Finally, suppose that $\theta = m$ and $m < n$. In this case the sum in (17.46) amounts to the net flow rate of solution into barrels associated with the terminal strong-linkage class Λ from barrels associated with the set of nonterminal complexes, denoted Λ'. Because Λ is terminal, there cannot be a flow from a Λ barrel to a Λ' barrel. Because the sole linkage class containing Λ is identical with $\mathscr{C}$ and because Λ is smaller than $\mathscr{C}$ there must be a flow from at least one Λ' barrel to a Λ barrel. For these reasons, the sum in (17.46) is positive, as was claimed in Assertion 17.1.16. That the sum is zero in the case $\theta = m$, $m = n$ is a consequence of Remark 16.1.4 and the fact that g is a member of $\operatorname{im} A_\kappa \subset \operatorname{span}(\Delta)$.

What we deduced on intuitive grounds from the salt-barrel picture for a single-linkage- class network having a single terminal strong-linkage class we now state as our Second Salt Theorem, and for a less restricted class of networks.[4] The symbol $\omega_{\mathscr{C}}$ retains its usual meaning, given by (17.28). Proof of the theorem is contained in Appendix 17.A.

---

[4] We replace the symbol κ with the more neutral α to free the theorem statement from the interpretation given to κ in the preceding discussion.

Theorem 17.1.17 (Second Salt Theorem [76]**).** *For reaction network* $\{\mathscr{S},\mathscr{C},\mathscr{R}\}$, *let* α *be an element of* $\mathbb{R}_+^{\mathscr{R}}$, *let* $A_\alpha : \mathbb{R}^{\mathscr{C}} \to \mathbb{R}^{\mathscr{C}}$ *be the linear transformation defined by*

$$A_\alpha x := \sum_{y\to y' \in \mathscr{R}} \alpha_{y\to y'} x_y (\omega_{y'} - \omega_y), \tag{17.47}$$

and let

$$g := A_\alpha \omega_{\mathscr{C}}. \tag{17.48}$$

Moreover, let $\Lambda \subset \mathscr{C}$ *be a terminal strong-linkage class, and let* $b \in \overline{\mathbb{R}}_+^{\mathscr{C}}$ *be an element of* $\ker A_\alpha$ *with* $\operatorname{supp} b = \Lambda$. *If the complexes of* Λ *are arranged in a sequence* $y(l), y(2), \ldots, y(m)$ *so that*

$$b_{y(l)} \geq b_{y(2)} \geq b_{y(3)} \geq \cdots \geq b_{y(m)}, \tag{17.49}$$

then, for $\theta = 1, 2, \ldots, m$,

$$\sum_{i=1}^{\theta} g_{y(i)} \geq 0. \tag{17.50}$$

In fact, for each $1 \leq \theta < m$, *inequality holds in* (17.50) *if* $b_{y(\theta)}$ *is greater than* $b_{y(\theta+1)}$. *When* $\theta = m$, *inequality holds in* (17.50) *if and only if* Λ *is smaller than the linkage class containing it.*

17.1.7 Proof of Proposition 17.1.10 (and the Deficiency One Theorem) in the Single-Linkage-Class Case

We now return to proof of Proposition 17.1.10 in the single-linkage-class case. In particular we want to show that when the complexes are arranged in a sequence such that the components of nonzero $b \in \ker A_\kappa \cap \overline{\mathbb{R}}_+^{\mathscr{C}}$ admit the inequalities

$$b_{y(1)} \geq b_{y(2)} \geq \ldots \geq b_{y(n)} \tag{17.51}$$

then there can be no $\mu \in \mathbb{R}^{\mathscr{S}}$ such that either

$$y(1)\cdot\mu \geq y(2)\cdot\mu \geq \ldots \geq y(n)\cdot\mu, \tag{17.52}$$

or

$$y(1)\cdot\mu \leq y(2)\cdot\mu \leq \ldots \leq y(n)\cdot\mu, \tag{17.53}$$

with strict inequality holding precisely where strict inequality holds in (17.51).

Remark 17.1.18. It is enough to preclude the first possibility, for if μ satisfies (17.53), then $-\mu$ satisfies (17.52). With this in mind, our focus will be on (17.52).

By supposition, $g := A_\kappa \omega_{\mathscr{C}}$ is not zero. Because $YA_\kappa \omega_{\mathscr{C}} = 0$, g is contained in $\ker Y$. And, because the image of A_κ is contained in $\operatorname{span}(\Delta)$, where Δ is given by (16.8), we can assert that g is a member of $\ker Y \cap \operatorname{span}(\Delta)$. From Remark 16.2.5, then, we get the following conditions that the components of g must satisfy:

$$\sum_{y \in \mathscr{C}} g_y = 0 \quad \text{and} \quad \sum_{y \in \mathscr{C}} g_y y = 0. \tag{17.54}$$

As a consequence of (17.54), we can write

$$\begin{aligned} & g_{y(1)}(y(1) - y(2)) + (g_{y(1)} + g_{y(2)})(y(2) - y(3)) \\ & + (g_{y(1)} + g_{y(2)} + g_{y(3)})(y(3) - y(4)) + \ldots + (\sum_{i=1}^{n-1} g_{y(i)})(y(n-1) - y(n)) = 0. \end{aligned} \tag{17.55}$$

Taking the scalar product with an arbitrary $\mu \in \mathbb{R}^{\mathscr{S}}$, we obtain

$$\sum_{\theta=1}^{n-1} (\sum_{i=1}^{\theta} g_{y(i)}) \left(y(\theta) \cdot \mu - y(\theta+1) \cdot \mu \right) = 0. \tag{17.56}$$

Now suppose that μ complies with the inequality string (17.52), with strict inequality holding precisely where strict inequality holds in (17.51). Suppose first that the network is weakly reversible—that is, $\Lambda = \mathscr{C}$ and $m = n$. In this case it follows from the Second Salt Theorem (Theorem 17.1.17) that each term is nonnegative. Moreover, because strict inequality must hold somewhere in (17.51) (Remark 17.1.15), at least one term is positive. In this case, we have a contradiction.

Suppose, on the other hand, that Λ is smaller than $\mathscr{C}$ (and $m < n$). From (17.34) $b_{y(m+1)} = b_{y(m+2)} = \cdots = b_{y(n)} = 0$. Thus, we have

$$(y(\theta) - y(\theta+1)) \cdot \mu = 0, \quad \theta = m+1, \ldots, n-1. \tag{17.57}$$

In this case, then, (17.56) can be replaced by

$$\sum_{\theta=1}^{m} (\sum_{i=1}^{\theta} g_{y(i)}) \left(y(\theta) \cdot \mu - y(\theta+1) \cdot \mu \right) = 0. \tag{17.58}$$

The Second Salt Theorem ensures that each term of the m terms is nonnegative and at least one (in particular the last) is positive. Again there is a contradiction.

This completes the proof of Proposition 17.1.10 in the single-linkage-class case. Thus, for a single linkage class, the (differently phrased) Proposition 17.1.7 is also proved, as is the Deficiency One Theorem (apart from the existence of a positive equilibrium when the network is weakly reversible). □

17.1.8 Proof of Proposition 17.1.7 (and the Deficiency One Theorem) with Multiple Linkage Classes

Here we extend the proof of the "only if" part of Proposition 17.1.7 to the case of multiple linkage classes. In this we will draw on ideas invoked in the proof of the "if" part given in Section 17.1.3.

Recall that we denote by $\mathscr{L}^1, \mathscr{L}^2, \ldots, \mathscr{L}^\ell$ the linkage classes of the network under consideration, and we denote by $\mathscr{R}^\theta$ the reactions associated with linkage class $\mathscr{L}^\theta$. That is,

$$\mathscr{R}^\theta := \{y \to y' \in \mathscr{R} \,:\, y \in \mathscr{L}^\theta,\ y' \in \mathscr{L}^\theta\}. \tag{17.59}$$

Then $\{\mathscr{R}^1, \mathscr{R}^2, \ldots, \mathscr{R}^\ell\}$ amounts to a partition of $\mathscr{R}$ into disjoint subsets, which in the sense of Appendix 6.A induces a decomposition of the network $\{\mathscr{S}, \mathscr{C}, \mathscr{R}\}$ into subnetworks $\{\mathscr{S}, \mathscr{L}^\theta, \mathscr{R}^\theta\}_{\theta=1,2,\ldots,\ell}$, with stoichiometric subspaces

$$S_\theta = \text{span}\,\{y' - y : y \to y' \in \mathscr{R}^\theta\}, \quad \theta = 1,2,\ldots,\ell. \tag{17.60}$$

Recall also that, because of condition *(ii)* in Theorem 17.1.1 , the stoichiometric subspace for the entire network is the direct sum of the stoichiometric subspaces associated with the linkage classes:

$$S = S_1 \oplus S_2 \oplus \cdots \oplus S_\ell. \tag{17.61}$$

Now suppose that, as in the statement of Proposition 17.1.7, $\kappa \in \mathbb{R}_+^{\mathscr{R}}$ is such that

$$\sum_{y \to y' \in \mathscr{R}} \kappa_{y \to y'} (y' - y) = 0, \tag{17.62}$$

and that $\mu \in \mathbb{R}^{\mathscr{S}}$ is such that

$$\sum_{y \to y' \in \mathscr{R}} \kappa_{y \to y'}\ \phi(y \cdot \mu)\ (y' - y) = 0. \tag{17.63}$$

Because of direct summing, these are equivalent to

$$\sum_{y \to y' \in \mathscr{R}^\theta} \kappa_{y \to y'} (y' - y) = 0, \quad \theta = 1,2,\ldots,\ell \tag{17.64}$$

and

$$\sum_{y \to y' \in \mathscr{R}^\theta} \kappa_{y \to y'}\, \phi(y \cdot \mu)\ (y' - y) = 0, \quad \theta = 1,2,\ldots,\ell. \tag{17.65}$$

We need to show that, when the hypotheses of Theorem 17.1.1 are satisfied, these equations can be satisfied only if μ is a member of $S^\perp$. Suppose that (17.64) and (17.65) hold with $\mu \notin S^\perp$. Then $(y' - y) \cdot \mu \neq 0$ for some reaction $y \to y'$. If $y \to y'$ is a member of the reaction subset $\mathscr{R}^j$, then we have

$$\sum_{y \to y' \in \mathscr{R}^j} \kappa_{y \to y'} (y' - y) = 0, \tag{17.66}$$

$$\sum_{y \to y' \in \mathscr{R}^j} \kappa_{y \to y'}\, \phi(y \cdot \mu)\ (y' - y) = 0, \tag{17.67}$$

and

$$\mu \notin (S_j)^\perp. \tag{17.68}$$

Because network $\{\mathscr{S},\mathscr{C},\mathscr{R}\}$ satisfies conditions *(i)* and *(iii)* of Theorem 17.1.1, so does the single-linkage-class network $\{\mathscr{S},\mathscr{L}^j,\mathscr{R}^j\}$. Thus, (17.66)–(17.68) stand in contradiction to the single-linkage-class version of Proposition 17.1.7 proved in the preceding subsection.

This completes the proof of Proposition 17.1.7 and of the Deficiency One Theorem (apart from the existence of a positive equilibrium in the weakly reversible case). □

Remark 17.1.19. Scrutiny of the proof given here will indicate that the role of condition *(ii)* in the Deficiency One Theorem is to ensure direct summing of the stoichiometric subspaces for the subnetworks associated with the linkage classes. A similar theorem obtains for any partition of the network $\{\mathscr{S},\mathscr{C},\mathscr{R}\}$ into *independent* subnetworks (in the sense of Appendix 6.A), each of which satisfies conditions *(i)* and *(iii)*. For a generalization of Proposition 17.1.7, see Proposition 6.3.1 in [76].

17.2 The Deficiency One Algorithm

In this section our interest is in reaction networks of deficiency one that do not satisfy condition *(ii)* of the Deficiency One Theorem. For such networks each linkage class has a deficiency of zero. We will restrict our attention further to *regular* deficiency one networks. Recall from Section 8.5 that a regular reaction network satisfies three mild conditions: First, each linkage class contains precisely one terminal strong-linkage class. Second, the reaction vectors are positively dependent (a condition necessary for the existence of a positive equilibrium). Third, each direct link of two *terminal* complexes is a cut-link (Definition 6.7.1).

In Section 8.5 we indicated by means of examples why, for the class of regular deficiency one networks, we cannot expect theorems such as the Deficiency Zero or Deficiency One Theorems to deny the capacity for multiple positive equilibria in bold, sweeping strokes. The fact is that, within the regular deficiency one class, seemingly inconsequential differences in network structure can serve to deny or engender the existence of mass action rate constants for which multiple stoichiometrically compatible positive equilibria exist.

In place of bold, sweeping theorems, the *Deficiency One Algorithm* [74, 77] provides means to decide unequivocally, *either positively or negatively*, a regular deficiency one network's capacity for multiple positive equilibria or for the existence of a degenerate positive equilibrium—seemingly *nonlinear* problems—through a systematic study of *linear* inequality systems. In Section 8.5 we gave cursory illustrations of how the algorithm works. Here we will go into far more depth.

17.2.1 Getting Started: Two Useful Propositions

To prepare for the reasoning behind the Deficiency One Algorithm, we will state two propositions connected with the capacity of a network, taken with mass action kinetics, to admit certain phenomena (for *some* assignment of rate constants): first, its capacity to admit multiple stoichiometrically compatible positive equilibria and, second, its capacity to admit a degenerate positive equilibrium.

We will need to recall Definition 8.5.1: For a reaction network $\{\mathscr{S},\mathscr{C},\mathscr{R}\}$ with stoichiometric subspace S, we say that a vector $\mu \in \mathbb{R}^{\mathscr{S}}$ is *sign-compatible with S* if there exists $\sigma \in S$ such that $\operatorname{sign}\mu_s = \operatorname{sign}\sigma_s$ for all $s \in \mathscr{S}$. It will be helpful to remember that $\mu \in \mathbb{R}^{\mathscr{S}}$ is sign-compatible with S if and only if there exist $\sigma \in S$ and a positive vector $p \in \mathbb{R}^{\mathscr{S}}_+$ such that $\mu = p\sigma$.

Proposition 17.2.1. *For a reaction network $\{\mathscr{S},\mathscr{C},\mathscr{R}\}$ with stoichiometric subspace S, the following are equivalent:*

(i) There exists $k \in \mathbb{R}^{\mathscr{R}}_+$ such that the mass action system $\{\mathscr{S},\mathscr{C},\mathscr{R},k\}$ admits two distinct positive equilibria that are stoichiometrically compatible.

(ii) There exist $\kappa \in \mathbb{R}^{\mathscr{R}}_+$ and a nonzero element $\mu \in \mathbb{R}^{\mathscr{S}}$ that is sign-compatible with S such that

$$\sum_{y\to y' \in \mathscr{R}} \kappa_{y\to y'}(y'-y) = 0 \tag{17.69}$$

and

$$\sum_{y\to y' \in \mathscr{R}} \kappa_{y\to y'} \exp(y\cdot\mu)\,(y'-y) = 0. \tag{17.70}$$

Proof. To prove that *(i)* implies *(ii)*, we suppose that c^* and c^{**}, both positive, are distinct equilibria of the mass action system $\{\mathscr{S},\mathscr{C},\mathscr{R},k\}$. That is,

$$\sum_{y\to y' \in \mathscr{R}} k_{y\to y'}(c^*)^y(y'-y) = 0 \tag{17.71}$$

and

$$\sum_{y\to y' \in \mathscr{R}} k_{y\to y'}(c^{**})^y(y'-y) = 0. \tag{17.72}$$

We suppose also that c^* and c^{**} are stoichiometrically compatible, which is to say that

$$\sigma := c^{**} - c^* \tag{17.73}$$

is a member of S. Now let

$$\mu := \ln c^{**} - \ln c^*. \tag{17.74}$$

Because $\operatorname{sgn}\mu_s = \operatorname{sgn}\sigma_s$ for all $s \in \mathscr{S}$, μ is sign-compatible with S (and nonzero). If we let

$$\kappa_{y\to y'} := k_{y\to y'}(c^*)^y, \quad \forall y\to y' \in \mathscr{R}, \tag{17.75}$$

then (17.69) and (17.70) follow from (17.71) and (17.72).

To prove that *(ii)* implies *(i)*, we suppose that $\kappa \in \mathbb{R}^{\mathscr{R}}_+$ and nonzero $\mu \in \mathbb{R}^{\mathscr{S}}$ satisfy equations (17.69) and (17.70). We suppose also that μ is sign-compatible with S, which implies the existence of a nonzero vector $\sigma \in S$ such that, for all $s \in \mathscr{S}$, $\operatorname{sgn}\mu_s = \operatorname{sgn}\sigma_s$. This, in turn, implies that $\operatorname{sgn}[\exp(\mu_s) - 1] = \operatorname{sgn}\sigma_s$. Thus, there is a positive vector $c^* \in \mathbb{R}^{\mathscr{S}}_+$ such that

$$\sigma = c^*(e^\mu - 1). \tag{17.76}$$

Now let $c^{**} \in \mathbb{R}_+^{\mathscr{S}}$ be defined by

$$c^{**} := c^* e^\mu. \tag{17.77}$$

From these it follows that $c^{**} - c^* = \sigma$, which in turn is a nonzero member of S. Thus c^{**} and c^* are stoichiometrically compatible and distinct. If we choose $k \in \mathbb{R}_+^{\mathscr{R}}$ to satisfy (17.75), then the equilibrium equations (17.71) and (17.72) become consequences of (17.69) and (17.70). That is, c^* and c^{**} are distinct stoichiometrically compatible positive equilibria of the mass action system $\{\mathscr{S},\mathscr{C},\mathscr{R},k\}$. □

Proposition 17.2.2. *For a reaction network $\{\mathscr{S},\mathscr{C},\mathscr{R}\}$ with stoichiometric subspace S, the following are equivalent:*

(i) There exists $k \in \mathbb{R}_+^{\mathscr{R}}$ such that the mass action system $\{\mathscr{S},\mathscr{C},\mathscr{R},k\}$ admits a degenerate positive equilibrium.

(ii) There exist $\kappa \in \mathbb{R}_+^{\mathscr{R}}$ and a nonzero element $\mu \in \mathbb{R}^{\mathscr{S}}$ that is sign-compatible with S such that

$$\sum_{y \to y' \in \mathscr{R}} \kappa_{y \to y'}(y' - y) = 0 \tag{17.78}$$

and

$$\sum_{y \to y' \in \mathscr{R}} \kappa_{y \to y'}\,(y \cdot \mu)\,(y' - y) = 0. \tag{17.79}$$

Proof. To prove that *(i)* implies *(ii)*, we suppose that c^* is a positive equilibrium of the mass action system $\{\mathscr{S},\mathscr{C},\mathscr{R},k\}$, which is to say that (17.78) obtains, with $\kappa \in \mathbb{R}_+^{\mathscr{R}}$ again given by (17.75). We suppose also that c^* is a degenerate equilibrium (i.e., at c^* the derivative of the species-formation-rate function is singular). Thus, there is a nonzero $\sigma \in S$ such that[5]

$$\sum_{y \to y' \in \mathscr{R}} \kappa_{y \to y'}\,(y \cdot \frac{\sigma}{c^*})\,(y' - y) = 0. \tag{17.80}$$

Now let $\mu := \sigma / c^*$. Because σ lies in S and c^* is positive, μ is sign-compatible with S. From (17.80) we obtain (17.79).

Proof that *(ii)* implies *(i)* is similar. □

17.2.2 A Key Overarching Question

In the two propositions of the preceding section, the capacity of a network $\{\mathscr{S},\mathscr{C},\mathscr{R}\}$ for multiple positive equilibria and its capacity for a degenerate positive equilibrium were, in each case, shown to be equivalent to the existence of vectors $\kappa \in \mathbb{R}_+^{\mathscr{R}}$ and $\mu \in \mathbb{R}^{\mathscr{S}}$ that jointly satisfy certain algebraic conditions, those conditions differing only slightly in the two cases. For this reason we can, as we did in proof of the Deficiency One Theorem, examine a network's capacity for multiple positive equilibria or for a degenerate positive equilibrium more or less simultaneously through consideration of the following question:

[5] Recall Section 3.6.

Question 17.2.3 *Let $\{\mathscr{S},\mathscr{C},\mathscr{R}\}$ be a reaction network with stoichiometric subspace S, and let $\phi : \mathbb{R} \to \mathbb{R}$ be continuous and strictly monotonically increasing. Do there exist a $\kappa \in \mathbb{R}_+^{\mathscr{R}}$ and a nonzero $\mu \in \mathbb{R}^{\mathscr{S}}$ that is sign-compatible with S such that*

$$\sum_{y \to y' \in \mathscr{R}} \kappa_{y \to y'} (y' - y) = 0 \tag{17.81}$$

and

$$\sum_{y \to y' \in \mathscr{R}} \kappa_{y \to y'} \, \phi(y \cdot \mu) \, (y' - y) = 0? \tag{17.82}$$

Because we will want to draw on the power of propositions and theorems in Chapter 16, it will be advantageous to rephrase Question 17.2.3 differently, along lines similar to those in Section 17.1.4:

Question 17.2.4 *Let $\{\mathscr{S},\mathscr{C},\mathscr{R}\}$ be a reaction network with stoichiometric subspace S, and let $\phi : \mathbb{R} \to \mathbb{R}$ be continuous and strictly monotonically increasing. Do there exist a $\kappa \in \mathbb{R}_+^{\mathscr{R}}$ and a nonzero $\mu \in \mathbb{R}^{\mathscr{S}}$ that is sign-compatible with S such that*

$$\omega_{\mathscr{C}} \in \ker Y A_\kappa, \tag{17.83}$$

and

$$\sum_{y \in \mathscr{C}} \phi(y \cdot \mu) \, \omega_y \in \ker Y A_\kappa ? \tag{17.84}$$

Remark 17.2.5. For κ such that $\omega_{\mathscr{C}}$ is not only in $\ker Y A_\kappa$ but also in $\ker A_\kappa$, there can be no μ with the specified properties: When $\omega_{\mathscr{C}}$ lies in $\ker A_\kappa$, (17.84) can be satisfied only if μ lies in $S^\perp$. (See Remarks 17.1.9 and 17.1.11.) In this case μ cannot be both nonzero and sign-compatible with S. Recall that $\omega_{\mathscr{C}} \in \ker A_\kappa$ corresponds to the case in which complex balancing obtains at the putative equilibrium c^*.

17.2.3 The Key Question, Posed for a Regular Deficiency One Network

In the preceding two subsections, we placed no restrictions on the network under consideration. Hereafter we will suppose (unless stated otherwise) that $\{\mathscr{S},\mathscr{C},\mathscr{R}\}$ is a particular regular deficiency one network. Because the network is regular, each linkage class contains precisely one terminal strong-linkage class ($t = \ell$). With this in mind, we denote by $\mathscr{L}^1, \mathscr{L}^2, \ldots, \mathscr{L}^\ell$ the network's linkage classes and by $\Lambda^1, \Lambda^2, \ldots, \Lambda^\ell$ its terminal strong-linkage classes, with $\Lambda^\theta \subset \mathscr{L}^\theta$, $\theta = 1,2,\ldots,\ell$.

Furthermore, we will suppose that, for the network under consideration and for a specified strictly increasing continuous function ϕ, Question 17.2.4 has an affirmative answer. *That is, we suppose that there exist specific values of κ and μ satisfying the question's requirements.* This supposition will lead to conditions on network structure that are *necessary* for an affirmative answer. It happens that those conditions will be sufficient as well.

Because $t = \ell$, because $\delta = 1$, and because (17.83) ensures that ker $Y A_\kappa$ contains a positive vector, Proposition 17.84 tells us that for the network

$$\dim \ker YA_\kappa = 1 + \ell.$$

Moreover, the First Salt Theorem (Theorem 16.4.2) tells us that there is a basis $\{b^1, b^2, \ldots, b^\ell\} \subset \overline{\mathbb{R}}_+^{\mathscr{C}}$ for $\ker A_\kappa$ with $\operatorname{supp} b^\theta = \Lambda^\theta$, $\theta = 1, 2, \ldots, \ell$. From Remark 17.2.5 it follows that

$$\omega_{\mathscr{C}} \notin \operatorname{span}\{b^1, b^2, \ldots, b^\ell\}. \tag{17.85}$$

Thus,

$$\{\omega_{\mathscr{C}}, b^1, b^2, \ldots, b^\ell\} \tag{17.86}$$

is a basis for $\ker YA_\kappa$. The inclusion (17.84) then requires the existence of numbers $\lambda_0, \lambda_1, \lambda_2, \ldots, \lambda_\ell$ such that

$$\sum_{y \in \mathscr{C}} \phi(y \cdot \mu)\, \omega_y = \lambda_0\, \omega_{\mathscr{C}} + \lambda_1 b^1 + \lambda_2 b^2 + \cdots + \lambda_\ell b^\ell. \tag{17.87}$$

In terms of components, we can write

$$\phi(y \cdot \mu) = \lambda_0 + \lambda_\theta b_y^\theta \ \text{ if } y \in \Lambda^\theta. \tag{17.88}$$

Recall that a nonterminal complex is one that does not reside in a terminal strong-linkage class. Thus, we can also write

$$\phi(y \cdot \mu) = \lambda_0 \ \text{ if } y \text{ is nonterminal.} \tag{17.89}$$

We will need a little more vocabulary: By a *trivial terminal strong-linkage class*, we mean a terminal strong-linkage class having just one complex. By a *reactive complex*, we mean a complex that is a reactant complex for at least one reaction. Thus, $y \in \mathscr{C}$ is reactant if there is a reaction $y \to y' \in \mathscr{R}$. (A complex is not reactive when it is a member of a trivial terminal strong-linkage class.)

We now partition the set of reactive complexes into three subsets, according to the signs of the λ_θ:

$$U := \bigcup_{\theta=1}^{\ell} \{\Lambda^\theta : \lambda_\theta > 0,\ \Lambda^\theta \text{ is nontrivial}\} \tag{17.90}$$

$$L := \bigcup_{\theta=1}^{\ell} \{\Lambda^\theta : \lambda_\theta < 0,\ \Lambda^\theta \text{ is nontrivial}\} \tag{17.91}$$

$$M := \{y \in \mathscr{C} : y \text{ is reactive},\ y \notin U,\ y \notin L\}. \tag{17.92}$$

Note that M is the union of the set of nonterminal complexes and the set

$$\bigcup_{\theta=1}^{\ell} \{\Lambda^\theta : \lambda_\theta = 0,\ \Lambda^\theta \text{ is nontrivial}\}. \tag{17.93}$$

The numbers $\{\lambda_\theta\}_{\theta=1,\ldots,\ell}$ in (17.87) gave rise to a partition of the reactive complexes into three subsets U, M, and L we shall find convenient to call the *upper*, *middle*, and *lower* parts. This provides motivation for the following definition, which will find broader use later on.

Definition 17.2.6. *An* **upper-middle-lower partition** *for a reaction network is a partition of its set of reactive complexes into three parts, U, M, and L, called the upper, middle, and lower parts, such that*

(i) all nonterminal complexes reside in M,

(ii) all complexes in the same nontrivial terminal strong-linkage class reside in the same part.

Remark 17.2.7. With respect to a given upper-middle-lower partition $P = \{U, M, L\}$, we will say that *complex y lies above complex y'* if

(i) y lies in U and y' lies in either M or L.

(ii) y lies in M and y' lies in L.

Until we say otherwise, it will be understood that the upper-middle-lower partition under consideration is the particular one induced by (17.87) in the sense of equations (17.90)–(17.92).

17.2.4 Three Observations Resulting in Some Linear Equations and Inequalities

In this section we will begin to see how the partition $P = \{U, M, L\}$ induces a system of equalities and inequalities joining linear forms in the set

$$\{y \cdot \mu : y \in \mathscr{C} \text{ is reactive}\}. \tag{17.94}$$

As we proceed, it will be useful to keep in mind that, for each b^θ, the number b^θ_y is positive for y in terminal strong-linkage class Λ^θ and is zero otherwise. The following observations are direct consequences of equations (17.88) and (17.89), the definitions of U, M and L, and the fact that $\phi(\cdot)$ is strictly increasing.

Observation 1: If y and y' are both members of M, then

$$y \cdot \mu = y' \cdot \mu. \tag{17.95}$$

Observation 2: If y' and y are reactive complexes and y' lies above y, then

$$y' \cdot \mu > y \cdot \mu. \tag{17.96}$$

Observation 3 (tentative): If y and y' are both members of nontrivial terminal strong-linkage class Λ^θ, then

$$\text{sgn}\,(y \cdot \mu - y' \cdot \mu) = \text{sgn}\,(b^\theta_y - b^\theta_{y'}) \ \text{ if } \Lambda^\theta \subset U \tag{17.97}$$

and

$$\text{sgn}\,(y \cdot \mu - y' \cdot \mu) = -\text{sgn}\,(b^\theta_y - b^\theta_{y'}) \ \text{ if } \Lambda^\theta \subset L. \tag{17.98}$$

In these three observations we are beginning to see that, when there is a $\{\kappa,\mu\}$ pair that satisfies the requirements of Questions 17.2.3 and 17.2.4, it is necessary that certain systems of linear equations and inequalities admit a solution, in particular a nonzero solution that is sign-compatible with the stoichiometric subspace. Although the equations and inequalities described in Observations 1 and 2 are expressed in terms of the reactive complexes alone—entities discernible from inspection of the reaction network—that does not appear to be true of the relations in Observation 3: When y' an y are members of terminal strong-linkage class $\Lambda^{\theta} \not\subset M$, relations joining $y'\cdot\mu$ and $y\cdot\mu$ are determined by the relative sizes of of $b^{\theta}_{y'}$ and b^{θ}_{y}. Because b^{θ} derives its identity only as a member of $\ker A_{\kappa}$, the ordering of the components of b^{θ} seems tied in an awkward way to the value of κ.

In fact, we will show in the next section that, for the regular deficiency one network under consideration, *the relative sizes of components of* b^{θ} *are fixed by the network itself and can be determined solely from network attributes.* This will permit us to replace the tentative Observation 3 with a far more useful one in which (17.97) and (17.98) are expressed more directly in terms of reaction network properties, with no reference at all to the specific value of κ.

17.2.5 Digression: Another (Minor) Salt Theorem

In this subsection we will present another salt theorem which, while not so substantive as the others, has an important bearing on Observation 3. So that we can be a little more general, we will temporarily step away from the context of the immediately preceding subsections and let $\{\mathscr{S},\mathscr{C},\mathscr{R}\}$ be an arbitrary reaction network.

Recall from Section 6.7 that a direct link $y \leftrightarrow y'$ in a reaction network is a *cut-link* if the linkage class containing y and y' can be partitioned into two sets of complexes, $\mathscr{W}(y)$ and $\mathscr{W}(y')$, one containing y and the other containing y', such that there is no direct link connecting a complex of $\mathscr{W}(y)$ with a complex of $\mathscr{W}(y')$ apart from $y \leftrightarrow y'$. When $y \leftrightarrow y'$ is a cut-link and g is a member of $\mathbb{R}^{\mathscr{C}}$, we denote by $[g,\ y \leftrightarrow y', y]$ the sum of all components of g on the y-side of the partition:

$$[g,\ y \leftrightarrow y',\ y] := \sum_{\bar{y} \in \mathscr{W}(y)} g_{\bar{y}}. \tag{17.99}$$

Note that if g is a member of span(Δ) and $\mathscr{L}$ is the linkage class containing y and y', then, from Remark 16.1.4,

$$\sum_{\bar{y} \in \mathscr{L}} g_{\bar{y}} = \sum_{\bar{y} \in \mathscr{W}(y)} g_{\bar{y}} + \sum_{\bar{y} \in \mathscr{W}(y')} g_{\bar{y}} = 0. \tag{17.100}$$

From this it follows that, for $g \in$ span(Δ),

$$[g,\ y \leftrightarrow y', y'] = -[g,\ y \leftrightarrow y', y]. \tag{17.101}$$

In Theorem 17.2.8 below, A_α and $\omega_{\mathscr{C}}$ have their usual meanings. The intuition behind the theorem is guided by the same salt-barrel picture that guided Theorems 16.4.2 and 17.1.17. Corresponding to network $\{\mathscr{S},\mathscr{C},\mathscr{R}\}$ we again imagine that there are barrels, one for each complex, containing salt solutions of perhaps differing salt concentrations. For each reaction $y \to y' \in \mathscr{R}$, there is a flow of salt solution from barrel y to barrel y' at volumetric flow rate $\alpha_{y\to y'}$. The steady-state concentration of salt in barrel y is denoted b_y.

Note that if, in network $\{\mathscr{S},\mathscr{C},\mathscr{R}\}$, $y \leftrightarrow y'$ is a cut-link, with y and y' in the same terminal strong-linkage class, then the steady-state rate of salt flow to barrel y from barrel y' must be exactly balanced by the rate of salt flow from barrel y to barrel y'. That is, we must have

$$\alpha_{y'\to y} b_{y'} = \alpha_{y\to y'} b_y. \tag{17.102}$$

From this it follows that

$$\operatorname{sgn}(b_y - b_{y'}) \;=\; \operatorname{sgn}(\alpha_{y'\to y} - \alpha_{y\to y'}). \tag{17.103}$$

Note that $\alpha_{y'\to y} - \alpha_{y\to y'}$ is not only the net rate of solution flow to barrel y from barrel y'; because $y \leftrightarrow y'$ is a cut-link, it is also the net flow rate of solution to the barrel collection $\mathscr{W}(y)$ from the barrel collection $\mathscr{W}(y')$. These ideas help guide the proof of the following theorem:

Theorem 17.2.8 (Third Salt Theorem [77]**).** *Suppose that $y \leftrightarrow y'$ is a cut-link of reaction network $\{\mathscr{S},\mathscr{C},\mathscr{R}\}$, with y and y' residing in terminal strong-linkage class Λ. For $\alpha \in \mathbb{R}^{\mathscr{R}}_{+}$, let*

$$g := A_\alpha \omega_{\mathscr{C}}, \tag{17.104}$$

and let b be a member of $\ker A_\alpha$ *with* $\operatorname{supp} b = \Lambda$. *Then*

$$\operatorname{sgn}(b_y - b_{y'}) \;=\; \operatorname{sgn}\,[g,\, y \leftrightarrow y',\, y]. \tag{17.105}$$

Proof. From (17.104) and Lemma 17.A.2 in the appendix, we obtain

$$\sum_{\bar{y}\,\in\,\mathscr{W}(y)} g_{\bar{y}} \;=\; \sum_{\mathscr{W}(y')\to\mathscr{W}(y)} \alpha_{\bar{y}'\to\bar{y}} \;-\; \sum_{\mathscr{W}(y)\to\mathscr{W}(y')} \alpha_{\bar{y}\to\bar{y}'}. \tag{17.106}$$

From the same lemma and the fact that $A_\kappa b = 0$, we also obtain

$$0 \;=\; \sum_{\mathscr{W}(y')\to\mathscr{W}(y)} \alpha_{\bar{y}'\to\bar{y}} b_{\bar{y}'} \;-\; \sum_{\mathscr{W}(y)\to\mathscr{W}(y')} \alpha_{\bar{y}\to\bar{y}'} b_{\bar{y}}. \tag{17.107}$$

Because $y \leftrightarrow y'$ is a cut-link, $y \to y'$ and $y' \to y$ are the only reactions connecting $\mathscr{W}(y)$ and $\mathscr{W}(y')$. Thus, we actually have

$$[g,\, y \leftrightarrow y',\, y] \;:=\; \sum_{\bar{y}\,\in\,\mathscr{W}(y)} g_{\bar{y}} = \alpha_{y'\to y} - \alpha_{y\to y'} \tag{17.108}$$

and

$$0 = \alpha_{y'\to y} b_{y'} - \alpha_{y\to y'} b_y. \tag{17.109}$$

From these last two equations, we obtain (17.105). □

17.2.6 Observation 3 Rewritten

We return now to the regular deficiency one network $\{\mathscr{S},\mathscr{C},\mathscr{R}\}$ that was presumed to give an affirmative answer to Question 17.2.4. Our aim in this subsection is to recast Observation 3 yet again, this time in light of information given by our latest salt theorem.

We suppose once again that $\kappa \in \mathbb{R}_+^{\mathscr{R}}$ and $\mu \in \mathbb{R}^{\mathscr{S}}$ satisfy the requirements of Question 17.2.4 and that the reactive complexes have been partitioned into upper, middle, and lower sets as prescribed in Section 17.2.3. Our current version of Observation 3 tells us that if $y \leftrightarrow y'$ is a cut-link, with y and y' residing in terminal strong-linkage class Λ, and if $b \in \overline{\mathbb{R}}_+^{\mathscr{C}}$ is an element of $\ker A_\kappa$ such that $\operatorname{supp} b = \Lambda$, then μ must satisfy

$$\operatorname{sgn}(y\cdot\mu - y'\cdot\mu) = \operatorname{sgn}(b_y - b_{y'}) \text{ if } \Lambda \subset U \tag{17.110}$$

and

$$\operatorname{sgn}(y\cdot\mu - y'\cdot\mu) = -\operatorname{sgn}(b_y - b_{y'}) \text{ if } \Lambda \subset L. \tag{17.111}$$

With the (presumed-nonzero) $g \in \mathbb{R}^{\mathscr{C}}$ given by

$$g := A_\kappa \omega_{\mathscr{C}} = \sum_{y\to y'\in\mathscr{R}} \kappa_{y\to y'}(\omega_{y'} - \omega_y), \tag{17.112}$$

Theorem 17.2.8 tells us that we can rewrite Observation 3 in the following way:

Observation 3: If $y \leftrightarrow y'$ is a cut-link, with y and y' residing in terminal strong-linkage class Λ, then

$$\operatorname{sgn}(y\cdot\mu - y'\cdot\mu) = \operatorname{sgn}[g,\, y\leftrightarrow y', y] \text{ if } \Lambda \subset U \tag{17.113}$$

and

$$\operatorname{sgn}(y\cdot\mu - y'\cdot\mu) = -\operatorname{sgn}[g,\, y\leftrightarrow y', y] \text{ if } \Lambda \subset L. \tag{17.114}$$

Recall that for a regular network—in particular the one under study—each direct link of two *terminal* complexes is a cut-link. Thus, Observation 3 provides, for each pair of adjacent terminal complexes $\{y,y'\}$, a linear relation of the form $y\cdot\mu > y'\cdot\mu$, $y\cdot\mu < y'\cdot\mu$, or $y\cdot\mu = y'\cdot\mu$ that μ must satisfy, the sense of the relation depending on the sign of $[g,\, y\leftrightarrow y', y]$. Our attention then turns naturally to ways in which that sign, for the various terminal cut-links, is fixed by (and can be determined directly from) reaction network attributes.

17.2.7 Confluence Vectors and Confluence Vector Orientations

From (17.112) it is evident that g is a member of $\operatorname{span}(\Delta)$, where Δ is given as usual by (16.8). Moreover, because $YA_\kappa\omega_{\mathscr{C}} = 0$, g is also a member of $\ker Y$. Thus g lies in $\ker Y \cap \operatorname{span}(\Delta)$, the dimension of which, from Proposition 16.2.1, is the deficiency of the network under study. In our case, then, $\dim \ker Y \cap \operatorname{span}(\Delta) = 1$.

From Remark 16.2.5, we know that g lies in $\ker Y \cap \operatorname{span}(\Delta)$ if and only if its components satisfy both of the following conditions:

$$\sum_{y \in \mathscr{L}^\theta} g_y = 0, \quad \theta = 1,2,\ldots,\ell, \tag{17.115}$$

$$\sum_{y \in \mathscr{C}} g_y\, y = 0. \tag{17.116}$$

Because $\dim \ker Y \cap \operatorname{span}(\Delta) = 1$, we know that these equations, taken together, admit a nonzero solution and that *all other solutions are scalar multiples of some fixed nonzero one*. Hereafter, we will suppose that g^* is some fixed nonzero solution of (17.115) and (17.116). *Note that these equations derive solely from knowledge of the network's complexes and how they are partitioned into linkage classes, nothing more. To find a nontrivial solution, g^* is a simple matter.* The particular solution g in which we have interest, defined by (17.112), is a nonzero multiple of g^*.

When the network under study is not weakly reversible, there is something else we know about g, apart from the fact that it lies in the line $\ker Y \cap \operatorname{span}(\Delta)$. We know something about the *direction* along that line in which g points:

For the network $\{\mathscr{S},\mathscr{C},\mathscr{R}\}$, we will say that a set of complexes $\mathscr{A} \subset \mathscr{C}$ is *absorptive* if $\mathscr{A}$ is not empty and there is no reaction from a complex in $\mathscr{A}$ to a complex not in $\mathscr{A}$; that is, $y \in \mathscr{A}$ and $y \to y' \in \mathscr{R}$ imply that y' is in $\mathscr{A}$. Suppose that $\mathscr{A}$ is absorptive and is not the union of linkage classes. If $\mathscr{A}'$ is the complement of $\mathscr{A}$ in $\mathscr{C}$ and if $\mathscr{A}' \to \mathscr{A}$ is the (nonempty) set of reactions from complexes in $\mathscr{A}'$ to complexes in $\mathscr{A}$, it follows from (17.112) and Lemma 17.A.2 in Appendix 17.A that

$$\sum_{y \in \mathscr{A}} g_y = \sum_{\mathscr{A}' \to \mathscr{A}} \kappa_{y' \to y} > 0. \tag{17.117}$$

Note that all terminal strong-linkage classes are absorptive. If the network $\{\mathscr{S},\mathscr{C},\mathscr{R}\}$ is not weakly reversible, then there is at least one terminal strong-linkage class Λ that is smaller than the linkage class $\mathscr{L}$ in which it resides (and is therefore not the union of linkage classes). In this case, we have

$$\sum_{y \in \Lambda} g_y > 0. \tag{17.118}$$

Thus, if $\{\mathscr{S},\mathscr{C},\mathscr{R}\}$ is not weakly reversible, then g can satisfy (17.112) only if its orientation along the line $\ker Y \cap \operatorname{span}(\Delta)$ is consistent with (17.118). When the network $\{\mathscr{S},\mathscr{C},\mathscr{R}\}$ under consideration is not weakly reversible, we will suppose that g^*, the basis for $\ker Y \cap \operatorname{span}(\Delta)$ described earlier, points in the direction consistent with (17.118). In this case, the vector g given by (17.112) is of the form λg^*, where λ is positive.

Our particular g, conforming to conditions (17.115)–(17.117), is an instance of what we shall call a *confluence vector* [77] for the network $\{\mathscr{S},\mathscr{C},\mathscr{R}\}$. The terminology is motivated by equation (17.41), which suggests that, for each $y \in \mathscr{C}$, the number g_y represents a net flow from the remaining complexes to complex y (a confluence at y). In the following definition, $\{\mathscr{S},\mathscr{C},\mathscr{R}\}$ indicates a general reaction network, not necessarily the particular one we are studying.

Definition 17.2.9. *A* **confluence vector** *for a reaction network* $\{\mathscr{S},\mathscr{C},\mathscr{R}\}$ *is a vector* $\bar{g} \in \mathbb{R}^{\mathscr{C}}$ *such that*

(i) For every linkage class $\mathscr{L}$

$$\sum_{y \in \mathscr{L}} \bar{g}_y = 0. \tag{17.119}$$

(ii)

$$\sum_{y \in \mathscr{C}} \bar{g}_y \, y = 0 \tag{17.120}$$

(iii) For every absorptive set $\mathscr{A} \subset \mathscr{C}$ *that is not the union of linkage classes*

$$\sum_{y \in \mathscr{A}} \bar{g}_y > 0. \tag{17.121}$$

Remark 17.2.10. For the particular regular deficiency one network under study, there is a nonzero confluence vector—our putative g is an example—and all others are multiples of some fixed one, g^*.

If the network is not weakly reversible, then, as we indicated earlier, all nonzero confluence vectors are *positive* multiples of g^*. In this case, it is easy to see that, *for each cut-link* $y \leftrightarrow y'$,

$$\text{sgn}\,[g,\, y \leftrightarrow y', y] = \text{sgn}\,[g^*,\, y \leftrightarrow y', y]. \tag{17.122}$$

Given any cut-link $y \leftrightarrow y'$, then, it is a simple matter to determine for our putative g (or for *any* nonzero confluence vector) the sign of $[g,\, y \leftrightarrow y', y]$: Find a nonzero confluence vector g^* for the network $\{\mathscr{S},\mathscr{C},\mathscr{R}\}$–a matter of straightforward linear-equation solving—and then determine sgn $[g^*,\, y \leftrightarrow y', y]$ from the definition (17.99):

$$\text{sgn}\ [g^*,\, y \leftrightarrow y',\, y] := \text{sgn} \sum_{\bar{y} \in \mathscr{W}(y)} g^*_{\bar{y}}, \tag{17.123}$$

where $\mathscr{W}(y)$ is the set of complexes on the y-side of the cut.

The situation is different when the network is weakly reversible. In this case, the regular deficiency one network under consideration must in fact be reversible. This is because, for a weakly reversible network, each terminal strong-linkage class coincides with the linkage class containing it, so all complexes are terminal. Regularity then requires that each link be a cut-link. That each reaction is reversible follows from Remark 6.7.3.

More important is the idea that, in the reversible case, there is no absorptive set that is not the union of linkage classes, so constraint (*iii*) in Definition 17.2.9 becomes no constraint at all. In this case, a confluence vector—in particular our putative nonzero g—can assume either one of *two* distinct orientations along the line $\ker Y \cap \text{span}\,(\Delta)$, pointing along either g^* or $-g^*$.

Motivated by these ideas, we posit the following definition:

Definition 17.2.11. *Two nonzero confluence vectors g' and g'' for a reaction network are* **similarly oriented** *(denoted $g' \cong g''$) if there is a positive number λ such that $g'' = \lambda g'$. The equivalence relation $\cong$ in the set of nonzero confluence vectors serves to partition them into equivalence classes called* **confluence vector orientations**.

Remark 17.2.12. If the regular deficiency one network under consideration is not reversible, then there is just one confluence vector orientation, characterized by g^*. If the network is reversible, then there are two, characterized by g^* and $-g^*$. In any case, our putative nonzero g is a member of one such orientation, and *the values of* $\text{sgn}\,[g,\, y \leftrightarrow y', y]$, *as $y \leftrightarrow y'$ ranges over all possible terminal cut-links, are completely determined by the orientation in which g resides*. This is to say that *the sense of all the inequalities imposed on μ in Observation 3 is determined by the confluence vector orientation in which g resides* and, of course, the U, M, L partition of reactive complexes given by (17.90)–(17.92).

17.2.8 The Inequality System Induced by a $\{U,M,L\}$ Partition and a Confluence Vector Orientation

For a regular deficiency one network $\{\mathscr{S},\mathscr{C},\mathscr{R}\}$, not necessarily the particular one under consideration so far, we will say what we mean by *the inequality system induced by* $\{P, g^\dagger\}$, where $P = \{U, M, L\}$ indicates a choice of upper-middle-lower partition of the reactive complexes and $g^\dagger$ indicates a choice of confluence vector orientation. (The vector $g^\dagger \in \ker Y \cap \text{span}\,(\Delta)$ is any representative member of the chosen orientation. If the network is not reversible, there is only one orientation; otherwise, there are two.)

We denote by $\mathscr{C}_\rightarrow$ the set of reactive complexes. The inequality system induced by $\{P, g^\dagger\}$ will amount to a collection of relations among linear forms in the set

$$\{y \cdot \mu\}_{y \in \mathscr{C}_\rightarrow}, \tag{17.124}$$

where μ is understood to be an unspecified element of $\mathbb{R}^{\mathscr{S}}$. Each relation is either an inequality or an equality. Thus, by an "inequality system," we embrace the possibility that some (or even all) of the relations might in fact be equations.

Example: To make the construction of an inequality system concrete, it will be helpful to have available an example to call upon as the procedure unfolds. For this purpose we will invoke the regular deficiency one network (17.125), which in Section 12.7 described a reactor driven by a classical mechanism of metal catalysis.

$$\begin{array}{ccc} A+S \rightleftarrows AS & & B+S \rightleftarrows BS \\ AS+BS & \rightarrow & C+2S \\ A \rightleftarrows & 0 & \leftrightarrows B \\ & \uparrow & \\ & C & \end{array} \tag{17.125}$$

In this case

$$\mathscr{C}_{\rightarrow} = \{A+S,\ AS,\ B+S,\ BS,\ AS+BS,\ A,\ B,\ C,\ 0\} \tag{17.126}$$

and the linear forms to be joined by equality or inequality signs are

$$\{\mu_A + \mu_S,\ \mu_{AS},\ \mu_B + \mu_S,\ \mu_{BS},\ \mu_{AS} + \mu_{BS},\ \mu_A,\ \mu_B,\ \mu_C,\ 0\}. \tag{17.127}$$

For network (17.125) there is only one choice of a confluence vector orientation, with a representative member $g^\dagger \in \mathbb{R}^{\mathscr{C}}$ given by

$$\begin{aligned} g_A^\dagger = 1,\quad g_B^\dagger = 1,\quad g_C^\dagger = -1,\quad g_0^\dagger = -1,\\ g_{A+S}^\dagger = -1,\quad g_{AS}^\dagger = 1,\\ g_{B+S}^\dagger = -1,\quad g_{BS}^\dagger = 1,\\ g_{AS+BS}^\dagger = -1,\quad g_{C+2S}^\dagger = 1. \end{aligned} \tag{17.128}$$

For the purposes of our example, we will construct the inequality system induced by $g^\dagger$ and the particular upper-middle-lower partition

$$U := \{A,\ 0,\ B,\ A+S,\ AS\},\ M := \{C,\ AS+BS\},\ L := \{B+S,\ BS\}. \tag{17.129}$$

The construction of the inequality system induced by a choice of upper-middle-lower partition and a choice of confluence vector orientation proceeds by the invocation of three rules, motivated by Observations 1–3 made earlier. We state the rules below and illustrate each by means of network (17.125), implemented for the partition shown in (17.129). Each rule gives rise to a new set of relations that μ must satisfy.

Rule 1: *If y and y' are members of M, then $\mu \in \mathbb{R}^{\mathscr{S}}$ must satisfy the relation*

$$y \cdot \mu = y' \cdot \mu.$$

Example (Rule 1): For our choice of M, application of Rule 1 results in just one relation,

$$\mu_C = \mu_{AS} + \mu_{BS}.$$

Rule 2: *If, with respect to the partition $\{U, M, L\}$, complex y lies above complex y' (Remark 17.2.7), then μ must satisfy the relation*

$$y \cdot \mu > y' \cdot \mu.$$

Example (Rule2): For our choice of an upper-middle-lower partition, application of Rule 2 results in a collection of inequalities represented schematically by

$$\{\mu_A,\ 0,\ \mu_B,\ \mu_A + \mu_S,\ \mu_{AS}\} > \{\mu_C,\ \mu_{AS} + \mu_{BS}\} > \{\mu_B + \mu_S,\ \mu_{BS}\}. \tag{17.130}$$

Rule 3: *Let $g^\dagger$ be any fixed confluence vector of the chosen orientation. If $\{y,\ y'\}$ is a pair of adjacent terminal complexes (in which case $y \leftrightarrow y'$ is a cut-link), then*

(i) for $\{y, y'\} \subset U$, μ must satisfy the relation

$$\begin{aligned} y\cdot\mu &> y'\cdot\mu \text{ if } [g^\dagger, y \leftrightarrow y', y] > 0,\\ y\cdot\mu &= y'\cdot\mu \text{ if } [g^\dagger, y \leftrightarrow y', y] = 0,\\ y\cdot\mu &< y'\cdot\mu \text{ if } [g^\dagger, y \leftrightarrow y', y] < 0; \end{aligned} \tag{17.131}$$

(ii) for $\{y, y'\} \subset L$, μ must satisfy the relation

$$\begin{aligned} y\cdot\mu &< y'\cdot\mu \text{ if } [g^\dagger, y \leftrightarrow y', y] > 0,\\ y\cdot\mu &= y'\cdot\mu \text{ if } [g^\dagger, y \leftrightarrow y', y] = 0,\\ y\cdot\mu &> y'\cdot\mu \text{ if } [g^\dagger, y \leftrightarrow y', y] < 0. \end{aligned} \tag{17.132}$$

Example (Rule 3): In network (17.125) consider the terminal adjacent complex pair $\{0, A\}$. For $g^\dagger$ given by (17.128),

$$[g^\dagger, 0 \leftrightarrow A, A] = g_A^\dagger = 1 > 0 \tag{17.133}$$

and

$$[g^\dagger, 0 \leftrightarrow A, 0] = g_0^\dagger + g_B^\dagger + g_C^\dagger = -1 + 1 + (-1) = -1 < 0. \tag{17.134}$$

Note that, in our chosen $\{U,M,L\}$ partition, 0 and A lie in U. By virtue of either (17.133) or (17.134), Rule 3 gives rise to the relation

$$\mu_A > 0. \tag{17.135}$$

Similarly, the terminal adjacent complex pairs $\{0, B\}$ and $\{A+S, AS\}$, both pairs residing in U, give rise to the relations

$$\mu_B > 0 \quad \text{and} \quad \mu_{AS} > \mu_A + \mu_S. \tag{17.136}$$

The situation for the adjacent terminal complex pair $\{B+S, BS\}$ is a little different. In this case

$$[g^\dagger, B+S \leftrightarrow BS, BS] = g_{BS}^\dagger = 1 > 0, \tag{17.137}$$

and, for our chosen $\{U,M,L\}$ partition, the complexes $\{B+S, BS\}$ reside in L. Rule 3 then gives rise to the relation

$$\mu_{BS} < \mu_B + \mu_{BS}. \tag{17.138}$$

Example Summary: For network (17.125) the inequality system induced by the upper-middle-lower partition (17.129) and the confluence vector orientation represented by (17.128) is expressed in the following relations:[6]

$$\begin{gathered} \{\mu_A, \mu_B\} > 0 > \mu_C = \mu_{AS} + \mu_{BS} > \mu_B + \mu_S > \mu_{BS} \\ \mu_{AS} > \mu_A + \mu_S > \mu_C. \end{gathered} \tag{17.139}$$

6 These represent a streamlined logically equivalent summary of the raw relations derived from Rules 1–3.

Remark 17.2.13. For network (17.125) there is only one confluence vector orientation, represented by (17.128). However, there are several upper-middle-lower partitions of the reactive complexes, and each will give rise to its own inequality system. For example, the partition

$$U := \{A,\ 0,\ B\},\ M := \{C,\ AS+BS\},\ L := \{\ A+S,\ AS,\ B+S,\ BS\} \tag{17.140}$$

gives rise to the relations

$$\begin{aligned} \{\mu_A, \mu_B\} \ > \ 0 \ > \ \mu_C \ = \ \mu_{AS} + \mu_{BS} \ > \ \mu_B + \mu_S \ > \ \mu_{BS} \\ \mu_{AS} + \mu_{BS} \ > \ \mu_A + \mu_S \ > \ \mu_{AS}. \end{aligned} \tag{17.141}$$

17.2.9 A Theorem Underlying the Deficiency One Algorithm

It will be useful to review the situation for the general regular deficiency one network considered earlier, the network for which we supposed an affirmative answer to Question 17.2.3 or, equivalently, to Question 17.2.4. For that network we deduced the existence of a confluence vector orientation and an upper-middle-lower partition (derived from $\lambda_0, \lambda_1, \lambda_2, \ldots, \lambda_\ell$) that induced a linear inequality system (derived from Observations 1–3) that the putative μ of Question 17.2.3 must satisfy. Thus, for an affirmative answer to Question 17.2.3, it is *necessary* that there exist a nonzero μ, sign-compatible with the stoichiometric subspace, that satisfies the inequality system induced by *some* confluence vector orientation and *some* upper-middle-lower partition of the reactive complexes.

The following proposition asserts that, for an arbitrary regular deficiency one network, this condition is not only necessary for an affirmative answer to Question 17.2.3, it is also sufficient.

Proposition 17.2.14 (A key to the Deficiency One Algorithm [74, 77]**).** *Let $\{\mathscr{S}, \mathscr{C}, \mathscr{R}\}$ be a regular deficiency one reaction network, let $\phi : \mathbb{R} \to \mathbb{R}$ be continuous and strictly monotonically increasing, and let $\mu \in \mathbb{R}^{\mathscr{S}}$ be nonzero and sign-compatible with the stoichiometric subspace. The following are equivalent:*

(i) There is an element $\kappa \in \mathbb{R}^{\mathscr{R}}_{+}$ such that

$$\sum_{y \to y' \in \mathscr{R}} \kappa_{y \to y'} (y' - y) = 0, \tag{17.142}$$

$$\sum_{y \to y' \in \mathscr{R}} \kappa_{y \to y'} \ \phi\,(y \cdot \mu)\ (y' - y) = 0. \tag{17.143}$$

(ii) There is a confluence vector orientation and an upper-middle-lower partition for which μ satisfies the induced linear inequality system.

About the proof: That *(i)* implies *(ii)* is a consequence of our discussion of requirements that μ must satisfy if Question 17.2.3 is to have an affirmative answer. That *(ii)* implies *(i)* follows from a construction of κ given in [77]. An earlier construction, in the special case $\phi(x) \equiv e^x$, was given in [74].[7] □

Proposition 17.2.14, in conjunction with Propositions 17.2.1 and 17.2.2, gives rise to the following theorem:

Theorem 17.2.15 (Foundations of the Deficiency One Algorithm [74, 77]**).** *For a regular deficiency one reaction network* $\{\mathscr{S},\mathscr{C},\mathscr{R}\}$ *the following are equivalent:*

(i) There is a rate constant assignment $k \in \mathbb{R}_+^{\mathscr{R}}$ *such that the mass action system* $\{\mathscr{S},\mathscr{C},\mathscr{R},k\}$ *admits a pair of distinct stoichiometrically compatible positive equilibria.*

(ii) There is a rate constant assignment $k' \in \mathbb{R}_+^{\mathscr{R}}$ *such that the mass action system* $\{\mathscr{S},\mathscr{C},\mathscr{R},k'\}$ *admits a degenerate positive equilibrium.*

(iii) There is a confluence vector orientation and an upper-middle-lower partition for which the induced linear inequality system is satisfied by a nonzero $\mu \in \mathbb{R}^{\mathscr{S}}$ *that is stoichiometrically compatible with the stoichiometric subspace.*

17.2.10 Algorithm Overview: Converting Questions About Nonlinear Equations into Questions About Linear Inequalities

Theorem 17.2.15 tells us that the apparently *nonlinear* problems of assessing a regular deficiency one network's capacity for multiple stoichiometrically compatible positive equilibria or for a degenerate positive equilibrium can be decided *unequivocally*—either positively or negatively—by examining whether *even one* of several *linear* systems of inequalities admits a solution.

For a particular network of interest, there will be one or perhaps two possible confluence vector orientations, depending on whether the network is reversible, and there will be several possible upper-middle-lower partitions of the reactive complexes. For each possibility there will be an induced system of linear inequalities that can be checked for a nonzero[8] solution that is sign-compatible with the network's stoichiometric subspace. Procedures for assessing the existence of solutions to linear inequality systems are now readily available and rapid in execution; they are routinely invoked in connection with standard linear programming implementations.

The Deficiency One Algorithm is, in effect, a procedure for assessing the existence of condition *(iii)* solutions for the variety of linear inequality systems that emerge from various choices of a confluence vector orientation and an upper-middle-lower partition. A step-by-step elaboration of the algorithm, with examples, is provided in [74].

[7] Equation (4.7) in [77] should read $\sigma = p(e^{\mu} - 1)$, which is the same as its counterpart (7.2) in [74].

[8] When the inequality system contains even one strict inequality, any solution will of necessity be nonzero. However, there can be crucial instances of upper-middle-lower partitions for which the induced linear inequality system contains only equations. Network 5.3 provides an example.

In fact, the Deficiency One Algorithm is implemented completely in the Chemical Reaction Network Toolbox [62]. In response to a user-supplied reaction network, the Toolbox will generate, internally, the several systems of inequalities that need be examined, and it will test each for a nonzero stoichiometrically-sign-compatible solution. When there is no solution for any such system, the Toolbox will say so. When there is a solution, the Toolbox will construct a sample rate constant assignment for which multiple stoichiometrically compatible positive equilibria exist; it will also construct rate constants for which there is a positive degenerate equilibrium. In each case the equilibria are reported along with a complete set of eigenvalues associated with them.

Remark 17.2.16 (Upper-middle-lower partitions that needn't be examined). To test whether condition *(iii)* is satisfied, exploration of certain upper-middle-lower partitions will be redundant or otherwise unnecessary.

By the *inversion* of upper-middle-lower partition $\{U,M,L\}$, we mean the upper-middle-lower partition $\{U',M,L'\}$ in which the upper and lower parts are interchanged—that is, in which $U' = L$ and $L' = U$. Given a specified confluence orientation, it is not difficult to see that the inequality system induced by the inversion of a particular upper-middle-lower partition will be identical to the inequality system induced by the original partition, but with all inequality signs reversed. Thus, both inequality systems will admit a nonzero stoichiometrically sign-compatible solution if and only if one of them does. To see whether condition *(iii)* is satisfied, then, there is redundancy in examining *both* a partition and its inversion.

The following propositions tell us that, so long as the number of *trivial* terminal strong-linkage classes is either zero or one, there are certain upper-middle-lower partitions that need not be examined, for the inequality systems they induce cannot admit a nonzero solution that is sign-compatible with the stoichiometric subspace.

Proposition 17.2.17 ([74, 77]). *Suppose that for a regular deficiency one network, there is a confluence vector orientation and a reactive-complex upper-middle-lower partition $\{U,M,L\}$ such that the induced linear inequality system admits a nonzero solution sign-compatible with the stoichiometric subspace. Then*

(i) if the network has no trivial terminal strong-linkage classes, then neither U nor L is empty,

(ii) if the network has just one trivial terminal strong-linkage class, then U and L are not both empty.

In the first case, then, there is no need to examine any partition in which either the upper or lower part is empty, and in the second case, there is no need to examine a partition in which only the middle part is nonempty. It is interesting to note that for the regular deficiency one network

$$\begin{aligned} B &\to A \\ A + 2B &\to 3B, \end{aligned} \tag{17.144}$$

in which there are *two* trivial terminal strong-linkage classes, the only feasible upper-middle-lower reactive-complex partition is $U = \{\varnothing\}$, $M = \{B, A+2B\}$, $L = \{\varnothing\}$, a partition in which only the middle part is populated. That partition (taken with the sole confluence vector orientation) gives rise to an inequality system consisting of the single *equation*, $\mu_B = \mu_A + 2\mu_B$. There is a nonzero solution, $\mu_A = -1$, $\mu_B = 1$, that is sign-compatible with the network's stoichiometric subspace. Recall from Section 5.1.3 and, in particular, Figure 5.4 that there are indeed rate constants—any choice of rate constants will do!—that give rise to a pair of distinct stoichiometrically compatible positive equilibria. In this case, any choice of rate constants will also give rise to a degenerate positive equilibrium.

Remark 17.2.18 (The number of partitions that need be considered). Suppose that in a regular deficiency one network, there are m *nontrivial* terminal strong-linkage classes. Taken with a particular confluence vector orientation, there are at most $\frac{3^m+1}{2}$ upper-middle-lower reactive-complex partitions that need be considered, once inversions have been eliminated. If there are no *trivial* terminal strong-linkage classes, then partitions with an empty upper part (or an empty lower part) can also be eliminated, in which case there are at most $\frac{3^m+1}{2} - 2^m$ that need be examined. If there is just one *trivial* terminal strong-linkage class, then there are at most $\frac{3^m+1}{2} - 1$ partitions that require study, for the partition in which all reactive complexes are assigned to the middle part can be ignored.

17.2.11 Some Earlier Examples Reconsidered

Here we reconsider some examples encountered earlier in the book, this time in light of the Deficiency One Algorithm. In each example the network of interest is regular and has a deficiency of one. In some cases we exhibit only a few of the linear inequality systems that the algorithm generates.

17.2.11.1 Tiny Two-Species Examples Having Great Subtlety

It will be useful to recall the simple but highly delicate examples discussed in Section 8.5, where we provided a first glimpse of the Deficiency One Algorithm. We repeat as Tables 17.1 and 17.2 the two tables included in that section. The first shows three small two-species networks and their capacity for multiple stoichiometrically compatible positive equilibria. It is only for the second of these networks that there are rate constants for which such equilibria are admitted by the resulting mass action differential equations. Yet the second network differs from the first only in the addition of species A to both sides of a single reversible reaction. And the second differs from the third only in a permutation of complexes 0 and A.

The Deficiency One Algorithm, then, must be sufficiently delicate as to distinguish correctly between networks as subtly different as these. In the second table, we exhibit for each network the linear inequality systems the algorithm delivers. (For each network there are two nontrivial terminal strong-linkage classes and no trivial

Table 17.1. Three simple deficiency one networks and their capacity for multiple equilibria

Network	Multiple Equilibria?
1. $A+B \rightleftarrows 2A$ $A \rightleftarrows 0 \rightleftarrows B$	No
2. $2A+B \rightleftarrows 3A$ $A \rightleftarrows 0 \rightleftarrows B$	Yes
3. $2A+B \rightleftarrows 3A$ $0 \rightleftarrows A \rightleftarrows B$	No

Table 17.2. Three simple deficiency one networks and their inequality systems

Network	Inequality Systems	Comment
1. $A+B \rightleftarrows 2A$ $A \rightleftarrows 0 \rightleftarrows B$	(a) $\mu_A > 0 > \mu_B > 2\mu_A > \mu_A + \mu_B$ (b) $\mu_B > 0 > \mu_A > \mu_A + \mu_B > 2\mu_A$	No solution for (a) or (b) Multiple equilibria impossible
2. $2A+B \rightleftarrows 3A$ $A \rightleftarrows 0 \rightleftarrows B$	(a) $\mu_A > 0 > \mu_B > 3\mu_A > 2\mu_A + \mu_B$ (b) $\mu_B > 0 > \mu_A > 2\mu_A + \mu_B > 3\mu_A$	No solution for (a) Solution for (b): $\mu_A = -2,\ \mu_B = 1$
3. $2A+B \rightleftarrows 3A$ $0 \rightleftarrows A \rightleftarrows B$	(a) $0 = \mu_A > \mu_B > 3\mu_A > 2\mu_A + \mu_B$ (b) $\mu_B > \mu_A = 0 > 2\mu_A + \mu_B > 3\mu_A$	No solution for (a) or (b) Multiple equilibria impossible

ones. Thus, from Remark 17.2.18 the number of inequality systems that need be considered for each confluence vector orientation is $\frac{3^2+1}{2} - 2^2 = 1$. Because each network is reversible, there are two orientations, so each network gives rise to two inequality systems.) It is easy to see that the *only* inequality system in the second table that admits a solution is 2(b).[9] Thus, the second network in the table is correctly identified by the Deficiency One Algorithm as the only network of three very similar ones which, when taken with mass action kinetics, has the capacity for multiple stoichiometrically compatible positive equilibria.

That the Deficiency One Algorithm is capable of such fine discrimination was already evident from the same tables presented earlier in Section 8.5. Now, however, we are in a position to explore the delicate machinery that gives rise to the subtly different inequality systems in Table 17.2. For this purpose, it will suffice to consider how inequality systems for the second and third networks come about. (The generation of inequality systems for the first network differs very little from that for the second network.)

For the purpose of finding a nonzero confluence vector for the second and third networks, the equations that Definition 17.2.9 requires us to study in the two cases turn out to be precisely the same:

$$g_{2A+B}(2A+B) + g_{3A}(3A) + g_A(A) + g_0(0) + g_B(B) = 0 \tag{17.145}$$

$$g_{2A+B} + g_{3A} = 0 \quad g_A + g_0 + g_B = 0. \tag{17.146}$$

A solution g^* is given by

$$g_A^* = 1,\ g_0^* = 0,\ g_B^* = -1,\ g_{2A+B}^* = 1,\ g_{3A}^* = -1. \tag{17.147}$$

By virtue of Remark 17.2.16, we need only consider, in each case, the partition

$$U = \{A, 0, B\}, M = \{\varnothing\}, L = \{2A+B, 3A\}. \tag{17.148}$$

Rule 2, implemented for Networks 2 and 3, gives rise to the same inequalities in both cases, which we indicate schematically by

$$\{\mu_A, 0, \mu_B\} > \{2\mu_A + \mu_B, 3\mu_A\}. \tag{17.149}$$

With respect to Rule 3, there are two opposite confluence vector orientations that must be considered. For each orientation, we must implement Rule 3 for Networks 2 and 3 separately, for the results are different.

Consider first Network 2. There are three cut-links: $A \leftrightarrow 0$, $0 \leftrightarrow B$, and $2A+B \leftrightarrow 3A$. For the particular orientation represented by the g^* given in (17.147), we have

$$\begin{aligned} [g^*, A \leftrightarrow 0, A] &= g_A^* = 1 > 0 \\ [g^*, 0 \leftrightarrow B, B] &= g_B^* = -1 < 0, \\ [g^*, 2A+B \leftrightarrow 3A, 2A+B] &= g_{2A+B}^* = 1 > 0. \end{aligned} \tag{17.150}$$

Because $U = \{A, 0, B\}$ and $L = \{2A+B, 3A\}$, we obtain from Rule 3 and (17.150) the relations

[9] As we indicated in Section 8.5, the stoichiometric subspace in each of these examples coincides with $\mathbb{R}^{\mathscr{S}}$, where $\mathscr{S} = \{A, B\}$. For each network, then, *any* vector $\mu \in \mathbb{R}^{\mathscr{S}}$ is sign-compatible with the stoichiometric subspace; the issue of sign compatibility is moot.

$$\mu_A > 0, \quad \mu_B < 0, \quad 2\mu_A + \mu_B < 3\mu_A. \tag{17.151}$$

Taken together, then, the relations that emerge from Rules 2 and 3, applied for the confluence vector orientation given by (17.147), are

$$\mu_A > 0 > \mu_B > 3\mu_A > 2\mu_A + \mu_B, \tag{17.152}$$

which is the first of the inequality systems shown for Network 2 in Table 17.2. For the reverse confluence vector orientation given by $-g^*$, the sense of the Rule 3 inequalities in (17.152) is reversed. Taken with the Rule 2 inequalities (17.149) (which do not become reversed), the reversed Rule 3 inequalities give the second inequality system shown for Network 2 in Table 17.2.

Now consider Network 3, for which there are subtly different cut-links, $A \leftrightarrow 0$, $A \leftrightarrow B$, and $2A + B \leftrightarrow 3A$, and, therefore, different Rule 3 relations that emerge. For the confluence vector orientation g^* given by (17.147), we obtain

$$\begin{aligned} [g^*, A \leftrightarrow 0, A] &= g_A^* + g_B^* = 1 + (-1) = 0 \\ [g^*, 0 \leftrightarrow B, B] &= g_B^* = -1 < 0, \\ [g^*, 2A + B \leftrightarrow 3A, 2A + B] &= g_{2A+B}^* = 1 > 0. \end{aligned} \tag{17.153}$$

As before, $U = \{A, 0, B\}$ and $L = \{2A + B, 3A\}$. We obtain from Rule 3 and, this time (17.153), the relations

$$\mu_A = 0, \quad \mu_B < 0, \quad 2\mu_A + \mu_B < 3\mu_A. \tag{17.154}$$

Taken with the Rule 2 inequalities (17.149), these give rise to the first inequality system displayed for Network 3 in Table 17.2:

$$0 = \mu_A > \mu_B > 3\mu_A > 2\mu_A + \mu_B. \tag{17.155}$$

For the confluence vector orientation given by $-g^*$, the sense of inequalities in (17.154) is reversed. These reversals give rise to the second inequality system displayed for Network 3 in Table 17.2.

17.2.11.2 A CFSTR Driven by a Classical Mechanism of Metal Catalysis

Once again we consider the open-reactor catalytic network (17.156) first discussed in Section 12.7. Because the network is not reversible, there is only one confluence vector orientation. Recall that a typical confluence vector was given by (17.128).

$$\begin{array}{ccc} A + S \rightleftarrows AS & & B + S \rightleftarrows BS \\ AS + BS & \rightarrow & C + 2S \\ A \rightleftarrows & 0 & \leftrightarrows B \\ & \uparrow & \\ & C & \end{array} \tag{17.156}$$

Here there are three *nontrivial* terminal strong-linkage classes: $\{A+S, AS\}$, $\{B+S, BS\}$, and $\{A, 0, B\}$. There is one *trivial* terminal strong-linkage class: $\{C+2S\}$. Thus, Remark 17.2.18 tells us that there are $\frac{3^3+1}{2} - 1 = 13$ upper-middle-lower partitions that should be considered.[10] These are displayed in their entirety in [74] along with the inequality systems they induce. We will probe just two of these.

Recall that earlier we considered, as examples, inequality systems induced by just two of the partitions: The partition

$$U := \{A,\ 0,\ B\},\ M := \{C,\ AS+BS\},\ L := \{\ A+S,\ AS,\ B+S,\ BS\} \tag{17.157}$$

induced the relations

$$\begin{aligned} \{\mu_A, \mu_B\} > 0 > \mu_C = \mu_{AS} + \mu_{BS} > \mu_B + \mu_S > \mu_{BS} \\ \mu_{AS} + \mu_{BS} > \mu_A + \mu_S > \mu_{AS} \end{aligned} \tag{17.158}$$

while the partition

$$U := \{A,\ 0,\ B,\ A+S,\ AS\},\ M := \{C,\ AS+BS\},\ L := \{B+S,\ BS\}. \tag{17.159}$$

gave rise to the inequality system

$$\begin{aligned} \{\mu_A, \mu_B\} > 0 > \mu_C = \mu_{AS} + \mu_{BS} > \mu_B + \mu_S > \mu_{BS} \\ \mu_{AS} > \mu_A + \mu_S > \mu_C. \end{aligned} \tag{17.160}$$

The Deficiency One Algorithm requires that we examine inequality systems such as these for nonzero solutions that are sign-compatible with the stoichiometric subspace. For network (17.156) it is not hard to determine that μ is sign-compatible with the stoichiometric subspace if and only if, in the set $\{\mu_S, \mu_{AS}, \mu_{BS}\}$, either all numbers are zero or else there is both a positive and a negative number.[11] It is evident that the inequality system (17.158) can admit no solution of this kind.

The situation for (17.160) is different. That system admits the solution

$$\mu_A = 2,\ \mu_B = 1,\ \mu_C = -1,\ \mu_S = -2.5,\ \mu_{AS} = 1,\ \mu_{BS} = -2,$$

which is sign-compatible with the stoichiometric subspace. Theorem 17.2.15 then tells us that there are rate constants for network (17.156) such that the corresponding differential equations admit a pair of distinct stoichiometrically compatible positive equilibria. Such a set of rate constants is constructed in [74]. The theorem also tells us that there are rate constants that give rise to a degenerate positive equilibrium.

If the only concern is the existence of rate constants for which these phenomena are admitted, that question is settled by partition (17.159). There is no need to examine any of the remaining ones.

---

10 There is a high degree of symmetry in network (17.156), afforded by an interchange of A and B, so the number of partitions that need be considered is, in practical terms, fewer than 13.

11 See Remark 3.A in [74] for a discussion of this example and, more generally, of some means to determine, on an ad hoc basis, sign patterns consistent with the stoichiometric subspace. Questions regarding stoichiometric sign compatibility are, in any case, resolved by means of linear programming in the Chemical Reaction Network Toolbox [62].

17.2.11.3 Some Small Networks from the Toy Zoo

In light shed by the Deficiency One Algorithm, the reader might want to examine some simple regular deficiency one networks considered in Chapter 5. Repeated below, for example, are the rabbits-and-wolves Lotka network (17.161), the Edelstein network (17.162), and the symmetry-breaking network involving left- and right-handed molecules.

$$\begin{aligned} A &\to 2A \\ A+B &\to 2B \\ B &\to 0 \end{aligned} \tag{17.161}$$

$$\begin{gathered} A \rightleftarrows 2A \\ A+B \rightleftarrows C \rightleftarrows B \end{gathered} \tag{17.162}$$

$$\begin{gathered} L+2R+P \rightleftarrows 3R+Q \qquad R+2L+P \rightleftarrows 3L+Q \\ P \rightleftarrows 0 \leftrightarrows Q \end{gathered} \tag{17.163}$$

The last of these is more difficult to study than the others, as there are 12 inequality systems that require examination. However, symmetry in the network, involving interchange of L and R, makes study of some inequality systems largely redundant. Keep in mind that all of the examples are resolved quickly by the Chemical Reaction Network Toolbox [62].

17.2.12 A Brief Remark About Higher-Deficiency Mass Action Theory

Earlier in the book, in Section 8.6, we indicated that there are procedures, similar to those in the Deficiency One Algorithm and specific to mass action systems, that *sometimes* answer questions for higher-deficiency networks, also through the study of linear inequality systems [42, 59–61, 115]. Indeed, some of these procedures are implemented in the Chemical Reaction Network Toolbox [62]. They are effective and fairly robust in the range of networks for which they provide answers.

We also indicated in Section 8.6 that a discussion of these techniques is beyond the intended scope of this book. They are complicated and difficult to explicate neatly. Here, however, we are at least in a position to say something about what goes awry in our discussion of the Deficiency One Algorithm when the deficiency one requirement is abandoned. The difficulties arise at two crucial points:

Recall that, for a network in which each linkage class contains no more than one terminal strong-linkage class, $\dim \ker YA_\kappa = \delta + \ell$, where as usual δ is the deficiency and ℓ is the number of linkage classes. Thus, for a deficiency *one* network, $\dim \ker YA_\kappa = 1 + \ell$, so $\ker YA_\kappa$ has the graceful basis $B := \{\omega_{\mathscr{C}}, b^1, b^2, \dots, b^\ell\}$, where $\{b^1, b^2, \dots, b^\ell\}$ is the simply structured and computationally attractive basis for $\ker A_\kappa$ ensured by the First Salt Theorem. When the deficiency is higher, the set

B is no longer a basis for $\ker YA_\kappa$. It must be enlarged, usually by the addition of clumsy vectors unsuited to the elegance of B.

Recall also our supposition that $YA_\kappa\omega_{\mathscr{C}} = 0$. Letting $g := A_\kappa\omega_{\mathscr{C}}$, we deduced that g is a member of $\ker Y \cap \text{span}\,(\Delta)$, the dimension of which is the deficiency of the network under study. For a deficiency *one* network, then, there is a single *line* in $\mathbb{R}^{\mathscr{C}}$ along which g must point. Moreover, that line is deducible *solely* from network structure. Thus, all candidates for g are scalar multiples of some fixed nonzero one, discernible from the network itself. In the language of Definition 17.2.9, all confluence vectors align along a single direction that is readily computed. For networks of higher deficiency, this is no longer true: confluence vectors have the annoying freedom to point in a great variety of directions.

Appendix 17.A Proof of the Second Salt Theorem

In this appendix we provide a proof of the Second Salt Theorem, which we repeat below as Theorem 17.A.1.

Theorem 17.A.1 (Second Salt Theorem [76]**).** *For a reaction network* $\{\mathscr{S},\mathscr{C},\mathscr{R}\}$*, let* α *be an element of* $\mathbb{R}^{\mathscr{R}}_{+}$*, let* $A_\alpha : \mathbb{R}^{\mathscr{C}} \to \mathbb{R}^{\mathscr{C}}$ *be the linear transformation defined by*

$$A_\alpha x := \sum_{y\to y' \in \mathscr{R}} \alpha_{y\to y'} x_y (\omega_{y'} - \omega_y), \tag{17.A.1}$$

and let

$$g := A_\alpha \omega_{\mathscr{C}}. \tag{17.A.2}$$

Moreover, let $\Lambda \subset \mathscr{C}$ *be a terminal strong-linkage class, and let* $b \in \overline{\mathbb{R}}^{\mathscr{C}}_{+}$ *an element of* $\ker A_\alpha$ *with* $\text{supp}\, b = \Lambda$*. If the complexes of* Λ *are arranged in a sequence* $y(1), y(2), \ldots, y(m)$ *so that*

$$b_{y(1)} \geq b_{y(2)} \geq b_{y(3)} \geq \cdots \geq b_{y(m)}, \tag{17.A.3}$$

then, for $\theta = 1, 2, \ldots, m$*,*

$$\sum_{i=1}^{\theta} g_{y(i)} \geq 0. \tag{17.A.4}$$

In fact, for each $1 \leq \theta < m$*, inequality holds in* (17.A.4) *if* $b_{y(\theta)}$ *is greater than* $b_{y(\theta+1)}$*. When* $\theta = m$*, inequality holds in* (17.A.4) *if and only if* Λ *is smaller than the linkage class containing it.*

Hereafter we invoke the following notation: When, for a network $\{\mathscr{S},\mathscr{C},\mathscr{R}\}$, $\mathscr{P} \subset \mathscr{C}$ is a set of complexes, we denote, as before, the complement of $\mathscr{P}$ in $\mathscr{C}$ by $\mathscr{P}'$. When $\mathscr{P}$ and $\mathscr{Q}$ are sets of complexes, we denote by $\mathscr{P} \to \mathscr{Q}$ the set of reactions having reactant complex in $\mathscr{P}$ and product complex in $\mathscr{Q}$. That is,

$$\mathscr{P} \to \mathscr{Q} := \{y \to y' \in \mathscr{R} : y \in \mathscr{P},\ y' \in \mathscr{Q}\}. \tag{17.A.5}$$

In particular, $\mathscr{C} \to \{y\}$ and $\{y\} \to \mathscr{C}$ are what we have sometimes denoted by $\mathscr{R}_{\to y}$ and $\mathscr{R}_{y\to}$.

Lemma 17.A.2. *For a reaction network* $\{\mathscr{S},\mathscr{C},\mathscr{R}\}$, *let* α *be an element of* $\mathbb{R}^{\mathscr{R}}$, *and let* $A_\alpha : \mathbb{R}^{\mathscr{C}} \to \mathbb{R}^{\mathscr{C}}$ *be as in* (17.A.1). *If* x *and* z *are elements of* $\mathbb{R}^{\mathscr{C}}$ *related by*

$$z = A_\alpha x \tag{17.A.6}$$

then, for each $\mathscr{P} \subset \mathscr{C}$,

$$\sum_{y \in \mathscr{P}} z_y = \sum_{\mathscr{P}' \to \mathscr{P}} \alpha_{y' \to y} x_{y'} - \sum_{\mathscr{P} \to \mathscr{P}'} \alpha_{y \to y'} x_y. \tag{17.A.7}$$

Proof. From (17.A.1) it follows that

$$z_y = \sum_{\mathscr{C} \to \{y\}} \alpha_{y' \to y} x_{y'} - \sum_{\{y\} \to \mathscr{C}} \alpha_{y \to y'} x_y. \tag{17.A.8}$$

Summing over $\mathscr{P} \subset \mathscr{C}$ we obtain

$$\sum_{y \in \mathscr{P}} z_y = \sum_{\mathscr{C} \to \mathscr{P}} \alpha_{y' \to y} x_{y'} - \sum_{\mathscr{P} \to \mathscr{C}} \alpha_{y \to y'} x_y. \tag{17.A.9}$$

Note, however, that

$$\mathscr{C} \to \mathscr{P} = (\mathscr{P} \to \mathscr{P}) \cup (\mathscr{P}' \to \mathscr{P})$$
$$\mathscr{P} \to \mathscr{C} = (\mathscr{P} \to \mathscr{P}) \cup (\mathscr{P} \to \mathscr{P}').$$

From this and (17.A.9), equation (17.A.7) follows. □

We turn now to proof of the Second Salt Theorem.

Proof (of Theorem 17.A.1). Let $\mathscr{P}$ be a subset of $\mathscr{C}$. Because $A_\kappa \omega_{\mathscr{C}} = g$ and $A_\kappa b = 0$, Lemma 17.A.2 gives the following equations:

$$\sum_{\mathscr{P}' \to \mathscr{P}} \kappa_{y' \to y} - \sum_{\mathscr{P} \to \mathscr{P}'} \kappa_{y \to y'} = \sum_{y \in \mathscr{P}} g_y \tag{17.A.10}$$

$$\sum_{\mathscr{P}' \to \mathscr{P}} \kappa_{y' \to y} b_{y'} - \sum_{\mathscr{P} \to \mathscr{P}'} \kappa_{y \to y'} b_y = 0. \tag{17.A.11}$$

We will begin by proving Theorem 17.A.1 for the case $\theta = m$. In this case we identify $\mathscr{P}$ in equation (17.A.10) with $\Lambda = \{y(1), y(2), \ldots, y(m)\}$. Because Λ is a terminal strong-linkage class, $\Lambda \to \Lambda'$ is empty and (17.A.10) becomes

$$\sum_{\Lambda' \to \Lambda} \kappa_{y' \to y} = \sum_{i=1}^{m} g_{y(i)}. \tag{17.A.12}$$

If Λ is smaller than the linkage class containing it, $\Lambda' \to \Lambda$ is not empty, and the positivity of κ ensures that both sides of (17.A.12) are positive. If Λ coincides with the linkage class containing it, $\Lambda' \to \Lambda$ is empty, and both sides of (17.A.12) are zero. (That the right side is zero also follows from Remark 16.2.5 and the fact that g lies in $\operatorname{im} A_\kappa \subset \operatorname{span}(\Delta)$.)

We consider next the case $\theta < m$, for which we identify $\mathscr{P}$ in equations (17.A.10) and (17.A.11) with $\{y(1), y(2), \ldots, y(\theta)\}$, with θ some fixed positive integer less than m. From (17.A.3) and (17.A.11), it follows that

$$b_{y(\theta+1)}\Big(\sum_{\mathscr{P}'\to\mathscr{P}} \kappa_{y'\to y}\Big) - b_{y(\theta)}\Big(\sum_{\mathscr{P}\to\mathscr{P}'} \kappa_{y\to y'}\Big) \geq 0. \tag{17.A.13}$$

Because

$$b_{y(\theta)} \geq b_{y(\theta+1)}, \tag{17.A.14}$$

we have

$$\sum_{\mathscr{P}'\to\mathscr{P}} \kappa_{y'\to y} - \sum_{\mathscr{P}\to\mathscr{P}'} \kappa_{y\to y'} \geq 0. \tag{17.A.15}$$

with strict inequality holding in (17.A.15) when strict inequality holds in (17.A.14). From this and (17.A.10), we obtain the desired result (17.A.4), with strict inequality holding when strict inequality holds in (17.A.14).

This completes the proof of Theorem 17.A.1. □

18

Concentration Robustness Foundations

In this chapter we will discuss the mathematical foundations of theorems presented without proof in Chapter 9, the expository chapter on concentration robustness. For proof of the Deficiency One Concentration Robustness Theorem (Theorem 9.5.1), most of the work was already done in the preceding chapter; there are just a few details that remain. Again, Theorem 9.5.1 was based on work with Guy Shinar [155]. We will need to say much more about the two theorems—Theorems 9.6.1 and 9.6.6—that assert the irrelevance of certain rate constants to the robust species concentration. Finally, we will make just a few remarks about the robustness-thwarting theorems, Theorems 9.7.1 and 9.7.10. Proofs of those were provided in [154].

18.1 The Deficiency One Concentration Robustness Theorem

Recall that a kinetic system $\{\mathscr{S},\mathscr{C},\mathscr{R},\mathscr{K}\}$ has *absolute concentration robustness relative to a species* $s \in \mathscr{S}$ if the system has a positive equilibrium c^* and if, for every other positive equilibrium c^{**}, it is true that $c^{**}_s = c^*_s$. Recall also that two complexes y' and y *differ only in species* s if there is a nonzero number α such that $y' - y = \alpha s$. We repeat below the first theorem we want to prove.

Theorem 18.1.1 (Deficiency One Concentration Robustness Theorem [155]**).** *Consider a mass action system that admits a positive equilibrium, and suppose that the deficiency of the underlying reaction network is one. If, in the network, there are two nonterminal complexes that differ only in species* s*, then the system has absolute concentration robustness relative to* s.

This theorem will emerge as a simple corollary of a more general theorem, presented below as Theorem 18.1.2 and proved at the end of this section. The more general theorem has connections to the mathematics of what we called *quasi-thermostatic* kinetic systems. Recall that a kinetic system $\{\mathscr{S},\mathscr{C},\mathscr{R},\mathscr{K}\}$–in particular a mass action system $\{\mathscr{S},\mathscr{C},\mathscr{R},k\}$–is quasi-thermostatic if it admits a positive equilibrium c^* and if the full set of positive equilibria is given by

M. Feinberg, *Foundations of Chemical Reaction Network Theory*,
Applied Mathematical Sciences 202,
https://doi.org/10.1007/978-3-030-03858-8_18

$$E := \{c \in \mathbb{R}_+^{\mathscr{S}} : \ln c - \ln c^* \in S^\perp\}, \tag{18.1}$$

where S is the stoichiometric subspace of the network $\{\mathscr{S},\mathscr{C},\mathscr{R}\}$. From this it follows easily that if c^* is a positive equilibrium of a quasi-thermostatic kinetic system, then, *whenever complexes y and y' reside in the same linkage class*, every other positive equilibrium c^{**} must satisfy the equation

$$(y' - y) \cdot \ln \frac{c^{**}}{c^*} = 0. \tag{18.2}$$

Theorems earlier in the book told us that for any weakly reversible network that has a deficiency of zero, that is star-like, or that satisfies the conditions of the Deficiency One Theorem, every assignment of rate constants results in a mass action system that is quasi-thermostatic.

Recall that not every deficiency one network satisfies the conditions of the Deficiency One Theorem. Nevertheless, the following theorem asserts that for *every* deficiency one network taken with mass action kinetics, equation (18.2) must be satisfied by every pair of positive equilibria *whenever complexes y and y' are nonterminal, even if those complexes do not reside in the same linkage class*. Proof is provided at the end of this section.

Theorem 18.1.2. *Let $\{\mathscr{S},\mathscr{C},\mathscr{R}\}$ be a deficiency one reaction network, and suppose that the mass action system $\{\mathscr{S},\mathscr{C},\mathscr{R},k\}$ admits a positive equilibrium c^*. If y and y' are nonterminal complexes, then each positive equilibrium c^{**} satisfies the equation*

$$(y' - y) \cdot \ln \frac{c^{**}}{c^*} = 0. \tag{18.3}$$

Theorem 18.1.1, the robustness theorem we want to prove, is an immediate corollary of Theorem 18.1.2, for if there are two nonterminal complexes y' and y that differ only in species $\mathscr{s}$—that is, if $y' - y = \alpha \mathscr{s}$ for some nonzero α—then, for the positive equilibria c^* and c^{**},

$$(y' - y) \cdot \ln \frac{c^{**}}{c^*} = \alpha \mathscr{s} \cdot \ln \frac{c^{**}}{c^*} = \alpha \ln \frac{c^{**}_{\mathscr{s}}}{c^*_{\mathscr{s}}} = 0. \tag{18.4}$$

From this it follows that $c^{**}_{\mathscr{s}} = c^*_{\mathscr{s}}$.

There is a slightly more general corollary of Theorem 18.1.2 that is also easy to prove:

Corollary 18.1.3. *Let $\{\mathscr{S},\mathscr{C},\mathscr{R}\}$ be a deficiency one reaction network, and suppose that the mass action system $\{\mathscr{S},\mathscr{C},\mathscr{R},k\}$ admits a positive equilibrium. Moreover, suppose that species $\mathscr{s}$ satisfies the condition*

$$\mathscr{s} \in \operatorname{span}\{y' - y \in \mathbb{R}^{\mathscr{S}} : y' \text{ and } y \text{ are nonterminal}\}. \tag{18.5}$$

Then that mass action system has absolute concentration robustness in $\mathscr{s}$.

Remark 18.1.4 (Robust ratios of monomials). Suppose that $\{\mathscr{S},\mathscr{C},\mathscr{R}\}$ is a deficiency one network in which y and y' are nonterminal complexes, and suppose that, for the mass action system $\{\mathscr{S},\mathscr{C},\mathscr{R},k\}$, c^* and c^{**} are positive equilibria. From Theorem 18.1.2 it follows that

$$\frac{(c^{**})^{y'}}{(c^{**})^{y}} = \frac{(c^{*})^{y'}}{(c^{*})^{y}}. \tag{18.6}$$

Thus, when then kinetics is mass action, when the underlying network has a deficiency of one, and when y' and y are nonterminal complexes, the ratio of monomials $c^{y'}/c^{y}$ has precisely the same value at all positive equilibria. That is, the ratio is absolutely robust.

Remark 18.1.5. In Chapter 9 we discussed an interesting example due to Germán Enciso, repeated below as (18.7). No species satisfies the condition of Theorem 18.1.1 or even the weaker condition in Corollary 18.1.3. Yet, when the kinetics is mass action, there is concentration robustness in species C.

$$\begin{aligned} &A+2B \to 3B \\ &B \to A \\ &A+B \rightleftarrows C \end{aligned} \tag{18.7}$$

This somewhat surprising result derives from a combination of two theorems acting on two different parts of the network. Note that the subnetworks

$$\begin{aligned} &A+2B \to 3B \\ &B \to A \end{aligned} \tag{18.8}$$

and

$$A+B \rightleftarrows C \tag{18.9}$$

are independent in the sense of Appendix 6.A. Thus if, for a particular assignment of rate constants to network (18.7), c^* and c^{**} are positive equilibria, then they must also be positive equilibria of the mass action subsystems corresponding to networks (18.8) and (18.9). The first of these has a deficiency of one. Note that the nonterminal complexes $A+2B$ and B have the difference $A+B$. From Theorem 18.1.2, then, c^* and c^{**} must satisfy

$$(A+B)\cdot \ln\frac{c^{**}}{c^{*}} = 0. \tag{18.10}$$

On the other hand, network (18.9) is reversible and has a deficiency of zero. From Theorem 16.7.1 it is quasi-thermodynamic and therefore quasi-thermostatic. Thus, we also have

$$(C-(A+B))\cdot \ln\frac{c^{**}}{c^{*}} = 0. \tag{18.11}$$

From the last two equations, we have

$$C \cdot \ln \frac{c^{**}}{c^{*}} = \ln \frac{c_C^{**}}{c_C^{*}} = 0, \tag{18.12}$$

whereupon $c_C^{**} = c_C^{*}$. The example illustrates the way in which different theorems governing different parts of a mass action system can have unexpected consequences for concentration robustness. This is discussed somewhat more fully in the supplementary material provided with [160].

We turn now to the proof of Theorem 18.1.2. The important ideas are largely contained in arguments from Chapter 17 supporting the Deficiency One Algorithm. What we need, however, is sufficiently brief that we can give a self-contained proof here.

Proof (of Theorem 18.1.2). We suppose that the network $\{\mathscr{S},\mathscr{C},\mathscr{R}\}$ has a deficiency of one and that it contains complexes that are not terminal, whereupon the network is not weakly reversible. Moreover, we suppose that $\{\mathscr{S},\mathscr{C},\mathscr{R},k\}$ is a mass action system that admits the positive equilibria c^* and c^{**}. Thus, we have the equilibrium equations

$$\sum_{y \to y' \in \mathscr{R}} k_{y\to y'}(c^*)^y(y'-y) = 0 \tag{18.13}$$

and

$$\sum_{y \to y' \in \mathscr{R}} k_{y\to y'}(c^{**})^y(y'-y) = 0. \tag{18.14}$$

In the spirit of Chapter 17, we can again let $\kappa \in \mathbb{R}_+^{\mathscr{R}}$ be defined by

$$\kappa_{y\to y'} := k_{y\to y'}(c^*)^y, \quad \forall y \to y' \in \mathscr{R}. \tag{18.15}$$

Moreover, we let $\mu \in \mathbb{R}^{\mathscr{S}}$ be defined by

$$\mu := \ln c^{**} - \ln c^*. \tag{18.16}$$

Then equations (18.13) and (18.14) can be written as

$$YA_\kappa \omega_{\mathscr{C}} = 0 \tag{18.17}$$

and

$$YA_\kappa \sum_{y \in \mathscr{C}} e^{y\cdot\mu}\,\omega_y = 0. \tag{18.18}$$

Here, as in Chapter 17, $Y : \mathbb{R}^{\mathscr{C}} \to \mathbb{R}^{\mathscr{S}}$ is the stoichiometric map defined by

$$Y\omega_y = y, \quad \forall y \in \mathscr{C},$$

$A_\kappa : \mathbb{R}^{\mathscr{C}} \to \mathbb{R}^{\mathscr{C}}$ is the linear map

$$A_\kappa x := \sum_{y \to y' \in \mathscr{R}} \kappa_{y \to y'} x_y (\omega_{y'} - \omega_y), \tag{18.19}$$

and $\omega_{\mathscr{C}}$ is the vector of $\mathbb{R}^{\mathscr{C}}$ such that $(\omega_{\mathscr{C}})_y = 1$ for each $y \in \mathscr{C}$.

In light of equations (18.17) and (18.18), the kernel of YA_κ assumes some importance. Recall from Proposition 16.B.9 that $\dim \ker YA_\kappa = t + 1$, where t is the number of terminal strong-linkage classes in the network $\{\mathscr{S}, \mathscr{C}, \mathscr{R}\}$. Clearly $\ker A_\kappa$ is contained in $\ker YA_\kappa$. Recall that the First Salt Theorem (Theorem 16.4.2) ensures that $\ker A_\kappa$ has a particularly nice basis $\{b^1, b^2, \dots, b^t\} \subset \overline{\mathbb{R}}_+^{\mathscr{C}}$: if $\{\Lambda^1, \Lambda^2, \dots, \Lambda^t\}$ are the terminal strong-linkage classes, then $\operatorname{supp} b^\theta = \Lambda^\theta$, $\theta = 1, 2, \dots, t$. Note that from (18.17) it follows that $\omega_{\mathscr{C}}$ is a member of $\ker YA_\kappa$. Moreover, because $\operatorname{supp} \omega_{\mathscr{C}}$ contains nonterminal complexes, $\omega_{\mathscr{C}}$ cannot lie in the span of $\{b^1, b^2, \dots, b^t\}$. Thus,

$$\{\omega_{\mathscr{C}}, b^1, b^2, \dots, b^t\}$$

is a basis for $\ker YA_\kappa$.

From this and (18.18), it follows that there are numbers $\lambda_0, \lambda_1, \lambda_2, \dots, \lambda_t$ such that

$$\sum_{y \in \mathscr{C}} e^{y \cdot \mu} \, \omega_y = \lambda_0 \, \omega_{\mathscr{C}} + \lambda_1 \, b^1 + \lambda_2 \, b^2 + \dots + \lambda_t \, b^t. \tag{18.20}$$

Since each b^θ has support entirely within the terminal complexes, it follows that for any y that is nonterminal

$$e^{y \cdot \mu} = \lambda_0.$$

Thus, if y and y' are both nonterminal, we have $y' \cdot \mu = y \cdot \mu$ or, equivalently, $(y' - y) \cdot \mu = 0$. In light of (18.16), this is equivalent to (18.3) in the statement of Theorem 18.1.2. □

18.2 Proof of the First Rate-Constant-Independence Theorem

In this section we will prove Theorem 9.6.1, the first of our rate-constant-independence theorems in Chapter 9. That theorem is repeated below as Theorem 18.2.1. Recall that the reaction(s) directly connecting complexes y and y' constitute a *cut-link* if, after removal of the arrow(s) joining y and y', there is no longer a (not necessarily directed) sequence of reaction arrows that connects y and y'. Recall also that when we refer to the reaction network obtained by removing reaction(s) of a cut-link, we mean the reaction network obtained by deleting the reaction(s) of the cut-link and any complex which, after that deletion, is not a participant in any reaction.

Theorem 18.2.1. *Consider a mass action system that satisfies all the conditions of Theorem 18.1.1, in particular the presence of nonterminal complexes y and y' that differ only in species $\mathscr{s}$. Suppose that removal of reaction(s) in a cut-link directly connecting complexes $\bar{y}$ and $\bar{y}'$ results in a network that again has y and y' as nonterminal complexes. Then, for the original mass action system, the concentration of $\mathscr{s}$ is not only identical in all positive equilibria, it is also independent of the rate constant(s) for reaction(s) directly connecting complexes $\bar{y}$ and $\bar{y}'$.*

In fact, we will prove the following more general theorem, of which Theorem 18.2.1 is an immediate corollary.

Theorem 18.2.2. *Consider a deficiency one reaction network with distinct nonterminal complexes y and y'. Suppose also that $\bar{y} \leftrightarrow \bar{y}'$ is a cut-link such that, in the subnetwork obtained by removal of $\bar{y} \leftrightarrow \bar{y}'$, y and y' remain as nonterminal complexes. Moreover, let c^* and c^{**} be positive equilibria, respectively, of two mass action systems that derive from the original network and that differ at most in the rate constant(s) for reaction(s) in $\bar{y} \leftrightarrow \bar{y}'$. Then*

$$(y' - y) \cdot \ln \frac{c^{**}}{c^*} = 0. \tag{18.21}$$

To see that Theorem 18.2.1 is a consequence of Theorem 18.2.2, see the discussion immediately following Theorem 18.1.2. See also the following remark.

Remark 18.2.3 (Extension of Remark 18.1.4). Under the conditions of Theorem 18.2.2, it follows from (18.21) that

$$\frac{(c^{**})^{y'}}{(c^{**})^{y}} = \frac{(c^{*})^{y'}}{(c^{*})^{y}} \tag{18.22}$$

when y' and y are the nonterminal complexes of the theorem statement. This is to say that the ratio of monomials $c^{y'}/c^{y}$ not only has the same value at all positive equilibria (as indicated in Remark 18.1.4), that value is also independent of rate constants associated with reactions in the cut-link $\bar{y} \leftrightarrow \bar{y}'$.

18.2.1 Motivation for the Proof

Because the proof of Theorem 18.2.2 will employ a trick, it will be worthwhile to first provide some motivation for the trick. We begin with an example.

Consider the toy mass action system (18.23). Note that (18.23) satisfies all the conditions of the main robustness theorem, Theorem 18.1.1. In particular, the underlying network has a deficiency of one, and there are two nonterminal complexes, $A + B$ and B, that differ only in species A. Thus, we have absolute concentration robustness relative to A. We want to argue that the concentration of A is not only the same in all positive steady states but also independent of k^0 (which is what Theorem 18.2.1 indicates).

$$B \overset{\alpha}{\rightarrow} C \overset{k^0}{\rightarrow} A \tag{18.23}$$

$$A + B \overset{\beta}{\rightarrow} 2B$$

The differential equations corresponding to (18.23) are shown in (18.24), along with a symbolic set of initial conditions (18.25).

$$\begin{aligned} \dot{c}_A &= k^0 c_C - \beta c_A c_B \\ \dot{c}_B &= -\alpha c_B + \beta c_A c_B \\ \dot{c}_C &= \alpha c_B - k^0 c_C \end{aligned} \tag{18.24}$$

$$c_A(0) = c_A^0 \quad c_B(0) = c_B^0 \quad c_C(0) = c_C^0 \tag{18.25}$$

When the system (18.24)-(18.25) gives rise to a positive equilibrium, the mixture composition at that steady state will depend upon the initial condition (more precisely, on the sum $c_A^0 + c_B^0 + c_C^0$). What Theorem 18.1.1 tells us is that although the steady state molar concentrations of species B and C might be initial-condition-dependent, the molar concentration of A is not. That is, *the molar concentration of A is the same in every positive steady state, regardless of the initial conditions that gave rise to it.*

Now consider the different but closely related mass action system (18.26).

$$\begin{gathered} B \xrightarrow{\alpha} C \qquad C + K \xrightarrow{1} A + K \\ A + B \xrightarrow{\beta} 2B \end{gathered} \tag{18.26}$$

Again we will write the corresponding differential equations along with a symbolic specification of initial conditions. This time, however, it will be helpful to denote the molar concentration of species K not by c_K but, instead, by k. The result is shown in (18.27) and (18.28).

$$\begin{aligned} \dot{c}_A &= k c_C - \beta c_A c_B \\ \dot{c}_B &= -\alpha c_B + \beta c_A c_B \\ \dot{c}_C &= \alpha c_B - k c_C \\ \dot{k} &= 0 \end{aligned} \tag{18.27}$$

$$c_A(0) = c_A^0 \quad c_B(0) = c_B^0 \quad c_C(0) = c_C^0 \quad k(0) = k^0 \tag{18.28}$$

Clearly, for the system (18.27)-(18.28), k remains fixed for all time at the constant value k^0. Consequently, *for* (18.27)-(18.28) *the dynamics and steady states of c_A, c_B, and c_C are identical to those of the system* (18.24)-(18.25). Note that the mass action system (18.26) also satisfies the conditions of Theorem 18.1.1: The deficiency is again one, and we again have two nonterminal complexes, $A + B$ and B, that differ only in species A. For (18.26), then, we again have absolute concentration robustness relative to species A.

Here, however, the meaning of absolute concentration robustness is different from what it was for the mass action system (18.23). As was the case for (18.23), in any positive steady state of (18.26), the molar concentration of A is independent of initial conditions. But now, in particular, *in any positive steady state, the molar concentration of A is independent of k^0.*

This leads to the following conclusion *about the original mass action system* (18.23): Because the dynamics and steady states of c_A, c_B, and c_C for (18.24)-(18.25) are identical to those for (18.27)-(18.28), *it must be the case that in any positive steady state of* (18.23), *the concentration of A is independent of* k^0.

Remark 18.2.4. Although the argument just given was framed in terms of initial conditions, it could just as well have been framed, somewhat more abstractly, in terms of stoichiometric compatibility classes for the networks in (18.23) and (18.26). In both cases, the dimension of the stoichiometric subspace is two, but in the second case, the positive stoichiometric compatibility classes are two-dimensional manifolds residing in the four-dimensional "species space" $\mathbb{R}^{\{A,B,C,K\}}$. In each of these stoichiometric compatibility classes, the value of k is invariant (although that value might change from one stoichiometric compatibility class to another). Absolute concentration robustness for (18.26) implies that, in any positive steady state, the concentration of A will be independent of stoichiometric compatibility class and, therefore, of the particular value of k associated with that stoichiometric compatibility class.

It is not difficult to see that the argument technique given above can be extended broadly to prove at least part of Theorem 18.2.1. In particular, it is extensible when the cut-link in question consists of a single *irreversible* reaction, as was the case for the cut-link $C \to A$ in (18.23). To see the difficulties that might be encountered when a cut-link consists of a reversible reaction pair, consider the slightly different mass action system (18.29).

$$B \overset{\alpha}{\to} C \underset{\bar{k}^0}{\overset{k^0}{\rightleftarrows}} A \qquad A + B \overset{\beta}{\to} 2B \tag{18.29}$$

This system satisfies the conditions of Theorem 18.1.1, and we again have absolute concentration robustness relative to species A. The cut-link $C \rightleftarrows A$ satisfies the requirements of Theorem 18.2.1, so the concentration of A is not only the same in all positive steady states but also independent of values of k^0 and $\bar{k}^0$.

To prove this along lines used earlier, we might consider mass action systems such as those shown in (18.30) or (18.31). In both cases, however, the network deficiency is *two*, not one, so the appeal to Theorem 18.1.1 used earlier becomes problematic. A similar difficulty arises for an irreversible reaction or a reversible reaction pair that is not a cut-link.

$$B \overset{\alpha}{\to} C \overset{\bar{k}^0}{\leftarrow} A \quad C + K \overset{1}{\to} A + K \qquad A + B \overset{\beta}{\to} 2B \tag{18.30}$$

$$B \overset{\alpha}{\to} C \qquad C+\bar{K} \overset{1}{\leftarrow} A+\bar{K} \qquad C+K \overset{1}{\to} A+K \tag{18.31}$$

$$A+B \overset{\beta}{\to} 2B$$

Going forward, then, we can anticipate a fairly straightforward argument in support of Theorem 18.2.1 in the case of an irreversible cut-link but a more complicated one when the cut-link consists of a reversible reaction pair. In both cases, the trick just employed—remove a cut-link and replace it with something different—will find use.

The proof of Theorem 18.2.2 will be divided into three parts. First, we will discuss the effect of the removal of a cut-link on the number of complexes and the number of linkage classes. Then, we will prove Theorem 18.2.2 when the cut-link in question consists of a single irreversible reaction; this will amount to a simple use of Theorem 18.1.2. Finally, we will consider the more difficult case of a cut-link consisting of a pair of reversible reactions.

18.2.2 A Consequence of Cut-Link Removal

Consider a reaction network with n complexes and ℓ linkage classes. We want to argue that *if the subnetwork obtained by removing a cut-link has* $\bar{n}$ *complexes and* $\bar{\ell}$ *linkage classes, then*

$$\bar{n} - \bar{\ell} = n - \ell - 1. \tag{18.32}$$

There are three distinct cases to consider, all represented in the schematic network example displayed in (18.33). (The precise nature of the complexes in the example is unimportant, as are the details of the arrow structure within various cut-links.)

$$y_1 \leftrightarrow y_2 \leftrightarrow y_3 \leftrightarrow y_4 \tag{18.33}$$

$$y_5 \leftrightarrow y_6$$

First, suppose that $y \leftrightarrow y'$ is an *internal cut-link* (such as $y_2 \leftrightarrow y_3$ in (18.33))—that is, one in which *both* complexes are participants in reactions other than those contained in the cut-link $y \leftrightarrow y'$. In the subnetwork obtained by removing $y \leftrightarrow y'$, there will be as many complexes as before, but now there will be one more linkage class.

Next, suppose that $y \leftrightarrow y'$ is an *end cut-link* (such as $y_1 \leftrightarrow y_2$ in (18.33))—that is, one in which *precisely one* of the complexes, say y', is a participant in a reaction not in the cut-link $y \leftrightarrow y'$. In the subnetwork obtained by removing $y \leftrightarrow y'$, the number of linkage classes will be the same as before, but, because of the removal of y, there will be one less complex.

Finally, suppose that $y \leftrightarrow y'$ is an *isolated cut-link* (such as $y_5 \leftrightarrow y_6$ in (18.33))—that is, one in which *neither* y *nor* y' is a participant in a reaction not contained in the cut-link $y \leftrightarrow y'$. In the subnetwork obtained by removing $y \leftrightarrow y'$, the number of complexes will be two less than before (because of the removal of both y and y'), and the number of linkage classes will be one less than before (because of the removal of the linkage class containing y and y').

18.2.3 Proof of Theorem 18.2.2 When $\bar{y} \leftrightarrow \bar{y}'$ Consists of a Single Irreversible Reaction $\bar{y} \to \bar{y}'$

Let $\{\mathscr{S},\mathscr{C},\mathscr{R}\}$ be the deficiency one network of the theorem statement, and suppose that in this network the cut-link $\bar{y} \leftrightarrow \bar{y}'$ consists of the single reaction $\bar{y} \to \bar{y}'$. We denote by $\{\mathscr{S}^\dagger,\mathscr{C}^\dagger,\mathscr{R}^\dagger\}$ the reaction network obtained by replacing the reaction $\bar{y} \to \bar{y}'$ with $\bar{y}+K \to \bar{y}'+K$, where K is a "species" that does not appear in the reactions of $\mathscr{R}$. That is,

$$\mathscr{S}^\dagger = \mathscr{S} \cup \{K\}, \quad \mathscr{R}^\dagger = \{\mathscr{R} \setminus \{\bar{y} \to \bar{y}\}\} \cup \{\bar{y}+K \to \bar{y}'+K\},$$

and $\mathscr{C}^\dagger$ is the set of complexes involved in $\mathscr{R}^\dagger$.

Remark 18.2.5. In formal terms, the complexes of the original network are elements of the vector space $\mathbb{R}^{\mathscr{S}}$. On the other hand, seemingly identical complexes, viewed as entities in the newly constructed network, are members of $\mathbb{R}^{\mathscr{S}^\dagger}$. To avoid an unnecessarily fussy narrative, we shall hereafter suppose, without consequence, that the original network has in its species set, $\mathscr{S}$, the species K—a species that, with respect to reactions in $\mathscr{R}$, is an inert. In this case, $\mathscr{S}$ and $\mathscr{S}^\dagger$ are identical. Nevertheless, we will continue to use $\mathscr{S}^\dagger$ when we have in mind the network $\{\mathscr{S}^\dagger,\mathscr{C}^\dagger,\mathscr{R}^\dagger\}$.

We can regard $\{\mathscr{S}^\dagger,\mathscr{C}^\dagger,\mathscr{R}^\dagger\}$ to derive first from removal of reaction $\bar{y} \to \bar{y}'$ from $\{\mathscr{S},\mathscr{C},\mathscr{R}\}$ to obtain the subnetwork $\{\mathscr{S},\bar{\mathscr{C}},\bar{\mathscr{R}}\}$, followed by addition of the reaction $\bar{y}+K \to \bar{y}'+K$. If, as in Section 18.2.2, we denote by n and $\bar{n}$ the number of complexes in $\{\mathscr{S},\mathscr{C},\mathscr{R}\}$ and $\{\mathscr{S},\bar{\mathscr{C}},\bar{\mathscr{R}}\}$, and by ℓ and $\bar{\ell}$ the corresponding number of linkage classes, then $\bar{n}-\bar{\ell} = n-\ell-1$. Because $\{\mathscr{S}^\dagger,\mathscr{C}^\dagger,\mathscr{R}^\dagger\}$ has two more complexes and one more linkage class than $\{\mathscr{S},\bar{\mathscr{C}},\bar{\mathscr{R}}\}$ it follows that

$$n^\dagger - \ell^\dagger = n - \ell, \tag{18.34}$$

where $n^\dagger$ and $\ell^\dagger$ are the number of complexes and the number of linkage classes in $\{\mathscr{S}^\dagger,\mathscr{C}^\dagger,\mathscr{R}^\dagger\}$.

Note that the networks $\{\mathscr{S},\mathscr{C},\mathscr{R}\}$ and $\{\mathscr{S}^\dagger,\mathscr{C}^\dagger,\mathscr{R}^\dagger\}$ have essentially[1] the same reaction vectors. Thus, the ranks, s and $s^\dagger$, of the two networks are identical. Denoting the deficiencies of the two networks by δ and $\delta^\dagger$, we have

$$\delta^\dagger := n^\dagger - \ell^\dagger - s^\dagger = n - \ell - s = \delta. \tag{18.35}$$

In this case, then, *the network* $\{\mathscr{S}^\dagger,\mathscr{C}^\dagger,\mathscr{R}^\dagger\}$, *like the network* $\{\mathscr{S},\mathscr{C},\mathscr{R}\}$, *has a deficiency of one.*

Let $\{\mathscr{S},\mathscr{C},\mathscr{R},k^*\}$ be a mass action system derived from $\{\mathscr{S},\mathscr{C},\mathscr{R}\}$, and suppose that $c^* \in \mathbb{R}^{\mathscr{S}}_+$ is one of its equilibria. Moreover, let $\{\mathscr{S}^\dagger,\mathscr{C}^\dagger,\mathscr{R}^\dagger,k^\dagger\}$ be the mass action system derived from $\{\mathscr{S}^\dagger,\mathscr{C}^\dagger,\mathscr{R}^\dagger\}$ that has rate constants given by (18.36).

$$k^\dagger_{\bar{y}+K\to\bar{y}'+K} = 1 \quad \text{and} \quad k^\dagger_{y\to y'} = k^*_{y\to y'}, \ \forall y \to y' \neq \bar{y}+K \to \bar{y}'+K \tag{18.36}$$

[1] Recall Remark 18.2.5.

It is not difficult to see that $c^\dagger \in \mathbb{R}_+^{\mathscr{S}^\dagger}$ defined by (18.37) is an equilibrium of $\{\mathscr{S}^\dagger, \mathscr{C}^\dagger, \mathscr{R}^\dagger, k^\dagger\}$.

$$c_K^\dagger = k^*_{\bar{y}\to\bar{y}'} \quad \text{and} \quad c_s^\dagger = c_s^*, \ \forall s \in \mathscr{S}^\dagger \setminus \{K\} \tag{18.37}$$

Now let $\{\mathscr{S}, \mathscr{C}, \mathscr{R}, k^{**}\}$ be a mass action system that differs from $\{\mathscr{S}, \mathscr{C}, \mathscr{R}, k^*\}$ only in the rate constant for the reaction $\bar{y} \to \bar{y}'$. That is,

$$k^{**}_{y\to y'} = k^*_{y\to y'}, \ \forall y \to y' \neq \bar{y} \to \bar{y}'. \tag{18.38}$$

If $c^{**} \in \mathbb{R}_+^{\mathscr{S}}$ is an equilibrium of $\{\mathscr{S}, \mathscr{C}, \mathscr{R}, k^{**}\}$, then $c^{\dagger\dagger} \in \mathbb{R}_+^{\mathscr{S}^\dagger}$ defined by (18.39) is, like $c^\dagger$, an equilibrium of $\{\mathscr{S}^\dagger, \mathscr{C}^\dagger, \mathscr{R}^\dagger, k^\dagger\}$.

$$c_K^{\dagger\dagger} = k^{**}_{\bar{y}\to\bar{y}'} \quad \text{and} \quad c_s^{\dagger\dagger} = c_s^{**}, \ \forall s \in \mathscr{S}^\dagger \setminus \{K\} \tag{18.39}$$

Note that the mass action system $\{\mathscr{S}^\dagger, \mathscr{C}^\dagger, \mathscr{R}^\dagger, k^\dagger\}$ admits a positive equilibrium and that $\{\mathscr{S}^\dagger, \mathscr{C}^\dagger, \mathscr{R}^\dagger\}$ is a deficiency one network. Clearly, if two complexes y and y' are nonterminal in $\{\mathscr{S}, \bar{\mathscr{C}}, \bar{\mathscr{R}}\}$ (the network obtained by removing $\bar{y} \to \bar{y}'$ from $\{\mathscr{S}, \mathscr{C}, \mathscr{R}\}$), they are also nonterminal in $\{\mathscr{S}^\dagger, \mathscr{C}^\dagger, \mathscr{R}^\dagger\}$. Moreover, neither y nor y' can contain K in its support. Applying Theorem 18.1.2 to the mass action system $\{\mathscr{S}^\dagger, \mathscr{C}^\dagger, \mathscr{R}^\dagger, k^\dagger\}$ and, in particular, to its equilibria $c^\dagger$ and $c^{\dagger\dagger}$, we obtain

$$(y' - y) \cdot \ln \frac{c^{\dagger\dagger}}{c^\dagger} = (y' - y) \cdot \ln \frac{c^{**}}{c^*} = 0. \tag{18.40}$$

The second equation in (18.40) is what we were required to show. □

18.2.4 Proof of Theorem 18.2.2 When $\bar{\mathbf{y}} \leftrightarrow \bar{\mathbf{y}}'$ Consists of Reversible Reactions $\bar{\mathbf{y}} \rightleftarrows \bar{\mathbf{y}}'$

The proof will be largely repetitive of the irreversible cut-link case, but there are important differences that arise at various points along the way. Some of these will cause us to revisit the proof of our primary deficiency one robustness theorem, Theorem 18.1.2.

Again, let $\{\mathscr{S}, \mathscr{C}, \mathscr{R}\}$ be the network in the statement of Theorem 18.2.2, but now suppose that the cut-link $\bar{y} \leftrightarrow \bar{y}'$ consists of a reversible reaction pair $\bar{y} \rightleftarrows \bar{y}'$. This time we denote by $\{\mathscr{S}^\dagger, \mathscr{C}^\dagger, \mathscr{R}^\dagger\}$ the reaction network obtained by replacing the reactions $\bar{y} \rightleftarrows \bar{y}'$ with two reactions $\bar{y} + K \to \bar{y}' + K$ and $\bar{y} + \bar{K} \leftarrow \bar{y}' + \bar{K}$ where K and $\bar{K}$ are "species" that do not appear in the reactions of $\mathscr{R}$. That is,

$$\mathscr{S}^\dagger = \mathscr{S} \cup \{K, \bar{K}\}, \ \mathscr{R}^\dagger = \{\mathscr{R} \setminus \{\bar{y} \rightleftarrows \bar{y}\}\} \cup \{\bar{y} + K \to \bar{y}' + K, \ \bar{y} + \bar{K} \leftarrow \bar{y}' + \bar{K}\},$$

and $\mathscr{C}^\dagger$ is the set of complexes that appear in $\mathscr{R}^\dagger$.

Remark 18.2.6. For reasons given in Remark 18.2.5, we suppose that $\mathscr{S}$, the species set of the original network, contains as inerts the species K and $\bar{K}$, whereupon $\mathscr{S}^\dagger = \mathscr{S}$.

We regard $\{\mathscr{S}^\dagger,\mathscr{C}^\dagger,\mathscr{R}^\dagger\}$ to derive first from removal of reactions $\bar{y} \rightleftarrows \bar{y}'$ from $\{\mathscr{S},\mathscr{C},\mathscr{R}\}$ to obtain the subnetwork $\{\mathscr{S},\bar{\mathscr{C}},\bar{\mathscr{R}}\}$ studied in Section 18.2.2, followed by addition of the reactions $\bar{y}+K \to \bar{y}'+K$ and $\bar{y}+\bar{K} \leftarrow \bar{y}'+\bar{K}$. If, as in Section 18.2.2, we denote by n and $\bar{n}$ the number of complexes in $\{\mathscr{S},\mathscr{C},\mathscr{R}\}$and $\{\mathscr{S},\bar{\mathscr{C}},\bar{\mathscr{R}}\}$, and by ℓ and $\bar{\ell}$ the corresponding number of linkage classes, we again have $\bar{n}-\bar{\ell}=n-\ell-1$. Now, however, $\{\mathscr{S}^\dagger,\mathscr{C}^\dagger,\mathscr{R}^\dagger\}$ has *four* more complexes and *two* more linkage classes than $\{\mathscr{S},\bar{\mathscr{C}},\bar{\mathscr{R}}\}$. From this it follows that

$$n^\dagger - \ell^\dagger = n - \ell + 1, \tag{18.41}$$

where $n^\dagger$ and $\ell^\dagger$ are the number of complexes and the number of linkage classes in $\{\mathscr{S}^\dagger,\mathscr{C}^\dagger,\mathscr{R}^\dagger\}$.

Note that the networks $\{\mathscr{S},\mathscr{C},\mathscr{R}\}$and $\{\mathscr{S}^\dagger,\mathscr{C}^\dagger,\mathscr{R}^\dagger\}$have essentially the same reaction vectors. Thus, the ranks, s and $s^\dagger$, of the two networks are identical. Denoting the deficiencies of the two networks by δ and $\delta^\dagger$, we have

$$\delta^\dagger := n^\dagger - \ell^\dagger - s^\dagger = n - \ell + 1 - s = \delta + 1. \tag{18.42}$$

In this case the network $\{\mathscr{S}^\dagger,\mathscr{C}^\dagger,\mathscr{R}^\dagger\}$, unlike the network $\{\mathscr{S},\mathscr{C},\mathscr{R}\}$, has a deficiency of *two*. For this reason we cannot, as in Section 18.2.3, appeal directly to Theorem 18.1.2. Nevertheless, we can more or less repeat a substantial part of Section 18.2.3, this time invoking a modification of the argument underlying Theorem 18.1.2.

Let $\{\mathscr{S},\mathscr{C},\mathscr{R},k^*\}$ be a mass action system derived from $\{\mathscr{S},\mathscr{C},\mathscr{R}\}$, and suppose that $c^* \in \mathbb{R}_+^{\mathscr{S}}$ is one of its equilibria. Furthermore, let $\{\mathscr{S}^\dagger,\mathscr{C}^\dagger,\mathscr{R}^\dagger,k^\dagger\}$ be a mass action system derived from $\{\mathscr{S}^\dagger,\mathscr{C}^\dagger,\mathscr{R}^\dagger\}$ and having rate constants defined by (18.43).

$$\begin{gathered} k^\dagger_{\bar{y}+K\to\bar{y}'+K} = 1, \quad k^\dagger_{\bar{y}+\bar{K}\leftarrow\bar{y}'+\bar{K}} = 1, \\ k^\dagger_{y\to y'} = k^*_{y\to y'}, \ \forall y \to y' \notin \{\bar{y}+K \to \bar{y}'+K, \bar{y}+\bar{K} \leftarrow \bar{y}'+\bar{K}\} \end{gathered} \tag{18.43}$$

It is not difficult to see that $c^\dagger \in \mathbb{R}_+^{\mathscr{S}^\dagger}$ defined by (18.44) is an equilibrium of $\{\mathscr{S}^\dagger,\mathscr{C}^\dagger,\mathscr{R}^\dagger,k^\dagger\}$.

$$c_K^\dagger = k^*_{\bar{y}\to\bar{y}'}, \ c_{\bar{K}}^\dagger = k^*_{\bar{y}\leftarrow\bar{y}'}, \quad \text{and} \quad c_s^\dagger = c_s^*, \ \forall s \in \mathscr{S}^\dagger \setminus \{K,\bar{K}\} \tag{18.44}$$

Now let $\{\mathscr{S},\mathscr{C},\mathscr{R},k^{**}\}$ be a mass action system that differs from $\{\mathscr{S},\mathscr{C},\mathscr{R},k^*\}$ only in the rate constants for the reactions $\bar{y} \rightleftarrows \bar{y}'$. That is,

$$k^{**}_{y\to y'} = k^*_{y\to y'}, \ \forall y \to y' \notin \{\bar{y} \rightleftarrows \bar{y}'\}. \tag{18.45}$$

If $c^{**} \in \mathbb{R}_+^{\mathscr{S}}$ is an equilibrium of $\{\mathscr{S},\mathscr{C},\mathscr{R},k^{**}\}$, then $c^{\dagger\dagger} \in \mathbb{R}_+^{\mathscr{S}^\dagger}$ defined by (18.46) is, like $c^\dagger$, an equilibrium of $\{\mathscr{S}^\dagger,\mathscr{C}^\dagger,\mathscr{R}^\dagger,k^\dagger\}$.

$$c_K^{\dagger\dagger} = k^{**}_{\bar{y}\to\bar{y}'}, \ c_{\bar{K}}^{\dagger\dagger} = k^{**}_{\bar{y}\leftarrow\bar{y}'}, \quad \text{and} \quad c_s^{\dagger\dagger} = c_s^{**}, \ \forall s \in \mathscr{S}^\dagger \setminus \{K,\bar{K}\} \tag{18.46}$$

We can now invoke the argument underlying the proof of Theorem 18.1.2 almost verbatim, up through the paragraph containing equation (18.19), *this time for the* *mass action system* $\{\mathscr{S}^\dagger,\mathscr{C}^\dagger,\mathscr{R}^\dagger,k^\dagger\}$, and with $c^\dagger$ and $c^{\dagger\dagger}$ replacing c^* and c^{**}. In particular, with

$$\mu^\dagger := \ln c^{\dagger\dagger} - \ln c^\dagger \tag{18.47}$$

we have

$$\omega_{\mathscr{C}^\dagger} := \sum_{y\in\mathscr{C}^\dagger} \omega_y \in \ker Y^\dagger A^\dagger_{\kappa^\dagger} \quad \text{and} \quad \sum_{y\in\mathscr{C}^\dagger} e^{y\cdot\mu^\dagger}\omega_y \in \ker Y^\dagger A^\dagger_{\kappa^\dagger}. \tag{18.48}$$

Here $Y^\dagger$, $\kappa^\dagger$, and $A^\dagger_{\kappa^\dagger}$ are the counterparts of the corresponding objects in Section 18.1, this time constructed for the mass action system $\{\mathscr{S}^\dagger,\mathscr{C}^\dagger,\mathscr{R}^\dagger,k^\dagger\}$ and the equilibrium $c^\dagger$.

Because the deficiency of network $\{\mathscr{S}^\dagger,\mathscr{C}^\dagger,\mathscr{R}^\dagger\}$ is two, Proposition 16.B.9 in the appendix of Chapter 16 tells us that, this time,

$$\dim\,(\ker Y^\dagger A^\dagger_{\kappa^\dagger}) \le 2 + t^\dagger, \tag{18.49}$$

where $t^\dagger$ is the number of terminal strong-linkage classes in $\{\mathscr{S}^\dagger,\mathscr{C}^\dagger,\mathscr{R}^\dagger\}$.

Although the linearly independent set $\left\{\omega_{\mathscr{C}^\dagger}, b^1,\dots,b^{t^\dagger}\right\}$, analogous to the one in the proof of Theorem 18.1.2, again resides in $\ker Y^\dagger A^\dagger_{\kappa^\dagger}$, we can no longer assert that it contains enough elements to constitute a basis for $\ker Y^\dagger A^\dagger_{\kappa^\dagger}$. However, it is not difficult to see that the vector

$$p := (\kappa^\dagger_{\bar{y}+K\to\bar{y}'+K})^{-1}\omega_{\bar{y}+K} + (\kappa^\dagger_{\bar{y}+\bar{K}\leftarrow\bar{y}'+\bar{K}})^{-1}\omega_{\bar{y}'+\bar{K}} \tag{18.50}$$

is a member of $\ker Y^\dagger A^\dagger_{\kappa^\dagger}$. Because the supports of the vectors in $\{b^1,b^2,\dots,b^t\}$ reside in distinct terminal strong-linkage classes of $\{\mathscr{S}^\dagger,\mathscr{C}^\dagger,\mathscr{R}^\dagger\}$, because $\operatorname{supp} p$ consists of complexes, $\bar{y}+K$ and $\bar{y}'+\bar{K}$, that are nonterminal in $\{\mathscr{S}^\dagger,\mathscr{C}^\dagger,\mathscr{R}^\dagger\}$, and because $\operatorname{supp}\omega_{\mathscr{C}^\dagger}$ contains the nonterminal complexes y and y' described in the theorem statement (that are different from $\bar{y}+K$ and $\bar{y}'+\bar{K}$), it follows that

$$\left\{\omega_{\mathscr{C}^\dagger}, p, b^1,\dots,b^{t^\dagger}\right\} \tag{18.51}$$

is linearly independent. In fact, from (18.49) it follows that (18.51) is a basis for $\ker Y^\dagger A^\dagger_{\kappa^\dagger}$.

Thus, there are numbers $\lambda_0 > 0, \xi > 0, \lambda_1,\dots,\lambda_{t^\dagger}$, such that

$$\sum_{y\in\mathscr{C}^\dagger} e^{y\cdot\mu^\dagger}\omega_y = \lambda_0\omega_{\mathscr{C}^\dagger} + \xi p + \sum_{\theta=1}^{t^\dagger}\lambda_\theta b^\theta. \tag{18.52}$$

Now consider two complexes y and y' that are nonterminal in the network $\{\mathscr{S},\bar{\mathscr{C}},\bar{\mathscr{R}}\}$ —that is, in the network obtained by removing reactions $\bar{y} \rightleftarrows \bar{y}'$ from the original network $\{\mathscr{S},\mathscr{C},\mathscr{R}\}$. Then y and y' are nonterminal in $\{\mathscr{S}^\dagger,\mathscr{C}^\dagger,\mathscr{R}^\dagger\}$ and differ from $\bar{y}+K$ or $\bar{y}'+\bar{K}$. From this and (18.52), it follows that $y\cdot\mu^\dagger = y'\cdot\mu^\dagger = \ln\lambda_0$. Because neither y nor y' has species K or $\bar{K}$ in its support, we have

$$(y'-y)\cdot\mu^{\dagger} = (y'-y)\cdot\ln\frac{c^{\dagger\dagger}}{c^{\dagger}} = (y'-y)\cdot\ln\frac{c^{**}}{c^{*}} = 0. \tag{18.53}$$

The last equation in (18.53) is what we were required to show. □

18.3 Proof of the Second Rate-Constant-Independence Theorem

Our aim in this section is to prove Theorem 9.6.6 of Chapter 9, which we repeat below as Theorem 18.3.1.

Theorem 18.3.1. *Consider a mass action system that satisfies all the conditions of Theorem 18.1.1, in particular the presence of nonterminal complexes y and y' that differ only in species $\mathscr{s}$. Suppose that removal of reaction(s) directly connecting complexes $\bar{y}$ and $\bar{y}'$ results in a network that (i) again has y and y' as nonterminal complexes and (ii) has more terminal strong-linkage classes than the original network. Then, for the original mass action system, the concentration of $\mathscr{s}$ is not only identical in all positive steady states, it is also independent of the rate constant(s) for reaction(s) directly connecting complexes $\bar{y}$ and $\bar{y}'$.*

In fact, we will prove the following more general theorem of which Theorem 18.3.1 is an immediate consequence. (Recall the discussion immediately following the statement of Theorem 18.1.2.)

Theorem 18.3.2. *Consider a deficiency one reaction network with distinct nonterminal complexes y and y'. Suppose also that $\bar{y} \leftrightarrow \bar{y}'$ is a direct link such that, in the subnetwork obtained by removal of $\bar{y} \leftrightarrow \bar{y}'$, (i) there are more terminal strong-linkage classes than in the original network and (ii) y and y' remain nonterminal. Moreover, let c^* and c^{**} be positive equilibria of two mass action systems that derive from the original network and that differ at most in the rate constant(s) for reaction(s) in $\bar{y} \leftrightarrow \bar{y}'$. Then*

$$(y'-y)\cdot\ln\frac{c^{**}}{c^{*}} = 0. \tag{18.54}$$

Remark 18.3.3 (Extension of Remark 18.1.4). An extension of Remark 18.1.4, parallel to the one given in Remark 18.2.3, can be made here as well.

Remark 18.3.4. If the reaction(s) in $\bar{y} \leftrightarrow \bar{y}'$ constitute a cut-link, then the conclusion of Theorem 18.2.2 is already a consequence of Theorem 18.3.2. Hence *we assume hereafter in this section that $\bar{y} \leftrightarrow \bar{y}'$ is not a cut-link.*

The proof of Theorem 18.3.2 will be divided into two subsections, one for the case in which $\bar{y} \leftrightarrow \bar{y}'$ consists of a single irreversible reaction and another for the case in which $\bar{y} \leftrightarrow \bar{y}'$ is a reversible reaction pair. In advance of that, however, it will be helpful to record some observations that apply in both cases. We denote by $\{\mathscr{S},\mathscr{C},\mathscr{R}\}$ the deficiency one reaction network of the theorem statement. If we consider the subnetwork $\{\mathscr{S},\bar{\mathscr{C}},\bar{\mathscr{R}}\}$ obtained by removing from $\{\mathscr{S},\mathscr{C},\mathscr{R}\}$ the reaction(s) in the (non-cut-link) link $\bar{y} \leftrightarrow \bar{y}'$, it is not difficult to see that the removal will leave the number of complexes, the number of linkage classes, and the rank unchanged. Because all of these are the same for both $\{\mathscr{S},\mathscr{C},\mathscr{R}\}$ and $\{\mathscr{S},\bar{\mathscr{C}},\bar{\mathscr{R}}\}$, it follows that the deficiency of $\{\mathscr{S},\bar{\mathscr{C}},\bar{\mathscr{R}}\}$ is one. If we denote by t and $\bar{t}$ the number of terminal strong-linkage classes in, respectively, $\{\mathscr{S},\mathscr{C},\mathscr{R}\}$ and $\{\mathscr{S},\bar{\mathscr{C}},\bar{\mathscr{R}}\}$, then, from the hypothesis of Theorem 18.3.2, we have

$$t+1 \leq \bar{t}. \tag{18.55}$$

18.3.1 Proof of Theorem 18.3.2 When $\bar{\mathbf{y}} \leftrightarrow \bar{\mathbf{y}}'$ Consists of a Single Reaction $\bar{\mathbf{y}} \to \bar{\mathbf{y}}'$

We suppose in this subsection that $\bar{y} \leftrightarrow \bar{y}'$ consists of the single reaction $\bar{y} \to \bar{y}'$. Motivated by the argument given in Section 18.2.3, we denote by $\{\mathscr{S}^\dagger,\mathscr{C}^\dagger,\mathscr{R}^\dagger\}$ the reaction network obtained by replacing the reaction $\bar{y} \to \bar{y}'$ in $\{\mathscr{S},\mathscr{C},\mathscr{R}\}$ with $\bar{y}+K \to \bar{y}'+K$, where K is a "species" that does not appear in the reactions of $\mathscr{R}$.[2] This amounts to adding the reaction $\bar{y}+K \to \bar{y}'+K$ to the reaction network $\{\mathscr{S},\bar{\mathscr{C}},\bar{\mathscr{R}}\}$. Note that $\{\mathscr{S}^\dagger,\mathscr{C}^\dagger,\mathscr{R}^\dagger\}$ has two more complexes and one more linkage class than $\{\mathscr{S},\mathscr{C},\mathscr{R}\}$; moreover, both networks have the same rank. Therefore, the deficiency of $\{\mathscr{S}^\dagger,\mathscr{C}^\dagger,\mathscr{R}^\dagger\}$ is *two*.

Much, but not all, of the argument remaining in this subsection will resemble very closely what was done in Sections 18.2.3 and 18.2.4. We repeat some of that text, adapted to the situation at hand.

Let $\{\mathscr{S},\mathscr{C},\mathscr{R},k^*\}$ be a mass action system derived from $\{\mathscr{S},\mathscr{C},\mathscr{R}\}$, and suppose that $c^* \in \mathbb{R}_+^{\mathscr{S}}$ is one of its equilibria. Moreover, let $\{\mathscr{S}^\dagger,\mathscr{C}^\dagger,\mathscr{R}^\dagger,k^\dagger\}$ be the mass action system derived from $\{\mathscr{S}^\dagger,\mathscr{C}^\dagger,\mathscr{R}^\dagger\}$ that has rate constants given by (18.56).

$$k^\dagger_{\bar{y}+K\to\bar{y}'+K} = 1 \quad \text{and} \quad k^\dagger_{y\to y'} = k^*_{y\to y'},\ \forall y \to y' \neq \bar{y}+K \to \bar{y}'+K \tag{18.56}$$

It is not difficult to see that $c^\dagger \in \mathbb{R}_+^{\mathscr{S}^\dagger}$ defined by (18.57) is an equilibrium of $\{\mathscr{S}^\dagger,\mathscr{C}^\dagger,\mathscr{R}^\dagger,k^\dagger\}$.

$$c^\dagger_K = k^*_{\bar{y}\to\bar{y}'} \quad \text{and} \quad c^\dagger_s = c^*_s,\ \forall s \in \mathscr{S}^\dagger \setminus \{K\} \tag{18.57}$$

Now let $\{\mathscr{S},\mathscr{C},\mathscr{R},k^{**}\}$ be a mass action system that differs from $\{\mathscr{S},\mathscr{C},\mathscr{R},k^*\}$ only in the rate constant for the reaction $\bar{y} \to \bar{y}'$. That is,

$$k^{**}_{y\to y'} = k^*_{y\to y'},\ \forall y \to y' \neq \bar{y} \to \bar{y}'. \tag{18.58}$$

If $c^{**} \in \mathbb{R}_+^{\mathscr{S}}$ in an equilibrium of $\{\mathscr{S},\mathscr{C},\mathscr{R},k^{**}\}$, then $c^{\dagger\dagger} \in \mathbb{R}_+^{\mathscr{S}^\dagger}$ defined by (18.59) is, like $c^\dagger$, an equilibrium of $\{\mathscr{S}^\dagger,\mathscr{C}^\dagger,\mathscr{R}^\dagger,k^\dagger\}$.

$$c^{\dagger\dagger}_K = k^{**}_{\bar{y}\to\bar{y}'} \quad \text{and} \quad c^{\dagger\dagger}_s = c^{**}_s,\ \forall s \in \mathscr{S}^\dagger \setminus \{K\} \tag{18.59}$$

---

[2] Again, for the sake of a less fussy treatment, we suppose that K is an inert in the original species set $\mathscr{S}$.

As in Section 18.2.4, we can invoke the argument underlying the proof of Theorem 18.1.2 almost verbatim, up through the paragraph containing equation (18.19), *now adapted to the mass action system* $\{\mathscr{S}^\dagger,\mathscr{C}^\dagger,\mathscr{R}^\dagger,k^\dagger\}$ *of this subsection*, and with $c^\dagger$ and $c^{\dagger\dagger}$ replacing c^* and c^{**} of Section 18.1. In particular, with

$$\mu^\dagger := \ln c^{\dagger\dagger} - \ln c^\dagger \tag{18.60}$$

we have

$$\omega_{\mathscr{C}^\dagger} := \sum_{y\in\mathscr{C}^\dagger} \omega_y \in \ker Y^\dagger A^\dagger_{\kappa^\dagger} \quad \text{and} \quad \sum_{y\in\mathscr{C}^\dagger} e^{y\cdot\mu^\dagger}\omega_y \in \ker Y^\dagger A^\dagger_{\kappa^\dagger}. \tag{18.61}$$

As in Section 18.2.4, $Y^\dagger$, $\kappa^\dagger$, and $A^\dagger_{\kappa^\dagger}$ are the counterparts of the corresponding objects in Section 18.1, constructed for the mass action system $\{\mathscr{S}^\dagger,\mathscr{C}^\dagger,\mathscr{R}^\dagger,k^\dagger\}$ and the equilibrium $c^\dagger$ of this subsection.

Because the deficiency of network $\{\mathscr{S}^\dagger,\mathscr{C}^\dagger,\mathscr{R}^\dagger\}$ is two, Proposition 16.B.9 tells us that, as in Section 18.2.4,

$$\dim\,(\ker Y^\dagger A^\dagger_{\kappa^\dagger}) \le 2 + t^\dagger, \tag{18.62}$$

where $t^\dagger$ is the number of terminal strong-linkage classes in $\{\mathscr{S}^\dagger,\mathscr{C}^\dagger,\mathscr{R}^\dagger\}$. This time, however, we will need a different argument going forward. *In particular, we want to show that, as a result of the hypothesis of Theorem 18.3.2, we actually have*

$$\dim\,(\ker Y^\dagger A^\dagger_{\kappa^\dagger}) \le 1 + t^\dagger. \tag{18.63}$$

In fact, we will see that equality holds.

To begin, we note that by a standard theorem of linear algebra

$$\dim\,(\ker Y^\dagger A^\dagger_{\kappa^\dagger}) = \dim \mathbb{R}^{\mathscr{C}^\dagger} - \dim\,(\operatorname{im} Y^\dagger A^\dagger_{\kappa^\dagger}), \tag{18.64}$$

where im $Y^\dagger A^\dagger_{\kappa^\dagger}$ is the image of the map $Y^\dagger A^\dagger_{\kappa^\dagger} : \mathbb{R}^{\mathscr{C}^\dagger} \to \mathbb{R}^{\mathscr{S}^\dagger}$. Moreover, we can also write

$$\dim\,(\ker YA_{\kappa^*}) = \dim \mathbb{R}^{\mathscr{C}} - \dim\,(\operatorname{im} YA_{\kappa^*}), \tag{18.65}$$

where Y, κ^*, and A_{κ^*} are the counterparts of the corresponding objects in Section 18.1, constructed for the mass action system $\{\mathscr{S},\mathscr{C},\mathscr{R},k^*\}$ and the equilibrium c^*.

From Proposition 16.B.9 we know that $\dim\,(\ker YA_{\kappa^*}) = 1 + t$, where t is the number of terminal strong-linkage classes in the deficiency one network $\{\mathscr{S},\mathscr{C},\mathscr{R}\}$. It follows from (18.65) that

$$\dim\,(\operatorname{im} YA_{\kappa^*}) = n - t - 1, \tag{18.66}$$

where n is the number of complexes in $\mathscr{C}$. It is not difficult to see that $\operatorname{im} YA_{\kappa^*}$ is contained in $\operatorname{im} Y^\dagger A^\dagger_{\kappa^\dagger}$, so

$$\dim\,(\operatorname{im} YA_{\kappa^*}) = n - t - 1 \le \dim\,(\operatorname{im} Y^\dagger A^\dagger_{\kappa^\dagger}). \tag{18.67}$$

Because the number of complexes in $\mathscr{C}^\dagger$ is $n+2$, it follows from (18.64) and (18.67) that

$$\dim\,(\ker Y^\dagger A^\dagger_{\kappa^\dagger}) \leq t+3. \tag{18.68}$$

In fact, from (18.55) we can write

$$\dim\,(\ker Y^\dagger A^\dagger_{\kappa^\dagger}) \leq \bar{t}+2, \tag{18.69}$$

where, again, $\bar{t}$ is the number of terminal strong-linkage classes in $\{\mathscr{S},\bar{\mathscr{C}},\bar{\mathscr{R}}\}$, the subnetwork obtained by removing $\bar{y} \to \bar{y}'$ from $\{\mathscr{S},\mathscr{C},\mathscr{R}\}$. Now $\{\mathscr{S}^\dagger,\mathscr{C}^\dagger,\mathscr{R}^\dagger\}$ contains precisely one more terminal strong-linkage class than $\{\mathscr{S},\bar{\mathscr{C}},\bar{\mathscr{R}}\}$—i.e., the singleton $\{\bar{y}'+K\}$. Thus, $t^\dagger = \bar{t}+1$, so

$$\dim\,(\ker Y^\dagger A^\dagger_{\kappa^\dagger}) \leq 1+t^\dagger, \tag{18.70}$$

which is what we wanted to establish on our way to the final result.

With $\{b^1,b^2,\ldots,b^{t^\dagger}\}$ denoting the basis for $\ker A^\dagger_{\kappa^\dagger}$ ensured by the First Salt Theorem (Theorem 16.4.2), it follows easily from (18.61) and (18.70) that

$$\{\omega_{\mathscr{C}^\dagger}, b^1, b^2, \ldots, b^{t^\dagger}\}$$

is a basis for $\ker YA^\dagger_{\kappa^\dagger}$. Thus, there are numbers $\lambda_0 > 0, \lambda_1, \ldots, \lambda_{t^\dagger}$, such that

$$\sum_{y \in \mathscr{C}^\dagger} e^{y \cdot \mu^\dagger} \omega_y = \lambda_0 \omega_{\mathscr{C}^\dagger} + \sum_{\theta=1}^{t^\dagger} \lambda_\theta b^\theta. \tag{18.71}$$

Now consider two complexes y and y' that are nonterminal in the network $\{\mathscr{S},\bar{\mathscr{C}},\bar{\mathscr{R}}\}$–that is, in the network obtained by removing reaction $\bar{y} \to \bar{y}'$ from the original network $\{\mathscr{S},\mathscr{C},\mathscr{R}\}$. Then y and y' are nonterminal in $\{\mathscr{S}^\dagger,\mathscr{C}^\dagger,\mathscr{R}^\dagger\}$ and differ from $\bar{y}+K$ or $\bar{y}'+\bar{K}$. From this and (18.71), it follows that $y \cdot \mu^\dagger = y' \cdot \mu^\dagger = \ln \lambda_0$. Because neither y nor y' has species K or $\bar{K}$ in its support, we have

$$(y'-y)\cdot\mu^\dagger = (y'-y)\cdot \ln\frac{c^{\dagger\dagger}}{c^\dagger} = (y'-y)\cdot \ln\frac{c^{**}}{c^*} = 0. \tag{18.72}$$

The last equation in (18.72) is what we were required to show. □

18.3.2 Proof of Theorem 18.3.2 When $\bar{\mathbf{y}} \leftrightarrow \bar{\mathbf{y}}'$ Consists of a Reversible Reaction Pair $\bar{\mathbf{y}} \rightleftarrows \bar{\mathbf{y}}'$

The proof here begins very much like that in Section 18.2.4, up through (18.48). We won't repeat the text here, except to point out differences. In this case, the network $\{\mathscr{S},\bar{\mathscr{C}},\bar{\mathscr{R}}\}$ obtained by removing the reactions $\bar{y} \rightleftarrows \bar{y}'$ from $\{\mathscr{S},\mathscr{C},\mathscr{R}\}$ has a deficiency of one, as was indicated earlier in this section. The addition to $\{\mathscr{S},\bar{\mathscr{C}},\bar{\mathscr{R}}\}$ of the reactions $\bar{y}+K \to \bar{y}'+K$ and $\bar{y}+\bar{K} \leftarrow \bar{y}'+\bar{K}$ leaves the rank unchanged, but there are now four more complexes and two more linkage classes. Thus, the deficiency of $\{\mathscr{S}^\dagger,\mathscr{C}^\dagger,\mathscr{R}^\dagger\}$ is *three*.

In this case, Proposition 16.B.9 tells us that

$$\dim\,(\ker Y^{\dagger} A^{\dagger}_{\kappa^{\dagger}}) \le 3 + t^{\dagger}. \tag{18.73}$$

However, we can make an argument very much like that in the preceding section to get a better bound. Keeping in mind that the number of complexes in $\mathscr{C}^{\dagger}$ is now *four* more than the number in $\mathscr{C}$ and that the number of terminal strong-linkage classes in $\{\mathscr{S}^{\dagger}, \mathscr{C}^{\dagger}, \mathscr{R}^{\dagger}\}$ is now *two* more than in $\{\mathscr{S}, \bar{\mathscr{C}}, \bar{\mathscr{R}}\}$, we can modify the argument in Section 18.3.1 to get

$$\dim\,(\ker Y^{\dagger} A^{\dagger}_{\kappa^{\dagger}}) \le 2 + t^{\dagger}. \tag{18.74}$$

Thereafter the argument is identical to the one at the end of Section 18.2.4, subsequent to equation (18.49). □

18.4 About Reaction Network Architectures that Thwart Concentration Robustness

In Section 9.7 of Chapter 9, we discussed theorems about reaction network structures that thwart concentration robustness. Here we will say just a little more about the underlying ideas. In fact, we can provide a very simple proof of the most elementary of our robustness-thwarting theorems, Theorem 9.7.1, which is repeated below as Theorem 18.4.1.

Theorem 18.4.1. *Consider a mass action system that admits a positive equilibrium. Moreover, suppose that the underlying reaction network is conservative and satisfies at least one of the following conditions:*

(i) The network has a deficiency of zero.
(ii) The network satisfies the conditions of the Deficiency One Theorem.
(iii) The network is reversible and star-like.

Then there is no species with respect to which the mass action system exhibits absolute concentration robustness.

We are now in a better position to understand just what the three conditions listed in the theorem statement do for us: As a result of theorems earlier in the book, *any one of them ensures that the mass action system under consideration is quasi-thermostatic.*[3] In fact, this is the role that those same conditions play in Theorem 9.7.10. Indeed, *the conclusions of both theorems remain true for any kinetic system that is quasi-thermostatic.*

With this in mind, we can state below (and prove easily) a more general version of Theorem 18.4.1.

Theorem 18.4.2. *Consider a kinetic system that is quasi-thermostatic. If the underlying reaction network is conservative, then there is no species with respect to which the system exhibits absolute concentration robustness.*

[3] Conditions (i) and (iii) also ensure that the mass action system is quasi-thermodynamic.

Proof. Let $\{\mathscr{S},\mathscr{C},\mathscr{R},\mathscr{K}\}$ be the kinetic system of the theorem statement. Because it is quasi-thermostatic, there is a $c^* \in \mathbb{R}_+^{\mathscr{S}}$ such that the positive equilibrium set is given by

$$E := \{c \in \mathbb{R}_+^{\mathscr{S}} : \ln c - \ln c^* \in S^{\perp}\}, \tag{18.75}$$

where S is the stoichiometric subspace of the network $\{\mathscr{S},\mathscr{C},\mathscr{R}\}$. Because the network is conservative, there is a positive vector $M \in S^{\perp}$. Now, for $\varepsilon > 0$ let $c(\varepsilon)$ satisfy

$$\ln c(\varepsilon) - \ln c^* = \varepsilon M. \tag{18.76}$$

That is, for each $s \in \mathscr{S}$,

$$c_s(\varepsilon) = c_s^* e^{\varepsilon M_s}. \tag{18.77}$$

Note that $c(\varepsilon)$ is a member of E and is therefore a positive equilibrium. Because each M_s is positive, we have $c_s(\varepsilon) \neq c_s^*$ for all $s \in \mathscr{S}$. Thus, there cannot be absolute concentration robustness relative to any species. □

As we indicated earlier, in the hypothesis of the more sophisticated and far more difficult robustness-thwarting Theorem 9.7.10, we can replace the mass action system considered by a quasi-thermostatic kinetic system in which the underlying reaction network is constructive and conservative. The proof is contained in [154].

19

Species-Reaction Graph Foundations

The very powerful Theorem 11.6.1 told us that, when a nondegenerate network's Species-Reaction Graph satisfies certain mild conditions, behavior is severely constrained to be largely stable and dull, provided only that the kinetics associated with the network resides in the large and natural weakly monotonic class. The underlying idea was that, when those graphical conditions are satisfied, the network's fully open extension is concordant. Nondegeneracy of the original network then ensured that it too is concordant, in which case the original network inherits all the dynamical attributes that concordance mandates.

The connection between a network's Species-Reaction Graph and concordance of the network's fully open extension was the subject of Theorem 11.5.1. By its very nature, that theorem marries algebraic aspects of the network with graph-theoretical ones, and it does so in a highly delicate and nuanced way. Proof of the theorem is not simple, nor is it brief. In this chapter we will not do the proof in full.

Rather, our aim is to indicate how the proof of Theorem 11.5.1 proceeds, with detail provided often but not always. From this, we hope the reader can gain some understanding of how the algebraic idea of network concordance makes contact with the graph-theoretical character of a network, as expressed in its Species-Reaction Graph. We hope too that the reader can gain insights into natural questions such as these: *Why in the Species-Reaction Graph do cycles containing an even (but not odd) number of c-pairs play such an important role? Why is there a distinction between even cycles that are stoichiometrically expansive and those that aren't?*

For readers seeking still more detail about the proof of Theorem 11.5.1, here is a little guide[1] to references in which that detail can be found: Proof of a weaker version of the theorem was given in [158]. In subsequent work with Daniel Knight [117, 118], a very different and better proof, relying on parts of the earlier one, resulted in the substantially sharper theorem stated here as Theorem 11.5.1. It is the line of reasoning in that second proof that we will try to convey here. Occasionally, we will draw on some more or less standard graph-theoretical terminology and ideas. For these, [21], [24], and [176] are useful references.

[1] The same guide applies to the proof of the *strong*-concordance variant, Theorem 11.8.1.

M. Feinberg, *Foundations of Chemical Reaction Network Theory*,
Applied Mathematical Sciences 202,
https://doi.org/10.1007/978-3-030-03858-8_19

19.1 Getting Started

We repeat below as Theorem 19.1.1 the assertion we want to prove.

Theorem 19.1.1 ([117, 118, 158]). *Suppose that a reaction network's Species-Reaction Graph has the following properties:*

(i) No ***even*** *cycle admits a stoichiometrically expansive orientation.*
(ii) No two consistently oriented ***even*** *cycles have as their intersection a* ***single*** *directed path* ***originating at a species vertex and terminating at a reaction vertex****.*

Then the network's fully open extension is concordant.

It will be helpful to have nearby the definition of a *discordance*. Recall that for a reaction network $\{\mathscr{S},\mathscr{C},\mathscr{R}\}$ with stoichiometric subspace S, the linear map $L:\mathbb{R}^{\mathscr{R}} \to S$ is defined by

$$L\alpha = \sum_{y\to y'\in\mathscr{R}} \alpha_{y\to y'}(y'-y). \tag{19.1}$$

Definition 19.1.2. *The reaction network* $\{\mathscr{S},\mathscr{C},\mathscr{R}\}$ *is* **discordant** *if there exist an* $\alpha \in \ker L$ *and a nonzero* $\sigma \in S$ *having the following properties:*

(i) For each $y \to y'$ *such that* $\alpha_{y\to y'} \neq 0$, $\operatorname{supp} y$ *contains a species* $s \in \mathscr{S}$ *for which* $\operatorname{sgn}\sigma_s = \operatorname{sgn}\alpha_{y\to y'}$.

(ii) For each $y \to y'$ *such that* $\alpha_{y\to y'} = 0$, *either* $\sigma_s = 0$ *for all* $s \in \operatorname{supp} y$ *or else* $\operatorname{supp} y$ *contains species* s *and* s' *for which* $\operatorname{sgn}\sigma_s = -\operatorname{sgn}\sigma_{s'}$, *both not zero.*

The network is **concordant** *if it is not discordant. A* **discordance** *for the network is a pair* $\{\alpha,\sigma\}$, *with* $\alpha \in \ker L$ *and* σ *a nonzero member of S, that satisfies conditions (i) and (ii).*

19.2 A Useful Supposition

Theorem 19.1.1 (and a more general theorem that we will soon state) are of the kind that assert that a network's fully open extension is concordant provided that its Species-Reaction Graph does not contain certain disagreeable objects. Our proof strategy will be to presume that there is a discordance for the network's fully open extension and then argue that, in the network's Species-Reaction Graph, at least one of those disagreeable objects must be present.

In the given reaction network, some or perhaps all of its non-degradation reactions[2] might be reversible. For the proof it will be far less cumbersome to work with networks in which every non-degradation reaction is irreversible. The following lemma, proved in [158], makes this possible.

[2] Recall that a degradation reaction is one of the form $s \to 0$.

Lemma 19.2.1. *If a fully open network is discordant, it is possible to choose from each reversible pair of non-degradation reactions at least one (and sometimes both) of the reactions for removal such that the resulting fully open subnetwork is again discordant.*

The lemma tells us that every fully open discordant network contains a discordant fully open subnetwork in which no non-degradation reaction is reversible. Rather than work with the original presumed-discordant fully open network, we will instead work with the smaller discordant fully open subnetwork whose existence is guaranteed by the Lemma 19.2.1.

Apart from small changes in the labels within the reaction nodes (i.e., replacement of $y \rightleftarrows y'$ by $y \to y'$), the Species-Reaction Graph drawn for the discordant subnetwork is a subgraph of the one drawn for the original presumed-discordant parent network. It is in the Species-Reaction Graph of the discordant reduced network that, we will argue, one or more of those certain disagreeable objects must reside. From there it is not difficult to argue that the same disagreeable objects must also be present in the parent Species-Reaction Graph for the original network.

Hereafter, then, we will suppose that $\{\mathscr{S}, \mathscr{C}, \mathscr{R}\}$ *is a discordant fully open network in which each non-degradation reaction is irreversible. Furthermore, we will suppose that* $\{\alpha, \sigma\}$ *is a particular discordance for the network.*

This supposition will remain in force for the remainder of this chapter, unless there is a statement to the contrary. What should also be kept in mind is the very mild assumption (Remark 11.3.1) tacit in all assertions about the Species-Reaction Graph: Networks admitted for consideration are required to satisfy the condition that no species appears on both sides of any reaction. That is, for each $y \to y' \in \mathscr{R}$ and each $s \in \mathscr{S}$, if $y_s > 0$, then $y'_s = 0$.

19.3 Signed Species and Signed Reactions

Drawing on the putative discordance $\{\alpha, \sigma\}$, we associate signs with the species and with the reactions in the following way: *Species* $s \in \mathscr{S}$ *is positive, negative, or zero according to whether* σ_s *is positive, negative, or zero.* A *signed species* is a species that is either positive or negative. *Reaction* $y \to y' \in \mathscr{R}$ *is positive, negative, or zero according to whether* $\alpha_{y \to y'}$ *is positive, negative, or zero.* A *signed reaction* is one that is either positive or negative. Because of the definition of a discordance, there must be at least one signed species.

Remark 19.3.1. For any degradation reaction $s \to 0$, Definition 19.1.2 requires that $\operatorname{sgn} \alpha_{s \to 0} = \operatorname{sgn} \sigma_s$, so $\operatorname{sgn}(s \to 0) = \operatorname{sgn} s$ for every $s \in \mathscr{S}$. (A similar situation obtains for any reaction of the form $ns \to y$, where n is a positive number.) For any synthesis reaction $0 \to s$, Definition 19.1.2 requires that $\alpha_{0 \to s} = 0$, so all such reactions are unsigned.

19.4 The Sign-Causality Graph Associated with a Discordance

In this section we will indicate how the discordance $\{\alpha, \sigma\}$ for network $\{\mathscr{S}, \mathscr{C}, \mathscr{R}\}$ gives rise to a *Sign-Causality Graph*, which can be associated with a subgraph of the network's Species-Reaction Graph.

19.4.1 Causal Units

Because the network $\{\mathscr{S}, \mathscr{C}, \mathscr{R}\}$ under consideration is fully open, it has a complete supply of degradation reactions: $\{s \to 0 : s \in \mathscr{S}\}$. We denote by $\mathscr{R}^*$ all the remaining (non-degradation) reactions that are signed. By supposition α of our putative discordance is a member of $\ker L$, so it satisfies the equation

$$\sum_{y \to y' \in \mathscr{R}^*} \alpha_{y \to y'}(y' - y) - \sum_{s \in \mathscr{S}} \alpha_{s \to 0}\, s = 0. \tag{19.2}$$

From this it follows that, for a particular choice of $s \in \mathscr{S}$, we can write

$$\sum_{y \to y' \in \mathscr{R}^*} \alpha_{y \to y'}(y'_s - y_s) = \alpha_{s \to 0}. \tag{19.3}$$

We will suppose for the moment that the indicated species s in (19.3) is *positive*. From Remark 19.3.1, then, we have

$$\sum_{y \to y' \in \mathscr{R}^*} \alpha_{y \to y'}(y'_s - y_s) > 0. \tag{19.4}$$

Clearly, at least one term on the left must be positive. In fact, adapting language from [158], we say that at least one term on the left is *causal* for the inequality.[3] Such a positive term can arise in two ways:

(i) There is a *positive* reaction $y \to y'$ (i.e., $\alpha_{y \to y'}$ is positive) with s in $\operatorname{supp} y'$ (whereupon $y'_s > 0$ and $y_s = 0$). For $\alpha_{y \to y'}$ to be positive, however, the conditions for a discordance in Definition 19.1.2 require that there be a *positive* species s' in $\operatorname{supp} y$. In this case, we regard reaction $y \to y'$ to be "causal" for the sign of $\alpha_{s \to 0}$ or, equivalently, for the sign of species s. Moreover, we regard species s' to be "causal" for the sign of reaction $y \to y'$. With this in mind, we write

$$\overset{+}{s'} \quad \overset{y}{\rightsquigarrow} \quad \boxed{\overset{+}{y \to y'}} \quad \overset{y'}{\rightsquigarrow} \quad \overset{+}{s}. \tag{19.5}$$

The signs above the species and the reactions indicate their respective signs. The complex labels above the "causal" arrows ($\rightsquigarrow$) indicate the complex in whose support the adjacent species resides.

---

[3] When we say that A is causal for B, we do not mean to suggest that A by itself determines the outcome B, only that A abets or contributes to B.

(ii) There is a *negative* reaction $y \to y'$ (i.e., $\alpha_{y\to y'}$ is negative) with $s \in \text{supp}\, y$ (whereupon $y_s > 0$ and $y'_s = 0$). For $\alpha_{y\to y'}$ to be negative, the conditions of Definition 19.1.2 require that there be a *negative* species s' in $\text{supp}\, y$. As in case (i), we regard reaction $y \to y'$ to be causal for the sign of species s and species s' to be causal for the sign of reaction $y \to y'$. This is reflected in the diagram

$$\overset{-}{s'} \; \overset{y}{\rightsquigarrow} \; \overset{-}{\boxed{y \to y'}} \; \overset{y}{\rightsquigarrow} \; \overset{+}{s}. \tag{19.6}$$

Now we will suppose instead that the indicated species s in (19.3) is *negative.* This time, from (19.3) and Remark 19.3.1, we have

$$\sum_{y\to y' \in \mathscr{R}^*} \alpha_{y\to y'}(y'_s - y_s) < 0, \tag{19.7}$$

whereupon at least one term on the left must be negative. There are two ways in which this can happen:

(i) There is a *negative* reaction $y \to y'$ (i.e., $\alpha_{y\to y'}$ is negative) such that $s \in \text{supp}\, y'$ (whereupon $y'_s > 0$ and $y_s = 0$). For $\alpha_{y\to y'}$ to be negative, the conditions in Definition 19.1.2 require that there be a *negative* species s' in $\text{supp}\, y$. Hence, we write the causality diagram

$$\overset{-}{s'} \; \overset{y}{\rightsquigarrow} \; \overset{-}{\boxed{y \to y'}} \; \overset{y'}{\rightsquigarrow} \; \overset{-}{s}. \tag{19.8}$$

(ii) There is a *positive* reaction $y \to y'$ (i.e., $\alpha_{y\to y'}$ is positive) with $s \in \text{supp}\, y$ (whereupon $y_s > 0$ and $y'_s = 0$). For $\alpha_{y\to y'}$ to be positive, the conditions of Definition 19.1.2 require that there be a *positive* species s' in $\text{supp}\, y$. With this in mind, we write

$$\overset{+}{s'} \; \overset{y}{\rightsquigarrow} \; \overset{+}{\boxed{y \to y'}} \; \overset{y}{\rightsquigarrow} \; \overset{-}{s}. \tag{19.9}$$

The objects shown in (19.5)–(19.6) and (19.8)–(19.9), called *causal units*, will play roles as edge pairs in the Sign-Causality Graph.

19.4.2 How the Sign-Causality Graph Is drawn

The *Sign-Causality Graph* associated with discordance $\{\alpha, \sigma\}$ is a bipartite graph drawn in the following way: The vertices are the signed species and the signed (non-degradation[4]) reactions. An edge $\rightsquigarrow$ is drawn from a signed species s' to a signed reaction $y \to y'$ whenever s' is contained in $\text{supp}\, y$ and the two signs agree; the edge is then labeled with the complex y. An edge $\rightsquigarrow$ is drawn from a signed reaction $y \to y'$

[4] In the event that there are signed degradation-like reactions such as $ms \to 0$ with $m \neq 1$, then those too are omitted from the Sign-Causality Graph

to a signed species $\mathscr{s}$ in either of the following circumstances: (i) $\mathscr{s}$ is contained in $\operatorname{supp} y'$ and the sign of $\mathscr{s}$ agrees with the sign of the reaction, in which case the edge carries the label y', or (ii) $\mathscr{s}$ is contained in $\operatorname{supp} y$ and the sign of $\mathscr{s}$ disagrees with the sign of the reaction, in which case the edge carries the label y. The signed species and the signed reactions are all labeled by their corresponding signs.

A *c-pair* in the Sign-Causality Graph, like a c-pair in the Species-Reaction Graph, is a pair of edges adjacent to the same reaction vertex that carry the same complex label. Here again, the term c-pair is meant to be a mnemonic for *complex pair.*

As we indicated earlier, a *causal unit* is a directed two-edge subgraph of the Sign-Causality Graph of the kind $\mathscr{s}' \rightsquigarrow R \rightsquigarrow \mathscr{s}$, where R denotes a reaction. (The symbol R will sometimes be used to indicate a reaction when there is no need to specify the reactant and product complexes.) The species $\mathscr{s}'$ is the *initiator* of the causal unit $\mathscr{s}' \rightsquigarrow R \rightsquigarrow \mathscr{s}$, and $\mathscr{s}$ is its *terminator*. Because, by supposition, a species can appear on only one side of a reaction, the initiator and terminator of a causal unit are always different species.

There are four distinct kinds of causal units, the ones already drawn in (19.5), (19.6), (19.8), and (19.9). Note that in (19.5) and (19.8), there are distinct complex labels on the two edges. The situation is different in (19.6) and (19.9), in which there are identical labels on the edges. By a *c-pair causal unit*, we mean a causal unit in which the two edges carry identical complex labels.

We come now to three important remarks.

Remark 19.4.1 (Species signs change in c-pair causal units but not otherwise). Note that a *c-pair* causal unit results in a *change* of species sign as the edges are traversed from the initiator species to the terminator species. Otherwise, a causal unit is characterized by *retention* of the sign.

Remark 19.4.2 (The Sign-Causality Graph's counterpart in the Species-Reaction Graph). The Sign-Causality Graph induced by the discordance $\{\alpha, \sigma\}$ can be identified in the obvious way with a subgraph of the Species-Reaction Graph drawn for network $\{\mathscr{S}, \mathscr{C}, \mathscr{R}\}$. Their vertices and edges are identical (apart perhaps from edge directions discussed in the following remark). This subgraph we will sometimes call *the Sign-Causality Graph's counterpart in the Species-Reaction Graph.*

Remark 19.4.3. Every edge in the Sign-Causality Graph is directed by the $\rightsquigarrow$ relation. Recall from Section 11.3.2, however, that only certain edges of the Species-Reaction Graph are assigned fixed directions. In particular, only certain edges of the Sign-Causality Graph's counterpart in the Species-Reaction Graph will be fixed-direction edges. Nevertheless, it is not difficult to check that, *in the Species-Reaction Graph counterpart, every fixed-direction edge carries a direction consistent with the $\rightsquigarrow$-direction in the Sign-Causality Graph.*

19.5 Why Even Cycles Are Necessary for a Discordance

Because, for the discordance $\{\alpha,\sigma\}$, σ is not zero, there is at least one signed species, say $\mathscr{s}_1$. It follows from the discussion in Section 19.4.1 that $\mathscr{s}_1$ is the terminator of a causal unit $\mathscr{s}_2 \rightsquigarrow R_2 \rightsquigarrow \mathscr{s}_1$, where $\mathscr{s}_2$ is distinct from the initiator $\mathscr{s}_1$. Because $\mathscr{s}_2$ is also signed, it must be the terminator of a causal unit $\mathscr{s}_3 \rightsquigarrow R_3 \rightsquigarrow \mathscr{s}_2$, where the signed species $\mathscr{s}_3$ is different from $\mathscr{s}_2$. Continuing in this way, we can assert that there is a directed sequence of the form

$$\cdots \rightsquigarrow \mathscr{s}_n \rightsquigarrow R_n \rightsquigarrow \mathscr{s}_{n-1} \rightsquigarrow R_{n-1} \cdots \rightsquigarrow \mathscr{s}_3 \rightsquigarrow R_3 \rightsquigarrow \mathscr{s}_2 \rightsquigarrow R_2 \rightsquigarrow \mathscr{s}_1, \tag{19.10}$$

with $\mathscr{s}_{i+1} \neq \mathscr{s}_i$.

Because the species are finite in number, two nonconsecutive species in the sequence (19.10) must coincide. Such a coincidence implies that *the Sign-Causality Graph for the putative discordance contains a directed cycle.* In fact, we can infer further that *each vertex in the Sign-Causality Graph either resides in a directed cycle or, if not, then there is a directed cycle $\rightsquigarrow$-upstream from it.*

Using language already invoked for the Species-Reaction Graph, we say that a cycle in the Sign-Causality Graph is *even* if it contains an even number of c-pairs. The following lemma explains, to a great extent, just why *even* cycles play such a crucial role.

Lemma 19.5.1. *Every directed cycle in the Sign-Causality Graph is even.*

Proof. If, in the Sign-Causality Graph, we traverse a directed cycle beginning at a species s^* and count the number of changes of species signs when we have returned to s^*, that number clearly must be even. Note that the cycle is the union of causal units. It follows from Remark 19.4.1 that the number of species-sign changes is identical to the number of c-pair causal units the cycle contains. Thus, the number of c-pairs the cycle contains is even. □

Remark 19.5.2 (Why, for a discordance, the Species-Reaction Graph must contain an even cycle). We argued earlier that the Sign-Causality Graph for the putative concordance $\{\alpha,\sigma\}$ must contain a directed cycle. Lemma 19.5.1 tells us that such a cycle must, in fact, be even. The counterpart of that cycle in the Species-Reaction Graph must be orientable and contain the same even number of c-pairs. Thus, *for our network $\{\mathscr{S},\mathscr{C},\mathscr{R}\}$ to be discordant, it is necessary that its Species-Reaction Graph contain an orientable even cycle.*

19.6 Sources in the Sign-Causality Graph

As with every graph, the Sign-Causality Graph drawn for the discordance $\{\alpha,\sigma\}$ has connected components. Because the Sign-Causality Graph is $\rightsquigarrow$-directed, it also has strong components.[5] Each strong component is either trivial—it consists of a single vertex—or else has the property that each edge is in a directed cycle.

Recall our observation that each vertex in the Sign-Causality Graph either resides in a directed cycle or, if not, then there is a directed cycle $\rightsquigarrow$-upstream from it. From this (and the fact that there is a finite number of vertices), it follows that each connected component contains at least one "initial" *nontrivial* strong component—that is, a nontrivial strong component with the property that none of its vertices has an $\rightsquigarrow$-incoming edge originating at a vertex outside that strong component. Such a nontrivial strong component we call a *source* in the Sign-Causality Graph.

For us, an essential fact is that the Sign-Causality Graph drawn for the presumed discordance $\{\alpha, \sigma\}$ has at least one source—in fact, at least one source within each connected component. Our proof of Theorem 19.1.1 will amount to showing that, when the hypothesis of the theorem is satisfied, the presence of a source in the putative Sign-Causality Graph is impossible.

With this in mind, *we will hereafter focus on one supposed source in the Sign-Causality Graph drawn for the discordance* $\{\alpha, \sigma\}$. *We denote its species vertices by* $\mathscr{S}_0$ *and its reaction vertices by* $\mathscr{R}_0$.

19.7 Inequalities Associated with a Source

We will begin by modifying some inequalities that we met in Section 19.4.1, this time with focus on the source under consideration. Note that the inequality (19.4), written for a *positive* species $s \in \mathscr{S}_0$, can be rewritten in the form

$$\sum_{y\to y' \in \mathscr{R}_0} \alpha_{y\to y'}(y'_s - y_s) \;+\; \sum_{y\to y' \in \mathscr{R}^* \setminus \mathscr{R}_0} \alpha_{y\to y'}(y'_s - y_s) \;>\; 0. \qquad (19.11)$$

Consider a nonzero term in the second sum corresponding to reaction $\bar{y} \to \bar{y}'$. Because that reaction is not a member of $\mathscr{R}_0$, any edge in the Sign-Causality graph connecting $\bar{y} \to \bar{y}'$ and the source-species s must be directed from s to $\bar{y} \to \bar{y}'$. (Recall that no vertex in a source is adjacent to an incoming edge originating at a vertex not in the source.) Thus, the reaction $\bar{y} \to \bar{y}'$ is not causal for the (positive) sign of s, so the corresponding term in (19.4) must be negative.

In effect, we have argued that the second sum in (19.4) cannot be positive. Therefore, when s is a *positive* species of the source, we can write

$$\sum_{y\to y' \in \mathscr{R}_0} \alpha_{y\to y'}(y'_s - y_s) > 0. \qquad (19.12)$$

When s is a *negative* species in the source, we can argue in a similar way to write

$$\sum_{y\to y' \in \mathscr{R}_0} \alpha_{y\to y'}(y'_s - y_s) < 0. \qquad (19.13)$$

---

[5] A directed graph is *strongly connected* if, for each choice of two vertices, those vertices are mutually reachable along directed paths. A *strong component* of a directed graph is a maximal strongly connected subgraph. A graph is *nontrivial* if it has an edge.

Remark 19.7.1. For a given source-species $s \in \mathscr{S}_0$, a nonzero term in either (19.12) or (19.13) might not correspond to any edge of the Sign-Causality Graph. This can happen if, for a particular reaction $y \to y'$, species s belongs to $\operatorname{supp} y'$ but disagrees in sign with $\alpha_{y \to y'}$. (In this case reaction $y \to y'$ is not causal for the sign of species s, nor is s causal for the sign of $y \to y'$.) If s is a positive species, then the term in question is negative, in which case it can be removed from (19.12) without changing the sense of the inequality. On the other hand, if s is negative, the term in question is positive and can be removed from (19.13) without changing that inequality's sense.

Hereafter, we shall assume that such removals of terms in (19.12) *or* (19.13) *that do not correspond to Sign-Causality Graph edges have been made, so that every term in these inequalities can be associated with an edge in the Sign-Causality Graph.*

19.8 Stoichiometric Coefficients in the Sign-Causality Graph

Much as we associated a coefficient with each edge of the Species-Reaction Graph, we will do the same in the Sign-Causality Graph. For a species-to-reaction edge $s \rightsquigarrow R$ of the Sign-Causality Graph, we denote by $e_{s \rightsquigarrow R}$ the (positive) stoichiometric coefficient of species s in the corresponding edge-labeling complex. For a reaction-to-species edge $R \rightsquigarrow s$, we denote by $f_{R \rightsquigarrow s}$ the (positive) stoichiometric coefficient of species s in its edge-labeling complex.

With respect to the source under consideration, it will be helpful to have available the following notation: For each species $s \in \mathscr{S}_0$, we denote by $\mathscr{R}_0 \rightsquigarrow s$ the set of all edges of the source that are incoming to s and by $s \rightsquigarrow \mathscr{R}_0$ the set of all edges of the source that are outgoing from s. With Remark 19.7.1 kept in mind, some analysis will indicate that the inequalities displayed in (19.12) and (19.13) can be written as a single system:

$$\boxed{\sum_{\mathscr{R}_0 \rightsquigarrow s} f_{R \rightsquigarrow s} |\alpha_R| - \sum_{s \rightsquigarrow \mathscr{R}_0} e_{s \rightsquigarrow R} |\alpha_R| > 0, \quad \forall s \in \mathscr{S}_0.} \tag{19.14}$$

Our aim will be to show that this inequality system, written for the putative discordance $\{\alpha, \sigma\}$, stands in contradiction to the hypothesis of Theorem 19.1.1.

19.9 Motivation: Proof of Theorem 19.1.1 for a Simple Special Case

Consideration of a simple special case will afford useful motivation for the remainder of the proof of Theorem 19.1.1. In Remark 19.5.2 we argued that, for our network $\{\mathscr{S}, \mathscr{C}, \mathscr{R}\}$ to be discordant, it is necessary that its Species-Reaction Graph contain an orientable even cycle. *Suppose, then, that in the Species-Reaction Graph for* $\{\mathscr{S}, \mathscr{C}, \mathscr{R}\}$ *there is precisely one orientable even cycle and that its only orientation is given by*

$$R_1 \to s_1 \to R_2 \to s_2 \to \cdots \to R_n \to s_n \to R_1. \tag{19.15}$$

In this case network $\{\mathscr{S},\mathscr{C},\mathscr{R}\}$ satisfies condition (ii) of Theorem 19.1.1 vacuously, for there is only one even cycle. Let us suppose that condition (i) is also satisfied, which is to say that lone even Species-Reaction Graph cycle admits no stoichiometrically expansive orientation. This amounts to the requirement that the oriented cycle's stoichiometric coefficients satisfy the relation

$$\frac{f_{R_1\to\mathscr{s}_1} f_{R_2\to\mathscr{s}_2} \cdots f_{R_n\to\mathscr{s}_n}}{e_{\mathscr{s}_1\to R_2} e_{\mathscr{s}_2\to R_3} \cdots e_{\mathscr{s}_n\to R_1}} \leq 1. \tag{19.16}$$

When conditions (i) and (ii) are both satisfied, Theorem 19.1.1 tells us that our putative discordance $\{\alpha,\sigma\}$ is impossible. Let us see why this is so.

Proof (The discordance $\{\alpha,\sigma\}$ is impossible). The discordance $\{\alpha,\sigma\}$ gives rise to a Sign-Causality Graph, in which there must be at least one source. Each source, in turn, has a corresponding subgraph in the Species-Reaction Graph for network $\{\mathscr{S},\mathscr{C},\mathscr{R}\}$. Because each edge of a source must be contained in an even $\rightsquigarrow$-directed cycle, the Species-Reaction Graph counterpart of a source can only be the sole orientable Species-Reaction Graph cycle (19.15). This tells us that there can be only one source in the Sign-Causality Graph, the $\rightsquigarrow$-directed even cycle

$$R_1 \rightsquigarrow \mathscr{s}_1 \rightsquigarrow R_2 \rightsquigarrow \mathscr{s}_2 \rightsquigarrow \cdots \rightsquigarrow R_n \rightsquigarrow \mathscr{s}_n \rightsquigarrow R_1. \tag{19.17}$$

For this elementary source, the inequality system (19.14) takes an especially simple form:

$$\begin{aligned} f_{R_1\rightsquigarrow\mathscr{s}_1}|\alpha_{R_1}| - e_{\mathscr{s}_1\rightsquigarrow R_2}|\alpha_{R_2}| &> 0,\\ f_{R_2\rightsquigarrow\mathscr{s}_2}|\alpha_{R_2}| - e_{\mathscr{s}_2\rightsquigarrow R_3}|\alpha_{R_3}| &> 0,\\ &\vdots\\ f_{R_n\rightsquigarrow\mathscr{s}_n}|\alpha_{R_n}| - e_{\mathscr{s}_n\rightsquigarrow R_1}|\alpha_{R_1}| &> 0. \end{aligned} \tag{19.18}$$

It is easy to confirm that the system (19.18) can be satisfied only if[6]

$$\frac{f_{R_1\to\mathscr{s}_1} f_{R_2\to\mathscr{s}_2} \cdots f_{R_n\to\mathscr{s}_n}}{e_{\mathscr{s}_1\to R_2} e_{\mathscr{s}_2\to R_3} \cdots e_{\mathscr{s}_n\to R_1}} > 1. \tag{19.19}$$

This, however, contradicts (19.16). □

Remark 19.9.1. In a sense, what makes this single-orientable-even-cycle case so simple to study is the fact that in a cycle each species vertex is adjacent to just two reaction vertices. That is what gave rise to the rudimentary form of the inequality system (19.18). What is also attractive about a cycle, however, is that each reaction vertex is adjacent to precisely two species vertices. As we shall see, there will be considerable pleasantness when, in a source (and its Species-Reaction Graph counterpart), a particular reaction vertex has just two species as neighbors.

---

[6] It is easy to see that the stoichiometric coefficients $\{f_{R\to\mathscr{s}}\}$ and $\{e_{\mathscr{s}\to R}\}$ associated with edges in the oriented Species-Reaction Graph cycle are identical to their counterparts $\{f_{R\rightsquigarrow\mathscr{s}}\}$ and $\{e_{\mathscr{s}\rightsquigarrow R}\}$ in the Sign-Causality Graph source cycle.

19.10 The Reflection of a Source in the Species-Reaction Graph

The special-case proof of Theorem 19.1.1 in the preceding section employed a technique that we will use more generally: Articulate conditions that, given the presumed discordance, must be satisfied by the putative Sign-Causality Graph source under consideration, and then ask how those conditions are reflected in the source's Species-Reaction Graph counterpart. Finally, show that those conditions cannot be satisfied when the Species-Reaction Graph conforms to the hypothesis of Theorem 19.1.1. With this in mind, we will want to have available a general characterization of a source's Species-Reaction Graph counterpart.

In what follows, we will sometimes suppose that a subgraph of the Species-Reaction Graph has been given an *orientation*. This is to say that *all* edges in the subgraph has been given directions, even those that are not fixed-direction edges. *In such situations, it will be understood that the edge directions assigned are consistent with directions already carried by the fixed-direction edges.*

The following definition will help us describe the general structure of a source's counterpart in the Species-Reaction Graph.

Definition 19.10.1 ([118]). *An* **even cycle cluster** *is a nontrivial subgraph of a Species-Reaction Graph, taken with an orientation of the subgraph's edges, such that the resulting directed subgraph is strongly connected and all of its directed cycles are even. An even cycle cluster is* **complete** *if it is not a subgraph of a larger even cycle cluster having the same vertices.*

Definition 19.10.1 will be illustrated with the help of network (19.20).[7] This example, due to Daniel Knight, will turn out to be revealing in several ways.

$$\begin{array}{c} A+D \rightleftarrows B \rightleftarrows A+E \\ \downarrow \\ C+F \\ C \rightleftarrows A \end{array} \tag{19.20}$$

In Figure 19.1 (a) we show the network's Species-Reaction Graph. In (b) we show an even cycle cluster. That even cycle cluster is not complete, however, as it is a proper subgraph of a still larger (complete) even cycle cluster (c) having the same vertices. The subgraph (d) will be explained in the next section.

[7] Network (19.20) is identical in structure to the enzyme example (11.5) of Chapter 11.

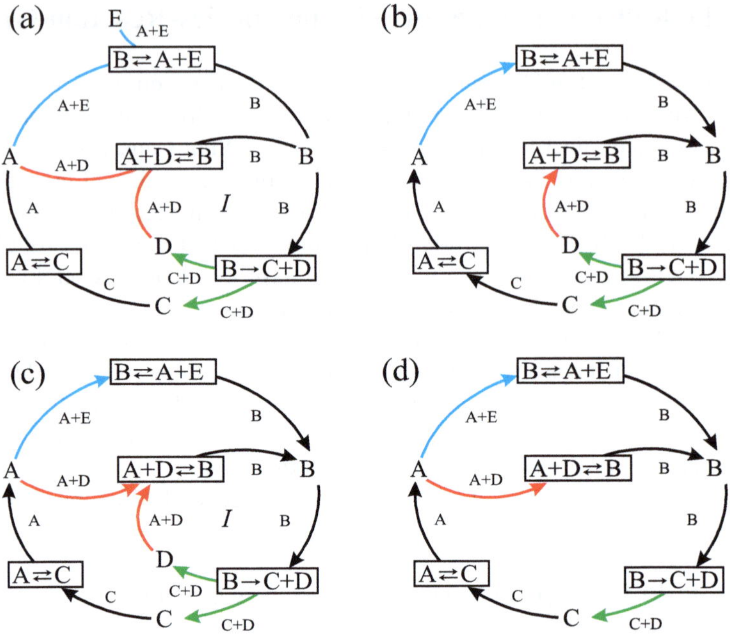

Fig. 19.1. (a) Species-Reaction Graph for network (19.20). (b) An incomplete even cycle cluster. (c) A complete even cycle cluster. (d) A simple core within (c). Adapted from [118]

The following proposition tells us that the counterpart of a source in the Species-Reaction Graph, when its edges are oriented as in the source, is invariably a complete even cycle cluster.

Proposition 19.10.2 ([118]). *Given a discordance for a reaction network, consider a source in the corresponding Sign-Causality Graph. Let G be the subgraph of the network's Species-Reaction Graph having the same vertices and edges as the source. Moreover, suppose that each edge in G is given a direction "→" identical to its ⇝-direction in the source. Then the resulting directed subgraph of the Species-Reaction Graph has edge directions consistent with the fixed-direction edges of the Species-Reaction Graph, and it is a complete even cycle cluster.*

Remark 19.10.3 (About the proof of Proposition 19.10.2). Because a source is a nontrivial strong component of the Sign-Causality Graph and because, from Lemma 19.5.1, all of its directed cycles are even, it is fairly evident that the source's oriented counterpart in the Species-Reaction Graph is an even cycle cluster. What is less evident is that the resulting even cycle cluster is complete. This is proved in [118].

Recall that we are supposing that the network $\{\mathscr{S},\mathscr{C},\mathscr{R}\}$ under study admits the discordance $\{\alpha,\sigma\}$, which in turn gives rise to a Sign-Causality Graph. Recall also that we are focusing on a particular source in the Sign-Causality Graph, one having species set $\mathscr{S}_0$ and reaction set $\mathscr{R}_0$. Proposition 19.10.2 tells us that we can identify that source with a subgraph of the Species-Reaction Graph, in particular a complete even cycle cluster in the Species-Reaction Graph with edge directions "→" congruent to the "⇝" directions in the source.

With the identification made in this way, *we will hereafter regard the putative source under study to be a complete even cycle cluster in the Species-Reaction Graph, oriented as described.* So that we can regard the inequality system (19.14) to be rooted in the Species-Reaction Graph, we will replace the "⇝" relation of the Sign-Causality Graph by the congruent Species-Reaction Graph relation "→".

$$\boxed{\sum_{\mathscr{R}_0 \to \mathcal{s}} f_{R\to \mathcal{s}}|\alpha_R| - \sum_{\mathcal{s} \to \mathscr{R}_0} e_{\mathcal{s}\to R}|\alpha_R| > 0, \quad \forall \mathcal{s} \in \mathscr{S}_0.} \tag{19.21}$$

The proof of Theorem 19.1.1 will amount to showing that, when the conditions of the theorem are satisfied, the system (19.21) cannot admit a solution $\{|\alpha_R|\}_{R\in\mathscr{R}_0}$. That is, we will show that the presumed discordance, giving rise to the Sign-Causality Graph source under consideration, cannot exist.

19.11 Proof Strategy Going Forward

The inequality system (19.21) for the putative source is one focused on the individual species of $\mathscr{S}_0$. To prove that (19.21) cannot obtain under the conditions of Theorem 19.1.1, we will show a contradiction in a dual inequality system focused instead on the individual reactions of $\mathscr{R}_0$.

Let M be an arbitrary nonzero member of $\overline{\mathbb{R}}_+^{\mathscr{S}}$ with support in $\mathscr{S}_0$. If (19.21) holds, then so must the single inequality (19.22):

$$\sum_{\mathcal{s}\in\mathscr{S}_0} M_{\mathcal{s}}\Big(\sum_{\mathscr{R}_0\to \mathcal{s}} f_{R\to \mathcal{s}}|\alpha_R| - \sum_{\mathcal{s}\to\mathscr{R}_0} e_{\mathcal{s}\to R}\,|\alpha_R|\Big) > 0. \tag{19.22}$$

For each reaction $R \in \mathscr{R}_0$, let $R \to \mathscr{S}_0$ denote the set of all edges of the source under consideration that are outgoing from R, and let $\mathscr{S}_0 \to R$ denote the set of all edges of the source that are incoming to R. With this notation in hand, we can rewrite (19.22) in the following way:

$$\sum_{R\in\mathscr{R}_0}\Big(\sum_{R\to\mathscr{S}_0} M_{\mathcal{s}} f_{R\to \mathcal{s}} - \sum_{\mathscr{S}_0\to R} M_{\mathcal{s}}\, e_{\mathcal{s}\to R}\Big)\,|\alpha_R| > 0. \tag{19.23}$$

If, for the putative source under consideration, we can find in $\overline{\mathbb{R}}_+^{\mathscr{R}}$ a nonzero M with support in $\mathscr{S}_0$ that satisfies (19.24), then we will have a contradiction of (19.23) and, therefore, of (19.21).

$$\sum_{R\to\mathscr{S}_0} M_{\mathcal{s}}\, f_{R\to \mathcal{s}} - \sum_{\mathscr{S}_0\to R} M_{\mathcal{s}}\, e_{\mathcal{s}\to R} \le 0, \quad \forall R\in\mathscr{R}_0 \tag{19.24}$$

Proof that there is such an M will be our goal. We will begin by showing that, when condition (ii) of Theorem 19.1.1 is satisfied, the putative source contains a simple subgraph that provides a congenial foothold for construction of an M having the desired properties.

19.12 When Condition (ii) of Theorem 19.1.1 Is Satisfied, Every Even Cycle Cluster Has a Simple Core

Condition (ii) of Theorem 19.1.1 is a prohibition against certain ways in which even cycles might be entangled in the Species-Reaction Graph. What we shall see in this section is that such a prohibition ensures that, within an even cycle cluster, there exists a "core" that has a simple structure. The nature of that core is described in the following definition.

Definition 19.12.1 ([118]). *An oriented subgraph, G, of the Species-Reaction Graph has a* **simple core** *if G contains a nontrivial subgraph G_*, consisting perhaps of G itself, for which the following are true:*

(i) G_* is strongly connected.
(ii) Every reaction vertex in G_* is, within G_*, adjacent to precisely two species vertices.
(iii) G contains no edge external to G_* that terminates in a species vertex of G_*.

Example 19.12.2. In Figure 19.1 the complete even cycle cluster (c) contains a simple core, exhibited in (d). By way of contrast, the reader can confirm that the even cycle cluster (b) contains no simple core.

The following proposition is proved in Appendix 19.A.

Proposition 19.12.3 ([118]). *If G is a nontrivial oriented strongly connected subgraph of the Species-Reaction Graph that has no simple core, then there exists in G a pair of directed cycles having as their intersection a single directed path originating at a species vertex and terminating at a reaction vertex.*

Example 19.12.4. Once again, we return to Figure 19.1. Recall that the even cycle cluster (b) contains no simple core. As Proposition 19.12.3 demands, two directed cycles in (b) have as their intersection a single species-to-reaction directed path, in this case the edge connecting species B to the reaction $B \to C + D$.

Recall too that the complete even cycle cluster (c) contains the simple core (d). Although (c), like (b), contains two directed cycles that intersect in a directed species-to-reaction edge, there is no contradiction of Proposition 19.12.3, which asserts a *necessary* condition for the absence of a simple core.

Proposition 19.12.5. *Suppose that a reaction network's Species-Reaction Graph satisfies condition (ii) of Theorem 19.1.1. Then every even cycle cluster contains a simple core.*

Proof. Suppose that a particular even cycle cluster lacks a simple core. Then, from Proposition 19.12.3, it contains two directed cycles that have as their intersection a single directed path originating at a species vertex and terminating at a reaction vertex. Because those cycles reside in an even cycle cluster, they must be even. This, however, contradicts condition (ii) of Theorem 19.1.1. □

19.13 Completion of the Proof of Theorem 19.1.1

In this section we will complete the proof of Theorem 19.1.1 (apart from proof of an intermediary proposition deferred to Appendix 19.B). Hereafter we shall suppose that, for the reaction network $\{\mathscr{S},\mathscr{C},\mathscr{R}\}$ under consideration, the Species-Reaction Graph satisfies conditions (i) and (ii) of Theorem 19.1.1.

We return to consideration of our putative source, having species set $\mathscr{S}_0$ and reaction set $\mathscr{R}_0$. Recall that the proof will be finished if we can show that there is an $M \in \overline{\mathbb{R}}_+^{\mathscr{S}}$, with support in $\mathscr{S}_0$, that satisfies the inequality system (19.24). Because the source under consideration is an even cycle cluster, Proposition 19.12.5 ensures that it contains a simple core. We can begin to build the required M by first considering those inequalities in the system (19.24) that correspond to reactions in the core. Toward that end, the following proposition is helpful. It is proved in Appendix 19.B.

Proposition 19.13.1 ([118, 158]). *Consider an oriented strongly connected subgraph G_* of the Species-Reaction Graph having species set $\mathscr{S}_*$ and reaction set $\mathscr{R}_*$. Suppose that no directed cycle in G_* is stoichiometrically expansive and that, within G_*, every reaction vertex is adjacent to precisely two species vertices. Then there is a set of positive numbers $\{M_s\}_{s \in \mathscr{S}_*}$ that satisfies the following system of inequalities:*

$$M_{s'}\, f_{R \to s'} - M_s\, e_{s \to R} \le 0, \quad \forall R \in \mathscr{R}_*. \tag{19.25}$$

Remark 19.13.2. The hypothesis of Proposition 19.13.1 requires that each reaction vertex in G_* has, within G_*, precisely one outgoing edge and one incoming edge. In (19.25) $f_{R \to s'}$ and $e_{s \to R}$ are the stoichiometric coefficients associated with the two edges in G_* adjacent to reaction vertex R.

Proposition 19.13.1 paves the way for the following proposition, which takes us just short of proof of Theorem 19.1.1.

Proposition 19.13.3 ([118]). *Suppose that a directed subgraph G_0 of the Species-Reaction Graph, with species set $\mathscr{S}_0$ and reaction set $\mathscr{R}_0$, has a simple core. Suppose also that no directed cycle within the simple core is stoichiometrically expansive. Then there is a set of nonnegative numbers $\{M_s\}_{s \in \mathscr{S}_0}$, not all zero, such that*

$$\sum_{R \to \mathscr{S}_0} M_s f_{R \to s} - \sum_{\mathscr{S}_0 \to R} M_s e_{s \to R} \le 0, \quad \forall R \in \mathscr{R}_0. \tag{19.26}$$

In fact, if $\mathscr{S}_*$ is the species set of the simple core, M_s can be taken positive for all $s \in \mathscr{S}_*$ and zero for all $s \notin \mathscr{S}_*$.

Proof. Let G_* be the simple core, and let $\mathscr{S}_*$ and $\mathscr{R}_*$ be its species and reaction sets. Because G_* is a simple core, it is strongly connected, and each of its reaction vertices is, within G_*, adjacent to precisely two species vertices of G_*. Note that, by hypothesis, no directed cycle in G_* is stoichiometrically expansive. Proposition 19.13.1 then ensures the existence of positive numbers $\{M_s^*\}_{s \in \mathscr{S}_*}$ that satisfy

$$M^*_{\mathcal{s}'} f_{R \to \mathcal{s}'} - M^*_{\mathcal{s}'} e_{\mathcal{s} \to R} \leq 0, \quad \forall R \in \mathcal{R}_*, \tag{19.27}$$

where, for each $R \in \mathcal{R}_*$, $\mathcal{s}'$ and $\mathcal{s}$ are the two species of $\mathcal{S}_*$ adjacent to R in G_*.

Next, we choose the set of nonnegative numbers $\{M_{\mathcal{s}}\}_{\mathcal{s} \in \mathcal{S}_0}$ in the following way: $M_{\mathcal{s}} = M^*_{\mathcal{s}}$ for all $\mathcal{s} \in \mathcal{S}_*$ and $M_{\mathcal{s}} = 0$ for all $\mathcal{s} \notin \mathcal{S}_*$. We want to show that this choice satisfies (19.26). Note that, because of the choice, (19.26) reduces to (19.28), in which it is understood that a sum over an empty set is zero.

$$\sum_{R \to \mathcal{S}_*} M^*_{\mathcal{s}} f_{R \to \mathcal{s}} - \sum_{\mathcal{S}_* \to R} M^*_{\mathcal{s}} e_{\mathcal{s} \to R} \leq 0, \quad \forall R \in \mathcal{R}_0 \tag{19.28}$$

To confirm that the inequality in (19.28) holds for a particular $R \in \mathcal{R}_0$, there are two possibilities that require examination: the first is that R is a reaction of the simple core G_* and the second is that it is not. These cases will be considered separately.

Suppose, then, that R is a reaction of the simple core G_*. From properties of a simple core, there are precisely two edges within G_* adjacent to R, one incoming (denoted $\mathcal{s}_* \to R$) and one outgoing (denoted $R \to \mathcal{s}'_*$). Thus, for the particular $R \in \mathcal{R}_*$ under consideration, the left side of (19.28) takes the form

$$M^*_{\mathcal{s}'_*} f_{R \to \mathcal{s}'_*} - M^*_{\mathcal{s}_*} e_{\mathcal{s}_* \to R}. \tag{19.29}$$

That this is nonpositive follows from (19.27).

Now suppose that R is not a reaction of the simple core. In this case, properties of a simple core ensure that $R \to \mathcal{S}_*$ is empty, whereupon the first sum in (19.28) is zero. For the particular $R \notin \mathcal{R}_*$ under consideration, the inequality in (19.28) is clearly satisfied. □

We are now coming to the end of the proof of Theorem 19.1.1. In particular, we return to consideration of our putative source in the Species-Reaction Graph, the source derived from the presumed discordance $\{\alpha, \sigma\}$ and having species set $\mathcal{S}_0$ and reaction set $\mathcal{R}_0$.

Suppose that conditions (i) and (ii) of Theorem 19.1.1 obtain. From Proposition 19.10.2 it follows that the source is an even cycle cluster. In this case condition (ii) and Proposition 19.12.3 ensure that the source contains a simple core. Condition (i) and Proposition 19.13.3 then ensure that there are nonnegative numbers $\{M_{\mathcal{s}}\}_{\mathcal{s} \in \mathcal{S}_0}$, not all zero, such that the inequality system (19.26) obtains. As explained in Section 19.11, the existence of such numbers precludes the possibility that the source inequality system (19.21) might be satisfied. From this it follows that there can be no source and, therefore, no discordance. This completes the proof of Theorem 19.1.1. □

19.14 A Species-Reaction Graph Theorem with a Weaker Hypothesis

In our proof of Theorem 19.1.1, condition (ii) exerted itself only to the extent that it ensured that the putative source under consideration—of necessity, an even cycle cluster—has a simple core. In the following theorem, we merely replace condition (ii) of Theorem 19.1.1 with the direct requirement that every possible candidate for a source in the Species-Reaction Graph—by Proposition 19.10.2, every *complete* even cycle cluster—have a simple core.

Theorem 19.14.1 (Simple Core Theorem [117, 118]**).** *Suppose that a reaction network's Species-Reaction Graph has the following properties:*

(i) No even cycle admits a stoichiometrically expansive orientation.
(ii) Every complete even cycle cluster has a simple core.

Then the network's fully open extension is concordant.

Remark 19.14.2. Recall from Definition 19.10.1 that an even cycle cluster is a subgraph of the Species-Reaction Graph, *taken with an orientation of its edges*, such that the resulting subgraph is strongly connected and all of its directed cycles are even. With this as background, condition (ii) of Theorem 19.14.1 should be understood in the following way: if any subgraph of the Species-Reaction Graph admits an orientation whereby the result is a complete even cycle, then that oriented subgraph is required to have a simple core.[8]

Remark 19.14.3. To see that the Simple Core Theorem gives information about networks for which Theorem19.1.1 stands silent, we return to network (19.20) and Figure 19.1, where the network's Species-Reaction Graph is drawn. Note that condition (ii) of Theorem 19.1.1 is not satisfied. When, in Figure 19.1(a), the large outer cycle and the small inner cycle traversing species B and D are given clockwise orientations, their intersection is the directed species-to-reaction edge connecting species B to reaction $B \to C+D$. For network (19.20), then, Theorem 19.1.1 says nothing.

On the other hand, both conditions of Theorem 19.14.1 are satisfied. Because every stoichiometric coefficient in the network is 1, no cycle admits a stoichiometrically expansive orientation. Moreover, every *complete* even cycle cluster has a simple core: We have already observed that the complete even cycle cluster in Figure 19.1(c) has the simple core shown in Figure 19.1(d). There are other complete even cycle clusters in Figure 19.1(a) that can be checked affirmatively. For example, the small inner cycle through species B and D, oriented clockwise, is a complete even cycle cluster, and it contains a simple core, the cycle itself. Recall that the even cycle cluster displayed in Figure 19.1 does not contain a simple core, but there is no violation of condition (ii) in Theorem 19.14.1. That even cycle cluster is not complete.

Indeed, the fully open extension of network (19.20) is concordant, as can be confirmed independently by means of the Chemical Reaction Network Toolbox [62].

Remark 19.14.4. Theorem 19.14.1, like Theorem 19.1.1, has deep ramifications for kinetic system behavior. If a *nondegenerate* network has a Species-Reaction Graph that satisfies conditions (i) and (ii) of Theorem 19.14.1, then the network itself—not just its fully open extension—is concordant (Theorem 10.6.17). So long as the

---

8 As indicated earlier, the orientation is required to be consistent with directions carried by the fixed-direction edges of the Species-Reaction Graph.

kinetics is weakly monotonic, therefore, the variety of behaviors the network can admit is severely constrained by Theorem 10.7.6.

Although Theorem 19.14.1 is more generous than Theorem 19.1.1 in the breadth of reaction networks for which it gives information, it should be said that its hypothesis is more difficult to check in particular instances. Moreover, although both theorems give highly subtle information about what is essential in a network for discordance (and certain exotic behaviors), the structural insights provided by Theorem 19.1.1 are far more tangible.

Appendix 19.A Proof of Proposition 19.12.3

Here we prove Proposition 19.12.3, repeated later in this appendix as Proposition 19.A.2.

Some additional vocabulary will lay the groundwork for a preliminary proposition. Let G be an oriented subgraph of the Species-Reaction Graph. Then $G_* \subset G$ will be called an *R-subgraph* of G if it satisfies the first two of the requirements for a simple core but not necessarily the third. That is, G_* is an R-subgraph of G if G_* is strongly connected and, within G_*, every reaction vertex is adjacent to precisely two species vertices. We say that G_* has an *R-to-S ear* in G if there exists in G a directed path that originates with a reaction vertex of G_* and terminates with a species vertex of G_* but contains no edge of G_*.

The following is a precursor to the proof of Proposition 19.12.3.

Proposition 19.A.1 ([118]). *If G is a nontrivial strongly connected directed subgraph of the Species-Reaction Graph that has no simple core, then there exists in G an R-subgraph having an R-to-S ear in G.*

Proof. We will show the existence of an R-subgraph with the desired R-to-S ear by means of an iterative construction. Because G is strongly connected, we know that it contains at least one R-subgraph—any directed cycle is an example—that we will designate G_0.

Now suppose that G_i is an R-subgraph of G, where i is a nonnegative integer index label that indicates the progress of our iteration. Because G has no simple core, G contains an edge that is not in G_i and that terminates at a species vertex of G_i. Let $R_* \to \mathcal{S}_*$ be such an edge. Since G is strongly connected, there is a directed path P in G that originates at a vertex of G_i, is edge-disjoint from G_i, and terminates in R_*. Let E be the union of P with the edge $R_* \to \mathcal{S}_*$. Note that E begins at a vertex of G_i, ends at the species vertex $\mathcal{S}_*$, and has no edge in common with G_i. If the originating vertex of E is a *reaction* vertex of G_i, then G_i is the required R-subgraph, having E as its required R-to-S ear.

If, instead, E originates at a *species* vertex of G_i, say $\mathcal{S}_{**}$ (which might be $\mathcal{S}_*$), then E is either a path (when $\mathcal{S}_{**} \neq \mathcal{S}_*$) or a cycle (when $\mathcal{S}_{**} = \mathcal{S}_*$), neither having an edge in common with G_i and whose only vertices in common with G_i are species vertices $\mathcal{S}_{**}$ and $\mathcal{S}_*$.

In this case, let $G_{i+1} = G_i \cup E$, which is a still larger R-subgraph of G. The construction applied above to G_i can be applied again to G_{i+i} and then sequentially thereafter if necessary. Because G has no simple core, the process will end in an R-subgraph having the required R-to-S ear. □

We are now in a position to prove Proposition 19.12.3, which we repeat below as Proposition 19.A.2.

Proposition 19.A.2 ([118]). *If G is a nontrivial oriented strongly connected subgraph of the Species-Reaction Graph that has no simple core, then there exists in G a pair of directed cycles having as their intersection a single directed path originating at a species vertex and terminating at a reaction vertex.*

Proof. Proposition 19.A.1 ensures the existence in G of an R-subgraph having an R-to-S ear. We denote by G_0 that R-subgraph and by E its R-to-S ear. Moreover, we denote by R_0 and $\mathscr{S}_0$ the reaction and species vertices at the ends of the ear. Because G_0 is strongly connected, G_0 contains within it a directed path P originating at R_0 and terminating at $\mathscr{S}_0$ and also a directed path Q originating at $\mathscr{S}_0$ and terminating at R_0. In Figure 19.2 we show some possibilities.

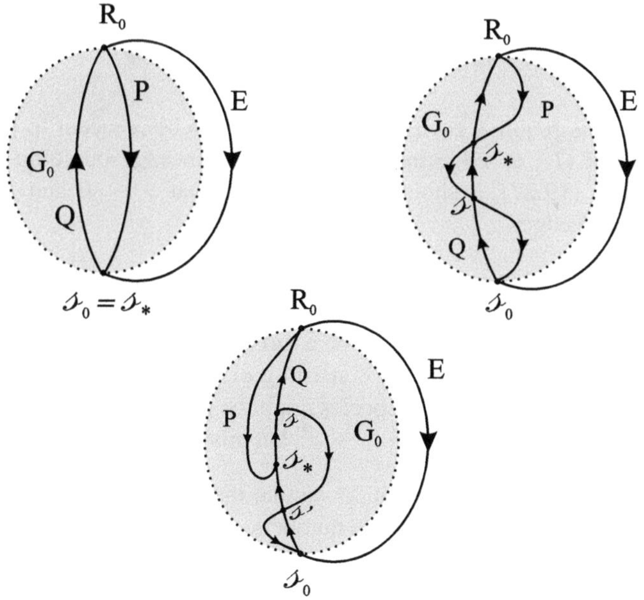

Fig. 19.2. Some possibilities in connection with Proposition 19.A.2. Adapted from [118]

Because G_0 is an R-subgraph of G, the first vertex along the path P that is also a vertex of Q must be a species vertex (which might be $\mathscr{S}_0$). We denote that first vertex by $\mathscr{S}_*$. By $R_0 P \mathscr{S}_*$ we mean the segment of P beginning at R_0 and ending at $\mathscr{S}_*$.

From the definition of $\mathscr{S}_*$, there can be no species vertex of $R_0 P \mathscr{S}_*$ that is also a vertex of Q (apart from $\mathscr{S}_*$). Thus, there can be no internal vertices common to the path $R_0 P \mathcal{s}_*$ and the directed segment of Q (denoted $\mathcal{s}_* Q R_0$) that begins at $\mathcal{s}_*$ and ends at R_0. Therefore, the union of the two paths, $R_0 P \mathcal{s}_* Q R_0$, is a directed cycle, denoted $C1$.

Note that the union of the ear E and the path Q is also a directed cycle, which can also be viewed as the union of E with the two complementary Q-segments $\mathcal{s}_0 Q \mathcal{s}_*$ and $\mathcal{s}_* Q R_0$. This second cycle, $C2 := E\mathcal{s}_0 Q \mathcal{s}_* Q R_0$, and the cycle $C1$ have as their intersection the single species-to-reaction path $\mathcal{s}_* Q R_0$. □

Appendix 19.B Proof of Proposition 19.13.1

Our goal in this appendix is to prove Proposition 19.13.1, which we repeat below as Proposition 19.B.1.

Proposition 19.B.1 ([118, 158]). *Consider an oriented strongly connected subgraph G_* of the Species-Reaction Graph having species set $\mathscr{S}_*$ and reaction set $\mathscr{R}_*$. Suppose that no directed cycle in G_* is stoichiometrically expansive and that, within G_*, every reaction vertex is adjacent to precisely two species vertices. Then there is a set of positive numbers $\{M_{\mathcal{s}}\}_{\mathcal{s} \in \mathscr{S}_*}$ that satisfies the following system of inequalities:*

$$M_{\mathcal{s}'}\, f_{R \to \mathcal{s}'} - M_{\mathcal{s}}\, e_{\mathcal{s} \to R} \le 0, \quad \forall R \in \mathscr{R}_*. \tag{19.B.1}$$

The hypothesis requires each reaction vertex R to be adjacent to precisely two directed edges of G_*, an edge incoming to R and an edge outgoing from R. For a particular R in (19.27), it should be understood that $\mathcal{s} \to R$ and $R \to \mathcal{s}'$ are, respectively, those edges.

Remark 19.B.2. Note that with each $R \in \mathscr{R}_*$ we can associate a unique causal unit $\mathcal{s} \to R \to \mathcal{s}'$ in G_*, and with each causal unit in G_* we can associate a unique R.[9] If we wish, then, we can regard G_* to be a directed strongly connected graph with vertices $\mathscr{S}_*$ and directed edge set $\mathscr{U}$ consisting of the causal units, one for each R. Note that, in such a graph, a pair of species might be joined by two different similarly directed edges, $\mathcal{s} \to R \to \mathcal{s}'$ and $\mathcal{s} \to \bar{R} \to \mathcal{s}'$, but this will cause no serious problems.

Proof. As in Remark 19.B.2, we let $\mathscr{U}$ denote the set of causal units in G_*. Our aim is to show the existence of positive numbers $\{M_{\mathcal{s}}\}_{\mathcal{s} \in \mathscr{S}_*}$ that satisfy (19.B.1) or, equivalently,

$$\ln M_{\mathcal{s}'} - \ln M_{\mathcal{s}} \le \ln \frac{e_{\mathcal{s} \to R}}{f_{R \to \mathcal{s}'}}, \quad \forall\, \mathcal{s} \to R \to \mathcal{s}' \in \mathscr{U}. \tag{19.B.2}$$

To prepare for a vector reformulation of (19.B.2), let $\{\omega_{\mathcal{s}}\}_{\mathcal{s} \in \mathscr{S}_*}$ be the standard basis for $\mathbb{R}^{\mathscr{S}_*}$, and let $\{\varepsilon_{\mathcal{s} \to R \to \mathcal{s}'}\}_{\mathcal{s} \to R \to \mathcal{s}' \in \mathscr{U}}$ be the standard basis for $\mathbb{R}^{\mathscr{U}}$. Now let $T : \mathbb{R}^{\mathscr{S}_*} \to \mathbb{R}^{\mathscr{U}}$ be the linear transformation defined by

[9] Here we are regarding the causal unit $\mathcal{s} \to R \to \mathcal{s}'$ to reside in the Species-Reaction Graph rather in the Sign-Causality Graph.

$$Tq := \sum_{\mathcal{s} \to R \to \mathcal{s}' \in \mathcal{U}} [(\omega_{\mathcal{s}'} - \omega_{\mathcal{s}}) \cdot q]\, \varepsilon_{\mathcal{s} \to R \to \mathcal{s}'}, \tag{19.B.3}$$

and let $z \in \mathbb{R}^{\mathcal{U}}$ be defined by

$$z := \sum_{\mathcal{s} \to R \to \mathcal{s}' \in \mathcal{U}} \ln \frac{e_{\mathcal{s} \to R}}{f_{R \to \mathcal{s}'}}\, \varepsilon_{\mathcal{s} \to R \to \mathcal{s}'}. \tag{19.B.4}$$

The existence of positive numbers $\{M_{\mathcal{s}}\}_{\mathcal{s} \in \mathcal{S}_*}$ satisfying (19.B.2) is equivalent to the existence of $q^* \in \mathbb{R}^{\mathcal{S}_*}$ satisfying

$$Tq^* \leq z, \tag{19.B.5}$$

because then we can choose $M_{\mathcal{s}} = \exp q^*_{\mathcal{s}}$, $\forall \mathcal{s} \in \mathcal{S}_*$ to satisfy (19.B.2).

From a theorem of Gale [87, p. 46], the existence of such a q^* is equivalent to the *nonexistence* of $p \in \ker T^T \cap \overline{\mathbb{R}}^{\mathcal{U}}_+$ satisfying

$$p \cdot z < 0, \tag{19.B.6}$$

where $T^T : \mathbb{R}^{\mathcal{U}} \to \mathbb{R}^{\mathcal{S}_*}$ is the transpose of T, given by

$$T^T x := \sum_{\mathcal{s} \to R \to \mathcal{s}' \in \mathcal{U}} x_{\mathcal{s} \to R \to \mathcal{s}'} (\omega_{\mathcal{s}'} - \omega_{\mathcal{s}}). \tag{19.B.7}$$

To prove the nonexistence, we will say that $c \in \overline{\mathbb{R}}^{\mathcal{U}}_+$ is a *directed cycle vector* if, in G_*, there is a directed cycle such that $c_{\mathcal{s} \to R \to \mathcal{s}'} = 1$ if $\mathcal{s} \to R \to \mathcal{s}'$ is a causal unit within the cycle and is zero otherwise. It is easy to see that every directed cycle vector is a member of $\ker T^T \cap \overline{\mathbb{R}}^{\mathcal{U}}_+$. In fact, Lemma 19.B.3 below tells us more.

Lemma 19.B.3. *Every nonzero member of* $\ker T^T \cap \overline{\mathbb{R}}^{\mathcal{U}}_+$ *is a positive linear combination of directed cycle vectors.*

Proof. A proof given in [158] is based on the idea that each positive vector in the pointed polyhedral cone $\ker T^T \cap \overline{\mathbb{R}}^{\mathcal{U}}_+$ has a representation as a nonnegative combination of the cone's extreme vectors. It is then argued that each nonzero extreme vector is a positive multiple of a directed cycle vector. A very different style of argument is essentially given in [24], where that book's Proposition 7.14 is, apart from vocabulary, almost the same as Lemma 19.B.3. □

Now suppose that $p \in \ker T^T \cap \overline{\mathbb{R}}^{\mathcal{U}}_+$ satisfies condition (19.B.6), and let

$$p = \sum_{\theta=1}^{k} \alpha_\theta c_\theta \tag{19.B.8}$$

be a representation of p as a positive linear combination of directed cycle vectors. Then, from (19.B.6),

$$p \cdot z = \sum_{\theta=1}^{k} \alpha_\theta (c_\theta \cdot z) < 0, \tag{19.B.9}$$

with all α_θ positive. However, (19.B.9) is contradicted by the fact that each $c_\theta \cdot z$ is nonnegative, as we now argue:

Suppose that, for a particular θ, the directed cycle corresponding to c_θ is

$$\mathcal{S}_1 \to R_1 \to \mathcal{S}_2 \to R_2 \cdots \to \mathcal{S}_n \to R_n \to \mathcal{S}_1. \tag{19.B.10}$$

Then

$$c_\theta \cdot z = \sum_{i=1}^{n} \ln \frac{e_{\mathcal{S}_i \to R_i}}{f_{R_i \to \mathcal{S}_{i+1}}} = -\ln \prod_{i=1}^{n} \frac{f_{R_i \to \mathcal{S}_{i+1}}}{e_{\mathcal{S}_i \to R_i}} \geq 0, \tag{19.B.11}$$

with the last inequality coming from the fact that no directed cycle in G_* is stoichiometrically expansive.

This completes the proof of Proposition 19.B.1. □

Correction to: Foundations of Chemical Reaction Network Theory

Correction to:
M. Feinberg, *Foundations of Chemical Reaction Network Theory*, Applied Mathematical Sciences 202, https://doi.org/10.1007/978-3-030-03858-8

The original version of this book has been revised because it was inadvertently published with a few errors.

Chapter 3

Page 24: In Definition 3.1.1., "$(y, y') \in \mathscr{R}$" has been updated to "$(y', y) \in \mathscr{R}$."

Chapter 10

Page 178: In Example 10.6.18, "Network (10.23)" has been updated to "Network (10.22)"
Page 190: In Equation 10.39, "$t = 1$, and $\ell = 1$" has been updated to "$t = 3$, and $\ell = 3$"

The updated version of this book can be found at
https://doi.org/10.1007/978-3-030-03858-8_3
https://doi.org/10.1007/978-3-030-03858-8_10
https://doi.org/10.1007/978-3-030-03858-8_12
https://doi.org/10.1007/978-3-030-03858-8_13

M. Feinberg, *Foundations of Chemical Reaction Network Theory*,
Applied Mathematical Sciences 202,
https://doi.org/10.1007/978-3-030-03858-8_20

Chapter 12

Page 266: In the caption of Figure 12.13, "$c_{AS} - c_{BS} - c_C$" has been updated to "... in a $c_{AS} - c_{AB} - c_{B_2S}$"

Chapter 13

Page 278: On the right side of (13.11), "f" has been updated to "μ"

References

Page 444: Reference 62 updated to https://doi.org/10.5281/zenodo.5149266
Page 445: Reference 78 updated to https://doi.org/10.5281/zenodo.5167618

References

1. Alon, U.: An Introduction to Systems Biology: Design Principles of Biological Circuits. Chapman and Hall/CRC, Boca Raton, FL (2006)
2. Anderson, D.F.: Global asymptotic stability for a class of nonlinear chemical equations. SIAM Journal on Applied Mathematics **68**(5), 1464 (2008)
3. Anderson, D.F.: A proof of the global attractor conjecture in the single linkage class case. SIAM Journal on Applied Mathematics **71**(4), 1487–1508 (2011)
4. Anderson, D.F., Enciso, G.A., Johnston, M.D.: Stochastic analysis of biochemical reaction networks with absolute concentration robustness. Journal of the Royal Society Interface **11**(93), 20130,943 (2014)
5. Anderson, D.F., Kurtz, T.G.: Stochastic Analysis of Biochemical Systems. Springer, New York (2015)
6. Anderson, D.F., Shiu, A.: The dynamics of weakly reversible population processes near facets. SIAM Journal on Applied Mathematics **70**(6), 1840–1858 (2010)
7. Angeli, D., Leenheer, P.D., Sontag, E.D.: A petri net approach to the study of persistence in chemical reaction networks. Mathematical Biosciences **210**(2), 598–618 (2007)
8. Appleman, J., Beard, W., Delcamp, T., Prendergast, N., Freisheim, J., Blakley, R.: Unusual transient-and steady-state kinetic behavior is predicted by the kinetic scheme operational for recombinant human dihydrofolate reductase. Journal of Biological Chemistry **265**(5), 2740–2748 (1990)
9. Aris, R.: Introduction to the Analysis of Chemical Reactors. Prentice-Hall, Inc., Englewood Cliffs, NJ (1965)
10. Aris, R.: Prolegomena to the rational analysis of systems of chemical reactions. Archive for Rational Mechanics and Analysis **19**(2), 81–99 (1965)
11. Aris, R.: Prolegomena to the rational analysis of systems of chemical reactions II. Some addenda. Archive for Rational Mechanics and Analysis **27**(5), 356–364 (1968)
12. Arnold, V.I.: Ordinary Differential Equations. Springer, New York (1992)

M. Feinberg, *Foundations of Chemical Reaction Network Theory*,
Applied Mathematical Sciences 202,
https://doi.org/10.1007/978-3-030-03858-8

13. Axler, S.: Linear Algebra Done Right, 3rd edn. Springer, New York (2014)
14. Banaji, M., Craciun, G.: Graph-theoretic approaches to injectivity and multiple equilibria in systems of interacting elements. Communications in Mathematical Sciences **7**(4), 867–900 (2009)
15. Banaji, M., Craciun, G.: Graph-theoretic criteria for injectivity and unique equilibria in general chemical reaction systems. Advances in Applied Mathematics **44**(2), 168–184 (2010)
16. Banaji, M., Donnell, P., Baigent, S.: P-matrix properties, injectivity, and stability in chemical reaction systems. SIAM Journal on Applied Mathematics **67**(6), 1523 (2007)
17. Batchelor, E., Goulian, M.: Robustness and the cycle of phosphorylation and dephosphorylation in a two-component regulatory system. Proceedings of the National Academy of Sciences of the United States of America **100**(2), 691-696 (2003)
18. Belousov, B.P.: A periodic reaction and its mechanism. In: Collection of Short Papers on Radiation Medicine for 1958. Med. Publ. Moscow (1959)
19. Beretta, E.: Stability problems of chemical networks. In: Nonlinear Differential Equations: Invariance, Stability, and Bifurcation, pp. 11–27. Academic Press, Inc. (1981)
20. Beretta, E., Vetrano, F., Solimano, F., Lazzari, C.: Some results about nonlinear chemical systems represented by trees and cycles. Bulletin of Mathematical Biology **41**(5), 641–664 (1979)
21. Biggs, N.: Algebraic Graph Theory. Cambridge University Press (1994)
22. Bilous, O., Amundson, N.R.: Chemical reactor stability and sensitivity. AIChE Journal **1**(4), 513–521 (1955)
23. Blackmond, D.G., McMillan, C.R., Ramdeehul, S., Schorm, A., Brown, J.M.: Origins of asymmetric amplification in autocatalytic alkylzinc additions. Journal of the American Chemical Society **123**(41), 10,103–10,104 (2001)
24. Bondy, A., Murty, U.: Graph Theory. Springer, New York (2010)
25. Boros, B.: Notes on the deficiency-one theorem: multiple linkage classes. Mathematical Biosciences **235**(1), 110–122 (2012)
26. Boros, B.: On the dependence of the existence of the positive steady states on the rate coefficients for deficiency-one mass action systems: single linkage class. Journal of Mathematical Chemistry **51**(9), 2455–2490 (2013)
27. Boros, B.: On the existence of the positive steady states of weakly reversible deficiency-one mass action systems. Mathematical Biosciences **245**(2), 157–170 (2013)
28. Boros, B.: Existence of positive steady states for weakly reversible mass-action systems. arXiv:1710.04732 [math] (2017)
29. Bowen, R.M.: Thermochemistry of reacting materials. The Journal of Chemical Physics **49**(4), 1625–1637 (1968)
30. Brauer, F., Nohel, J.A.: The Qualitative Theory of Ordinary Differential Equations: An Introduction. Dover Publications (1989)
31. Bridgman, P.: Note on the principle of detailed balancing. Physical Review **31**(1), 101 (1928)
32. Chen, K.C., Calzone, L., Csikasz-Nagy, A., Cross, F.R., Novak, B., Tyson, J.J.: Integrative analysis of cell cycle control in budding yeast. Molecular Biology of the Cell **15**(8), 3841–3862 (2004)
33. Chen, K.C., Csikasz-Nagy, A., Gyorffy, B., Val, J., Novak, B., Tyson, J.J.: Kinetic analysis of a molecular model of the budding yeast cell cycle. Molecular Biology of the Cell **11**(1), 369–391 (2000)
34. Chicone, C.: Ordinary Differential Equations with Applications, 2nd edn. Springer, New York (2006)

35. Clarke, B.L.: Graph theoretic approach to the stability analysis of steady state chemical reaction networks. The Journal of Chemical Physics **60**(4), 1481–1492 (1974)
36. Clarke, B.L.: Theorems on chemical network stability. The Journal of Chemical Physics **62**(3), 773–775 (1975)
37. Clarke, B.L.: Stability of Complex Reaction Networks. In: I. Prigogine, S.A. Rice (eds.) Advances in Chemical Physics, vol. 43, pp. 1–215. Wiley, New York (1980)
38. Clarke, B.L.: Stoichiometric network analysis. Cell Biochemistry and Biophysics **12**(1), 237–253 (1988)
39. Coleman, B.D., Gurtin, M.E.: Thermodynamics with internal state variables. The Journal of Chemical Physics **47**(2), 597–613 (1967)
40. Coleman, B.D., Noll, W.: The thermodynamics of elastic materials with heat conduction and viscosity. Archive for Rational Mechanics and Analysis **13**(1), 167–178 (1963)
41. Conley, C.: Isolated invariant sets and the Morse index. CBMS Regional Conference Series in Mathematics, 38. American Mathematical Society, Providence, RI (1978)
42. Conradi, C., Flockerzi, D.: Multistationarity in mass action networks with applications to ERK activation. Journal of Mathematical Biology **65**(1), 107–156 (2011)
43. Cox, M.P., Ertl, G., Imbihl, R.: Spatial self-organization of surface structure during an oscillating catalytic reaction. Physical Review Letters **54**(15), 1725–1728 (1985)
44. Craciun, G.: Systems of nonlinear differential equations deriving from complex chemical reaction networks. Ph.D. thesis, The Ohio State University (2002)
45. Craciun, G.: Toric differential inclusions and a proof of the global attractor conjecture. arXiv:1501.02860 (2015)
46. Craciun, G., Feinberg, M.: Multiple equilibria in complex chemical reaction networks. I. The injectivity property. SIAM Journal on Applied Mathematics **65**(5), 1526–1546 (2005)
47. Craciun, G., Feinberg, M.: Multiple equilibria in complex chemical reaction networks: extensions to entrapped species models. IEE Proc. Syst. Biol **153**(4), 179–186 (2006)
48. Craciun, G., Feinberg, M.: Multiple equilibria in complex chemical reaction networks. II. The species-reaction graph. SIAM Journal on Applied Mathematics **66**(4), 1321–1338 (2006)
49. Craciun, G., Feinberg, M.: Multiple equilibria in complex chemical reaction networks: semi-open mass action systems. SIAM Journal on Applied Mathematics **70**(6), 1859–1877 (2010)
50. Craciun, G., Nazarov, F., Pantea, C.: Persistence and permanence of mass-action and power-law dynamical systems. SIAM Journal on Applied Mathematics **73**(1), 305–329 (2013)
51. Craciun, G., Pantea, C.: Identifiability of chemical reaction networks. Journal of Mathematical Chemistry **44**(1), 244–259 (2008)
52. Craciun, G., Tang, Y., Feinberg, M.: Understanding bistability in complex enzyme-driven reaction networks. Proceedings of the National Academy of Sciences **103**(23), 8697–8702 (2006)
53. De Kepper, P., Boissonade, J.: Theoretical and experimental analysis of phase diagrams and related dynamical properties in the Belousov–Zhabotinskii system. The Journal of Chemical Physics **75**(1), 189–195 (1981)
54. Delattre, P.: Evolution of Molecular Systems: Theoretical Foundations and Applications in Chemistry and Biology (in French). Maloine-Doin, Paris (1971)
55. Deng, J., Jones, C., Feinberg, M., Nachman, A.: On the steady states of weakly reversible chemical reaction networks. arXiv:1111.2386 [physics, q-bio] (2011)

56. Dickenstein, A., Pérez Millán, M.: How far is complex balancing from detailed balancing? Bulletin of Mathematical Biology **73**(4), 811–828 (2011)
57. Dirac, P.A.M.: The conditions for statistical equilibrium between atoms, electrons and radiation. Proceedings of the Royal Society of London, Series A **106**(739), 581–596 (1924)
58. Edelstein, B.B.: Biochemical model with multiple steady states and hysteresis. Journal of Theoretical Biology **29**(1), 57–62 (1970)
59. Ellison, P.: The advanced deficiency algorithm and its applications to mechanism discrimination. Ph.D. thesis, University of Rochester (1998)
60. Ellison, P., Feinberg, M.: How catalytic mechanisms reveal themselves in multiple steady-state data: I. Basic principles. Journal of Molecular Catalysis. A: Chemical **154**(1-2), 155–167 (2000)
61. Ellison, P., Feinberg, M., Yue, M., Saltsburg, H.: How catalytic mechanisms reveal themselves in multiple steady-state data: II. An ethylene hydrogenation example. Journal of Molecular Catalysis A: Chemical **154**, 169–184 (2000). [Corrigendum **260**, 306 (2006)]
62. Ellison, P., Ji, H., Knight, D., Feinberg, M.: The Chemical Reaction Network Toolbox, Version 2.3 (2014). Available at https://doi.org/10.5281/zenodo.5149266
63. Enciso, G.A.: Transient absolute robustness in stochastic biochemical networks. Journal of the Royal Society Interface **13**(121) (2016)
64. Epstein, I.R., Pojman, J.A.: An Introduction to Nonlinear Chemical Dynamics. Oxford University Press, New York (1998)
65. Epstein, I.R., Showalter, K.: Nonlinear chemical dynamics: oscillations, patterns, and chaos. The Journal of Physical Chemistry **100**(31), 13,132–13,147 (1996)
66. Érdi, P., Tóth, J.: Mathematical Models of Chemical Reactions: Theory and Applications of Deterministic and Stochastic Models. Princeton University Press, Princeton, NJ (1989)
67. Feinberg, M.: Complex balancing in general kinetic systems. Archive for Rational Mechanics and Analysis **49**(3), 187–194 (1972)
68. Feinberg, M.: On chemical kinetics of a certain class. Archive for Rational Mechanics and Analysis **46**(1), 1–41 (1972)
69. Feinberg, M.: Constitutive equations for ideal gas mixtures and ideal solutions as consequences of simple postulates. Chemical Engineering Science **32**(1), 75–78 (1977)
70. Feinberg, M.: Mathematical aspects of mass action kinetics. In: L. Lapidus, N. Amundson (eds.) Chemical Reactor Theory: A Review, pp. 1–78. Prentice Hall, Englewood Cliffs, N.J. (1977)
71. Feinberg, M.: Lectures on Chemical Reaction Networks (1979). Written version of lectures given at the Mathematical Research Center, University of Wisconsin, Madison, WI Available at http://crnt.osu.edu/LecturesOnReactionNetworks
72. Feinberg, M.: Chemical oscillations, multiple equilibria, and reaction network structure. In: W.E. Stewart, W.H. Ray, C. Conley (eds.) Dynamics and Modeling of Reactive Systems, Mathematics Research Center Symposia Series, pp. 59–130. Academic Press (1980). Available at http://crnt.osu.edu
73. Feinberg, M.: Chemical reaction network structure and the stability of complex isothermal reactors I. The deficiency zero and deficiency one theorems. Chemical Engineering Science **42**(10), 2229–2268 (1987)
74. Feinberg, M.: Chemical reaction network structure and the stability of complex isothermal reactors II. Multiple steady states for networks of deficiency one. Chemical Engineering Science **43**(1), 1–25 (1988)

75. Feinberg, M.: Necessary and sufficient conditions for detailed balancing in mass action systems of arbitrary complexity. Chemical Engineering Science **44**(9), 1819–1827 (1989)
76. Feinberg, M.: The existence and uniqueness of steady states for a class of chemical reaction networks. Archive for Rational Mechanics and Analysis **132**(4), 311–370 (1995)
77. Feinberg, M.: Multiple steady states for chemical reaction networks of deficiency one. Archive for Rational Mechanics and Analysis **132**(4), 371–406 (1995)
78. Feinberg, M., Ellison, P.: The Chemical Reaction Network Toolbox, Version 1.1 (1996). Available at https://doi.org/10.5281/zenodo.5167618
79. Feinberg, M., Horn, F.J.M.: Dynamics of open chemical systems and the algebraic structure of the underlying reaction network. Chemical Engineering Science **29**(3), 775–787 (1974)
80. Feinberg, M., Horn, F.J.M.: Chemical mechanism structure and the coincidence of the stoichiometric and kinetic subspaces. Archive for Rational Mechanics and Analysis **66**(1), 83–97 (1977). The definition of linkage class in this article should read as it does here.
81. Feinberg, M., Lavine, R.: Thermodynamics based on the Hahn-Banach theorem: the Clausius inequality. Archive for Rational Mechanics and Analysis **82**(3), 203–293 (1983)
82. Feinberg, M., Lavine, R.: Foundations of the Clausius-Duhem inequality. In: J. Serrin (ed.) New Perspectives in Thermodynamics, pp. 49–64. Springer, New York (1986). Also available at https://crnt.osu.edu and as Appendix 2A in Truesdell, C., Rational Thermodynamics, Springer (1984)
83. Feinberg, M., Terman, D.: Traveling composition waves on isothermal catalyst surfaces. Archive for Rational Mechanics and Analysis **116**(1), 35–69 (1991)
84. Feliu, E., Wiuf, C.: Preclusion of switch behavior in networks with mass-action kinetics. Applied Mathematics and Computation **219**(4), 1449–1467 (2012)
85. Fowler, R., Milne, E.: A note on the principle of detailed balancing. Proceedings of the National Academy of Sciences **11**(7), 400–402 (1925)
86. Frank, F.C.: On spontaneous asymmetric synthesis. Biochimica et Biophysica Acta **11**, 459–463 (1953)
87. Gale, D.: The Theory of Linear Economic Models. University of Chicago Press, Chicago (1960)
88. Gavalas, G.R.: Nonlinear Differential Equations of Chemically Reacting Systems, *Springer Tracts in Natural Philosophy*, vol. 17. Springer, New York (1968)
89. Geiseler, W., Bar-Eli, K.: Bistability of the oxidation of cerous ions by bromate in a stirred flow reactor. The Journal of Physical Chemistry **85**(7), 908–914 (1981)
90. Gnacadja, G.: Reachability, persistence, and constructive chemical reaction networks I: Reachability approach to the persistence of chemical reaction networks. Journal of Mathematical Chemistry **49**(10), 2117–2136 (2011)
91. Gnacadja, G.: Reachability, persistence, and constructive chemical reaction networks II: A formalism for species composition in chemical reaction network theory and application to persistence. Journal of Mathematical Chemistry **49**(10), 2137–2157 (2011)
92. Gnacadja, G.: Reachability, persistence, and constructive chemical reaction networks III: A mathematical formalism for binary enzymatic networks and application to persistence. Journal of Mathematical Chemistry **49**(10), 2158–2176 (2011)
93. Goentoro, L., Kirschner, M.W.: Evidence that fold-change, and not absolute level, of β-catenin dictates Wnt signaling. Molecular Cell **36**(5), 872–884 (2009)

94. Golubitsky, M., Langford, W.F.: Classification and unfoldings of degenerate Hopf bifurcations. Journal of Differential Equations **41**(3), 375–415 (1981)
95. Golubitsky, M., Schaeffer, D.: Singularities and Groups in Bifurcation Theory: Volume I. Springer, New York (1984)
96. Goulian, M.: Robust control in bacterial regulatory circuits. Current Opinion in Microbiology **7**(2), 198-202 (2004)
97. Graziani, K.R., Hudson, J.L., Schmitz, R.A.: The Belousov-Zhabotinskii reaction in a continuous flow reactor. The Chemical Engineering Journal **12**(1), 9–21 (1976)
98. Greub, W.H.: Linear Algebra, 4th edn. Springer, New York (1981)
99. Guckenheimer, J., Holmes, P.: Nonlinear Oscillations, Dynamical Systems, and Bifurcations of Vector Fields. Springer, New York (2002)
100. Gunawardena, J.: Multisite protein phosphorylation makes a good threshold but can be a poor switch. Proceedings of the National Academy of Sciences **102**(41), 14,617–14,622 (2005)
101. Gurtin, M.E., Vargas, A.S.: On the classical theory of reacting fluid mixtures. Archive for Rational Mechanics and Analysis **43**(3), 179–197 (1971)
102. Halmos, P.R.: Finite-Dimensional Vector Spaces. Springer, New York (2015)
103. Hárs, V., Tóth, J.: On the inverse problem of reaction kinetics. In: M. Farkas (ed.) Qualitative Theory of Ordinary Differential Equations, *Colloquia Mathematica Societatis János Bolyai*, vol. 30, pp. 363–379. North-Holland - János Bolyai Mathematical Society, Budapest (1981). URL www.math.bme.hu/~jtoth/pubtexts/SzegedInverse.pdf
104. Hirsch, M.W., Smale, S., Devaney, R.L.: Differential Equations, Dynamical Systems, and an Introduction to Chaos, Third Edition. Academic Press (2012)
105. Ho, P.Y., Li, H.Y.: Determination of multiple steady states in an enzyme kinetics involving two substrates in a CSTR. Bioprocess Engineering **22**(6), 557–561 (2000)
106. Hoffman, K., Kunze, R.: Linear Algebra, 2nd edn. Pearson India (2015)
107. Horn, F.: Necessary and sufficient conditions for complex balancing in chemical kinetics. Archive for Rational Mechanics and Analysis **49**(3), 172–186 (1972)
108. Horn, F.: The dynamics of open reaction systems. In: Mathematical Aspects of Chemical and Biochemical Problems and Quantum Chemistry, Proceedings of the SIAM-AMS Symposium on Applied. Mathematics, vol. 8, pp. 125–137 (1974)
109. Horn, F., Jackson, R.: General mass action kinetics. Archive for Rational Mechanics and Analysis **47**(2), 81–116 (1972)
110. Huang, C.Y., Ferrell, J.E.: Ultrasensitivity in the mitogen-activated protein kinase cascade. Proceedings of the National Academy of Sciences **93**(19), 10,078–10,083 (1996)
111. Hudson, J.L., Mankin, J.C.: Chaos in the Belousov - Zhabotinskii reaction. The Journal of Chemical Physics **74**(11), 6171–6177 (1981)
112. Hyver, C.: Valeurs propres des systèmes de transformation représentables par des graphes en arbres. Journal of Theoretical Biology **42**(3), 397–409 (1973)
113. Hyver, C.: Stability of transformation systems representable by tree graphs: extension to structures of biological importance. Journal of Theoretical Biology **82**(2), 187–204 (1980)
114. Israel, K.: Monotone behavior for equilibria of dynamical systems. Ph.D. thesis, Duke University (1984)
115. Ji, H.: Uniqueness of equilibria for complex chemical reaction networks. Ph.D. thesis, The Ohio State University (2011)
116. Joshi, B., Shiu, A.: Simplifying the Jacobian criterion for precluding multistationarity in chemical reaction networks. SIAM Journal on Applied Mathematics **72**(3), 857–876 (2012)

117. Knight, D.: Reactor behavior and its relation to chemical reaction network structure. Ph.D. thesis, The Ohio State University (2015)
118. Knight, D., Shinar, G., Feinberg, M.: Sharper graph-theoretical conditions for the stabilization of complex reaction networks. Mathematical Biosciences **262**(1), 10–27 (2015)
119. Krambeck, F.J.: The mathematical structure of chemical kinetics in homogeneous single-phase systems. Archive for Rational Mechanics and Analysis **38**(5), 317–347 (1970)
120. Lang, S.: Linear Algebra. Addison-Wesley, Reading, MA (1966)
121. LaPorte, D.C., Thorsness, P.E., Koshland, D.E.: Compensatory phosphorylation of isocitrate dehydrogenase. A mechanism for adaptation to the intracellular environment. Journal of Biological Chemistry **260**(19), 10,563-10,568 (1985)
122. Lee, E., Salic, A., Krüger, R., Heinrich, R., Kirschner, M.W.: The roles of APC and axin derived from experimental and theoretical analysis of the Wnt pathway. PLoS Biology **1**(1), e10 (2003)
123. Leib, T., Rumschitzki, D., Feinberg, M.: Multiple steady states in complex isothermal CFSTRs. I: General considerations. Chemical Engineering Science **43**(2), 321–328 (1988)
124. Lewis, G.: A new principle of equilibrium. Proceedings of the National Academy of Sciences **11**(3), 179–183 (1925)
125. Lorenz, E.N.: Deterministic nonperiodic flow. Journal of the Atmospheric Sciences **20**(2), 130–141 (1963)
126. Lotka, A.J.: Contribution to the theory of periodic reactions. The Journal of Physical Chemistry **14**(3), 271–274 (1909)
127. Markevich, N.I., Hoek, J.B., Kholodenko, B.N.: Signaling switches and bistability arising from multisite phosphorylation in protein kinase cascades. The Journal of Cell Biology **164**(3), 353–359 (2004)
128. Mincheva, M.: Oscillations in biochemical reaction networks arising from pairs of subnetworks. Bulletin of Mathematical Biology **73**(10), 2277–2304 (2011)
129. Mincheva, M., Roussel, M.R.: Graph-theoretic methods for the analysis of chemical and biochemical networks. I. Multistability and oscillations in ordinary differential equation models. Journal of mathematical biology **55**(1), 61 (2007)
130. Mincheva, M., Roussel, M.R.: Graph-theoretic methods for the analysis of chemical and biochemical networks. II. Oscillations in networks with delays. Journal of Mathematical Biology **55**(1), 87–104 (2007)
131. Mullins, M.E.: Hydrocarbon reactions over transition metals: observations of surface hydrogen. Ph.D. thesis, University of Rochester (1983)
132. Nailor, J.: Behavior of equilibria in quasi-thermodynamic chemical reaction networks with mass-action kinetics. Ph.D. thesis, Duke University (1991)
133. Nickerson, H.K., Spencer, D.C., Steenrod, N.E.: Advanced Calculus. D. Van Nostrand Company, Inc. (1959). Re-published by Dover Publications, Mineola, NY, (2011).
134. Noll, W.: On certain convex sets of measures and on phases of reacting mixtures. Archive for Rational Mechanics and Analysis **38**(1), 1–12 (1970)
135. Pantea, C.: On the persistence and global stability of mass-action systems. SIAM Journal on Mathematical Analysis **44**(3), 1636–1673 (2012)
136. Prigogine, I., Lefever, R.: Symmetry breaking instabilities in dissipative systems. II. The Journal of Chemical Physics **48**(4), 1695–1700 (1968)
137. Qiao, L., Nachbar, R.B., Kevrekidis, I.G., Shvartsman, S.Y.: Bistability and oscillations in the Huang-Ferrell model of MAPK signaling. PLoS Comput Biol **3**(9), e184 (2007)

138. Rawlings, J., Ekerdt, J.: Chemical Reactor Analysis and Design Fundamentals, 2nd edn. Nob Hill Publishing, Madison, Wisconsin (2013)
139. Razon, L.F., Schmitz, R.A.: Intrinsically unstable behavior during the oxidation of carbon monoxide on platinum. Catalysis Reviews - Science and Engineering **28**(1), 89–164 (1986)
140. Reid, C.: Hilbert, 2nd edn. Copernicus, New York (1996)
141. Reidl, J., Borowski, P., Sensse, A., Starke, J., Zapotocky, M., Eiswirth, M.: Model of calcium oscillations due to negative feedback in olfactory cilia. Biophysical Journal **90**(4), 1147–1155 (2006)
142. Ricci, F., Stillinger, F.H., Debenedetti, P.G.: Creation and persistence of chiral asymmetry in a microscopically reversible molecular model. The Journal of Physical Chemistry B **117**(2), 602–614 (2013)
143. Rockafellar, R.T.: Convex Analysis. Princeton University Press, Princeton, NJ (1996)
144. Rumschitzki, D.: On the theory of multiple steady states in isothermal CSTR's. Ph.D. thesis, University of California, Berkeley [Work performed at the University of Rochester] (1983)
145. Rumschitzki, D., Feinberg, M.: Multiple steady states in complex isothermal CFSTRs II. Homogeneous reactors. Chemical Engineering Science **43**(2), 329–337 (1988)
146. Savageau, M.: Biochemical Systems Analysis: A Study of Function and Design in Molecular Biology. Addison-Wesley, Reading, MA (1976)
147. Schlosser, P.M.: A graphical determination of the possibility of multiple steady states in complex isothermal CFSTRs. Ph.D. thesis, University of Rochester (1988)
148. Schlosser, P.M., Feinberg, M.: A theory of multiple steady states in isothermal homogeneous CFSTRs with many reactions. Chemical Engineering Science **49**(11), 1749–1767 (1994)
149. Schmitz, R.A., Graziani, K.R., Hudson, J.L.: Experimental evidence of chaotic states in the Belousov-Zhabotinskii reaction. The Journal of Chemical Physics **67**(7), 3040–3044 (1977)
150. Segel, I.H.: Enzyme Kinetics: Behavior and Analysis of Rapid Equilibrium and Steady-State Enzyme Systems. Wiley-Interscience, New York (1993)
151. Seydel, R.: Practical Bifurcation and Stability Analysis, 3rd edn. Springer, New York (2009)
152. Shapiro, A.: Statics and dynamics of multicell reaction systems. Ph.D. thesis, University of Rochester (1975)
153. Shapiro, A., Horn, F.J.M.: On the possibility of sustained oscillations, multiple steady states, and asymmetric steady states in multicell reaction systems. Mathematical Biosciences **44**(1–2), 19–39 (1979). [See Errata in **46**(2):157 (1979).]
154. Shinar, G., Alon, U., Feinberg, M.: Sensitivity and robustness in chemical reaction networks. SIAM Journal on Applied Mathematics **69**(4), 977–998 (2009)
155. Shinar, G., Feinberg, M.: Structural sources of robustness in biochemical reaction networks. Science **327**(5971), 1389–1391 (2010)
156. Shinar, G., Feinberg, M.: Design principles for robust biochemical reaction networks: What works, what cannot work, and what might almost work. Mathematical Biosciences **231**(1), 39–48 (2011)
157. Shinar, G., Feinberg, M.: Concordant chemical reaction networks. Mathematical Biosciences **240**(2), 92–113 (2012)
158. Shinar, G., Feinberg, M.: Concordant chemical reaction networks and the species-reaction graph. Mathematical Biosciences **241**(1), 1–23 (2013)

159. Shinar, G., Milo, R., Rodriguez-Martinez, M., Alon, U.: Input-output robustness in simple bacterial signaling systems. Proceedings of the National Academy of Sciences **104**(50), 19,931–19,935 (2007)
160. Shinar, G., Rabinowitz, J.D., Alon, U., Papin, J.A.: Robustness in glyoxylate bypass regulation. PLoS Computational Biology **5**(3), e1000,297 (2009)
161. Siegel, D., MacLean, D.: Global stability of complex balanced mechanisms. Journal of Mathematical Chemistry **27**(1), 89–110 (2000)
162. Sontag, E.D.: Structure and stability of certain chemical networks and applications to the kinetic proofreading model of t-cell receptor signal transduction. IEEE Transactions on Automatic Control **46**(7), 1028–1047 (2001)
163. Stoer, J., Witzgall, C.: Convexity and Optimization in Finite Dimensions I. Springer, New York (2012)
164. Strogatz, S.H.: Nonlinear Dynamics and Chaos: With Applications to Physics, Biology, Chemistry, and Engineering, Second Edition. Westview Press (2014)
165. Tolman, R.C.: The principle of microscopic reversibility. Proceedings of the National Academy of Sciences **11**(7), 436–439 (1925)
166. Tyson, J., Kauffman, S.: Control of mitosis by a continuous biochemical oscillation: synchronization; spatially inhomogeneous oscillations. Journal of Mathematical Biology **1**(4), 289–310 (1975)
167. Tyson, J.J., Chen, K., Novak, B.: Network dynamics and cell physiology. Nature Reviews Molecular Cell Biology **2**(12), 908–916 (2001)
168. Voet, D., Voet, J.G.: Biochemistry, 3rd edn. Wiley, New York (2004)
169. Voit, E.O., Martens, H.A., Omholt, S.W.: 150 years of the mass action law. PLoS Computational Biology **11**(1), e1004,012 (2015)
170. Vol'pert, A., Hudjaev, S.: Analysis in Classes of Discontinuous Functions and Equations of Mathematical Physics. Martinus Nijhoff Publishers, Leiden (1985)
171. Wegscheider, R.: Über simultane gleichgewichte und die beziehungen zwischen thermodynamik und reactionskinetik homogener systeme. Monatshefte für Chemie und verwandte Teile anderer Wissenschaften **22**(8), 849–906 (1901)
172. Wegscheider, R.: Über simultane gleichgewichte und die beziehungen zwischen thermodynamic und reaktionskinetik homogener systeme. Zeitschrift fur Physikalische Chemie **39**, 257–303 (1902)
173. Wei, J.: Axiomatic treatment of chemical reaction systems. The Journal of Chemical Physics **36**(6), 1578–1584 (1962)
174. Wei, J.: Structure of complex chemical reaction systems. Industrial & Engineering Chemistry Fundamentals **4**(2), 161–167 (1965)
175. Wei, J., Prater, C.D.: The structure and analysis of complex reaction systems. Advances in Catalysis **13**, 203–392 (1962)
176. Wilson, R.J.: Introduction to Graph Theory. Addison-Wesley, Reading, MA (1996)
177. Wiuf, C., Feliu, E.: Power-law kinetics and determinant criteria for the preclusion of multistationarity in networks of interacting species. SIAM Journal on Applied Dynamical Systems **12**(4), 1685–1721 (2013)
178. Zhabotinsky, A.M.: Periodical oxidation of malonic acid in solution: a study of the Belousov reaction kinetics. Biofizika **9**, 306–11 (1964)

Index

M. Feinberg, *Foundations of Chemical Reaction Network Theory*,
Applied Mathematical Sciences 202,
https://doi.org/10.1007/978-3-030-03858-8

MIX
Papier aus verantwortungsvollen Quellen
Paper from responsible sources
FSC® C105338

If you have any concerns about our products,
you can contact us on
ProductSafety@springernature.com

In case Publisher is established outside the EU,
the EU authorized representative is:
Springer Nature Customer Service Center GmbH
Europaplatz 3, 69115 Heidelberg, Germany

Printed by Libri Plureos GmbH
in Hamburg, Germany